Foundations of Antenna Radiation Theory

Foundations of Antenna Radiation Theory

Eigenmode Analysis

Wen Geyi
Waterloo, Canada

IEEE Antennas and Propagation Society

IEEE Press Series on Electromagnetic Wave Theory and Applications
Douglas H. Werner, Series Editor

IEEE PRESS
WILEY

Published by John Wiley & Sons, Inc., Hoboken, New Jersey.
Published simultaneously in Canada.

Library of Congress Cataloging-in-Publication Data
Names: Wen, Geyi, author.
Title: Foundations of antenna radiation theory : eigenmode analysis / Wen Geyi.
Description: Hoboken, New Jersey : Wiley-IEEE Press, [2023] | Includes index.
Identifiers: LCCN 2022059830 (print) | LCCN 2022059831 (ebook) | ISBN 9781394170852 (cloth) | ISBN 9781394170869 (adobe pdf) | ISBN 9781394170876 (epub)
Subjects: LCSH: Antennas (Electronics) | Antenna radiation patterns.
Classification: LCC TK7871.6 .W46 2023 (print) | LCC TK7871.6 (ebook) | DDC 621.382/4–dc23/eng/20230111
LC record available at https://lccn.loc.gov/2022059830
LC ebook record available at https://lccn.loc.gov/2022059831

Cover Design and Image: Wiley

Set in 9.5/12.5pt STIXTwoText by Straive, Pondicherry, India

Contents

About the Author

Wen Geyi (Fellow, IEEE) was born in Pingjiang, Hunan, China, in 1963. He received the B.Eng., M.Eng., and Ph.D. degrees in electrical engineering from Xidian University, Xi'an, China, in 1982, 1984, and 1987, respectively. From 1988 to 1990, he was a lecturer at the Radio Engineering Department, Southeast University, Nanjing, China. From 1990 to 1992, he was an associate professor at the Institute of Applied Physics, University of Electronic Science and Technology of China (UESTC), Chengdu, China. From 1992 to 1993, he was a visiting researcher at the Department of Electrical and Computer Engineering, University of California at Berkeley, Berkeley, CA, United States. From 1993 to 1998, he was a full professor at the Institute of Applied Physics, UESTC. He was a visiting professor at the Electrical Engineering Department, University of Waterloo, Waterloo, ON, Canada, from February 1998 to May 1998. From 1996 to 1997, he was the vice chairman of the Institute of Applied Physics, UESTC, where he was the chairman of the institute from 1997 to 1998. From 1998 to 2007, he was with Blackberry Ltd., Waterloo, ON, Canada, first as a senior scientist with the Radio Frequency Department and then the director of the Advanced Technology Department. Since 2010, he has been a National Distinguished Professor with Fudan University, Shanghai, China, and the Nanjing University of Information Science and Technology (NUIST), Nanjing, where he is currently the director of the Research Center of Applied Electromagnetics. He has authored over 100 journal publications and *Foundations for Radio Frequency Engineering* (World Scientific, 2015), *Foundations of Applied Electrodynamics* (Wiley, 2010), *Advanced Electromagnetic Field Theory* (China: National Defense Publishing House, 1999), and *Modern Methods for Electromagnetic Computations* (China: Henan Science and Technology Press, 1994). He holds more than 40 patents.

Preface

Wireless technologies have revolutionized many different fields in industry as well as in our daily lives. As a vital device in wireless systems, antennas play an important role in boosting overall system performance. The demand on various types of antennas for different wireless applications is growing rapidly, which raises many challenges for antenna designers. For example, wireless terminals have become smaller, and antennas must be squeezed into an even smaller space. At the same time, multiple antenna systems and antennas covering multiple frequency bands are being deployed to wireless terminals to meet the increasing demand for new services and to improve the communication quality. To overcome these challenges, antenna designers need a better understanding of antenna theory.

Antenna theory usually contains three different but related subjects: generic properties of antenna, antenna analysis, and antenna synthesis. The generic properties of antenna are meant to be valid for all antennas, and they are the fundamentals of antenna design. For historical or technical reasons, many of the generic properties of antenna discovered in the last few decades have not yet been reflected in most antenna books. To include these new results in a book, one has to introduce a number of concepts that are barely touched in many antenna books, such as the stored field energy around antenna, the radiation quality factor, and the spherical vector wave functions. Antenna analysis examines the radiation properties of antenna with a specified current distribution, of which the radiated field is conventionally expressed as an integration. Such a process is, however, not always the most efficient since the integration must be carried out for each observation point in order to find the field distribution outside the source region. The antenna synthesis, also called pattern synthesis, is the opposite process of analysis, in which the current distribution or type of antenna, including the geometry and feeding mechanism, is determined in an optimal way so that a prescribed field distribution in the far- or near-field region can be achieved. Since a continuous current distribution is not easy to realize in practice, it must be discretized and then realized by an antenna array. For this reason, various antenna synthesis methods are primarily developed for antenna array. The conventional array synthesis methods are dependent on the array factor, which is no longer effective when the array elements are not identical, the surrounding environment is too complicated, or the inter-element spacing becomes very small. New array synthesis methods based on eigenmode analysis have been developed in recent years and can overcome the existing problems associated with the array factor, but have not yet been incorporated into textbooks therefore limiting accessibility to students and researchers.

The main theme of this book is eigenmode analysis and its applications in antenna theory and design. The free space can be considered as a spherical waveguide. An antenna may therefore be viewed as a waveguide junction that connects the feeding line and the spherical waveguide, transforming the guided modes into spherical modes in transmitting mode or converting the spherical modes into guided modes in receiving mode. For this reason, it is possible to build a theory for antennas that parallels the theory for waveguides. The eigenmode analysis is the foundation of waveguide theory, and its importance in physics and engineering cannot be overstressed. An eigenmode is a possible state of a system when it is free of excitation, and the corresponding eigenvalue often represents an important quantity of the system, for example the total energy of the system (such as in quantum mechanics) or the natural oscillation frequency (such as in a metal cavity resonator). An arbitrary state of the system can be expressed as a linear combination of the eigenmodes. If only one or a few eigenmodes dominate in the linear combination, this will significantly simplify the analysis of the problem. In the eigenmode expansion of a field, the expansion coefficients are expressed as the integrals over the source region and the integrations are only performed once. After the expansion coefficients are determined, the evaluation of the field distribution outside the source region only involves the sum of series, which decreases the computational burden and simplifies the numerical treatment most of the time as compared to the conventional integral representation.

There have been several modal theories for studying electromagnetic (EM) radiation and scattering problems. The singularity expansion method (SEM) is based on the analysis in complex frequency domain and formulated by electric field integral equation. The resonant frequencies and the modes in SEM are complex, which significantly increases the computational time and the difficulty in numerical implementations. The eigenmode expansion method (EEM) uses the eigenfunctions of an integral operator. Same with the SEM, the eigenvalues and the eigenmodes in EEM are complex numbers. In addition, the EEM lacks a solid mathematical foundation. The characteristic mode (CM) analysis is another interesting modal theory and is carried out in the real frequency domain, of which the characteristic values (eigenvalues) and CMs are all real. It is noted that all the modes involved in the CM, SEM, and EEM formulations depend not only on the properties of the scatterer but also on the working frequency.

This book contains the new developments in antenna theory, with the goal to address the aforementioned problems and challenges in the best possible way and is hoped to be a useful alternative to the traditional approaches. The antenna radiation problems in both closed and open region are treated in a unified manner in terms of the eigenmodes available from the systems. The eigenmodes are derived from waveguides, cavity resonators, and spherical waveguide and are independent of frequency, and can therefore be used to expand the fields in either frequency or time domain. The organization and treatment of the proposed book is quite different from the previous books on similar topics. The method of eigenmodes, similar to the Fourier series expansion in signal analysis, is used throughout the book. The antenna analysis problems are treated by combining the method of separation of variables, Green's function, and variational method. The variational method establishes the complete set of eigenmodes and their properties, and the method of separation of variable is used to find the eigenmodes for simple geometries. The radiated field is then expanded by using the eigenmodes, from which dyadic Green's functions can be determined, avoiding the problem caused by the inappropriate selection of the eigenmodes for the expansion of a point source. When the dyadic Green's functions are applied to the integral equation formulation for an antenna, a significant computational burden can be reduced and the numerical treatment can be simplified. The array synthesis problems are also treated as an eigenvalue problem with the method of maximum power transmission efficiency (MMPTE). The variational expression is established for the power transmission efficiency (PTE) between the antenna array under design and a testing array. An algebraic eigenvalue problem resulting from the variational principle is then solved, and the eigenvector corresponding to the maximum eigenvalue is selected as the distribution of excitations for the array under design.

The contents of the book are selected for their fundamentality and importance, and many of them are formulated in terms of eigenmode theory and appear in book form for the first time. The book not only discusses the antenna radiation problems in open space but also those in waveguide and cavity resonator, and it consists of six chapters. Chapter 1 describes the basics of EM field equations and their solution methods and provides the necessary background information for later chapters. It begins with the introduction of Maxwell equations, the wave equations, and the theorems for EM fields. Three analytical tools for the solution of boundary value problems are introduced, and they are the separation of variables, Green's function, and the variational method. The main focus of this chapter is the treatment of eigenvalue problems arising in matrix theory, scalar and vector fields, and they are fundamental to our later discussions. By means of the Rayleigh quotient (a variational expression for the eigenvalue problem), the eigenmodes of the

Laplacian operator acting on a scalar or a vector field are treated in a similar manner, and a complete set of eigenmodes is constructed by the variational analysis of the Rayleigh quotient. In order to understand how a vector field is decomposed into longitudinal, transverse, and harmonic components, the Helmholtz theorems for the vector fields defined in finite or infinite region are presented. As a generalization of the Helmholtz theorem, the eigenfunctions of the curl operator are also explored, in terms of which the plane-wave expansions for the fields and the dyadic Green's functions are obtained.

Chapter 2 investigates the radiation problems in waveguide. The eigenvalue problems in waveguide are approached in transverse field for its generality. Various dyadic Green's functions for waveguide are derived directly from the field expansions in terms of the vector modal functions, which avoids a problem caused by the incompleteness of the eigenfunctions selected to expand a dyadic point source in the conventional study of dyadic Green's functions in waveguides. By the equivalence principle, three common waveguide discontinuity problems, the excitation of waveguide, obstacles in waveguide, and the coupling between waveguides, are analyzed and treated as a radiation problem and compactly reformulated by using the dyadic Green's functions. The radiated field in time domain is approached by the vector modal functions, and the transient processes in the waveguide are studied. For reference, the vector modal functions in typical waveguides are summarized.

Chapter 3 deals with the radiation problems in metal cavity resonators. In particular, the vector modal functions in the waveguide cavity resonator are derived from the waveguide modes. The dyadic Green's functions of electric and magnetic type for a cavity resonator are established from the modal expansions of the fields. Like the waveguide theory, all the cavity-related problems are treated as a radiation problem through the use of equivalence principle. The circuit parameters for the cavity with multiple waveguide ports are evaluated by the modal analysis. The vector modal functions for typical waveguide cavity resonators are derived. It is demonstrated that the dyadic Green's functions for the waveguide cavity reduce to those for the waveguide if the two ends of the waveguide cavity are extended to infinity. The time-domain fields generated by the sources in the waveguide cavity are expanded in terms of the vector modal functions in waveguide, and the transient responses in the cavity resonator are examined.

Chapter 4 discusses the generic properties of antenna. Typical antenna parameters are summarized. Complete set of vector modal functions for the spherical waveguide is rigorously constructed from spherical harmonics, and the modal expansions of the dyadic Green's functions are derived from the field expansions. A general definition of the stored field energy of antenna is proposed by means of a conservation law for the stored field energies in an arbitrary medium. Two methods for evaluating the radiation quality factor are elucidated. One is from the input impedance of antenna and the other is via the current distribution. The modal quality factors are thoroughly examined, and their finite power series expansions are obtained. The upper bounds on the product of gain and bandwidth for both directional and omnidirectional antenna are presented, and their applications in small antenna design are demonstrated. The upper bounds answer a common question of how much space should be requested to accommodate an antenna to realize a specified performance. The radiated fields from a transient source are studied through the field expansions in terms of the vector modal functions for the spherical waveguide.

In antenna analysis, the induced current distribution on antenna is either given or to be determined from an impressed source. In many cases, one must resort to numerical techniques to solve an integral equation or a set of differential equations derivable from Maxwell equations with boundary conditions to find the induced current distribution. Chapter 5 is devoted to the modal analysis of typical antennas. The free space is considered as a spherical waveguide, and the radiated field is expressed as a linear combination of spherical vector wave functions. The integral equations for an antenna consisting of composite materials are derived by the modal expansions of dyadic Green's functions. Instead of using the integral representation of the fields in conventional antenna analysis, typical antennas, including dipole, loop, aperture, and patch antenna, are all analyzed by the spherical wave functions or the eigenmodes in both near- and far-field regions. An arbitrary scatterer is said to be resonant if its stored electric field energy is equal to the stored magnetic field energy. Based on this definition, a method for computing the resonant modes is proposed and applied to the antenna design.

Antenna synthesis involves using well-organized optimization methods to find the current distribution so as to achieve a specified field distribution in the near- or far-field region. A single antenna is often around one wavelength in size, and its radiation pattern covers a wide angle and thus exhibits poor directivity. In order to enhance the directivity and increase the flexibility of shaping the radiation pattern, one must use an antenna array. The performances of antenna array are controlled by the relative positioning of elements and the distribution of excitations. To achieve a desired field pattern, a performance index (target function) must be properly chosen and optimized. For a wireless system planned for the transmission of either information or power, a natural performance index is the PTE between the transmitting (Tx) and receiving (Rx) antennas, which is defined as the ratio of the power delivered to the load of the receiver to the input power of the transmitter. To attain the best possible quality of wireless communication or power transfer, the PTE must be maximized. Motivated by the fact that antennas must be designed to enhance the PTE for all wireless systems, the PTE can thus be adopted as a performance index for the design of antennas. The optimization procedure provides a powerful and universal technique for the synthesis of antenna arrays of all types and can overcome the existing challenges with the conventional array synthesis methods using array factors. The technique is based on eigenmode analysis and called the method of maximum power transmission efficiency (MMPTE), which can achieve various field patterns in any complicated environment in the near- or far-field region. The conventional methods of antenna synthesis largely depend on field theory, while the MMPTE reduces the field synthesis problem into a circuit analysis problem so that circuit theory may be applied to solve the original field problem. This feature makes the design process of antenna array more accessible for those who are not very familiar with EM field theory. The circuit parameters can be acquired by simulation or measurement, and therefore the MMPTE is applicable to any complicated problem. Whenever the simulation is beyond the capability of a state-of-the-art computer, one can resort to measurement to find the circuit parameters. Another important feature of the MMPTE is that it contains the information of the environment between Tx and Rx arrays, and therefore can be made adaptive to complicated environment, guaranteeing the best possible performance of the antenna array. The MMPTE has been verified to be superior to most existing array design methods in terms of simplicity, applicability, generality, and design accuracy. It generates an optimized distribution of excitation for the antenna array to assure that the gain and efficiency of the array is maximized for a fixed array configuration and is equally applicable for both near- and far-field synthesis problems. Chapter 6 summarizes several typical array synthesis methods based on the array factors, including Schelkunoff unit circle method, Dolph–Chebyshev method, and the Fourier transform method. The main part of this chapter is the formulations of MMPTE, including the unconstrained MMPTE, weighted MMPTE, constrained MMPTE, and extended MMPTE (EMMPTE). The EMMPTE is a field method but follows a procedure similar to MMPTE. A number of applications of MMPTE and EMMPTE are demonstrated.

For the convenience of readers, three Appendices A, B, and C are included to provide the fundamentals of vector analysis, dyadic analysis, and the SI unit system. A unified theory for fields (UTF) is explored in Appendix D. The UTF unveils that an arbitrary static field (either scalar or vectorial, called an ontological field) in an inertial system (static system) will merge as two vector fields in an inertial system moving relative to the static system, which satisfy Maxwell-like equations. Therefore, the Maxwell equations are valid not only for describing the EM fields but also for any physical fields for which an ontological field exists. The UTF is based on the theory of special relativity and the Helmholtz theorem. In order to find how the vector field changes in different inertial systems, one only needs to examine how the curl and divergence of the vector field transform. As a demonstration, the Maxwell-like equations for the gravitational field are derived from the UTF, and they are also derived from the Einstein field equations in the theory of general relativity. Some universal laws of nature are shown to be derivable from the UTF.

The book can be used either for undergraduate or graduate courses on "Advanced Antenna Theory," or as a reference for researchers and engineers in the areas of microwave, antenna, and EM compatibility. The prerequisites for the book are advanced calculus and linear algebra. After reading the book, the readers should be able to better understand antenna radiation theory and antenna analysis and synthesis from a different perspective in terms of eigenmode analysis. The SI units are used throughout the book. A $e^{j\omega t}$ time variation is assumed for time-harmonic fields. A special symbol "□" is used to indicate the end of a theorem, a remark, or an example.

The author is grateful to his family. Without their constant support and encouragement, the book would never have been completed.

Wen Geyi
Waterloo, Ontario, Canada

1

Eigenvalue Theory

Science is spectral analysis. Art is light synthesis.
– Karl Kraus (Austrian writer and journalist, 1874–1936)

The study of eigenvalue problems can be traced back to the eighteenth century, when Swiss mathematician and physicist Leonhard Euler (1707–1783) investigated the rotational motion of a rigid body. The word "eigen" is from German and means "own" or "belonging to," and was first used by German mathematician David Hilbert (1862–1943) to characterize eigenvalues and eigenvectors in 1904. Eigenvalue problems often arise in mathematics, physics, and engineering sciences. In linear algebra, an **eigenvector** of a linear transformation is a nonzero vector that changes by a scalar factor when the linear transformation acts on it. The scalar factor is called the **eigenvalue** corresponding to the eigenvector. Geometrically, this implies that the eigenvector is not rotated after transformation. The eigenvalue problem for a differential operator often results from the boundary value problems defined in a finite region. When the defining region is unbounded, the discrete eigenvalues become a continuum. A very useful technique for studying the eigenvalue problem is to establish the Rayleigh quotient for the eigenvalues and then use the calculus of variations to investigate the properties of eigenmodes. In physics, an **eigenmode** of a system is a possible state when the system is free of excitation, which might exist in the system on its own under certain conditions, and is also called an **eigenstate** of the system. The **method of eigenfunctions** is very similar to the Fourier series expansion in signal analysis, and will be used throughout this book. The method is based on the solution of an eigenvalue problem available from the system. An arbitrary state of the system can be expressed as a linear combination of the eigenmodes, and the expansion coefficients can then be determined from the source conditions or the initial values of the system. If only one or a few eigenmodes dominate in the linear combination, this will significantly simplify the analysis of the problem.

Foundations of Antenna Radiation Theory: Eigenmode Analysis, First Edition. Wen Geyi.

Published 2023 by John Wiley & Sons, Inc.

The modal theory for a scatterer plays an important role in antenna theory and designs. The basic idea behind the modal theory is to introduce the fundamental field patterns, called **modes**, so that the fields outside the scatterer can be expanded into a linear combination of these modes. There have been several modal theories for studying electromagnetic (EM) radiation and scattering problems (exterior boundary value problems). The **singularity expansion method** (SEM) is based on the analysis in complex frequency domain and formulated by electric field integral equation [1, 2]. The **natural resonant frequencies** arise from the requirement that a nontrivial current distribution exists on a conducting scatterer free of incident fields. The corresponding field patterns are called **natural resonant modes**. The natural resonant frequencies and the modes in SEM are complex, which significantly increases the computational time and the difficulty in numerical implementations. The **eigenmode expansion method** (EEM) expands the currents and the radiated fields in terms of the eigenmodes of an integral operator [3, 4]. Same with the SEM, the eigenvalues and the eigenmodes in EEM are complex numbers. The EEM is based on the eigenfunctions of integral equations and lacks a solid mathematical foundation. The integral operator involved in EEM is not symmetric, and it is therefore hard to prove the existence and completeness of the eigenfunctions. A more useful method for the study of scattering problem is the **singular function expansion**, which was first introduced by the German mathematician Erhard Schmidt (1876–1959) in 1907 [5], and has been applied to study various scattering problems [6, 7]. The theory of **characteristic mode** is another interesting modal notion and is carried out in the real frequency domain [8–11], of which the characteristic values (eigenvalues) and the corresponding characteristic modes are all real. In general, the characteristic values range from $-\infty$ to $+\infty$, among which those of the smallest magnitudes are the most important for radiation and scattering problems. The external resonant modes correspond to the zero characteristic values, and can be determined approximately by sweeping the frequency. It is noted that all the abovementioned modal formulations depend not only on the properties of the scatterer but also on the operating frequency.

The eigenvalue problems discussed in this book are derived from waveguide, cavity resonator, and spherical waveguide, whose eigenfunctions are independent of frequency and can thus be used to expand the fields in either frequency or time domain. The importance of eigenvalue theory in mathematics and physics cannot be overstated. There have been various methods developed to calculate eigenvalues and eigenfunctions, with the most important one being the variational method based on the Rayleigh quotient [12]. This chapter provides the necessary background information for later chapters. The Maxwell equations and the solution methods for partial differential equations (PDEs) are briefly introduced. The emphasis is upon the eigenvalue theory for operators, including the matrix and the

Laplacian on scalar and vector fields. The properties of eigenfunctions are derived from the Rayleigh quotient, and the Ritz method for the numerical solution of the Rayleigh quotient is demonstrated. Also included in this chapter is the Helmholtz theorem, which states that any vector field can be decomposed into the sum of an irrotational vector field and a solenoidal vector field. Such a decomposition has interesting applications in the modal expansion of fields and is the theoretical basis of introducing scalar and vector potentials. The Helmholtz theorem indicates that a vector field is fully determined by its divergence and curl. Indeed, Maxwell equations are nothing but a couple of rules that regulate the divergences and the curls of electric and magnetic fields according to impressed and induced sources. As a generalization of Helmholtz theorem, the eigenfunctions of curl operator are discussed, in terms of which the plane-wave expansions for the fields as well as the dyadic Green's functions can be obtained.

1.1 Maxwell Equations

Maxwell equations are a set of PDEs that unify electricity and magnetism and describe how electric and magnetic fields, as the functions of space and time, are generated by charges and currents and altered by each other. They have been proved to be very successful in explaining and predicting a variety of macroscopic EM phenomena.

1.1.1 Wave Equations

The **generalized Maxwell equations** that include both electric and magnetic sources consist of two vector equations and two scalar equations:

$$\begin{aligned} \nabla \times \mathbf{H}(\mathbf{r},t) &= \frac{\partial \mathbf{D}(\mathbf{r},t)}{\partial t} + \mathbf{J}(\mathbf{r},t), \\ \nabla \times \mathbf{E}(\mathbf{r},t) &= -\frac{\partial \mathbf{B}(\mathbf{r},t)}{\partial t} - \mathbf{J}_m(\mathbf{r},t), \\ \nabla \cdot \mathbf{D}(\mathbf{r},t) &= \rho(\mathbf{r},t), \\ \nabla \cdot \mathbf{B}(\mathbf{r},t) &= \rho_m(\mathbf{r},t). \end{aligned} \tag{1.1}$$

In the above, $\mathbf{r}$ is the observation point of the fields in meter (m) and t is the time in second (s), $\mathbf{H}$ is the **magnetic field intensity** measured in amperes per meter (A/m), $\mathbf{B}$ is the **magnetic induction intensity** measured in tesla (Wb/m^2), $\mathbf{E}$ is **electric field intensity** measured in volts per meter (V/m), $\mathbf{D}$ is the **electric induction intensity** measured in coulombs per square meter (C/m^2), $\mathbf{J}$ is **electric current density** measured in amperes per square meter (A/m^2), ρ is the **electric charge density** measured in coulombs per cubic meter (C/m^3), $\mathbf{J}_m$

is **magnetic current density** in volts per square meter (V/m^2), and ρ_m is **magnetic charge density** in webers per cubic meter (Wb/m^3). The first equation is **Ampère's law**, and it describes how the electric field changes according to the current density and magnetic field. The positive sign in the first equation indicates that the directions of the magnetomotive force and the electric current are related by the right-hand rule. The term $\partial \mathbf{D}/\partial t$ was introduced by Maxwell in 1861 and is called **displacement current**, which is necessary for the existence of wave solutions. The second equation is **Faraday's law**, and it characterizes how the magnetic field varies according to the electric field and equivalent magnetic current density. The minus sign in the second equation indicates that the directions of electromotive force and the magnetic current are related by the left-hand rule, which is required by **Lenz's law**. In other words, when an electromotive force is generated by a change of magnetic flux, the polarity of the induced electromotive force is such that it produces a current whose magnetic field opposes the change, which produces it. The third equation is **Coulomb's law**, and it says that the electric field depends on the charge distribution and obeys the inverse square law. The last equation shows that the magnetic field also obeys the inverse square law and depends on the equivalent magnetic charge distribution. It should be understood that none of the experiments had anything to do with waves at the time when Maxwell derived his equations. Maxwell equations imply more than the experimental facts. The **continuity equation** can be derived from (1.1):

$$\nabla \cdot \mathbf{J} = -\frac{\partial \rho}{\partial t}. \tag{1.2}$$

The electric charge density ρ and the electric current density $\mathbf{J}$ in Maxwell equations are free charge density and currents and they exclude charges and currents forming part of the structure of atoms and molecules. The bound charges and currents are regarded as material, which are not included in ρ and $\mathbf{J}$. The current density usually consists of two parts: $\mathbf{J} = \mathbf{J}_{con} + \mathbf{J}_{imp}$. Here, $\mathbf{J}_{imp}$ is referred to as external or **impressed current source**, which is independent of the fields and delivers energy to electric charges in a system. The impressed current source can be of electric and magnetic type as well as of non-electric or nonmagnetic origin. $\mathbf{J}_{con} = \sigma \mathbf{E}$, where σ is the **conductivity** of the medium in siemens per meter (S/m), denotes the **conduction current** induced by the impressed source $\mathbf{J}_{imp}$. Sometimes it is convenient to introduce an **impressed electric field** $\mathbf{E}_{imp}$ defined by $\mathbf{J}_{imp} = \sigma \mathbf{E}_{imp}$. In more general situation, one may write $\mathbf{J} = \mathbf{J}_{ind} + \mathbf{J}_{imp}$, where $\mathbf{J}_{ind}$ is the **induced current** by the impressed current $\mathbf{J}_{imp}$. The continuity equation for the magnetic current $\mathbf{J}_m$ and magnetic charges ρ_m can be derived from (1.1):

$$\nabla \cdot \mathbf{J}_m = -\frac{\partial \rho_m}{\partial t}. \tag{1.3}$$

The inclusions of magnetic sources $\mathbf{J}_m$ and ρ_m make Maxwell equations more symmetric although there has been no evidence that the magnetic current and charge are physically present (Appendix D gives an explanation of why the magnetic charge does not exist). The validity of introducing such concepts in Maxwell equations is justified by the equivalence principle, i.e. they are introduced as a mathematical equivalent to EM fields. For the time-harmonic (sinusoidal) fields with a single frequency ω, the generalized Maxwell equations (1.1) reduce to

$$\begin{aligned} \nabla \times \mathbf{H}(\mathbf{r}) &= j\omega \mathbf{D}(\mathbf{r}) + \mathbf{J}(\mathbf{r}), \\ \nabla \times \mathbf{E}(\mathbf{r}) &= -j\omega \mathbf{B}(\mathbf{r}) - \mathbf{J}_m(\mathbf{r}), \\ \nabla \cdot \mathbf{D}(\mathbf{r}) &= \rho(\mathbf{r}), \\ \nabla \cdot \mathbf{B}(\mathbf{r}) &= \rho_m(\mathbf{r}), \end{aligned} \tag{1.4}$$

where all the field quantities denote the **complex amplitudes** (**phasors**), defined by

$$\mathbf{E}(\mathbf{r}, t) = \operatorname{Re}\left[\mathbf{E}(\mathbf{r})e^{j\omega t}\right], \quad \text{etc.}$$

For brevity, the same notations will be used for both time-domain and frequency-domain quantities. The **boundary conditions** on the surface between two different media can be easily obtained as follows:

$$\begin{aligned} \mathbf{u}_n \times (\mathbf{H}_1 - \mathbf{H}_2) &= \mathbf{J}_s, \\ \mathbf{u}_n \times (\mathbf{E}_1 - \mathbf{E}_2) &= -\mathbf{J}_{ms}, \\ \mathbf{u}_n \cdot (\mathbf{D}_1 - \mathbf{D}_2) &= \rho_s, \\ \mathbf{u}_n \cdot (\mathbf{B}_1 - \mathbf{B}_2) &= \rho_{ms}, \end{aligned} \tag{1.5}$$

where $\mathbf{u}_n$ is the unit normal of the boundary directed from medium 2 to medium 1; $\mathbf{J}_s$ and ρ_s are the **surface current density** and **surface charge density**, respectively; $\mathbf{J}_{ms}$ and ρ_{ms} are the **surface magnetic current density** and **surface magnetic charge density**, respectively.

Maxwell equations (without magnetic sources) are a set of 7 equations involving 16 unknowns (i.e. five vectors **E, H, B, D, J** and one scalar ρ and the last equation of (1.1) is not independent). To determine the fields, nine more equations are needed, and they are given by the **generalized constitutive relations**:

$$\mathbf{D} = f(\mathbf{E}, \mathbf{H}), \quad \mathbf{B} = g(\mathbf{E}, \mathbf{H})$$

together with the **generalized Ohm's law**

$$\mathbf{J} = h(\mathbf{E}, \mathbf{H})$$

if the medium is conducting. The constitutive relations establish the connections between field quantities and reflect the properties of the medium, and they are totally independent of the Maxwell equations. For time-harmonic fields, the constitutive relations for a **bi-anisotropic medium** are defined by

$$\mathbf{D} = \overleftrightarrow{\boldsymbol{\varepsilon}} \cdot \mathbf{E} + \overleftrightarrow{\boldsymbol{\xi}} \cdot \mathbf{H},$$
$$\mathbf{B} = \overleftrightarrow{\boldsymbol{\zeta}} \cdot \mathbf{E} + \overleftrightarrow{\boldsymbol{\mu}} \cdot \mathbf{H},$$

where $\overleftrightarrow{\boldsymbol{\mu}}$, $\overleftrightarrow{\boldsymbol{\varepsilon}}$, $\overleftrightarrow{\boldsymbol{\xi}}$, and $\overleftrightarrow{\boldsymbol{\zeta}}$ are dyadics. The medium is called **anisotropic** if $\overleftrightarrow{\boldsymbol{\xi}} = \overleftrightarrow{\boldsymbol{\zeta}} = 0$. The medium is called **isotropic** if $\overleftrightarrow{\boldsymbol{\xi}} = \overleftrightarrow{\boldsymbol{\zeta}} = 0$ and $\overleftrightarrow{\boldsymbol{\mu}}$ and $\overleftrightarrow{\boldsymbol{\varepsilon}}$ are, respectively, degenerated to $\overleftrightarrow{\boldsymbol{\mu}} = \mu\overleftrightarrow{\mathbf{I}}$ and $\overleftrightarrow{\boldsymbol{\varepsilon}} = \varepsilon\overleftrightarrow{\mathbf{I}}$, where $\overleftrightarrow{\mathbf{I}}$ is the identity dyadic; μ and ε are referred to as **permittivity** and **permeability** of the medium, respectively.

The EM wave equations are second-order PDEs that describe the propagation of EM waves through a medium. On elimination of **E** or **H** in the generalized Maxwell equations (1.4), the **wave equations** for the time-harmonic fields in an inhomogeneous and anisotropic medium are

$$\begin{aligned} \nabla \times \overleftrightarrow{\boldsymbol{\mu}}^{-1} \cdot \nabla \times \mathbf{E}(\mathbf{r}) - \omega^2 \overleftrightarrow{\boldsymbol{\varepsilon}} \cdot \mathbf{E}(\mathbf{r}) &= -j\omega \mathbf{J}(\mathbf{r}) - \nabla \times \overleftrightarrow{\boldsymbol{\mu}}^{-1} \cdot \mathbf{J}_m, \\ \nabla \times \overleftrightarrow{\boldsymbol{\varepsilon}}^{-1} \cdot \nabla \times \mathbf{H}(\mathbf{r}) - \omega^2 \overleftrightarrow{\boldsymbol{\mu}} \cdot \mathbf{H}(\mathbf{r}) &= -j\omega \mathbf{J}_m(\mathbf{r}) + \nabla \times \overleftrightarrow{\boldsymbol{\varepsilon}}^{-1} \cdot \mathbf{J}. \end{aligned} \tag{1.6}$$

If the medium is homogeneous and isotropic, μ and ε are constants, and the wave equations (1.6) are simplified to

$$\begin{aligned} \nabla \times \nabla \times \mathbf{E} - k^2 \mathbf{E} &= -j\omega\mu \mathbf{J} - \nabla \times \mathbf{J}_m, \\ \nabla \times \nabla \times \mathbf{H} - k^2 \mathbf{H} &= -j\omega\varepsilon \mathbf{J}_m + \nabla \times \mathbf{J}, \end{aligned} \tag{1.7}$$

where $k = \omega\sqrt{\mu\varepsilon}$ is the **wavenumber**. It can be seen that the source terms on the right-hand side of (1.7) are very complicated. To simplify the analysis, the EM potential functions can be introduced. The wave equations may be used to solve the following three different field problems:

1) EM fields in source-free region: wave propagations in space and waveguides, wave oscillation in cavity resonators, etc.
2) EM fields generated by known source distributions: antenna radiations, excitations in waveguides and cavity resonators, etc.
3) Interaction of fields and sources: wave propagation in plasma, coupling between electron beams and propagation mechanism, etc.

1.1.2 Properties of Electromagnetic Fields

A number of theorems can be derived from Maxwell equations [13–16], and they usually bring deep physical insight into the EM field problems. When applied properly, these theorems can simplify the problems dramatically.

1.1.2.1 Superposition Theorem

Superposition theorem applies to all linear systems. Suppose that the impressed current source $\mathbf{J}_{imp}$ can be expressed as a linear combination of independent impressed current sources $\mathbf{J}_{imp}^k$ $(k = 1, 2, ..., n)$:

$$\mathbf{J}_{imp} = \sum_{k=1}^{n} a_k \mathbf{J}_{imp}^k,$$

where a_k $(k = 1, 2, ..., n)$ are arbitrary constants. If $\mathbf{E}^k$ and $\mathbf{H}^k$ are the fields produced by the source $\mathbf{J}_{imp}^k$, the **superposition theorem** for EM fields asserts that the fields $\mathbf{E} = \sum\limits_{k=1}^{n} a_k \mathbf{E}^k$ and $\mathbf{H} = \sum\limits_{k=1}^{n} a_k \mathbf{H}^k$ are a solution of Maxwell equations produced by the source $\mathbf{J}_{imp}$.

1.1.2.2 Conservation of Electromagnetic Field Energy

The law of **conservation of EM field energy** is known as the **Poynting theorem**, named after the English physicist John Henry Poynting (1852–1914). It can be found from (1.1) that

$$-\mathbf{J}_{imp} \cdot \mathbf{E} - \mathbf{J}_{ind} \cdot \mathbf{E} = \nabla \cdot \mathbf{S} + \mathbf{E} \cdot \frac{\partial \mathbf{D}}{\partial t} + \mathbf{H} \cdot \frac{\partial \mathbf{B}}{\partial t}, \tag{1.8}$$

where $\mathbf{J} = \mathbf{J}_{imp} + \mathbf{J}_{ind}$ has been assumed. In a region V bounded by S, the integral form of (1.8) is

$$-\int_V \mathbf{J}_{imp} \cdot \mathbf{E} dV = \int_V \mathbf{J}_{ind} \cdot \mathbf{E} dV + \int_S \mathbf{S} \cdot \mathbf{u}_n dS + \int_V \left(\mathbf{E} \cdot \frac{\partial \mathbf{D}}{\partial t} + \mathbf{H} \cdot \frac{\partial \mathbf{B}}{\partial t} \right) dV, \tag{1.9}$$

where $\mathbf{u}_n$ is the unit outward normal of S, and $\mathbf{S} = \mathbf{E} \times \mathbf{H}$ is the **Poynting vector** representing the EM power-flow density measured in watts per square meter (W/m^2). It is assumed that this explanation holds for all media. Thus, the left-hand side of the above equation stands for the power supplied by the impressed current source. The first term on the right-hand side is the work done per second by the electric field to maintain the current in the conducting part of the system. The second term denotes the EM power flowing out of S. The last term can be interpreted as the work done per second by the impressed source to establish the fields. The total field energy density w of the EM fields may be defined as follows:

$$dw = \left(\mathbf{E} \cdot \frac{\partial \mathbf{D}}{\partial t} + \mathbf{H} \cdot \frac{\partial \mathbf{B}}{\partial t} \right) dt. \tag{1.10}$$

If all the sources and fields are zero at $t = -\infty$, the total field energy density can be written as

$$w = w_e + w_m, \tag{1.11}$$

where w_e and w_m are the **electric field energy density** and **magnetic field energy density**, respectively,

$$\begin{aligned} w_e &= \frac{1}{2}\mathbf{E}\cdot\mathbf{D} + \int_{-\infty}^{t} \frac{1}{2}\left(\mathbf{E}\cdot\frac{\partial \mathbf{D}}{\partial t} - \mathbf{D}\cdot\frac{\partial \mathbf{E}}{\partial t}\right)dt, \\ w_m &= \frac{1}{2}\mathbf{H}\cdot\mathbf{B} + \int_{-\infty}^{t} \frac{1}{2}\left(\mathbf{H}\cdot\frac{\partial \mathbf{B}}{\partial t} - \mathbf{B}\cdot\frac{\partial \mathbf{H}}{\partial t}\right)dt. \end{aligned} \tag{1.12}$$

Equation (1.9) can thus be written as

$$-\int_V \mathbf{J}_{imp}\cdot\mathbf{E}dV = \int_V \mathbf{J}_{ind}\cdot\mathbf{E}dV + \int_S \mathbf{S}\cdot\mathbf{u}_n dS + \frac{\partial}{\partial t}\int_V (w_e + w_m)dV. \tag{1.13}$$

In general, the field energy density w does not represent the stored field energy density in the fields: the energy temporarily located in the fields and completely recoverable when the fields are reduced to zero. The field energy density w given by (1.11) can be considered as the stored field energy density only if the medium is lossless (i.e. $\nabla\cdot\mathbf{S} = 0$). If the medium is isotropic and time-invariant, (1.12) reduces to

$$w_e = \frac{1}{2}\mathbf{E}\cdot\mathbf{D}, \quad w_m = \frac{1}{2}\mathbf{H}\cdot\mathbf{B}. \tag{1.14}$$

For the time-harmonic fields, the **time averages** of Poynting vector, the field energy densities in (1.14) over one period of the sinusoidal wave $e^{j\omega t}$, denoted T, are

$$\begin{aligned} \frac{1}{T}\int_0^T \mathbf{E}\times\mathbf{H}dt &= \frac{1}{2}\,\mathrm{Re}\left(\mathbf{E}\times\overline{\mathbf{H}}\right), \\ \frac{1}{T}\int_0^T \frac{1}{2}\mathbf{E}\cdot\mathbf{D}dt &= \frac{1}{4}\,\mathrm{Re}\left(\mathbf{E}\cdot\overline{\mathbf{D}}\right), \\ \frac{1}{T}\int_0^T \frac{1}{2}\mathbf{H}\cdot\mathbf{B}dt &= \frac{1}{4}\,\mathrm{Re}\left(\mathbf{H}\cdot\overline{\mathbf{B}}\right). \end{aligned}$$

The energy balance relations for the time-harmonic fields can be derived by using complex variable analysis [17]. Let $s = \alpha + j\omega$ denote the complex frequency. For an arbitrary analytic function $f(\mathbf{r}, s)$, the Cauchy–Riemann conditions imply

$$\frac{\partial f(\mathbf{r}, s)}{\partial \alpha} = -j\frac{\partial f(\mathbf{r}, s)}{\partial \omega}. \tag{1.15}$$

For sufficiently small α, the analytic function has the first-order expansion

$$\begin{aligned} f(\mathbf{r},s) &= f(\mathbf{r},\alpha + j\omega) \approx f(\mathbf{r},s)\big|_{\alpha=0} + \alpha\frac{\partial f(\mathbf{r},s)}{\partial \alpha}\bigg|_{\alpha=0} \\ &= f(\mathbf{r},j\omega) - j\alpha\frac{\partial f(\mathbf{r},j\omega)}{\partial \omega}, \end{aligned} \tag{1.16}$$

where (1.15) has been used. If the Laplace transform

$$\mathbf{E}(\mathbf{r}, s) = \int_{-\infty}^{\infty} \mathbf{E}(\mathbf{r}, t)e^{-st}dt$$

is applied to the first two equations of the Maxwell equations (1.1) (with $\mathbf{J}_m = 0$), one may find

$$\begin{aligned} \nabla \times \mathbf{H}(\mathbf{r}, s) &= \mathbf{J}(\mathbf{r}, s) + s\mathbf{D}(\mathbf{r}, s), \\ \nabla \times \mathbf{E}(\mathbf{r}, s) &= -s\mathbf{B}(\mathbf{r}, s). \end{aligned} \tag{1.17}$$

From the vector identity $\nabla \cdot (\mathbf{a} \times \mathbf{b}) = \mathbf{b} \cdot \nabla \times \mathbf{a} - \mathbf{a} \cdot \nabla \times \mathbf{b}$ and (1.17), one may obtain

$$\begin{aligned} \nabla \cdot \left[\mathbf{E}(\mathbf{r}, s) \times \overline{\mathbf{H}}(\mathbf{r}, s)\right] = &-\mathbf{E}(\mathbf{r}, s) \cdot \overline{\mathbf{J}}(\mathbf{r}, s) \\ &-\alpha\left[\overline{\mathbf{H}}(\mathbf{r}, s) \cdot \mathbf{B}(\mathbf{r}, s) + \mathbf{E}(\mathbf{r}, s) \cdot \overline{\mathbf{D}}(\mathbf{r}, s)\right] \\ &-j\omega\left[\overline{\mathbf{H}}(\mathbf{r}, s) \cdot \mathbf{B}(\mathbf{r}, s) - \mathbf{E}(\mathbf{r}, s) \cdot \overline{\mathbf{D}}(\mathbf{r}, s)\right], \end{aligned} \tag{1.18}$$

where the bar denotes complex conjugate. By use of the expansion (1.16), the first-order expansion for the electric field can be expressed by

$$\begin{aligned} \mathbf{E}(\mathbf{r}, s) &= \mathbf{E}(\mathbf{r}, \alpha + j\omega) \approx \mathbf{E}(\mathbf{r}, j\omega) - j\alpha\frac{\partial \mathbf{E}(\mathbf{r}, j\omega)}{\partial \omega} \\ &= \mathbf{E}(\mathbf{r}) - j\alpha\frac{\partial \mathbf{E}(\mathbf{r})}{\partial \omega}, \end{aligned} \tag{1.19}$$

where $\mathbf{E}(\mathbf{r})$ denotes the phasor for the time-harmonic electric field. By introducing the first-order expansions for all the field quantities into (1.18), one may arrive at

$$
\begin{aligned}
&\nabla\cdot\left[\mathbf{E}(\mathbf{r})\times\overline{\mathbf{H}}(\mathbf{r})\right]+j\alpha\nabla\cdot\left[\mathbf{E}(\mathbf{r})\times\frac{\partial\overline{\mathbf{H}}(\mathbf{r})}{\partial\omega}-\frac{\partial\mathbf{E}(\mathbf{r})}{\partial\omega}\times\overline{\mathbf{H}}(\mathbf{r})\right]\\
&\quad=-\mathbf{E}(\mathbf{r})\cdot\overline{\mathbf{J}}(\mathbf{r})-j\alpha\left[\mathbf{E}(\mathbf{r})\cdot\frac{\partial\overline{\mathbf{J}}(\mathbf{r})}{\partial\omega}-\frac{\partial\mathbf{E}(\mathbf{r})}{\partial\omega}\cdot\overline{\mathbf{J}}(\mathbf{r})\right]\\
&\quad-j\omega\left[\mathbf{B}(\mathbf{r})\cdot\overline{\mathbf{H}}(\mathbf{r})-\mathbf{E}(\mathbf{r})\cdot\overline{\mathbf{D}}(\mathbf{r})\right]\\
&\quad-\alpha\left[\mathbf{B}(\mathbf{r})\cdot\overline{\mathbf{H}}(\mathbf{r})+\mathbf{E}(\mathbf{r})\cdot\overline{\mathbf{D}}(\mathbf{r})\right]\\
&\quad-\alpha\omega\left[\overline{\mathbf{H}}(\mathbf{r})\cdot\frac{\partial\mathbf{B}(\mathbf{r})}{\partial\omega}-\mathbf{B}(\mathbf{r})\cdot\frac{\partial\overline{\mathbf{H}}(\mathbf{r})}{\partial\omega}\right]\\
&\quad-\alpha\omega\left[\mathbf{E}(\mathbf{r})\cdot\frac{\partial\overline{\mathbf{D}}(\mathbf{r})}{\partial\omega}-\overline{\mathbf{D}}(\mathbf{r})\cdot\frac{\partial\mathbf{E}(\mathbf{r})}{\partial\omega}\right].
\end{aligned}
\tag{1.20}
$$

Comparing the coefficients of similar terms containing α on both sides of (1.20) yields the energy balance relation

$$
\begin{aligned}
&\frac{1}{4}\mathbf{E}\cdot\overline{\mathbf{D}}+\frac{1}{4}\omega\left(\mathbf{E}\cdot\frac{\partial\overline{\mathbf{D}}}{\partial\omega}-\overline{\mathbf{D}}\cdot\frac{\partial\mathbf{E}}{\partial\omega}\right)+\frac{1}{4}\mathbf{B}\cdot\overline{\mathbf{H}}+\frac{1}{4}\omega\left(\overline{\mathbf{H}}\cdot\frac{\partial\mathbf{B}}{\partial\omega}-\mathbf{B}\cdot\frac{\partial\overline{\mathbf{H}}}{\partial\omega}\right)\\
&\quad=-j\frac{1}{4}\nabla\cdot\left(\mathbf{E}\times\frac{\partial\overline{\mathbf{H}}}{\partial\omega}-\frac{\partial\mathbf{E}}{\partial\omega}\times\overline{\mathbf{H}}\right)-j\frac{1}{4}\left(\mathbf{E}\cdot\frac{\partial\overline{\mathbf{J}}}{\partial\omega}-\frac{\partial\mathbf{E}}{\partial\omega}\cdot\overline{\mathbf{J}}\right),
\end{aligned}
\tag{1.21}
$$

and the well-known Poynting theorem for time-harmonic fields

$$
-\frac{1}{2}\mathbf{E}\cdot\overline{\mathbf{J}}=\nabla\cdot\frac{1}{2}(\mathbf{E}\times\overline{\mathbf{H}})+j2\omega\left(\frac{1}{4}\mathbf{B}\cdot\overline{\mathbf{H}}-\frac{1}{4}\mathbf{E}\cdot\overline{\mathbf{D}}\right).
\tag{1.22}
$$

Consequently, the complex analysis produces two energy balance relations simultaneously. In this sense, the complex analysis implies more than the real analysis. It should be noted that the Poynting theorem (1.9) in time domain and the Poynting theorem (1.22) in frequency domain are independent. This property can be used to find the stored field energies of small antenna [18]. The physical implication of the energy conservation law (1.21) becomes clear if it is decomposed into the real and imaginary parts. The real part is given by

$$
w_e+w_m=\nabla\cdot\mathrm{Im}\frac{1}{4}\left(\mathbf{E}\times\frac{\partial\overline{\mathbf{H}}}{\partial\omega}-\frac{\partial\mathbf{E}}{\partial\omega}\times\overline{\mathbf{H}}\right)+\mathrm{Im}\frac{1}{4}\left(\mathbf{E}\cdot\frac{\partial\overline{\mathbf{J}}}{\partial\omega}-\frac{\partial\mathbf{E}}{\partial\omega}\cdot\overline{\mathbf{J}}\right),
\tag{1.23}
$$

where w_e and w_m are defined by

$$
w_e=\frac{1}{4}\,\mathrm{Re}\,\mathbf{E}\cdot\overline{\mathbf{D}}+\frac{1}{4}\omega\,\mathrm{Re}\left(\mathbf{E}\cdot\frac{\partial\overline{\mathbf{D}}}{\partial\omega}-\overline{\mathbf{D}}\cdot\frac{\partial\mathbf{E}}{\partial\omega}\right),
\tag{1.24}
$$

$$w_m = \frac{1}{4}\operatorname{Re}\mathbf{H}\cdot\overline{\mathbf{B}} + \frac{1}{4}\omega\operatorname{Re}\left(\mathbf{H}\cdot\frac{\partial\overline{\mathbf{B}}}{\partial\omega} - \overline{\mathbf{B}}\cdot\frac{\partial\mathbf{H}}{\partial\omega}\right). \tag{1.25}$$

The expressions (1.24) and (1.25) were first derived by the author in [17], and they are shown to, respectively, represent the (time averaged) **stored electric field energy density** and the **stored magnetic field energy density** in an arbitrary medium. In Chapter 4, a new narrow-band approach will be introduced to rederive energy expressions (1.24) and (1.25). The stored field energies can be decomposed into the sum of two distinct parts: the dominant (nondispersive) parts defined by

$$w_e^{dom} = \frac{1}{4}\operatorname{Re}\mathbf{E}\cdot\overline{\mathbf{D}}, \tag{1.26}$$

$$w_m^{dom} = \frac{1}{4}\operatorname{Re}\mathbf{H}\cdot\overline{\mathbf{B}}, \tag{1.27}$$

and the dispersive parts defined by

$$w_e^{dis} = \frac{1}{4}\operatorname{Re}\omega\left(\mathbf{E}\cdot\frac{\partial\overline{\mathbf{D}}}{\partial\omega} - \overline{\mathbf{D}}\cdot\frac{\partial\mathbf{E}}{\partial\omega}\right), \tag{1.28}$$

$$w_m^{dis} = \frac{1}{4}\operatorname{Re}\omega\left(\mathbf{H}\cdot\frac{\partial\overline{\mathbf{B}}}{\partial\omega} - \overline{\mathbf{B}}\cdot\frac{\partial\mathbf{H}}{\partial\omega}\right), \tag{1.29}$$

which are caused by the dispersion of materials. The imaginary part of (1.21) gives

$$\begin{aligned} w_{ed} - w_{md} = & -\operatorname{Im}\pi\omega\left(\mathbf{E}\cdot\frac{\partial\overline{\mathbf{D}}}{\partial\omega} - \overline{\mathbf{D}}\cdot\frac{\partial\mathbf{E}}{\partial\omega}\right) - \operatorname{Im}\pi\omega\left(\overline{\mathbf{H}}\cdot\frac{\partial\mathbf{B}}{\partial\omega} - \mathbf{B}\cdot\frac{\partial\overline{\mathbf{H}}}{\partial\omega}\right) \\ & -\nabla\cdot\operatorname{Re}\pi\left(\mathbf{E}\times\frac{\partial\overline{\mathbf{H}}}{\partial\omega} - \frac{\partial\mathbf{E}}{\partial\omega}\times\overline{\mathbf{H}}\right) - \operatorname{Re}\pi\left(\mathbf{E}\cdot\frac{\partial\overline{\mathbf{J}}}{\partial\omega} - \frac{\partial\mathbf{E}}{\partial\omega}\cdot\overline{\mathbf{J}}\right), \end{aligned} \tag{1.30}$$

where w_{ed} and w_{md} are the (average) **dissipated electric field energy density** and **dissipated magnetic field energy density**, respectively, defined by

$$w_{ed} = \operatorname{Im}\pi\mathbf{E}\cdot\overline{\mathbf{D}}, \tag{1.31}$$

$$w_{md} = \operatorname{Im}\pi\mathbf{H}\cdot\overline{\mathbf{B}}. \tag{1.32}$$

As a result, the energy balance relation (1.21) gives two expressions: one is for the sum of stored electric and magnetic field energies and the other is for the difference of the dissipated electric and magnetic field energies, both expressions being valid in an arbitrary medium. It will be shown in Chapter 4 that (1.23) logically gives the definition of stored field energy of antenna in an arbitrary medium.

Remark 1.1 The quantities $\frac{1}{4}\,\mathrm{Re}\,\mathbf{H}\cdot\overline{\mathbf{B}}$ and $\frac{1}{4}\,\mathrm{Re}\,\mathbf{E}\cdot\overline{\mathbf{D}}$ in the Poynting theorem (1.22) only represent the nondispersive parts (1.26) and (1.27). For this reason, the Poynting theorem (1.22) cannot be considered as an energy balance relation for the time-harmonic fields in a general medium. □

1.1.2.3 Equivalence Theorem

It is known that there is no answer to the question of whether field or source is primary. The equivalence theorem just indicates that the distinction between the field and source is kind of blurred. Let V be an arbitrary region bounded by S, as shown in Figure 1.1. Two sources that produce the same fields inside a region are said to be **equivalent** within that region. Similarly, two EM fields $\{\mathbf{E}_1, \mathbf{D}_1, \mathbf{H}_1, \mathbf{B}_1\}$ and $\{\mathbf{E}_2, \mathbf{D}_2, \mathbf{H}_2, \mathbf{B}_2\}$ are said to be equivalent inside a region if they both satisfy the Maxwell equations and are equal in that region. The main application of the equivalence theorem is to find equivalent sources to replace the influences of substance (the medium is homogenized), so that the formulae for retarding potentials can be used. The equivalent sources may be located inside S (equivalent volume sources) or on S (equivalent surface sources). One of the important equivalence theorems is found by American mathematician Sergei Alexander Schelkunoff (1897–1992) and English mathematician Augustus Edward Hough Love (1863–1940), whose derivation can be found in [16].

Schelkunoff-Love Equivalence Theorem: Let $\{\mathbf{E}, \mathbf{D}, \mathbf{H}, \mathbf{B}\}$ be the EM fields with source confined in S. The following surface sources on S

$$\mathbf{J}_s = \mathbf{u}_n \times \mathbf{H}, \quad \mathbf{J}_{ms} = -\mathbf{u}_n \times \mathbf{E}, \tag{1.33}$$

produce the same fields $\{\mathbf{E}, \mathbf{D}, \mathbf{H}, \mathbf{B}\}$ outside S and a zero field inside S. □

Since the sources in (1.33) produce a zero field inside S, the interior of S may be filled with a perfect electric conductor. By use of the Lorentz reciprocity theorem [see (1.36)], it can be shown that the surface electric current pressed tightly on the

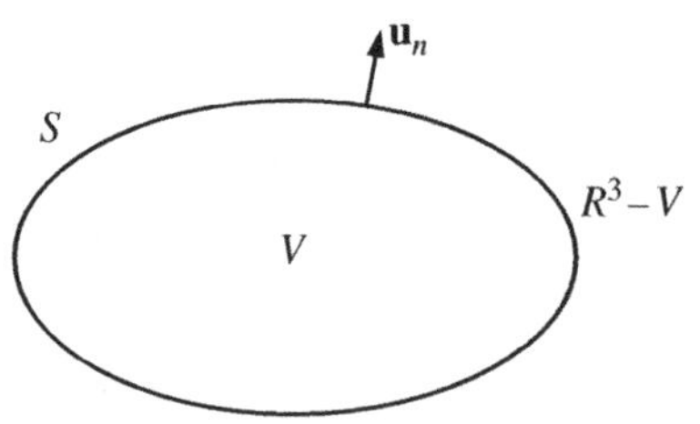

Figure 1.1 Equivalence theorem.

perfect conductor does not produce fields. As a result, only the surface magnetic current is needed in (1.33). Similarly, the interior of S may be filled with a perfect magnetic conductor, and in this case the surface magnetic current does not produce fields and only the surface electric current is needed in (1.33). In both cases, one cannot directly apply the vector potential formula even if the medium outside S is homogeneous.

1.1.2.4 Reciprocity

A linear system is said to be **reciprocal** if the response of the system with a particular load and a source is the same as the response when the source and the load are interchanged. Consider two sets of time-harmonic sources, $\mathbf{J}_1$, $\mathbf{J}_{m1}$ and $\mathbf{J}_2$, $\mathbf{J}_{m2}$, of the same frequency in the same linear medium. The fields produced by the two sources are, respectively, denoted by $\mathbf{E}_1$, $\mathbf{H}_1$ and $\mathbf{E}_2$, $\mathbf{H}_2$. The reciprocity can be stated as

$$\begin{aligned}\int_V (\mathbf{E}_2 \cdot \mathbf{J}_1 - \mathbf{H}_2 \cdot \mathbf{J}_{m1}) dV &= \int_V (\mathbf{E}_1 \cdot \mathbf{J}_2 - \mathbf{H}_1 \cdot \mathbf{J}_{m2}) dV \\ &\quad + \int_S (\mathbf{E}_1 \times \mathbf{H}_2 - \mathbf{E}_2 \times \mathbf{H}_1) \cdot \mathbf{u}_n dS,\end{aligned} \tag{1.34}$$

where V is a finite region bounded by S. If both sources are outside S, the surface integral in (1.34) is zero. If both sources are inside S, it can be shown that the surface integral is also zero by using the radiation condition. Therefore, one obtains the **Lorentz form of reciprocity**

$$\int_S (\mathbf{E}_1 \times \mathbf{H}_2 - \mathbf{E}_2 \times \mathbf{H}_1) \cdot \mathbf{u}_n dS = 0 \tag{1.35}$$

and the **Rayleigh–Carson form of reciprocity**

$$\int_V (\mathbf{E}_2 \cdot \mathbf{J}_1 - \mathbf{H}_2 \cdot \mathbf{J}_{m1}) dV = \int_V (\mathbf{E}_1 \cdot \mathbf{J}_2 - \mathbf{H}_1 \cdot \mathbf{J}_{m2}) dV. \tag{1.36}$$

If the surface S only contains the sources $\mathbf{J}_1(\mathbf{r})$ and $\mathbf{J}_{m1}(\mathbf{r})$, (1.34) becomes

$$\int_V (\mathbf{E}_2 \cdot \mathbf{J}_1 - \mathbf{H}_2 \cdot \mathbf{J}_{m1}) dV = \int_S (\mathbf{E}_2 \cdot \mathbf{u}_n \times \mathbf{H}_1 - \mathbf{H}_2 \cdot \mathbf{E}_1 \times \mathbf{u}_n) dS.$$

This is the familiar form of **Huygens' principle**. The EM reciprocity theorem (1.35) can also be generalized to an anisotropic medium.

1.2 Methods for Partial Differential Equations

Various analytic and numerical methods for the solution of PDEs have been developed [19–21]. Linear PDEs are generally solved by means of the **method of separation of variables**, the **method of Green's function**, named after the British mathematician George Green (1793–1841), and the **variational method**. Some usual trinities for PDEs are summarized in Table 1.1.

1.2.1 Method of Separation of Variables

The study of eigenvalue problems has its roots in the **method of separation of variables** or series solutions of PDEs. The basic idea of separation of variables is to seek a solution of the form of a product of functions, each of which depends on one variable only, so that the solution of original PDEs may reduce to the solution of ordinary differential equations. The latter is usually solved by the power series methods, resulting in various special functions. The method of separation of variables, also called Fourier method, was first introduced by Swiss mathematician Johann Bernoulli (1667–1748) between the years 1694 and 1697. The Helmholtz equation will be used to illustrate the procedure. The **Helmholtz equation**, named after the German physicist Hermann Ludwig Ferdinand von

Table 1.1 Some trinities for PDEs.

Trinity	Description
Three types of PDEs	Elliptical, hyperbolic, and parabolic.
Three types of problems	Boundary value problems, initial value problems, and eigenvalue problems.
Three types of boundary conditions	Dirichlet boundary condition, named after the German mathematician Johann Peter Gustav Lejeune Dirichlet (1805–1859); Neumann boundary condition, named after the German mathematician Carl Gottfried Neumann (1832–1925); and Robin boundary condition, named after the French mathematician Victor Gustave Robin (1855–1897).
Three mathematical tools	Divergence theorem, inequalities, and convergence theorems.
Three analytical methods	Method of separation of variables, method of Green's function, and variational method.
Three numerical methods	Finite element method, finite difference method, and moment method.

Helmholtz (1821–1894), also called **reduced wave equation**, is the time-independent form of wave equation, and is defined by

$$(\nabla^2 + k^2)u = 0, \tag{1.37}$$

where k is a constant. When k is zero, the Helmholtz equation reduces to the Laplace equation, named after the French mathematician Pierre-Simon marquis de Laplace (1749–1827). The Helmholtz equation is separable in 11 orthogonal coordinate systems [22]. The separated solutions form a subset of all solutions of (1.37) and can be served as a basis in terms of which all solutions of (1.37) can be expressed as a linear combination of the separated solutions.

1.2.1.1 Rectangular Coordinate System

In rectangular coordinate system, Helmholtz equation (1.37) becomes

$$\frac{\partial^2 u}{\partial x^2} + \frac{\partial^2 u}{\partial y^2} + \frac{\partial^2 u}{\partial z^2} + k^2 u = 0. \tag{1.38}$$

One may seek a solution in the form of product of three functions of one coordinate each

$$u = X(x)Y(y)Z(z). \tag{1.39}$$

If this is substituted into (1.38), one obtains

$$\frac{1}{X}\frac{d^2X}{dx^2} + \frac{1}{Y}\frac{d^2Y}{dy^2} + \frac{1}{Z}\frac{d^2Z}{dz^2} + k^2 = 0. \tag{1.40}$$

Since k is a constant and each term depends on one variable only and can change independently, the left-hand side of (1.40) can sum to zero for all coordinate values only if each term is a constant. Thus,

$$\begin{aligned}
&\frac{d^2X}{dx^2} + k_x^2 X = 0,\\
&\frac{d^2Y}{dy^2} + k_y^2 Y = 0,\\
&\frac{d^2Z}{dz^2} + k_z^2 Z = 0,
\end{aligned} \tag{1.41}$$

where k_x, k_y, and k_z are **separation constants** and satisfy

$$k_x^2 + k_y^2 + k_z^2 = k^2. \tag{1.42}$$

The solutions of (1.41) are harmonic functions, denoted by $X(k_x x)$, $Y(k_y y)$, and $Z(k_z z)$, and they are any linear combination of the following independent **harmonic functions**:

$$e^{ik_\alpha \alpha}, e^{-ik_\alpha \alpha}, \cos k_\alpha \alpha, \sin k_\alpha \alpha \; (\alpha = x, y, z). \tag{1.43}$$

Consequently, the solution (1.39) may be expressed as

$$u = X(k_x x)Y(k_y y)Z(k_z z). \tag{1.44}$$

The separation constants k_x, k_y, and k_z are also called **eigenvalues**, and they are determined by the boundary conditions. The corresponding solutions (1.44) are called **eigenfunctions** or **elementary wavefunctions**. The general solution of (1.38) can be expressed as a linear combination of the eigenfunctions. For the solutions defined in finite regions, only discrete spectra of eigenvalues are involved. The discrete spectra become a continuum for the solutions defined in infinite regions. The harmonic functions should be properly selected according to the physical properties that the solutions must have. Note that the exponential functions in (1.43) represent a travelling wave while the sine and cosine functions represent a standing wave. Also note that the separation constants are a complex number for the waves propagating in a lossy medium.

1.2.1.2 Cylindrical Coordinate System

In a cylindrical coordinate system, (1.37) can be written as

$$\frac{1}{\rho}\frac{\partial}{\partial \rho}\left(\rho \frac{\partial u}{\partial \rho}\right) + \frac{1}{\rho^2}\frac{\partial^2 u}{\partial \varphi^2} + \frac{\partial^2 u}{\partial z^2} + k^2 u = 0. \tag{1.45}$$

By the method of separation of variables, the solutions may be assumed to be of the form

$$u = R(\rho)\Phi(\varphi)Z(z). \tag{1.46}$$

Introducing (1.46) into (1.45) yields

$$\begin{aligned} &\frac{d^2 R}{d\rho^2} + \frac{1}{\rho}\frac{dR}{d\rho} + \left(\mu^2 - \frac{p^2}{\rho^2}\right)R = 0, \\ &\frac{d^2\Phi}{d\varphi^2} + p^2\Phi = 0, \\ &\frac{d^2 Z}{dz^2} + \beta^2 Z = 0, \end{aligned} \tag{1.47}$$

where μ, p, and β are separation constants and satisfy

$$\beta^2 + \mu^2 = k^2. \tag{1.48}$$

The first equation of (1.47) is **Bessel equation**, named after the German mathematician Friedrich Wilhelm Bessel (1784–1846), whose solutions are **Bessel functions**:

$$J_p(\mu\rho), N_p(\mu\rho), H_p^{(1)}(\mu\rho), H_p^{(2)}(\mu\rho),$$

where $J_p(\mu\rho)$ and $N_p(\mu\rho)$ are the Bessel functions of the first and second kind, $H_p^{(1)}(\mu\rho)$ and $H_p^{(2)}(\mu\rho)$ are the Bessel functions of the third and fourth kind, also called **Hankel functions** of first and second kind, respectively, named after German mathematician Hermann Hankel (1839–1873). The Bessel function of the first kind is defined by

$$J_p(\mu z) = \sum_{m=0}^{\infty} \frac{(-1)^m}{\Gamma(m+1)\Gamma(p+m+1)} \left(\frac{\mu z}{2}\right)^{p+2m}, \tag{1.49}$$

where $\Gamma(\alpha)$ is the gamma function defined by

$$\Gamma(\alpha) = \int_0^{\infty} x^{\alpha-1} e^{-x} dx, \alpha > 0.$$

If p is not an integer, a second independent solution is $J_{-p}(\mu z)$. If $p = n$ is an integer, $J_{-n}(\mu z)$ is related to $J_n(\mu z)$ by $J_{-n}(z) = (-1)^n J_n(z)$. The Bessel function of the second kind is defined by

$$N_p(\mu z) = \frac{\cos p\pi J_p(\mu z) - J_{-p}(\mu z)}{\sin p\pi}, \tag{1.50}$$

and the Bessel functions of the third (Hankel function of the first kind) and fourth kind (Hankel function of the second kind) are defined by

$$\begin{aligned} H_p^{(1)}(\mu z) &= J_p(\mu z) + jN_p(\mu z), \\ H_p^{(2)}(\mu z) &= J_p(\mu z) - jN_p(\mu z). \end{aligned} \tag{1.51}$$

The solutions of second and third equation of (1.47) are harmonic functions. Note that only $J_p(\mu\rho)$ is finite at $\rho = 0$. The separation constants μ and p are determined by the boundary conditions. For example, if the field u is finite and satisfies homogeneous Dirichlet boundary condition $u = 0$ at $\rho = a$, the separation constant μ is determined by $J_p(\mu\rho) = 0$. If the cylindrical region contains all φ from 0 to 2π, the separation constant p is usually determined by the requirement that the field is single-valued, i.e. $\Phi(0) = \Phi(2\pi)$. In this case, p must be integers. If the cylindrical region only contains a circular sector, p will be fractional numbers.

Let $R_p(\mu z) = AJ_p(\mu z) + BN_p(\mu z)$, where A and B are constants. Some typical recurrence relations for the linear combination of the Bessel functions are listed below:

$$\frac{2p}{\mu z} R_p(\mu z) = R_{p-1}(\mu z) + R_{p+1}(\mu z),$$

$$\frac{1}{\mu} \frac{d}{dz} R_p(\mu z) = \frac{1}{2} \left[R_{p-1}(\mu z) - R_{p+1}(\mu z)\right],$$

$$z \frac{d}{dz} R_p(\mu z) = pR_p(\mu z) - \mu z R_{p+1}(\mu z),$$

$$\frac{d}{dz} \left[z^p R_p(\mu z)\right] = \mu z^p R_{p-1}(\mu z),$$

$$\frac{d}{dz} \left[z^{-p} R_p(\mu z)\right] = -\mu z^{-p} R_{p+1}(\mu z).$$

The Bessel functions have the orthogonality property

$$\int_0^1 x J_p\left(\chi_{pm} x\right) J_p\left(\chi_{pn} x\right) dx = \begin{cases} 0, & m \neq n \\ \frac{1}{2} \left[J'_p\left(\chi_{pm}\right)\right]^2, & m = n \end{cases}, \tag{1.52}$$

where χ_{pn} denotes the nth positive zero of the Bessel function J_p, i.e. $J_p(\chi_{pn}) = 0$, and the prime denotes the derivative of the Bessel function with respect to its argument. Let $C[a, b]$ denote the set of continuous functions defined on the closed interval $[a, b]$. An arbitrary continuous function in $C[0, 1]$ can be expanded in terms of the Bessel functions.

Theorem 1.1 (Fourier–Bessel Expansion)

If the function $f(x) \in C[0, 1]$ and the integral $\int_0^1 \sqrt{t} f(t) dt$ exist, it has the expansion

$$f(x) = \sum_n a_n J_p\left(\chi_{pn} x\right), \quad p > -1, \tag{1.53}$$

where the expansion coefficients can be determined from (1.52):

$$a_n = \frac{2}{\left[J'_p\left(\chi_{pn}\right)\right]^2} \int_0^1 x f(x) J_p\left(\chi_{pn} x\right) dx = \frac{2}{\left[J_{p+1}\left(\chi_{pn}\right)\right]^2} \int_0^1 x f(x) J_p\left(\chi_{pn} x\right) dx.$$

□

1.2.1.3 Spherical Coordinate System

In spherical coordinate system, (1.37) can be expressed as

$$\frac{1}{r^2}\frac{\partial}{\partial r}\left(r^2\frac{\partial u}{\partial r}\right)+\frac{1}{r^2\sin\theta}\frac{\partial}{\partial\theta}\left(\sin\theta\frac{\partial u}{\partial\theta}\right)+\frac{1}{r^2\sin\theta}\frac{\partial^2 u}{\partial\varphi^2}+k^2u=0. \quad (1.54)$$

By means of the separation of variables, one may assume

$$u=R(r)\Theta(\theta)\Phi(\varphi). \quad (1.55)$$

Substitution of (1.55) into (1.54) leads to

$$\begin{aligned}&\frac{1}{R}\frac{d}{dr}\left(r^2\frac{dR}{dr}\right)+k^2r^2=\beta^2,\\&\frac{1}{\Theta\sin\theta}\frac{d}{d\theta}\left(\sin\theta\frac{d\Theta}{d\theta}\right)-\frac{m^2}{\sin^2\theta}=-\beta^2,\\&\frac{d^2\Phi}{d\varphi^2}+m^2\Phi=0,\end{aligned} \quad (1.56)$$

where β and m are separation constants. Let $x=\cos\theta$ and $P(x)=\Theta(\theta)$. The second equation of (1.56) becomes

$$(1-x^2)\frac{d^2P}{dx^2}-2x\frac{dP}{dx}+\left(\beta^2-\frac{m^2}{1-x^2}\right)P=0. \quad (1.57)$$

This is called **Legendre equation**, named after the French mathematician Adrien-Marie Legendre (1752–1833). The points $x=\pm1$ are singular. Equation (1.57) has two linearly independent solutions and can be expressed as a power series at $x=0$. In general, the series solution diverges at $x=\pm1$. But if one lets $\beta^2=n(n+1)$, $n=0,1,2\ldots$, the series will be finite at $x=\pm1$ and have finite terms. Thus, the separation constant β is determined naturally and (1.56) can be rewritten as

$$\begin{aligned}&\frac{d}{dr}\left(r^2\frac{dR}{dr}\right)+\left[k^2r^2-n(n+1)\right]R=0,\\&(1-x^2)\frac{d^2P}{dx^2}-2x\frac{dP}{dx}+\left[n(n+1)-\frac{m^2}{1-x^2}\right]P=0,\\&\frac{d^2\Phi}{d\varphi^2}+m^2\Phi=0.\end{aligned} \quad (1.58)$$

The solutions of the first equation of (1.58) are **spherical Bessel functions** of the first and second kinds, **spherical Hankel functions** of the first and second kinds, respectively, defined by

$$
\begin{aligned}
j_n(kr) &= \sqrt{\frac{\pi}{2kr}} J_{n+1/2}(kr),\\
n_n(kr) &= \sqrt{\frac{\pi}{2kr}} N_{n+1/2}(kr),\\
h_n^{(1)}(kr) &= \sqrt{\frac{\pi}{2kr}} H_{n+1/2}^{(1)}(kr),\\
h_n^{(2)}(kr) &= \sqrt{\frac{\pi}{2kr}} H_{n+1/2}^{(2)}(kr).
\end{aligned}
\tag{1.59}
$$

Let $z_n(kr) = Aj_n(kr) + Bn_n(kr)$, where A and B are constants. The recurrence relations for the linear combination of the spherical Bessel functions are summarized as follows:

$$
\begin{aligned}
&\frac{2n+1}{kr} z_n(kr) = z_{n-1}(kr) + z_{n+1}(kr),\\
&\frac{2n+1}{k}\frac{d}{dr} z_n(kr) = nz_{n-1}(kr) - (n+1)z_{n+1}(kr),\\
&\frac{d}{dr}\left[r^{n+1} z_n(kr)\right] = kr^{n+1} z_{n-1}(kr),\\
&\frac{d}{dr}\left[r^{-n} z_n(kr)\right] = -kr^{-n} z_{n+1}(kr).
\end{aligned}
$$

The solutions of the second equation of (1.58) are **associated Legendre functions** of first and second kinds defined, respectively, by

$$
P_n^m(x) = \frac{(1-x^2)^{m/2}}{2^n n!}\frac{d^{m+n}}{dx^{m+n}}\left(x^2-1\right)^n, m \le n,
\tag{1.60}
$$

and

$$
Q_n^m(x) = \left(1-x^2\right)^{\frac{m}{2}}\frac{d^m}{dx^m} Q_n(x), m \le n,
\tag{1.61}
$$

where

$$
Q_n(x) = \frac{1}{2}P_n^0(x)\ln\frac{1+x}{1-x} - \sum_{r=1}^{n}\frac{1}{r}P_{r-1}^0(x)P_{n-r}^0(x)
$$

is the **Legendre function of the second kind**. The following integrations on orthogonality are useful:

1. $$\int_{-1}^{1} \frac{P_n^m(x)P_n^k(x)}{1-x^2}dx = \frac{1}{m}\frac{(n+m)!}{(n-m)!}\delta_{mk}.$$
2. $$\int_{-1}^{1} P_l^m(x)P_n^m(x)dx = \frac{2}{2l+1}\frac{(l+m)!}{(l-m)!}\delta_{ln}.$$
3. $$\int_0^{\pi}\left[\frac{dP_n^m(\cos\theta)}{d\theta}\frac{dP_l^m(\cos\theta)}{d\theta} + \frac{m^2}{\sin^2\theta}P_n^m(\cos\theta)P_l^m(\cos\theta)\right]\sin\theta d\theta$$
$$= \frac{2}{2n+1}\frac{(n+m)!}{(n-m)!}n(n+1)\delta_{nl}.$$

In the above, $\delta_{mn} = \begin{cases} 1, m = n \\ 0, m \neq n \end{cases}$. The solutions of the third equation of (1.58) are harmonic functions. Note that the separation constants are not related in the spherical coordinate system.

Theorem 1.2 (Completeness of Associated Legendre Functions)

Any function $f(x) \in C[-1, 1]$ satisfying the boundary conditions $f(-1) = f(1) = 0$ has the expansion

$$f(x) = \sum_{m \le n} a_n P_n^m(x),$$

where

$$a_n = \frac{2n+1}{2}\frac{(n-m)!}{(n+m)!}\int_{-1}^{1} f(x)P_n^m(x)dx.$$

□

1.2.2 Method of Green's Function

A source may be divided into a number of elementary sources. The superposition theorem indicates that the field generated by the source can be expressed as the sum of the fields generated by the elementary sources. The concept of the Green's function was first developed by the British mathematician George Green in the 1820s. Physically, Green's function represents the field produced by a point source (an elementary source). The well-known impulse response of a linear system is a typical Green's function with specified initial conditions. Green's function provides a general method to solve PDEs, by means of which the solution of a PDE can be represented by an integral defined over the source region or on a closed surface enclosing the source. Consider a PDE defined in a source region V contained in R^3 (the three-dimensional (3D) Euclidian space)

$$\hat{L}u(\mathbf{r}) = f(\mathbf{r}), \tag{1.62}$$

where $\hat{L}$ is a differential operator, $\mathbf{r} = (x, y, z) \in R^3$, and f is a known source function. The solution of the above equation can be expressed by

$$u(\mathbf{r}) = \hat{L}^{-1} f(\mathbf{r}),$$

where $\hat{L}^{-1}$ stands for the inverse of $\hat{L}$ and is often represented by an integral operator with a kernel function $G(\mathbf{r}, \mathbf{r}')$:

$$\hat{L}^{-1} f(\mathbf{r}) = -\int_V G(\mathbf{r}, \mathbf{r}') f(\mathbf{r}') dV(\mathbf{r}'). \tag{1.63}$$

If $\hat{L}$ is applied to both sides of the above equation and use is made of $\hat{L}\hat{L}^{-1} = \hat{I}$, where $\hat{I}$ is the **identity operator**, one obtains

$$\hat{L}\hat{L}^{-1} f(\mathbf{r}) = f(\mathbf{r}) = -\int_V \hat{L}G(\mathbf{r}, \mathbf{r}') f(\mathbf{r}') dV(\mathbf{r}').$$

This implies that the kernel function G satisfies

$$\hat{L}G(\mathbf{r}, \mathbf{r}') f(\mathbf{r}) = -\delta(\mathbf{r} - \mathbf{r}'), \tag{1.64}$$

where $\delta(\mathbf{r})$ is the 3D **Dirac delta function** defined symbolically by

$$\int_{R^3} \delta(\mathbf{r} - \mathbf{r}') \phi(\mathbf{r}') dV(\mathbf{r}') = \phi(\mathbf{r}) \tag{1.65}$$

with $\phi(\mathbf{r})$ being an arbitrary smooth function defined in R^3. The kernel function G is called the **fundamental solution** or **Green's function** of Eq. (1.62).

1.2.2.1 Green's Functions for Helmholtz Equation

Let $\boldsymbol{\rho} = (x, y)$ denote a point in the two-dimensional (2D) Euclidian space R^2 and $d\Omega = dxdy$ the differential area element. The fundamental solutions of Laplace and Helmholtz equations are summarized in Table 1.2. The 2D Dirac delta function $\delta(\boldsymbol{\rho})$ is defined by

$$\int_{R^2} \delta(\boldsymbol{\rho} - \boldsymbol{\rho}') \phi(\boldsymbol{\rho}') d\Omega(\boldsymbol{\rho}') = \phi(\boldsymbol{\rho}),$$

Table 1.2 Green's functions.

Equations	Green's functions
2D Laplace equation: $\nabla^2 G(\boldsymbol{\rho}, \boldsymbol{\rho}') = -\delta(\boldsymbol{\rho}-\boldsymbol{\rho}')$	$G(\boldsymbol{\rho}, \boldsymbol{\rho}') = -\frac{1}{2\pi}\ln\|\boldsymbol{\rho}-\boldsymbol{\rho}'\|$
3D Laplace equation: $\nabla^2 G(\mathbf{r}, \mathbf{r}') = -\delta(\mathbf{r}-\mathbf{r}')$	$G(\mathbf{r}, \mathbf{r}') = \frac{1}{4\pi\|\mathbf{r}-\mathbf{r}'\|}$
2D Helmholtz equation: $(\nabla^2 + k^2)G(\boldsymbol{\rho}, \boldsymbol{\rho}') = -\delta(\boldsymbol{\rho}-\boldsymbol{\rho}')$	$G(\boldsymbol{\rho}, \boldsymbol{\rho}') = \frac{1}{4j}H_0^{(2)}(k\|\boldsymbol{\rho}-\boldsymbol{\rho}'\|)$
3D Helmholtz equation: $(\nabla^2 + k^2)G(\mathbf{r}, \mathbf{r}') = -\delta(\mathbf{r}-\mathbf{r}')$	$G(\mathbf{r}, \mathbf{r}') = \frac{e^{-jk\|\mathbf{r}-\mathbf{r}'\|}}{4\pi\|\mathbf{r}-\mathbf{r}'\|}$

where $\phi(\mathbf{r})$ is an arbitrary smooth function defined in R^2. It can be seen that the Green's functions are symmetric $G(\mathbf{r}, \mathbf{r}') = G(\mathbf{r}', \mathbf{r})$.

Example 1.1 The Green's function for one-dimensional Helmholtz equation satisfies

$$\begin{cases} \dfrac{d^2G(z,z')}{dz^2} + k^2G(z,z') = -\delta(z-z'), \\ \lim\limits_{z\to\pm\infty}\left(\dfrac{dG}{dz} \pm jkG\right) = 0, \end{cases} \tag{1.66}$$

where the one-dimensional Dirac delta function $\delta(z)$ is defined by

$$\int_{-\infty}^{\infty} \delta(z-z')\phi(z')dz' = \phi(z)$$

and $\phi(z)$ is an arbitrary smooth function defined in the real axis R. The second equation in (1.66) denotes the radiation condition at infinity. To solve (1.66), one may let

$$G(z,z') = \begin{cases} G_1(z,z'), z < z' \\ G_2(z,z'), z > z' \end{cases}.$$

The functions G_1 and G_2 satisfy the homogeneous Helmholtz equation in the regions $z < z'$ and $z > z'$, respectively, and can be written as

$$G_1(z,z') = a_1e^{-jk(z-z')} + b_1e^{jk(z-z')},$$
$$G_2(z,z') = a_2e^{-jk(z-z')} + b_2e^{jk(z-z')},$$

where a_1, b_1, a_2, b_2 are constants to be determined. On account of the radiation conditions at $z = \pm\infty$, one may easily find $a_1 = b_2 = 0$. Thus,

$$\begin{aligned} G_1(z,z') &= b_1 e^{jk(z-z')}, \\ G_2(z,z') &= a_2 e^{-jk(z-z')}. \end{aligned} \tag{1.67}$$

The first equation of (1.66) implies that G must be continuous while its first derivative has a jump discontinuity at the source point $z = z'$:

$$\begin{aligned} &G_1(z,z') = G_2(z,z'), \\ &\frac{dG_2(z,z')}{dz} - \frac{dG_1(z,z')}{dz} = -1. \end{aligned}$$

Introducing (1.67) into the above equations yields

$$a_2 = b_1 = \frac{1}{j2k}.$$

The Green's function for one-dimensional Helmholtz equation is thus given by

$$G(z,z') = \begin{cases} \dfrac{1}{j2k} e^{jk(z-z')}, z < z' \\ \dfrac{1}{j2k} e^{-jk(z-z')}, z > z' \end{cases} = \frac{1}{j2k} e^{-jk|z-z'|}. \tag{1.68}$$

□

1.2.2.2 Dyadic Green's Functions and Integral Representations

The Green's function often appears as the kernel of an integral operator. For a scalar field, the kernel is also a scalar; but for a vector field, the kernel must be a dyadic. A dyadic is a second-order tensor formed by putting two vectors side by side. Its manipulation rules are analogous to that for matrix algebra (see Appendix B). Dyadic notation was first established by Josiah Willard Gibbs (1839–1903) in 1884. The application of dyadic Green's function in solving EM boundary value problem can be traced back to Schwinger's work in the early 1940s. Levine and Schwinger applied the dyadic Green's function to investigate the diffraction problem by an aperture in an infinite plane conducting screen [23]. In 1953, Morse and Feshbach discussed various applications of dyadic Green's functions [20]. A systematic study of dyadic Green's functions and their applications in EM engineering can be found in [24]. One of the advantages of using dyadic Green's functions is that it affords a compact formulation or solution for the field problems. Consider an electric current element in the direction of $\alpha(\alpha = x, y, z)$ located at $\mathbf{r}'$:

$$\mathbf{J}^{(\alpha)}(\mathbf{r}) = -\frac{1}{j\omega\mu}\delta(\mathbf{r}-\mathbf{r}')\mathbf{u}_\alpha,$$

which produces EM fields $\mathbf{E}^{(\alpha)}(\mathbf{r})$ and $\mathbf{H}^{(\alpha)}(\mathbf{r})$ at $\mathbf{r}$. Let

$$\begin{aligned} \mathbf{G}_e^{(\alpha)}(\mathbf{r},\mathbf{r}') &= \mathbf{E}^{(\alpha)}(\mathbf{r}), \\ \mathbf{G}_m^{(\alpha)}(\mathbf{r},\mathbf{r}') &= -j\omega\mu\mathbf{H}^{(\alpha)}(\mathbf{r}). \end{aligned} \tag{1.69}$$

Here $\mathbf{G}_e^{(\alpha)}(\mathbf{r},\mathbf{r}')$ and $\mathbf{G}_m^{(\alpha)}(\mathbf{r},\mathbf{r}')$ are, respectively, referred to as **electric and magnetic Green's functions** along direction α in free space. It follows from Maxwell equations that

$$\begin{aligned} \nabla\times\mathbf{G}_e^{(\alpha)}(\mathbf{r},\mathbf{r}') &= \mathbf{G}_m^{(\alpha)}(\mathbf{r},\mathbf{r}'), \\ \nabla\times\mathbf{G}_m^{(\alpha)}(\mathbf{r},\mathbf{r}') &= \mathbf{u}_\alpha\delta(\mathbf{r}-\mathbf{r}') + k^2\mathbf{G}_e^{(\alpha)}(\mathbf{r},\mathbf{r}'). \end{aligned}$$

The dyadic functions defined by

$$\begin{aligned} \overleftrightarrow{\mathbf{G}}_e(\mathbf{r},\mathbf{r}') &= \sum_{\alpha=x,y,z}\mathbf{G}_e^{(\alpha)}(\mathbf{r},\mathbf{r}')\mathbf{u}_\alpha, \\ \overleftrightarrow{\mathbf{G}}_m(\mathbf{r},\mathbf{r}') &= \sum_{\alpha=x,y,z}\mathbf{G}_m^{(\alpha)}(\mathbf{r},\mathbf{r}')\mathbf{u}_\alpha, \end{aligned}$$

are, respectively, called **electric and magnetic dyadic Green's functions** in free space. Apparently,

$$\begin{aligned} \nabla\times\overleftrightarrow{\mathbf{G}}_e(\mathbf{r},\mathbf{r}') &= \overleftrightarrow{\mathbf{G}}_m(\mathbf{r},\mathbf{r}'), \\ \nabla\times\overleftrightarrow{\mathbf{G}}_m(\mathbf{r},\mathbf{r}') &= \overleftrightarrow{\mathbf{I}}\delta(\mathbf{r}-\mathbf{r}') + k^2\overleftrightarrow{\mathbf{G}}_e(\mathbf{r},\mathbf{r}'), \end{aligned} \tag{1.70}$$

where $\overleftrightarrow{\mathbf{I}}$ is the identity dyadic. Note that $\overleftrightarrow{\mathbf{G}}_e(\mathbf{r},\mathbf{r}')$ is a symmetric while $\overleftrightarrow{\mathbf{G}}_m(\mathbf{r},\mathbf{r}')$ is antisymmetric upon interchange of $\mathbf{r}$ and $\mathbf{r}'$:

$$\begin{aligned} \overleftrightarrow{\mathbf{G}}_e(\mathbf{r},\mathbf{r}') &= \overleftrightarrow{\mathbf{G}}_e(\mathbf{r}',\mathbf{r}), \\ \overleftrightarrow{\mathbf{G}}_m(\mathbf{r},\mathbf{r}') &= -\overleftrightarrow{\mathbf{G}}_m(\mathbf{r}',\mathbf{r}). \end{aligned} \tag{1.71}$$

It follows from (1.70) that

$$\begin{aligned} \nabla\times\nabla\times\overleftrightarrow{\mathbf{G}}_e(\mathbf{r},\mathbf{r}') - k^2\overleftrightarrow{\mathbf{G}}_e(\mathbf{r},\mathbf{r}') &= \overleftrightarrow{\mathbf{I}}\delta(\mathbf{r}-\mathbf{r}'), \\ \nabla\times\nabla\times\overleftrightarrow{\mathbf{G}}_m(\mathbf{r},\mathbf{r}') - k^2\overleftrightarrow{\mathbf{G}}_m(\mathbf{r},\mathbf{r}') &= \nabla\times\left[\overleftrightarrow{\mathbf{I}}\delta(\mathbf{r}-\mathbf{r}')\right]. \end{aligned} \tag{1.72}$$

The free space electric dyadic Green's function $\overleftrightarrow{\mathbf{G}}_e(\mathbf{r},\mathbf{r}')$ may be represented by free space Green's function $G(\mathbf{r},\mathbf{r}')$:

$$\overleftrightarrow{\mathbf{G}}_e(\mathbf{r},\mathbf{r}') = \left(\overleftrightarrow{\mathbf{I}} + \frac{1}{k^2}\nabla\nabla\right)G(\mathbf{r},\mathbf{r}'). \tag{1.73}$$

In fact, the first equation of (1.72) may be written as

$$-\nabla^2\overleftrightarrow{\mathbf{G}}_e(\mathbf{r},\mathbf{r}') - k^2\overleftrightarrow{\mathbf{G}}_e(\mathbf{r},\mathbf{r}') + \nabla\nabla\cdot\overleftrightarrow{\mathbf{G}}_e(\mathbf{r},\mathbf{r}') = \overleftrightarrow{\mathbf{I}}\delta(\mathbf{r}-\mathbf{r}'). \tag{1.74}$$

Taking the divergence of the first equation of (1.72) yields

$$\nabla\cdot\overleftrightarrow{\mathbf{G}}_e(\mathbf{r},\mathbf{r}') = -\frac{1}{k^2}\nabla\cdot\left[\overleftrightarrow{\mathbf{I}}\delta(\mathbf{r}-\mathbf{r}')\right] = -\frac{1}{k^2}\nabla\delta(\mathbf{r}-\mathbf{r}').$$

Insertion of the above into (1.74) gives

$$\nabla^2\overleftrightarrow{\mathbf{G}}_e(\mathbf{r},\mathbf{r}') + k^2\overleftrightarrow{\mathbf{G}}_e(\mathbf{r},\mathbf{r}') = -\left(\overleftrightarrow{\mathbf{I}} + \frac{1}{k^2}\nabla\nabla\right)\delta(\mathbf{r}-\mathbf{r}').$$

Obviously, expression (1.73) for the free space electric dyadic Green's function satisfies the above equation. The free space magnetic dyadic Green's function may be expressed as

$$\overleftrightarrow{\mathbf{G}}_m(\mathbf{r},\mathbf{r}') = \nabla\times\overleftrightarrow{\mathbf{G}}_e(\mathbf{r},\mathbf{r}') = \nabla\times\left[G(\mathbf{r},\mathbf{r}')\overleftrightarrow{\mathbf{I}}\right] = \nabla\times\left[\overleftrightarrow{\mathbf{G}}_0(\mathbf{r},\mathbf{r}')\right],$$

where $\overleftrightarrow{\mathbf{G}}_0 = G\overleftrightarrow{\mathbf{I}}$ satisfies Helmholtz equation

$$(\nabla^2 + k^2)\overleftrightarrow{\mathbf{G}}_0(\mathbf{r},\mathbf{r}') = -\overleftrightarrow{\mathbf{I}}\delta(\mathbf{r}-\mathbf{r}'). \tag{1.75}$$

Consider the scattering problem of an incident field by an obstacle bounded by S as illustrated in Figure 1.2. Let V be the region bounded by $S + S_\infty$, where S_∞ is a surface enclosing the source and the obstacle. By letting $\mathbf{P} = \mathbf{E}$, $\mathbf{Q} = \mathbf{G}_e^{(\alpha)}$ in the vector Green's identity

$$\int_V (\mathbf{P}\cdot\nabla\times\nabla\times\mathbf{Q} - \mathbf{Q}\cdot\nabla\times\nabla\times\mathbf{P})dV = -\int_S (\mathbf{u}_n\times\mathbf{Q}\cdot\nabla\times\mathbf{P} - \mathbf{u}_n\times\mathbf{P}\cdot\nabla\times\mathbf{Q})dS$$
$$+\int_{S_\infty} (\mathbf{u}_n\times\mathbf{Q}\cdot\nabla\times\mathbf{P} - \mathbf{u}_n\times\mathbf{P}\cdot\nabla\times\mathbf{Q})dS,$$

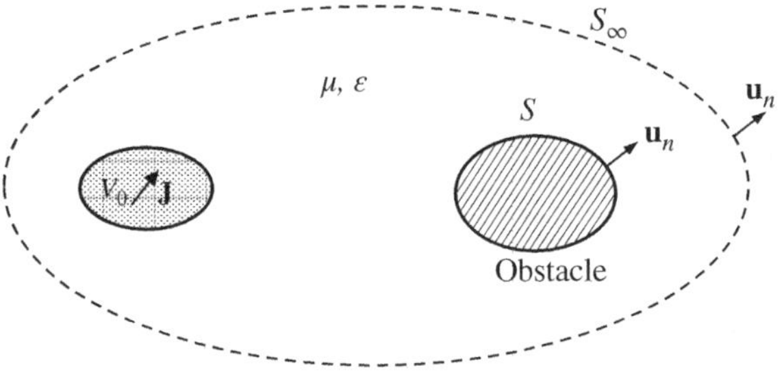

Figure 1.2 Scattering by an obstacle.

and by making use of the first Eq. (1.72) and the wave equation for the electric field

$$\nabla \times \nabla \times \mathbf{E}(\mathbf{r}) - k^2\mathbf{E}(\mathbf{r}) = -j\omega\mu\mathbf{J}(\mathbf{r}) - \nabla \times \mathbf{J}_m(\mathbf{r}), \tag{1.76}$$

one may find the integral expression for the electric field

$$\begin{aligned}\mathbf{E}(\mathbf{r}) &= -j\omega\mu \int_{V_0} \overleftrightarrow{\mathbf{G}}_e(\mathbf{r},\mathbf{r}') \cdot \mathbf{J}(\mathbf{r}')dV(\mathbf{r}') - \int_{V_0} \overleftrightarrow{\mathbf{G}}_m(\mathbf{r},\mathbf{r}') \cdot \mathbf{J}_m(\mathbf{r}')dV(\mathbf{r}') \\ &\quad - j\omega\mu \int_S \overleftrightarrow{\mathbf{G}}_e(\mathbf{r},\mathbf{r}') \cdot \mathbf{J}_s(\mathbf{r}')dS(\mathbf{r}') - \int_S \overleftrightarrow{\mathbf{G}}_m(\mathbf{r},\mathbf{r}') \cdot \mathbf{J}_{ms}(\mathbf{r}')dS(\mathbf{r}'),\end{aligned} \tag{1.77}$$

where $\mathbf{J}_s = \mathbf{u}_n \times \mathbf{H}$ and $\mathbf{J}_{ms} = -\mathbf{u}_n \times \mathbf{E}$ are, respectively, the equivalent surface electric and magnetic currents. Similarly, one may find the integral expression for the magnetic field

$$\begin{aligned}\mathbf{H}(\mathbf{r}) &= -j\omega\varepsilon \int_{V_0} \overleftrightarrow{\mathbf{G}}_e(\mathbf{r},\mathbf{r}') \cdot \mathbf{J}_m(\mathbf{r}')dV(\mathbf{r}') + \int_{V_0} \overleftrightarrow{\mathbf{G}}_m(\mathbf{r},\mathbf{r}') \cdot \mathbf{J}(\mathbf{r}')dV(\mathbf{r}') \\ &\quad - j\omega\varepsilon \int_S \overleftrightarrow{\mathbf{G}}_e(\mathbf{r},\mathbf{r}') \cdot \mathbf{J}_{ms}(\mathbf{r}')dS(\mathbf{r}') + \int_S \overleftrightarrow{\mathbf{G}}_m(\mathbf{r},\mathbf{r}') \cdot \mathbf{J}_s(\mathbf{r}')dS(\mathbf{r}').\end{aligned} \tag{1.78}$$

In the derivation of (1.77) and (1.78), the surface integral over S_∞ has been ignored since it approaches to zero due to the radiation condition as S_∞ expands to infinity. The volume integrals in (1.77) and (1.78) represent the incident fields while the surface integrals denote the scattered fields produced by the induced currents on the obstacle.

Remark 1.2 Care must be exercised when the observation point $\mathbf{r}$ is inside the source region V_0. The dyadic Green's functions in (1.77) and (1.78) become infinite when $\mathbf{r}$ approaches $\mathbf{r}'$. This singular behavior is often treated with the principal volume method by introducing an exclusion volume around $\mathbf{r}$ [25–27]. □

Remark 1.3 For many applications discussed in this book, the dyadic Green's functions may be expanded in terms of eigenfunctions of the Helmholtz equation for the vector fields. In these cases, the integral representations (1.77) and (1.78) become an infinite series. □

1.2.3 Variational Method

Variational method or **calculus of variations** is a generalization of calculus and can be traced back to the early 1730s when Swiss mathematician Leonhard Euler

(1707–1783) and French mathematician Joseph-Louis Lagrange (1736–1813) elaborated the subject. Instead of finding the extrema of a function in calculus, the calculus of variations deals with maximizing or minimizing functionals, which are often expressed as definite integrals involving functions and their derivatives. The extremal functions that make the functional stationary (i.e. attain a maximum or minimum value) can be obtained by assuming that the rate of change of the functional is zero. Let F be a map from a linear space consisting of functions into the real axis. Such a map is called a **functional**. Let v be an arbitrary function in the space. The **gradient** or **functional derivative** of F at u, denoted by $\delta F(u)/\delta u$ and used to describe the rate of change of the functional, is defined by [16]

$$\left(\frac{\delta F(u)}{\delta u}, v\right) = \frac{d}{d\varepsilon}F(u+\varepsilon v)\bigg|_{\varepsilon=0}, \tag{1.79}$$

where $(\cdot,\cdot)$ is an **inner product** defined by the following rules:

1) Positive definiteness: $(u, u) \geq 0$ and $(u, u) = 0$ if and only if $u = 0$.
2) Hermitian property: $(u, v) = \overline{(v, u)}$.
3) Homogeneity: $(\alpha u, v) = \alpha(u, v)$.
4) Additivity: $(u + v, w) = (u, w) + (v, w)$.

Here, u, v, z are functions and α is a number. A linear space equipped with an inner product is called an **inner product space**.

Extremum Theorem: A necessary condition for a functional F to have an extremum at u is that its functional derivative vanishes at u:

$$\frac{\delta F(u)}{\delta u} = 0. \tag{1.80}$$

This is referred to as **Euler–Lagrangian equation**. □

Example 1.2 An **operator** (or **transformation**) $\hat{L}$ on an inner product space is defined as a map from the inner product space to itself. An operator $\hat{L}$ is called **self-adjoint** if it satisfies

$$\left(\hat{L}u, v\right) = \left(u, \hat{L}v\right).$$

The most important functional to be dealt with in this book is the **Rayleigh quotient**, named after the English scientist Lord Rayleigh (1942–1919), defined by

$$\lambda(u) = \frac{\left(\hat{L}u, u\right)}{(u, u)}. \tag{1.81}$$

Since $\hat{L}$ is self-adjoint, λ can be shown to be real. For an arbitrary v, the functional derivative of the Rayleigh quotient can be determined from (1.79):

$$\left(\frac{\delta\lambda}{\delta u}, v\right) = \frac{d}{d\varepsilon}\lambda(u+\varepsilon v)\bigg|_{\varepsilon=0} = \frac{d}{dt}\frac{(\hat{L}(u+\varepsilon v), u+\varepsilon v)}{(u+\varepsilon v, u+\varepsilon v)}\bigg|_{\varepsilon=0}$$
$$= \frac{1}{(u,u)} 2\mathrm{Re}\left(\hat{L}u - \lambda u, v\right).$$

If $\delta\lambda/\delta u = 0$, one obtains the eigenvalue equation

$$\hat{L}u = \lambda u. \tag{1.82}$$

This is the Lagrangian equation for the Rayleigh quotient (1.81). □

1.3 Eigenvalue Problem for Hermitian Matrix

When the dimension of the underlying vector space of a linear transformation is finite, the linear transformation is reduced to a matrix under a fixed basis. A complex square matrix $[A]$ is called **Hermitian**, named after the French mathematician Charles Hermite (1822–1901), if $[A] = [A]^H$, where the superscript H denotes the conjugate transpose of a matrix. Hermitian matrices are the complex extension of real symmetric matrices and are fundamental to mathematics, physics, and engineering sciences.

1.3.1 Properties

Let $[A]$ denote an $N \times N$ Hermitian matrix. If an $N \times 1$ matrix $[x]$ (also called a **column vector** or **vector**) and a scalar λ satisfy the equation

$$[A][x] = \lambda[x], \tag{1.83}$$

then λ is called an **eigenvalue** of $[A]$, $[x]$ is called an **eigenvector** corresponding to the eigenvalue λ, and $(\lambda, [x])$ is called an **eigenpair** of $[A]$. The eigenvalue problem (1.83) for a Hermitian matrix has the following properties, which can be found in a standard text book on matrix [e.g. 28]:

1) All the eigenvalues are real and they can be ordered such that

   $$\lambda_1 \le \lambda_2 \le \cdots \le \lambda_N,$$

   and the corresponding eigenvectors are denoted by $[x_1]$, $[x_2]$, ..., $[x_N]$.
2) The eigenfunctions corresponding to different eigenvalues are orthogonal.

3) Assume that the eigenvectors $[x_1], [x_2], \ldots, [x_N]$ are orthonormal. The matrix $[A]$ has the **spectral decomposition**

$$[A] = [U][\Lambda][U]^H = \sum_{j=1}^{N} \lambda_j [x_j] [x_j]^H,$$

where

$$[U] = [[x_1], [x_2], \ldots, [x_N]],$$
$$[\Lambda] = \mathrm{diag}(\lambda_1, \lambda_2, \ldots, \lambda_N).$$

The matrix $[U]$ is unitary $[U]^H[U] = [I]$, where $[I]$ denotes identity matrix.

The following statements are equivalent.

1) The Hermitian matrix $[A]$ is **positive**, i.e. $[x]^H[A][x] > 0$ for all $[x] \neq 0$.
2) All eigenvalues of $[A]$ are positive.
3) There exists a nonsingular $n \times n$ matrix $[R]$, such that $[A] = [R]^H[R]$.
4) There exists a nonsingular $n \times n$ matrix $[R]$, such that $[R]^H[A][R]$ is positive.

A positive matrix $[A]$ (not necessarily Hermitian) has the following properties:

1) The diagonal elements of $[A]$ are all positive.
2) $\det[A] > 0$.
3) $[A]^{-1}$ is positive.
4) The element of the largest absolute value is a diagonal element.

1.3.2 Rayleigh Quotient

The **Rayleigh quotient** for an $n \times n$ Hermitian matrix $[A]$ is defined by

$$\lambda = \frac{[x]^H [A][x]}{[x]^H [x]}. \tag{1.84}$$

Any vector $[x]$ can be expressed as a linear combination of the eigenvectors

$$[x] = \sum_{j=1}^{N} c_j [x_j] = [U][c_j], \tag{1.85}$$

where $[c_j] = [c_1, c_2, \ldots, c_N]^T$. Assume that the eigenvectors $[x_1], [x_2], \ldots, [x_N]$ of $[A]$ are orthonormal. Introducing the above expansion into (1.84), one may find

$$\begin{aligned}\lambda &= \frac{[x]^H[A][x]}{[x]^H[x]} = \frac{[c_j]^H[U]^H[A][U][c_j]}{[c_j]^H[U]^H[U][c_j]} = \frac{[c_j]^H[\Lambda][c_j]}{[c_j]^H[c_j]} \\ &= \frac{\lambda_1|c_1|^2 + \lambda_2|c_2|^2 + \cdots + \lambda_n|c_N|^2}{|c_1|^2 + |c_2|^2 + \cdots + |c_N|^2},\end{aligned} \tag{1.86}$$

which implies

$$\lambda_1 \le \frac{[x]^H[A][x]}{[x]^H[x]} \le \lambda_N. \tag{1.87}$$

As a result, the following optimization problems are established:

$$\begin{aligned}\lambda_1 &= \min \frac{[x]^H[A][x]}{[x]^H[x]}, \\ \lambda_N &= \max \frac{[x]^H[A][x]}{[x]^H[x]},\end{aligned} \tag{1.88}$$

where $[x] \in C^N$ is called **trial vector** and C^N stands for the set consisting of all complex column vectors of dimension N. To determine other eigenvalues, the trial vector $[x]$ will be assumed to be orthogonal to $[x_1]$. Thus, one finds $c_1 = 0$ in the expansion (1.85), and (1.86) reduces to

$$\lambda = \frac{\lambda_2|c_2|^2 + \cdots + \lambda_N|c_N|^2}{|c_2|^2 + \cdots + |c_N|^2} \ge \lambda_2. \tag{1.89}$$

The equality holds for $[x] = [x_2]$. The above relation suggests

$$\lambda_2 = \min_{[x]\perp[x_1]} \frac{[x]^H[A][x]}{[x]^H[x]}. \tag{1.90}$$

The second eigenvalue λ_2 can thus be found by minimizing (1.84) with the supplementary constraint that the trial vector $[x]$ must be orthogonal to the first eigenvector $[x_1]$. In general, one may write

$$\lambda_n = \min_{[x]\in[[x_1],[x_2],\ldots,[x_{n-1}]]^\perp} \frac{[x]^H[A][x]}{[x]^H[x]}, \quad 2 \le n \le N. \tag{1.91}$$

Similarly,

$$\lambda_n = \max_{[x]\in[[x_{n+1}],\ldots,[x_N]]^\perp} \frac{[x]^H[A][x]}{[x]^H[x]}, 1 \le n \le N-1. \tag{1.92}$$

1.4 Eigenvalue Problems for the Laplace Operator on Scalar Field

If the dimension of the underlying vector space of a linear transformation is infinite, the linear transformation is usually a differential operator defined on a subspace, or an integral operator defined on the whole space. In 1894, the French mathematician Jules Henri Poincaré (1854–1912) established the existence of an infinite sequence of eigenvalues and the corresponding eigenfunctions for the Laplace operator under Dirichlet boundary condition. This key result extends the eigenvector and eigenvalue theory of a square matrix and has played an important role in mathematical physics.

1.4.1 Rayleigh Quotient

Let $L^2(\Omega)$ denote the set of all functions defined on a region Ω bounded by Γ in 2D or 3D space such that $\int_\Omega |u|^2 d\Omega < \infty$, where $d\Omega$ denotes the area element (2D) or volume element (3D). The **inner product** and the **norm** in $L^2(\Omega)$ are, respectively, defined by

$$(u, v) = \int_\Omega u\bar{v}d\Omega, \quad \|u\| = (u, u)^{1/2}. \tag{1.93}$$

Two functions are said to be **orthogonal** if their inner product is zero. Consider the eigenvalue problems for the **Laplace operator** (also called **Laplacian**) $-\nabla^2$ acting on the scalar fields with three different boundary conditions:

1) **Dirichlet problem**:

$$\begin{cases} -\nabla^2 u(\mathbf{r}) = \lambda u(\mathbf{r}), \mathbf{r} \in \Omega \\ u(\mathbf{r}) = 0, \mathbf{r} \in \Gamma \end{cases}. \tag{1.94}$$

2) **Neumann problem**:

$$\begin{cases} -\nabla^2 u(\mathbf{r}) = \lambda u(\mathbf{r}), \mathbf{r} \in \Omega \\ \dfrac{\partial u(\mathbf{r})}{\partial n} = 0, \mathbf{r} \in \Gamma \end{cases}, \tag{1.95}$$

where $\partial/\partial n$ denotes the derivative in the direction normal to the boundary Γ.

3) **Robin problem**:

$$\begin{cases} -\nabla^2 u(\mathbf{r}) = \lambda u(\mathbf{r}), \mathbf{r} \in \Omega \\ \dfrac{\partial u(\mathbf{r})}{\partial n} + a(\mathbf{r})u(\mathbf{r}) = 0, \mathbf{r} \in \Gamma \end{cases}, \tag{1.96}$$

where $a(\mathbf{r})$ is a continuous function.

For any functions u_1 and u_2 satisfying one of the above boundary conditions, the Laplacian $-\nabla^2$ is **symmetric**

$$\begin{aligned}(-\nabla^2 u_1, u_2) &= \int_\Omega -\bar{u}_2 \nabla^2 u_1 d\Omega = \int_\Omega \nabla u_1 \nabla \bar{u}_2 d\Omega - \int_\Gamma \bar{u}_2 \frac{\partial u_1}{\partial n} d\Gamma \\ &= \int_\Omega \nabla u_1 \nabla \bar{u}_2 d\Omega = (u_1, -\nabla^2 u_2),\end{aligned} \tag{1.97}$$

where use has been made of the **Green's first identity**

$$\int_\Omega (u\nabla^2 v + \nabla u \nabla v) d\Omega = \int_\Gamma u \frac{\partial v}{\partial n} d\Gamma. \tag{1.98}$$

The symmetric property of the Laplacian is also easily seen from the **Green's second identity**

$$\int_\Omega (u\nabla^2 v - v\nabla^2 u) d\Omega = \int_\Gamma \left(u\frac{\partial v}{\partial n} - v\frac{\partial u}{\partial n} \right) d\Gamma. \tag{1.99}$$

Apparently,

$$(-\nabla^2 u, u) = \int_\Omega \|\nabla u\|^2 d\Omega \geq 0, \tag{1.100}$$

which implies that the Laplacian $-\nabla^2$ is **non-negative**. The three eigenvalue problems (1.94)–(1.96) have the properties:

1) All the eigenvalues are real and the corresponding eigenfunctions can be chosen to be real.
2) The eigenfunctions corresponding to different eigenvalues are orthogonal.
3) All the eigenvalues are positive for Dirichlet problem (1.94). All the eigenvalues are non-negative (positive or zero) for Neumann and Robin problems (1.95) and (1.96).

The **Rayleigh quotient** for the Laplacian $-\nabla^2$ can be obtained by taking the inner product of the first equation of (1.94) with u:

$$\lambda = \frac{(-\nabla^2 u, u)}{(u, u)}. \tag{1.101}$$

After applying Green's first identity to (1.101), the Rayleigh quotient can be written as

$$\lambda = \frac{1}{(u, u)} \left[(\nabla u, \nabla u) - \int_\Gamma \bar{u} \frac{\partial u}{\partial n} d\Gamma \right]. \tag{1.102}$$

If u satisfies the Dirichlet boundary condition, the boundary integral in (1.102) vanishes and the Rayleigh quotient reduces to

$$\lambda = \frac{(\nabla u, \nabla u)}{(u, u)}. \tag{1.103}$$

Consider the minimization of the Rayleigh quotient

$$\lambda = \min_{h|_\Gamma = 0} \frac{(\nabla h, \nabla h)}{(h, h)}, \tag{1.104}$$

where h is a smooth function (called **trial function**), satisfying the Dirichlet boundary condition $h|_\Gamma = 0$.

Theorem 1.3 If u_1 is a solution of the minimum problem (1.104) and λ_1 is the value of the minimum (i.e. the Rayleigh quotient reaches minimum value λ_1 when $h = u_1$), then λ_1 is the smallest eigenvalue of the Dirichlet problem (1.94) and u_1 its corresponding eigenfunction. □

Proof. Let v be an arbitrary function and ε be a small number. Since λ_1 is a minimum of the Rayleigh quotient (1.103), the function

$$\lambda(\varepsilon) = \frac{(\nabla(u_1 + \varepsilon v), \nabla(u_1 + \varepsilon v))}{(u_1 + \varepsilon v, u_1 + \varepsilon v)} \tag{1.105}$$

has a minimum at $\varepsilon = 0$ with

$$\lambda_1 = \frac{(\nabla u_1, \nabla u_1)}{(u_1, u_1)}. \tag{1.106}$$

By ordinary calculus, this implies $\lambda'(0) = 0$. Thus,

$$\lambda'(0) = \frac{1}{(u_1, u_1)^2}[2(\nabla u_1, \nabla v)(u_1, u_1) - 2(u_1, v)(\nabla u_1, \nabla u_1)] = 0.$$

It follows that

$$(\nabla u_1, \nabla v) - (u_1, v)\frac{(\nabla u_1, \nabla u_1)}{(u_1, u_1)} = 0. \tag{1.107}$$

By using Green's first identity (1.98), (1.106), and the Dirichlet boundary condition, one may obtain

$$(\nabla^2 u_1, v) + \lambda_1(u_1, v) = 0.$$

Since v is arbitrary, it may be concluded that λ_1 is an eigenvalue of (1.94) and its eigenfunction is u_1. It is easy to verify that λ_1 is the smallest eigenvalue of (1.94). The proof is completed.

Once the first eigenvalue λ_1 and its eigenfunction u_1 are determined, the second eigenvalue λ_2 can be found by minimizing (1.104) with the supplementary constraint that the trial function must be orthogonal to the first eigenfunction u_1. In general, u_n is the nth eigenfunction that minimizes the Rayleigh quotient (1.104) under the conditions

$$(u_n, u_1) = (u_n, u_2) = \cdots (u_n, u_{n-1}) = 0,$$

and the corresponding eigenvalues satisfy $0 < \lambda_1 \le \lambda_2 \le \cdots$.

Consider now the Neumann eigenvalue problem (1.95). Instead of using (1.104), the following minimization problem for the Rayleigh quotient will be introduced:

$$\lambda = \min \frac{(\nabla h, \nabla h)}{(h, h)}, \tag{1.108}$$

where the trial function h is not required to satisfy any boundary conditions.

Theorem 1.4 If u_1 is a solution of the minimum problem (1.108) and λ_1 is the value of the minimum, then λ_1 is the smallest eigenvalue of the Neumann problem (1.95) and u_1 its corresponding eigenfunction. □

Proof. The steps of the proof is similar to that of Theorem 1.3 until (1.107), which, after applying Green's first identity, can be written as

$$(\nabla^2 u_1, v) + \lambda_1(u_1, v) = \int_\Gamma v \frac{\partial u_1}{\partial n} d\Gamma. \tag{1.109}$$

Since v is an arbitrary function, one can first choose v arbitrarily inside Ω and $v = 0$ on the boundary Γ. Then, (1.109) implies $\nabla^2 u_1 + \lambda_1 u_1 = 0$ inside Ω. Thus,

$$\int_\Gamma v \frac{\partial u_1}{\partial n} d\Gamma = 0$$

holds for an arbitrary function v. Now selecting $v = \partial u_1/\partial n$ yields $\partial u_1/\partial n = 0$, which is the Neumann boundary condition. The proof is completed.

The above discussion indicates that the Neumann boundary condition is naturally met after minimization. For this reason, it is called **natural boundary condition** or **free boundary condition**. It is noted that the first eigenvalue λ_1 for Neumann problem (1.95) can be zero and the corresponding eigenfunction becomes a constant. Other eigenvalues and their eigenfunctions can be determined by (1.108) in the same way as discussed for Dirichlet problem.

Remark 1.4 It can be shown that the Dirichlet boundary condition in (1.94) becomes a natural boundary condition if one uses the following Rayleigh quotient

$$\lambda = \min \frac{(\nabla h, \nabla h) - 2\int_{\Gamma} h\frac{\partial h}{\partial n} d\Gamma}{(h, h)}, \tag{1.110}$$

where the trial functions are not required to satisfy any boundary conditions. □

1.4.2 Properties of Eigenvalues

Let λ_j and λ_j', respectively, denote Dirichlet and Neumann eigenvalues. The Rayleigh quotients in (1.104) and (1.108) have the same expressions except that the trial functions for λ_j satisfy extra constraints. Some important properties of eigenvalues are summarized below [29]:

1) $\lambda_j' < \lambda_j (j = 1, 2, \ldots)$.
2) As the domain increases, the eigenvalues λ_n or λ_n' decreases.
3) If Ω is a plane domain in a 2D space, the eigenvalues for the Dirichlet problem (1.94) satisfy $\lim\limits_{n\to\infty} \frac{\lambda_n}{n} = \frac{4\pi}{A}$, where A is the area of Ω. If Ω is a solid domain in a 3D space, the eigenvalues for the Dirichlet problem (1.94) satisfy $\lim\limits_{n\to\infty} \frac{\lambda_n^{3/2}}{n} = \frac{6\pi^2}{V}$, where V is the volume of Ω.
4) $\lim\limits_{n\to\infty} \lambda_n = \infty$ and $\lim\limits_{n\to\infty} \lambda_n' = \infty$.

The first property results from the consideration that additional constraints will increase the value of minimum. In waveguide theory, this property implies that the dominant mode of a hollow metal waveguide is always a TE mode. The second property is simply because the set of trial functions defined in a larger domain includes those defined in a smaller domain contained in the larger domain. In waveguide theory, this property implies that the cutoff wavenumber of a hollow metal waveguide decreases as the cross section of the waveguide increases. As the cross section becomes infinitely large, the discrete cutoff wavenumbers will become closer together and merge into the positive axis. Only the proof of the last property will be given here. To this end, one needs the following **Rellich's theorem**, named after the Austrian-German mathematician Franz Rellich (1906–1955).

Rellich's Theorem: Any sequence $\{f_n\}$ that satisfies

$$\|f_n\|^2 = \int_\Omega f_n^2 d\Omega \leq c_1, \quad \|\nabla f_n\|^2 = \int_\Omega (\nabla f_n)^2 d\Omega \leq c_2,$$

where c_1 and c_2 are constants, has a subsequence, still denoted by $\{f_n\}$, such that

$$\lim_{m,n\to\infty} \int_\Omega (f_m - f_n)^2 d\Omega = 0.$$

□

According to the solution procedures of the eigenfunctions described by (1.104) or (1.108), one may write

$$\lambda_n = \frac{(\nabla u_n, \nabla u_n)}{(u_n, u_n)}. \tag{1.111}$$

If λ_n is assumed to be finite as $n \to \infty$, the sequence $\{u_n|(u_n, u_n) = 1\}$ satisfies the conditions stated in Rellich's theorem. So, one can choose a subsequence of $\{u_n|(u_n, u_n) = 1\}$, still denoted by $\{u_n|(u_n, u_n) = 1\}$, such that

$$\lim_{m,n\to\infty} \int_\Omega (u_m - u_n)^2 d\Omega = 0.$$

This contradicts the fact that $\int_\Omega (u_m - u_n)^2 d\Omega = 2$. Therefore, the assumption that λ_n remains finite as $n \to \infty$ is wrong. The proof is completed.

Remark 1.5 For the eigenvalue problems defined in 2D or 3D space, the eigenvalues are numbered multiple integers. For example, if Ω is a rectangle $[0, a] \times [0, b]$ in the plane, one has

$$\lambda_n = \left(\frac{p\pi}{a}\right)^2 + \left(\frac{q\pi}{b}\right)^2, p = 1, 2, \ldots, q = 1, 2, \ldots. \tag{1.112}$$

In this case, n stands for multiple indices (p, q) and it is difficult to see how (1.112) is related to property 3. To solve this problem, one can introduce the **enumeration function** $N(\lambda)$, defined as the number of eigenvalues that do not exceed λ. Clearly $N(\lambda)$ is the number of lattice points (p, q) satisfying

$$\left(\frac{p}{a}\right)^2 + \left(\frac{q}{b}\right)^2 \leq \frac{\lambda}{\pi^2}.$$

If the eigenvalues (1.112) are arranged in increasing order, it is easy to see $N(\lambda) = n$. □

1.4.3 Completeness of Eigenfunctions

A set of eigenfunctions is said to be **complete** in $L^2(\Omega)$ if any function in $L^2(\Omega)$ can be expressed as a linear combination of the eigenfunctions.

Theorem 1.5 The eigenfunctions resulted from (1.104) (Dirichlet boundary condition) and (1.108) (Neumann boundary condition) are complete in $L^2(\Omega)$. In other words, an arbitrary function $f \in L^2(\Omega)$ can be represented as a linear combination of the eigenfunctions in the sense that

$$\left\| f - \sum_{n=1}^{N} c_n u_n \right\| \to 0, \tag{1.113}$$

as $N \to \infty$. In the above, $u_n (n = 1, 2, ...)$ are the eigenfunctions obtained either from (1.104) or (1.108), which are assumed to be mutually orthogonal, and

$$c_n = \frac{(f, u_n)}{(u_n, u_n)}, \quad (n = 1, 2, ...) \tag{1.114}$$

are the (Fourier) expansion coefficients. □

Proof. Let f be a trial function. The remainder of the expansion of f

$$r_N = f - \sum_{n=1}^{N} c_n u_n$$

is also a trial function. From the orthogonality of the eigenfunctions, one may easily verify that $(r_N, u_j) = 0$ for $j \leq N$, which implies

$$\lambda_N \leq \frac{(\nabla r_N, \nabla r_N)}{(r_N, r_N)} \tag{1.115}$$

by the construction of the eigenfunctions. After expanding the numerator, one may find

$$(\nabla r_N, \nabla r_N) = (\nabla f, \nabla f) - \sum_{n=1}^{N} c_n^2 \lambda_n (u_n, u_n) \leq \|\nabla f\|^2. \tag{1.116}$$

It follows from (1.115) and (1.116) that

$$\|r_N\|^2 \leq \frac{\|\nabla f\|^2}{\lambda_N}. \tag{1.117}$$

Since $\lambda_N \to \infty$ as $N \to \infty$, one may find $\|r_N\| \to 0$ as $N \to \infty$. Hence, (1.113) is valid for a trial function f. It can be shown that an arbitrary function in $L^2(\Omega)$ can be approximated by a trial function. Thus, the proof is completed.

Remark 1.6 The convergence in the function space $L^2(\Omega)$ as demonstrated in (1.113) is usually nonuniform. Great care must be taken in performing term-by-term differentiation, integration, and limiting process of the series. In many cases, as will be seen later in the theory of waveguide and cavity resonator, the function and its derivative must be expanded separately. □

Remark 1.7 If the set of eigenfunctions $\{u_n\}$ is **orthonormal**

$$(u_m, u_n) = \delta_{mn}, \tag{1.118}$$

and satisfy the homogeneous Dirichlet condition in (1.94) or Neumann boundary condition in (1.95), one can construct a set of vector field $\{\mathbf{E}_n\}$, defined by $\mathbf{E}_n = (1/\lambda_n)\nabla u_n$, such that

$$\int_\Omega \mathbf{E}_m \cdot \mathbf{E}_n d\Omega = \delta_{mn}. \tag{1.119}$$

Equation (1.119) is a direct consequence of the Green's first identity (1.98). □

1.4.4 Differential Equations with Variable Coefficients

The main results obtained in previous sections can be generalized to the differential equations with variable coefficients

$$\frac{1}{w(\mathbf{r})}[-\nabla\cdot p(\mathbf{r})\nabla + q(\mathbf{r})]u(\mathbf{r}) = \lambda u(\mathbf{r}), \mathbf{r}\in\Omega, \tag{1.120}$$

where p and w are assumed to be positive, and q is a continuous function. Three boundary value problems similar to (1.94)–(1.96) may be introduced with the operator $-\nabla^2$ replaced by $(-\nabla\cdot p\nabla + q)/w$. The inner product and the norm are now, respectively, defined by

$$(u, v)_w = \int_\Omega w(\mathbf{r})u(\mathbf{r})\bar{v}(\mathbf{r})d\Omega, \ \ \|u\|_w = \sqrt{(u, u)_w}.$$

Let $L_w^2(\Omega)$ denote the set of all functions defined on the region Ω in 2D or 3D space such that $\|u\|_w < \infty$. Then, all the eigenvalues of (1.120) under the three boundary conditions are real and the corresponding eigenfunctions can be chosen to be real, and the eigenfunctions corresponding to different eigenvalues are orthogonal. The Rayleigh quotient for the eigenvalue problem (1.120) is given by

$$\lambda = \frac{\left(\frac{1}{w}(-\nabla\cdot p\nabla + q)u, u\right)_w}{(u, u)_w}. \tag{1.121}$$

By use of the identity $\nabla \cdot (\phi \mathbf{a}) = \phi \nabla \cdot \mathbf{a} + \mathbf{a} \cdot \nabla \phi$, the above quotient can be written as

$$\lambda = \frac{\int\limits_{\Omega} (p|\nabla u|^2 + qu^2) d\Omega - \int\limits_{\Gamma} p \frac{\partial u}{\partial n} u d\Gamma}{(u, u)_w}. \tag{1.122}$$

For the Dirichlet and Neumann problems, the boundary term in (1.122) vanishes and the quotient reduces to

$$\lambda = \frac{\int\limits_{\Omega} (p|\nabla u|^2 + qu^2) d\Omega}{(u, u)_w}. \tag{1.123}$$

During the minimization process for determining the eigenvalues and the eigenfunctions, the trial functions in the quotient (1.123) must satisfy the Dirichlet boundary condition for Dirichlet problem while they are free for Neumann problem. For the Robin problem, (1.122) becomes

$$\lambda = \frac{\int\limits_{\Omega} (p|\nabla u|^2 + qu^2) d\Omega + \int\limits_{\Gamma} pau^2 d\Gamma}{(u, u)_w}. \tag{1.124}$$

In this case, the trial functions are also free.

As a special case in one dimension, let us consider the most important eigenvalue problem in mathematical physics, the **Sturm–Liouville type**

$$\frac{1}{w(x)} \left[-\frac{d}{dx} p(x) \frac{d}{dx} + q(x) \right] u(x) = \lambda u(x), x \in (a, b), \tag{1.125}$$

where the functions p and w are smooth and positive, and q is real piecewise continuous function, all defined over a finite interval $[a, b]$. Then, the following properties can be established for the eigenvalue problem (1.125):

1) The eigenvalues are real and all the corresponding eigenfunctions can be assumed to be real.
2) The eigenfunctions of different eigenvalues are orthogonal.
3) The eigenvalue problem (1.125) has an infinite set of eigenvalues $\lambda_1 \leq \lambda_2 \leq \cdots \leq \lambda_n \cdots$, and $\lambda_n \to \infty$ as $n \to \infty$. In addition, the corresponding eigenfunctions $\{u_n\}$ constitute a complete system in $L_w^2(a, b)$.

1.4.5 Green's Function and Spectral Representation

The eigenvalue theory can be used to solve the operator equation of the type

$$(\hat{L} - \xi)g(\mathbf{r}, \xi) = f(\mathbf{r}), \tag{1.126}$$

where $\hat{L}$ is a self-adjoint operator, ξ is a complex parameter, f is a known source function, and g is the unknown field. Suppose that $\{u_n\}$ is the complete set of the normalized eigenfunctions of the eigenvalue problem, which satisfy

$$\hat{L}u_n = \lambda_n u_n.$$

Then, both the source and the field may have the expansions

$$f = \sum_n f_n u_n, \quad g = \sum_n g_n u_n,$$

with $f_n = (f, u_n)$, $g_n = (g, u_n)$. Substituting these expansions into (1.126) gives

$$g = -\sum_n \frac{f_n}{\xi - \lambda_n} u_n. \tag{1.127}$$

Let C_R be a circle of radius R at the origin in the complex ξ-plane. Then,

$$\int_{C_R} g d\xi = -\sum_n f_n u_n \int_{C_R} \frac{d\xi}{\xi - \lambda_n}, \tag{1.128}$$

where the sum is over those eigenvalues λ_n contained within the circle. The singularities of the integrand are simple poles with residue of unity at all $\xi = \lambda_n$ within the contour. Since $\hat{L}$ is self-adjoint, the poles must lie on the real axis in the ξ-plane. Taking the limit as $R \to \infty$ and using the residue theorem, one may find

$$\lim_{R\to\infty} \int_{C_R} g d\xi = -2\pi j \sum_n f_n u_n,$$

where the sum is now over all of the eigenfunctions. Therefore, there exists a relationship between the source and the field

$$f = -\frac{1}{2\pi j} \int_C g d\xi, \tag{1.129}$$

where C is a circle at infinity obtained in the limit operation. As a special case, one may consider the Green's function defined by

$$(\hat{L} - \xi)G(\mathbf{r}, \mathbf{r}'; \xi) = \delta(\mathbf{r} - \mathbf{r}'). \tag{1.130}$$

From (1.129), it is easy to see

$$\delta(\mathbf{r}-\mathbf{r}') = -\frac{1}{2\pi j}\int_C G(\mathbf{r},\mathbf{r}';\xi)d\xi. \tag{1.131}$$

This is called the **spectral representation of the delta function** for the operator $\hat{L}$. From (1.127), the Green's function has the expansion

$$G(\mathbf{r},\mathbf{r}';\xi) = -\sum_n \frac{u_n(\mathbf{r})\bar{u}_n(\mathbf{r}')}{\xi-\lambda_n}. \tag{1.132}$$

Combining (1.131) and (1.132) yields

$$\delta(\mathbf{r}-\mathbf{r}') = \sum_n u_n(\mathbf{r})\bar{u}_n(\mathbf{r}'). \tag{1.133}$$

Equation (1.133) is the **completeness identity** of the eigenfunctions and it should be taken as a symbolic equality.

Example 1.3 Let us consider the Green's function defined by

$$\begin{cases} -\left(\dfrac{d^2}{dx^2}+\xi\right)G(x,x';\xi) = \delta(x-x') \\ G(0,x';\xi) = G(a,x';\xi) = 0 \end{cases}.$$

The Green's function is easily found to be [30]

$$G(x,x';\xi) = \frac{\sin\sqrt{\xi}x\sin\sqrt{\xi}(a-x')}{\sqrt{\xi}\sin\sqrt{\xi}a}H(x'-x) + \frac{\sin\sqrt{\xi}(a-x)\sin\sqrt{\xi}x'}{\sqrt{\xi}\sin\sqrt{\xi}a}H(x-x').$$

Considering

$$\begin{aligned}\frac{1}{2\pi j}\int_C G(x,x';\xi)d\xi &= \frac{1}{2\pi j}\int_C \frac{\sin\sqrt{\xi}x\sin\sqrt{\xi}(a-x')}{\sqrt{\xi}\sin\sqrt{\xi}a}H(x'-x)d\xi \\ &\quad + \frac{1}{2\pi j}\int_C \frac{\sin\sqrt{\xi}(a-x)\sin\sqrt{\xi}x'}{\sqrt{\xi}\sin\sqrt{\xi}a}H(x-x')d\xi\end{aligned}$$

and

$$\begin{aligned}\operatorname*{Res}_{\xi=(n\pi/a)^2}\frac{\sin\sqrt{\xi}x\sin\sqrt{\xi}(a-x')}{\sqrt{\xi}\sin\sqrt{\xi}a} &= \frac{2\sin\sqrt{\xi}x\sin\sqrt{\xi}(a-x')}{\xi^{-1/2}\sin\sqrt{\xi}a + a\cos\sqrt{\xi}a}\Big|_{\xi=(n\pi/a)^2} \\ &= -\frac{2}{a}\sin\frac{n\pi x}{a}\sin\frac{n\pi x'}{a} \\ &= \operatorname*{Res}_{\xi=(n\pi/a)^2}\frac{\sin\sqrt{\xi}(a-x)\sin\sqrt{\xi}x'}{\sqrt{\xi}\sin\sqrt{\xi}a},\end{aligned}$$

where $\operatorname{Res}_{\xi=a} f(\xi)$ denotes the residue of $f(\xi)$ at $\xi = a$, one may find

$$\frac{1}{2\pi j}\int_C G(x,x';\xi)d\xi = -\frac{2}{a}\sum_{n=1}^{\infty} \sin\frac{n\pi x}{a}\sin\frac{n\pi x'}{a}$$

from residue theorem. It follows from (1.131) that

$$\delta(x-x') = \sum_{n=1}^{\infty}\sqrt{\frac{2}{a}}\sin\frac{n\pi x}{a}\sqrt{\frac{2}{a}}\sin\frac{n\pi x'}{a}. \tag{1.134}$$

By shifting the origin to $a/2$, the above equation can be written as

$$\begin{aligned}\delta(x-x') &= \frac{2}{a}\sum_{n=1}^{\infty}\sin\frac{n\pi}{a}\left(x-\frac{a}{2}\right)\sin\frac{n\pi}{a}\left(x'-\frac{a}{2}\right)\\ &= \frac{2}{a}\sum_{n=1}^{\infty}\left(\sin\frac{n\pi x}{a}\sin\frac{n\pi x'}{a}\cos^2\frac{n\pi}{2}+\cos\frac{n\pi x}{a}\cos\frac{n\pi x'}{a}\sin^2\frac{n\pi}{2}\right)\\ &= \frac{2}{a}\sum_{n=1}^{\infty}\sin\frac{2n\pi x}{a}\sin\frac{2n\pi x'}{a}+\frac{2}{a}\sum_{n=1}^{\infty}\cos\frac{(2n-1)\pi x}{a}\cos\frac{(2n-1)\pi x'}{a}.\end{aligned} \tag{1.135}$$

As a becomes very large, the eigenvalues $\lambda_n = (n\pi/a)^2$ become closer and closer together until they merge in the limit into a continuous spectrum (positive axis). Setting

$$\tau_n = \frac{n\pi}{a}, \Delta\tau_n = \frac{\pi}{a},$$

and letting $a \to \infty$, (1.134) can be written as

$$\begin{aligned}\delta(x-x') &= \lim_{a\to\infty}\frac{2}{\pi}\sum_{n=1}^{\infty}\sin\frac{n\pi x}{a}\sin\frac{n\pi x'}{a}\frac{\pi}{a}\\ &= \lim_{\Delta\tau_n\to\infty}\frac{2}{\pi}\sum_{n=1}^{\infty}\sin\tau_n x\sin\tau_n x'\Delta\tau_n = \frac{2}{\pi}\int_0^{\infty}\sin\tau x\sin\tau x' d\tau.\end{aligned} \tag{1.136}$$

Similarly, if a approaches to infinity, (1.135) becomes

$$\delta(x-x') = \frac{1}{\pi}\int_0^{\infty}\cos\tau(x-x')d\tau = \frac{1}{2\pi}\int_{-\infty}^{\infty}e^{j\tau(x-x')}d\tau. \tag{1.137}$$

This is the **Fourier integral representation**. □

Example 1.4 Consider the eigenvalue problem

$$-\nabla^2 u(\mathbf{r}) = \lambda u(\mathbf{r}), \mathbf{r} = (x, y, z) \in R^3.$$

The above equation has normalized plane-wave solutions

$$u_{\mathbf{k}}(\mathbf{r}) = \frac{1}{(2\pi)^{3/2}} e^{j\mathbf{k}\cdot\mathbf{r}}, \ \ \mathbf{k} = (k_x, k_y, k_z),$$

which satisfy the orthonormal condition

$$\int_{R^3} u_{\mathbf{k}}(\mathbf{r})\bar{u}_{\mathbf{k}'}(\mathbf{r})dxdydz = \frac{1}{(2\pi)^3}\int_{R^3} e^{j(\mathbf{k}-\mathbf{k}')\cdot\mathbf{r}}dxdydz = \delta(\mathbf{k}-\mathbf{k}').$$

The corresponding eigenvalue is given by $\lambda = k^2 = |\mathbf{k}|^2$. The Green's function defined by

$$(\nabla^2 + k_0^2)G(\mathbf{r}, \mathbf{r}') = -\delta(\mathbf{r}-\mathbf{r}'), \mathbf{r}, \mathbf{r}' \in R^3$$

is easily found to be $e^{-jk_0|\mathbf{r}-\mathbf{r}'|}/4\pi|\mathbf{r}-\mathbf{r}'|$. In the expansion (1.132), one may let $\xi = k_0^2$, replace the discrete summation index n by the continuous vector $\mathbf{k}$, u_n by $u_{\mathbf{k}}$, λ_n by k^2, and the discrete summation by the integral over the whole space, to find

$$\frac{e^{-jk_0|\mathbf{r}-\mathbf{r}'|}}{4\pi|\mathbf{r}-\mathbf{r}'|} = \frac{1}{(2\pi)^3}\int_{R^3} \frac{e^{j\mathbf{k}\cdot(\mathbf{r}-\mathbf{r}')}}{k^2 - k_0^2} dk_x dk_y dk_z. \tag{1.138}$$

The above expression is the plane-wave expansion for the Green's function $e^{-jk_0|\mathbf{r}-\mathbf{r}'|}/4\pi|\mathbf{r}-\mathbf{r}'|$. □

1.5 Eigenvalue Problems for the Laplace Operator on Vector Field

A vector field can be decomposed into three scalar field components and a vector field problem can always be reduced to scalar field problems. But this procedure is not always the most effective. One can also express a vector field as the sum of the gradient of a scalar potential and the curl of a vector potential. Since the vector potential must satisfy the gauge condition, only two components of the vector potential are independent. A vector field is usually required to satisfy certain boundary conditions, which must be imposed to the scalar field components that represent the original vector field. The imposition of the boundary conditions to the scalar fields is very complicated in many cases. Therefore, one is forced to adopt a direct approach to solve vector field equation [20].

1.5.1 Rayleigh Quotient

Consider the eigenvalue problems for the Laplace operator on the vector field with the **boundary condition of electric type** (BCE)

$$\begin{cases} -\nabla^2\mathbf{e}(\boldsymbol{\rho}) = \lambda\mathbf{e}(\boldsymbol{\rho}), \boldsymbol{\rho} \in \Omega \\ \text{BCE} : \mathbf{u}_n \times \mathbf{e}(\boldsymbol{\rho}) = 0, \nabla \cdot \mathbf{e}(\boldsymbol{\rho}) = 0, \boldsymbol{\rho} \in \Gamma \end{cases}, \tag{1.139}$$

and that with the **boundary condition of magnetic type** (BCM)

$$\begin{cases} -\nabla^2\mathbf{h}(\boldsymbol{\rho}) = \lambda\mathbf{h}(\boldsymbol{\rho}), \boldsymbol{\rho} \in \Omega \\ \text{BCM} : \mathbf{u}_n \cdot \mathbf{h}(\boldsymbol{\rho}) = 0, \mathbf{u}_n \times \nabla \times \mathbf{h}(\boldsymbol{\rho}) = 0, \boldsymbol{\rho} \in \Gamma \end{cases}, \tag{1.140}$$

where the operator ∇^2 is defined by $\nabla^2\mathbf{a} = -\nabla\times\nabla\times\mathbf{a} + \nabla\nabla\cdot\mathbf{a}$. For convenience, the domain Ω will be assumed to be a finite region in 2D space and is bounded by Γ. Let $[L^2(\Omega)]^2$ denote the **product space** $L^2(\Omega)\times L^2(\Omega)$. The **inner product** and the **norm** in $[L^2(\Omega)]^2$ are, respectively, defined by

$$(\mathbf{a}_1, \mathbf{a}_2) = \int_\Omega \mathbf{a}_1 \cdot \bar{\mathbf{a}}_2 d\Omega, \quad \|\mathbf{a}\| = (\mathbf{a}, \mathbf{a})^{1/2}, \quad \mathbf{a}_1, \mathbf{a}_2, \mathbf{a} \in \left[L^2(\Omega)\right]^2. \tag{1.141}$$

It is noted that the BCE implies that the tangential component as well as the normal derivative of the normal component of the vector field $\mathbf{e}$ vanish on the boundary while the BCM implies that the normal component and the normal derivative of the tangential component of the vector field $\mathbf{h}$ are zero on the boundary. For any two vector fields $\mathbf{a}_1$ and $\mathbf{a}_2$ satisfying the BCE or BCM, the Laplacian $-\nabla^2$ is symmetric

$$\begin{aligned} \left(-\nabla^2\mathbf{a}_1, \mathbf{a}_2\right) &= \int_\Omega \nabla \times \mathbf{a}_1 \cdot \nabla \times \bar{\mathbf{a}}_2 d\Omega + \int_\Omega \nabla \cdot \mathbf{a}_1 \nabla \cdot \bar{\mathbf{a}}_2 d\Omega \\ &\quad - \int_\Gamma (\mathbf{u}_n \times \bar{\mathbf{a}}_2) \cdot \nabla \times \mathbf{a}_1 d\Gamma - \int_\Gamma (\mathbf{u}_n \cdot \bar{\mathbf{a}}_2) \nabla \cdot \mathbf{a}_1 d\Gamma \\ &= \int_\Omega (\nabla \times \mathbf{a}_1 \cdot \nabla \times \bar{\mathbf{a}}_2 + \nabla \cdot \mathbf{a}_1 \nabla \cdot \bar{\mathbf{a}}_2) d\Omega = \left(\mathbf{a}_1, -\nabla^2\mathbf{a}_2\right) \end{aligned} \tag{1.142}$$

after applying integration by parts. The Laplacian $-\nabla^2$ is **non-negative**

$$\left(-\nabla^2\mathbf{a}, \mathbf{a}\right) = \int_\Omega \left(|\nabla \times \mathbf{a}|^2 + |\nabla \cdot \mathbf{a}|^2\right) d\Omega \geq 0.$$

Similar to the eigenvalue problems for scalar fields, the eigenvalue problems (1.139) and (1.140) for vector fields have the properties:

1) All the eigenvalues are real and the corresponding eigenfunctions can be chosen to be real.
2) The eigenfunctions corresponding to different eigenvalues are orthogonal.
3) All the eigenvalues are positive or zero.

From now on, it will be assumed that all the eigenfunctions are real, and a shorthand notation $\mathbf{a}^2 = \mathbf{a} \cdot \mathbf{a}$ will be used. The **Rayleigh quotient** for the Laplacian $-\nabla^2$is

$$\lambda = \frac{(-\nabla^2\mathbf{a}, \mathbf{a})}{(\mathbf{a}, \mathbf{a})}$$
$$= \frac{\int_\Omega (\nabla \times \mathbf{a})^2 d\Omega + \int_\Omega (\nabla \cdot \mathbf{a})^2 d\Omega - \int_\Gamma (\mathbf{u}_n \times \mathbf{a}) \cdot \nabla \times \mathbf{a} d\Gamma - \int_\Gamma (\mathbf{u}_n \cdot \mathbf{a})\nabla \cdot \mathbf{a} d\Gamma}{\int_\Omega \mathbf{a}^2 d\Omega}. \tag{1.143}$$

The line integrals vanish if **a** satisfies the BCE or BCM. One may thus introduce the minimization problem

$$\lambda = \min_{\text{BCE or BCM}} \frac{\int_\Omega [(\nabla \times \mathbf{g})^2 + (\nabla \cdot \mathbf{g})^2] d\Omega}{\int_\Omega \mathbf{g}^2 d\Omega}, \tag{1.144}$$

where **g** is the trial function and is assumed to satisfy the BCE or BCM.

Theorem 1.6 If $\mathbf{a}_1$ is a solution of the minimum problem (1.144) and λ_1 is the value of the minimum, then λ_1 is the smallest eigenvalue of the Rayleigh quotient (1.144) and $\mathbf{a}_1$ is its corresponding eigenfunction. □

Proof Let **d** be an arbitrary function that satisfies the BCE or BCM, and ε be a small number. Since λ_1 is a minimum of (1.144), the expression

$$\lambda(\varepsilon) = \frac{\int_\Omega \{[\nabla \times (\mathbf{a}_1 + \varepsilon\mathbf{d})]^2 + [\nabla \cdot (\mathbf{a}_1 + \varepsilon\mathbf{d})]^2\} d\Omega}{\int_\Omega (\mathbf{a}_1 + \varepsilon\mathbf{d})^2 d\Omega} \tag{1.145}$$

has a minimum at $\varepsilon = 0$ with

$$\lambda_1 = \frac{\int\limits_{\Omega} \left[(\nabla \times \mathbf{a}_1)^2 + (\nabla \cdot \mathbf{a}_1)^2 \right] d\Omega}{\int\limits_{\Omega} \mathbf{a}_1^2 d\Omega}. \tag{1.146}$$

By ordinary calculus, this requires $\lambda'(0) = 0$. A simple calculation leads to

$$\begin{aligned} \lambda'(0) \;=\; & \frac{2\int\limits_{\Omega} \mathbf{a}_1^2 d\Omega \int\limits_{\Omega} \nabla \times \mathbf{a}_1 \cdot \nabla \times \mathbf{d} d\Omega - 2\int\limits_{\Omega} \mathbf{a}_1 \cdot \mathbf{d} d\Omega \int\limits_{\Omega} (\nabla \times \mathbf{a}_1)^2 d\Omega}{\left(\int\limits_{\Omega} \mathbf{a}_1^2 d\Omega \right)^2} \\ & + \frac{2\int\limits_{\Omega} \mathbf{a}_1^2 d\Omega \int\limits_{\Omega} \nabla \cdot \mathbf{a}_1 \nabla \cdot \mathbf{d} d\Omega - 2\int\limits_{\Omega} \mathbf{a}_1 \cdot \mathbf{d} d\Omega \int\limits_{\Omega} (\nabla \cdot \mathbf{a}_1)^2 d\Omega}{\left(\int\limits_{\Omega} \mathbf{a}_1^2 d\Omega \right)^2}. \end{aligned}$$

The condition that $\lambda'(0)$ must be zero implies

$$\begin{aligned} & \int\limits_{\Omega} \mathbf{a}_1^2 d\Omega \int\limits_{\Omega} \nabla \times \mathbf{a}_1 \cdot \nabla \times \mathbf{d} d\Omega - \int\limits_{\Omega} \mathbf{a}_1 \cdot \mathbf{d} d\Omega \int\limits_{\Omega} (\nabla \times \mathbf{a}_1)^2 d\Omega \\ & + \int\limits_{\Omega} \mathbf{a}_1^2 d\Omega \int\limits_{\Omega} \nabla \cdot \mathbf{a}_1 \nabla \cdot \mathbf{d} d\Omega - \int\limits_{\Omega} \mathbf{a}_1 \cdot \mathbf{d} d\Omega \int\limits_{\Omega} (\nabla \cdot \mathbf{a}_1)^2 d\Omega = 0. \end{aligned} \tag{1.147}$$

By the vector identities

$$\begin{aligned} & \int\limits_{\Omega} \nabla \times \mathbf{a} \cdot \nabla \times \mathbf{b} d\Omega = \int\limits_{\Omega} \mathbf{a} \cdot \nabla \times \nabla \times \mathbf{b} d\Omega + \int\limits_{\Gamma} \nabla \times \mathbf{b} \cdot (\mathbf{u}_n \times \mathbf{a}) d\Gamma, \\ & \int\limits_{\Omega} \nabla \cdot \mathbf{a} \nabla \cdot \mathbf{b} d\Omega = -\int\limits_{\Omega} \mathbf{a} \cdot \nabla \nabla \cdot \mathbf{b} d\Omega + \int\limits_{\Gamma} \nabla \cdot \mathbf{b} (\mathbf{u}_n \cdot \mathbf{a}) d\Gamma, \end{aligned} \tag{1.148}$$

(1.147) may be rewritten as

$$\int_{\Omega} \mathbf{d}\cdot\nabla\times\nabla\times\mathbf{a}_1 d\Omega - \int_{\Omega} \mathbf{d}\cdot\nabla\nabla\cdot\mathbf{a}_1 d\Omega - \int_{\Omega} \mathbf{a}_1\cdot\mathbf{d}d\Omega \frac{\int_{\Omega}\left[(\nabla\times\mathbf{a}_1)^2+(\nabla\cdot\mathbf{a}_1)^2\right]d\Omega}{\int_{\Omega}\mathbf{a}_1^2 d\Omega}$$
$$+\int_{\Gamma}(\mathbf{u}_n\times\mathbf{d})\cdot\nabla\times\mathbf{a}_1 d\Gamma + \int_{\Gamma}(\mathbf{u}_n\cdot\mathbf{d})\nabla\cdot\mathbf{a}_1 d\Gamma = 0. \tag{1.149}$$

On account of (1.146) and the same boundary conditions that both $\mathbf{d}$ and $\mathbf{e}_1$ must satisfy, the line integrals in (1.149) vanish. Thus,

$$\int_{\Omega} \mathbf{d}\cdot(\nabla\times\nabla\times\mathbf{a}_1 - \nabla\nabla\cdot\mathbf{a}_1 - \lambda_1\mathbf{a}_1)d\Omega = 0. \tag{1.150}$$

Since $\mathbf{d}$ is arbitrary, the above equation indicates that λ_1 is an eigenvalue of (1.139) and its eigenfunction is $\mathbf{a}_1$. It is easy to verify that λ_1 is the smallest eigenvalue of (1.139) or (1.140). The proof is completed.

1.5.2 Completeness of Vector Modal Functions

A procedure similar to the study of eigenvalue problems for scalar fields produces a set of orthogonal eigenfunctions $\{\mathbf{a}_1, \mathbf{a}_2, ...\}$, and the corresponding eigenvalues satisfy $0 \le \lambda_1 \le \lambda_2 \le \cdots$. The eigenfunction $\mathbf{a}_n$ is called nth **vector modal function**. From now on, it will be assumed that the vector modal functions are orthonormal

$$\int_{\Omega} \mathbf{a}_m\cdot\mathbf{a}_n d\Omega = \delta_{mn}. \tag{1.151}$$

Theorem 1.7

$$\lim_{n\to\infty}\lambda_n = \infty.$$

□

Proof. Since $\mathbf{a}_n$ is normalized, one may write

$$\lambda_n = \int_{\Omega}\left[(\nabla\times\mathbf{a}_n)^2 + (\nabla\cdot\mathbf{a}_n)^2\right]d\Omega = \int_{\Omega}(\nabla\times\nabla\times\mathbf{a}_n - \nabla\nabla\cdot\mathbf{a}_n)\cdot\mathbf{a}_n d\Omega \tag{1.152}$$

from (1.143). Using the vector identity $\nabla \times \nabla \times \mathbf{a} - \nabla \nabla \cdot \mathbf{a} = -\nabla^2 \mathbf{a}$ and Green's first identity (1.98), the above equation may be reduced to

$$\begin{aligned}\lambda_n &= -\int_\Omega \mathbf{a}_n \cdot \nabla^2 \mathbf{a}_n d\Omega = -\int_\Omega \left(a_{xn} \nabla^2 a_{xn} + a_{yn} \nabla^2 a_{yn}\right) d\Omega \\ &= \int_\Omega \left[(\nabla a_{xn})^2 + \left(\nabla a_{yn}\right)^2\right] d\Omega - \int_\Gamma \mathbf{a}_n \cdot \frac{\partial \mathbf{a}_n}{\partial n} d\Gamma,\end{aligned} \tag{1.153}$$

where the decomposition $\mathbf{a}_n = a_{xn}\mathbf{u}_x + a_{yn}\mathbf{u}_y$ in the rectangular coordinate system has been used. The boundary integral term vanishes due to the boundary conditions BCE or BCM. In fact, one may write

$$\begin{aligned}\mathbf{a}_n \cdot \frac{\partial \mathbf{a}_n}{\partial n} &= [(\mathbf{a}_n \cdot \mathbf{u}_n)\mathbf{u}_n + (\mathbf{a}_n \cdot \mathbf{u}_t)\mathbf{u}_t] \cdot \left[\frac{\partial(\mathbf{a}_n \cdot \mathbf{u}_n)}{\partial n}\mathbf{u}_n + \frac{\partial(\mathbf{a}_n \cdot \mathbf{u}_t)}{\partial n}\mathbf{u}_t\right] \\ &= (\mathbf{a}_n \cdot \mathbf{u}_n)\frac{\partial(\mathbf{a}_n \cdot \mathbf{u}_n)}{\partial n} + (\mathbf{a}_n \cdot \mathbf{u}_t)\frac{\partial(\mathbf{a}_n \cdot \mathbf{u}_t)}{\partial n},\end{aligned} \tag{1.154}$$

where $\mathbf{u}_n$ and $\mathbf{u}_t$ are, respectively, the unit normal and unit tangent along the boundary Γ, and can be used to decompose the field $\mathbf{a}_n$, as illustrated in Figure 1.3.

If the field $\mathbf{a}_n$ satisfies the BCE or BCM, it is easy to show that

$$\mathbf{a}_n \cdot \mathbf{u}_t = 0, \frac{\partial(\mathbf{a}_n \cdot \mathbf{u}_n)}{\partial n} = 0, \boldsymbol{\rho} \in \Gamma, \tag{1.155}$$

and

$$\mathbf{a}_n \cdot \mathbf{u}_n = 0, \frac{\partial(\mathbf{a}_n \cdot \mathbf{u}_t)}{\partial n} = 0, \boldsymbol{\rho} \in \Gamma, \tag{1.156}$$

respectively, for the BCE and BCM. Substituting (1.155) or (1.156) into (1.154), one may find $\mathbf{a}_n \cdot \partial \mathbf{a}_n / \partial n = 0$ in both cases. Thus,

$$\lambda_n = \int_\Omega \left[(\nabla a_{xn})^2 + \left(\nabla a_{yn}\right)^2\right] d\Omega. \quad \left[L^2(\Omega)\right]^2. \tag{1.157}$$

If λ_n is assumed to be finite as $n \to \infty$, Rellich's theorem thus applies. As a result, one can choose a subsequence of $\{\mathbf{a}_n\}$, still denoted $\{\mathbf{a}_n\}$, such that

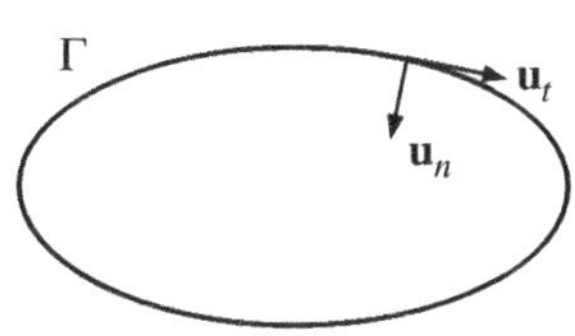

Figure 1.3 Decomposition of field along local rectangular coordinate system.

$\lim_{n,m\to\infty} \|a_{xn} - a_{xm}\| = 0$. Another subsequence can be chosen from this subsequence, such that $\lim_{n,m\to\infty} \|a_{yn} - a_{ym}\| = 0$. Then,

$$\lim_{n,m\to\infty} \|\mathbf{a}_n - \mathbf{a}_m\|^2 = \lim_{n,m\to\infty} \left(\|a_{xn} - a_{xm}\|^2 + \|a_{yn} - a_{ym}\|^2\right) = 0.$$

This contradicts the fact that $\|\mathbf{a}_n - \mathbf{a}_m\|^2 = 2$. Consequently, the assumption that λ_n remains finite as $n \to \infty$ is invalid. The proof is completed.

Theorem 1.8 The eigenfunctions determined from (1.144) are complete in $[L^2(\Omega)]^2$. In other words, an arbitrary vector function $\mathbf{F} \in [L^2(\Omega)]^2$ can be expressed as a linear combination of the eigenfunctions $\{\mathbf{a}_n\}$ in the sense that

$$\lim_{N\to\infty} \left\| \mathbf{F} - \sum_{n=1}^{N} c_n \mathbf{a}_n \right\| \to 0, \tag{1.158}$$

where $c_n = (\mathbf{F}, \mathbf{a}_n)$, $n = 1, 2, \ldots$, are the (Fourier) expansion coefficients. □

Proof. Let $\mathbf{F}$ be a trial function. The remainder of the expansion of $\mathbf{F}$

$$\mathbf{r}_N = \mathbf{F} - \sum_{n=1}^{N} c_n \mathbf{a}_n$$

is also a trial function. From the orthogonality of the eigenfunctions, it is easy to verify that $(\mathbf{r}_N, \mathbf{a}_j) = 0$ for $j \le N$, which implies

$$\lambda_N \le \frac{\int_\Omega \left(|\nabla \times \mathbf{r}_N|^2 + |\nabla \cdot \mathbf{r}_N|^2\right) d\Omega}{\int_\Omega |\mathbf{r}_N|^2 d\Omega} \tag{1.159}$$

by the construction of the eigenfunctions. Expanding the integrands in the numerator, one may find

$$\begin{aligned}
|\nabla \times \mathbf{r}_N|^2 &= |\nabla \times \mathbf{F}|^2 - 2\sum_{n=1}^{N} c_n [\mathbf{F} \cdot \nabla \times \nabla \times \mathbf{a}_n + \nabla \cdot (\mathbf{F} \times \nabla \times \mathbf{a}_n)] \\
&\quad + \sum_{m=1}^{N}\sum_{n=1}^{N} c_m c_n [\mathbf{a}_m \cdot \nabla \times \nabla \times \mathbf{a}_n + \nabla \cdot (\mathbf{a}_m \times \nabla \times \mathbf{a}_n)], \\
(\nabla \cdot \mathbf{r}_N)^2 &= (\nabla \cdot \mathbf{F})^2 - 2\sum_{n=1}^{N} c_n [\nabla \cdot (\mathbf{F} \nabla \cdot \mathbf{a}_n) - \mathbf{F} \cdot \nabla\nabla \cdot \mathbf{a}_n] \\
&\quad + \sum_{m=1}^{N}\sum_{n=1}^{N} c_m c_n [\nabla \cdot (\mathbf{a}_m \nabla \cdot \mathbf{a}_n) - \mathbf{a}_m \cdot \nabla\nabla \cdot \mathbf{a}_n].
\end{aligned}$$

Thus,

$$\int_{\Omega}(|\nabla \times \mathbf{r}_N|^2 + |\nabla \cdot \mathbf{r}_N|^2)d\Omega = \int_{\Omega}\left[|\nabla \times \mathbf{F}|^2 + |\nabla \cdot \mathbf{F}|^2\right]d\Omega - \sum_{n=1}^{N} c_n^2\lambda_n \leq \int_{\Omega}\left[|\nabla \times \mathbf{F}|^2 + |\nabla \cdot \mathbf{F}|^2\right]d\Omega. \tag{1.160}$$

It follows from (1.159) and (1.160) that

$$\|\mathbf{r}_N\|^2 \leq \frac{\int_{\Omega}\left[|\nabla \times \mathbf{F}|^2 + |\nabla \cdot \mathbf{F}|^2\right]d\Omega}{\lambda_N}. \tag{1.161}$$

Since $\lim_{N\to\infty} \lambda_N = \infty$, it is to see $\lim_{N\to\infty} \|\mathbf{r}_N\| = 0$. Hence (1.158) is valid for a trial function $\mathbf{F}$. It can be shown that an arbitrary function in $[L^2(\Omega)]^2$ can be approximated by a trial function. The proof is completed.

Remark 1.8 In the proof of the completeness of eigenfunctions for both scalar and vector fields, it has been stated that an arbitrary function in $L^2(\Omega)$ or $[L^2(\Omega)]^2$ can be approximated by a trial function without giving the details. Serious readers may find that a proof for the existence of eigenfunctions is also missing in the preceding discussion. To prove the existence and completeness of eigenfunctions, the concept of the generalized solutions for differential equations must be used, which is beyond the scope of this book. For a rigorous treatment, please refer to [16, 31]. □

Remark 1.9 Similar to (1.110), the BCE and BCM can be made free by, respectively, introducing the minimization problems [32]

$$\lambda = \min \frac{\int_{\Omega}\left[(\nabla \times \mathbf{g})^2 + (\nabla \cdot \mathbf{g})^2\right]d\Omega - 2\int_{\Gamma}(\mathbf{u}_n \times \mathbf{g}) \cdot \nabla \times \mathbf{g}d\Gamma}{\int_{\Omega}\mathbf{g}^2 d\Omega}, \tag{1.162}$$

$$\lambda = \min \frac{\int_{\Omega}\left[(\nabla \times \mathbf{g})^2 + (\nabla \cdot \mathbf{g})^2\right]d\Omega - 2\int_{\Gamma}(\mathbf{u}_n \cdot \mathbf{g})\nabla \cdot \mathbf{g}d\Gamma}{\int_{\Omega}\mathbf{g}^2 d\Omega}. \tag{1.163}$$

It is readily shown that the solution of (1.162) is an eigenfunction that satisfies (1.139). Let **d** be an arbitrary function and ε be a small number. Suppose that **a** is a solution of (1.162). Then, the expression

$$\lambda(\varepsilon)=\frac{\int_\Omega\{[\nabla\times(\mathbf{a}+\varepsilon\mathbf{d})]^2+[\nabla\cdot(\mathbf{a}+\varepsilon\mathbf{d})]^2\}d\Omega-2\int_\Gamma[\mathbf{u}_n\times(\mathbf{a}+\varepsilon\mathbf{d})]\cdot\nabla\times(\mathbf{a}+\varepsilon\mathbf{d})d\Gamma}{\int_\Omega(\mathbf{a}+\varepsilon\mathbf{d})^2d\Omega} \tag{1.164}$$

must satisfy $\lambda'(0) = 0$, which leads to

$$\begin{aligned}
&2\int_\Omega\mathbf{a}^2d\Omega\int_\Omega\nabla\times\mathbf{a}\cdot\nabla\times\mathbf{d}d\Omega-2\int_\Omega\mathbf{a}\cdot\mathbf{d}d\Omega\int_\Omega(\nabla\times\mathbf{a})^2d\Omega\\
&+2\int_\Omega\mathbf{a}^2d\Omega\int_\Omega\nabla\cdot\mathbf{a}\nabla\cdot\mathbf{d}d\Omega-2\int_\Omega\mathbf{a}\cdot\mathbf{d}d\Omega\int_\Omega(\nabla\cdot\mathbf{a})^2d\Omega\\
&-\int_\Omega\mathbf{a}^2d\Omega\left[2\int_\Gamma(\mathbf{u}_n\times\mathbf{a})\cdot\nabla\times\mathbf{d})d\Gamma+2\int_\Gamma(\mathbf{u}_n\times\mathbf{d})\cdot\nabla\times\mathbf{a}d\Gamma\right]\\
&+4\int_\Omega\mathbf{a}\cdot\mathbf{d}d\Omega\int_\Gamma(\mathbf{u}_n\times\mathbf{a})\cdot\nabla\times\mathbf{a}d\Gamma=0.
\end{aligned}$$

By means of the vector identities (1.148), one may find

$$\begin{aligned}
&\int_\Omega\mathbf{d}\cdot\nabla\times\nabla\times\mathbf{a}d\Omega-\int_\Omega\mathbf{d}\cdot\nabla\nabla\cdot\mathbf{a}d\Omega\\
&-\int_\Omega\mathbf{d}\cdot\mathbf{a}d\Omega\frac{\int_\Omega(\nabla\times\mathbf{a})^2d\Omega+\int_\Omega(\nabla\cdot\mathbf{a})^2d\Omega-2\int_\Gamma\mathbf{u}_n\times\mathbf{a}\cdot\nabla\times\mathbf{a}d\Gamma}{\int_\Omega\mathbf{a}^2d\Omega}\\
&+\int_\Gamma\nabla\cdot\mathbf{a}(\mathbf{u}_n\cdot\mathbf{d})d\Gamma-\int_\Gamma(\mathbf{u}_n\times\mathbf{a})\cdot\nabla\times\mathbf{d}d\Gamma=0.
\end{aligned}$$

This can be simplified to

$$\int_{\Omega} \mathbf{d} \cdot [\nabla \times \nabla \times \mathbf{a} - \nabla\nabla \cdot \mathbf{a} - \lambda \mathbf{a}] d\Omega + \int_{\Gamma} \nabla \cdot \mathbf{a}(\mathbf{u}_n \cdot \mathbf{d}) d\Gamma - \int_{\Gamma} (\mathbf{u}_n \times \mathbf{a}) \cdot \nabla \times \mathbf{d} d\Gamma = 0. \tag{1.165}$$

Since $\mathbf{d}$ is arbitrary, it can be first chosen arbitrarily inside Ω such that $\mathbf{u}_n \cdot \mathbf{d} = \mathbf{u}_n \times \nabla \times \mathbf{d} = 0$ on the boundary Γ. From Eq. (1.165), the following equation can be obtained:

$$\nabla \times \nabla \times \mathbf{a} - \nabla\nabla \cdot \mathbf{a} - \lambda \mathbf{a} = 0, \boldsymbol{\rho} \in \Omega.$$

Thus,

$$\int_{\Gamma} \nabla \cdot \mathbf{a}(\mathbf{u}_n \cdot \mathbf{d}) d\Gamma - \int_{\Gamma} (\mathbf{u}_n \times \mathbf{a}) \cdot \nabla \times \mathbf{d} d\Gamma = 0, \tag{1.166}$$

which is valid for an arbitrary $\mathbf{d}$. One can first choose $\mathbf{u}_n \cdot \mathbf{d} = 0$ in the above equation so that

$$\int_{\Gamma} (\mathbf{u}_n \times \mathbf{a}) \cdot \nabla \times \mathbf{d} d\Gamma = 0. \tag{1.167}$$

From the expansion for curl of $\mathbf{d}$

$$\begin{aligned} \nabla \times \mathbf{d} &= \left(\frac{\partial}{\partial n} \mathbf{u}_n + \frac{\partial}{\partial t} \mathbf{u}_t \right) \times [(\mathbf{d} \cdot \mathbf{u}_n)\mathbf{u}_n + (\mathbf{d} \cdot \mathbf{u}_t)\mathbf{u}_t] \\ &= -\mathbf{u}_z \frac{\partial(\mathbf{d} \cdot \mathbf{u}_t)}{\partial n} + \mathbf{u}_z \frac{\partial(\mathbf{d} \cdot \mathbf{u}_n)}{\partial t}, \end{aligned}$$

it is easy to see that $\nabla \times \mathbf{d}$ can still be selected arbitrarily even though $\mathbf{u}_n \cdot \mathbf{d} = 0$ has been assumed. Hence, (1.167) implies

$$\mathbf{u}_n \times \mathbf{a} = 0, \boldsymbol{\rho} \in \Gamma.$$

One can also first choose $\nabla \times \mathbf{d} = 0$ in (1.166) so that

$$\int_{\Gamma} \nabla \cdot \mathbf{a}(\mathbf{u}_n \cdot \mathbf{d}) d\Gamma = 0. \tag{1.168}$$

For $\mathbf{u}_n \cdot \mathbf{d}$ can still be chosen arbitrarily, one may find

$$\nabla \cdot \mathbf{a} = 0, \boldsymbol{\rho} \in \Gamma.$$

Therefore, the vector field $\mathbf{a}$ is an eigenfunction that satisfies (1.139). In a similar way, it can be shown that the solution of (1.163) is an eigenfunction that satisfies (1.140). □

1.5.3 Classification of Vector Modal Functions

The vector modal function $\mathbf{a}_n$ belongs to one of the following four types:

1. $\nabla \times \mathbf{a}_n = 0, \nabla \cdot \mathbf{a}_n = 0$,
2. $\nabla \times \mathbf{a}_n \neq 0, \nabla \cdot \mathbf{a}_n = 0$,
3. $\nabla \times \mathbf{a}_n = 0, \nabla \cdot \mathbf{a}_n \neq 0$,
4. $\nabla \times \mathbf{a}_n \neq 0, \nabla \cdot \mathbf{a}_n \neq 0$.

A complete set of eigenfunctions can be constructed from the first three types [32], which is implied by the Helmholtz theorem to be discussed later. Indeed, if $\mathbf{a}_n$ belongs to type 4, two new functions may be introduced through

$$\mathbf{a}'_n = c'\nabla \times \nabla \times \mathbf{a}_n, \quad \mathbf{a}''_n = c''\nabla\nabla \cdot \mathbf{a}_n, \tag{1.169}$$

where c' and c'' are two normalizing constants. It can be verified that both $\mathbf{a}'_n$ and $\mathbf{a}''_n$ are eigenfunctions that are mutually orthogonal and satisfy (1.139). Apparently, $\mathbf{a}'_n$ and $\mathbf{a}''_n$, respectively, belong to types 2 and 3. From (1.139) and (1.169), one may find

$$\mathbf{a}_n = \frac{1}{\lambda_n}\left(\frac{1}{c'}\mathbf{a}'_n - \frac{1}{c''}\mathbf{a}''_n\right).$$

As a result, the eigenfunctions belonging to type 4 can be expressed as a linear combination of the eigenfunctions in the first three types. A complete set of eigenfunctions can be established as follows. Suppose that the nth eigenfunction $\mathbf{a}_n$ happens to fall into type 4 during the minimization process with (1.144). Instead of using $\mathbf{e}_n$ as the nth eigenfunction, one may take $\mathbf{a}'_n$ and $\mathbf{a}''_n$ as the nth and $(n+1)$th eigenfunctions, respectively. This process guarantees that all the eigenfunctions obtained with (1.144) fall into the first three types. Based on the above analysis, an arbitrary vector field $\mathbf{F}$ can be expressed as a linear combination of the eigenfunctions in the first three types and therefore can be split up into three components

$$\mathbf{F} = \mathbf{F}_L + \mathbf{F}_T + \mathbf{F}_H, \tag{1.170}$$

where

$$\begin{cases} \nabla \cdot \mathbf{F}_L = \nabla \cdot \mathbf{F} \\ \nabla \times \mathbf{F}_L = 0 \end{cases}, \begin{cases} \nabla \cdot \mathbf{F}_T = 0 \\ \nabla \times \mathbf{F}_L = \nabla \times \mathbf{F} \end{cases}, \begin{cases} \nabla \cdot \mathbf{F}_H = 0 \\ \nabla \times \mathbf{F}_H = 0 \end{cases}. \tag{1.171}$$

The components $\mathbf{F}_L$, $\mathbf{F}_T$, and $\mathbf{F}_H$ are, respectively, called **longitudinal**, **transverse**, and **harmonic**. The decomposition is very useful if $\mathbf{F}$ is unknown but $\nabla \times \mathbf{F}$ and $\nabla \cdot \mathbf{F}$ are specified. Such a decomposition is usually referred to as Helmholtz theorem, which will be discussed in Section 1.7.

It is noted that all the results obtained for eigenvalue problems (1.139) and (1.140) in 2D space are also valid in 3D space.

1.6 Ritz Method for the Solution of Eigenvalue Problem

For the solution of the eigenvalue problem

$$\hat{L}u = \lambda u, \tag{1.172}$$

where $\hat{L}$ is a **positive-bounded-below operator** $(\hat{L}u, u) \geq c\|u\|^2$ with c being a constant, one may resort to seeking the solution of the minimization problem of the Rayleigh quotient for the operator $\hat{L}$:

$$\lambda = \min \frac{(\hat{L}u, u)}{(u, u)}. \tag{1.173}$$

This is equivalent to

$$\lambda = \min_{(u,u)=1} (\hat{L}u, u). \tag{1.174}$$

The minimization problem (1.174) can be solved numerically. Let $\{u_j | j = 1, 2, \ldots, N\}$ be a set of linearly independent functions in the domain of operator $\hat{L}$, called **basis** or **trial functions**. As an approximation, the unknown function u may be expanded as follows:

$$u = \sum_{j=1}^{N} a_j u_j, \tag{1.175}$$

where a_j are the expansion coefficients to be determined. In this case, (1.174) is equivalent to

$$\begin{cases} \lambda = \min \sum\limits_{i,j=1}^{N} (\hat{L}u_i, u_j) a_i a_j \\ \text{s.t.} \ \sum\limits_{i,j=1}^{N} (u_i, u_j) a_i a_j = 1. \end{cases} \tag{1.176}$$

The constrained problem (1.176) can be solved by the **method of Lagrangian multipliers**. The **Lagrangian function** for (1.176) is given by

$$L(a_i, \xi) = \sum_{i,j=1}^{N} (\hat{L}u_i, u_j) a_i a_j - \xi \sum_{i,j=1}^{N} (u_i, u_j) a_i a_j, \tag{1.177}$$

where ξ is the Lagrangian multiplier. The partial derivatives of the Lagrangian function with respect to the coefficients a_j must be zero, which results in the linear system of equations

$$\sum_{i=1}^{N} a_i \left[(\hat{L}u_i, u_j) - \xi (u_i, u_j) \right] = 0, \ \ j = 1, 2, \ldots, N. \tag{1.178}$$

The existence of a nonzero solution requires that the determinant of the system (1.178) is zero

$$\begin{bmatrix} (\hat{L}u_1,u_1)-\xi(u_1,u_1) & (\hat{L}u_2,u_1)-\xi(u_2,u_1) & \cdots & (\hat{L}u_N,u_1)-\xi(u_N,u_1) \\ (\hat{L}u_1,u_2)-\xi(u_1,u_2) & (\hat{L}u_2,u_2)-\xi(u_2,u_2) & \cdots & (\hat{L}u_N,u_2)-\xi(u_N,u_2) \\ \vdots & \vdots & \ddots & \vdots \\ (\hat{L}u_1,u_N)-\xi(u_1,u_N) & (\hat{L}u_2,u_N)-\xi(u_2,u_N) & \cdots & (\hat{L}u_N,u_N)-\xi(u_N,u_N) \end{bmatrix} = 0. \tag{1.179}$$

It is easy to see that (1.179) has N roots. Substituting the roots into (1.178), the coefficients a_i can be determined up to a constant multiplier c. One can then use the constraint $(u, u) = 1$ to determine the constant c. If the set $\{u_j | j = 1, 2, ..., N\}$ is orthonormal

$$(u_i, u_j) = \delta_{ij} = \begin{cases} 1, i = j \\ 0, i \neq j \end{cases}, \tag{1.180}$$

(1.179) reduces to

$$\begin{bmatrix} (\hat{L}u_1, u_1) - \xi & (\hat{L}u_2, u_1) & \cdots & (\hat{L}u_N, u_1) \\ (\hat{L}u_1, u_2) & (\hat{L}u_2, u_2) - \xi & \cdots & (\hat{L}u_N, u_2) \\ \vdots & \vdots & \ddots & \vdots \\ (\hat{L}u_1, u_N) & (\hat{L}u_2, u_N) & \cdots & (\hat{L}u_N, u_N) - \xi \end{bmatrix} = 0. \tag{1.181}$$

The procedure described above is called **Ritz method**, named after Swiss theoretical physicist Walther Heinrich Wilhelm Ritz (1878–1909) [33].

Example 1.5 Let $\hat{L} = -\left(d^2/dx^2\right)$. The domain of the operator $\hat{L}$ consists of the smooth functions which satisfy the boundary conditions $u(0) = u(1) = 0$. It is easy to find that the eigenvalues of $\hat{L}$ are $\lambda_n = (n\pi)^2$, $n = 1, 2, 3, ...$ The Ritz method can be used to estimate the first eigenvalue λ_1. In this case, the minimization problem (1.173) is of the form

$$\lambda = \min \frac{\int_0^1 u'^2 dx}{\int_0^1 u^2 dx}.$$

Choose the trial function $u_1 = x(1-x)$ in the domain of $\hat{L}$. Substituting $u = a_1 u_1$ into the above, one may obtain $\lambda = 10$, which is good approximation to the first eigenvalue $\lambda_1 = \pi^2$. □

1.7 Helmholtz Theorems

In 1905, German mathematician Blumenthal (1876–1944) showed that every continuously differentiable vector field that vanishes at infinity can be split into an irrotational (curl free) and a solenoidal (divergence free) part. It means that a vector field **F** can be decomposed into the sum of a gradient and a curl

$$\mathbf{F}(\mathbf{r}) = -\nabla\phi(\mathbf{r}) + \nabla \times \mathbf{A}(\mathbf{r}). \tag{1.182}$$

The component $\mathbf{F}^{\parallel} = -\nabla\phi$ generated by the scalar field ϕ is irrotational (longitudinal); the component $\mathbf{F}^{\perp} = \nabla \times \mathbf{A}$ is solenoidal (transverse). Apparently, the decomposition (1.182) can be carried out in an infinite number of ways. The vector potential **A** in (1.182) is determined to within a gradient. In order for the decomposition to be unique, certain restrictions have to be placed on the vector field **F**.

1.7.1 Helmholtz Theorem for the Field in Infinite Space

By use of the identity

$$\nabla^2 \frac{1}{4\pi R} = -\delta(\mathbf{r}-\mathbf{r}'), \tag{1.183}$$

the vector field **F** defined on the infinite space can be divided into the sum of two components

$$\mathbf{F}(\mathbf{r}) = \int_{R^3} \mathbf{F}(\mathbf{r}')\delta(\mathbf{r}-\mathbf{r}')dV(\mathbf{r}') = -\nabla^2 \int_{R^3} \frac{\mathbf{F}(\mathbf{r}')}{4\pi R} dV(\mathbf{r}') = \mathbf{F}^{\parallel} + \mathbf{F}^{\perp}, \tag{1.184}$$

where

$$\begin{aligned} \mathbf{F}^{\parallel}(\mathbf{r}) &= -\nabla\nabla\cdot \int_{R^3} \frac{\mathbf{F}(\mathbf{r}')}{4\pi R} dV(\mathbf{r}'), \\ \mathbf{F}^{\perp}(\mathbf{r}) &= \nabla \times \nabla \times \int_{R^3} \frac{\mathbf{F}(\mathbf{r}')}{4\pi R} dV(\mathbf{r}') \end{aligned} \tag{1.185}$$

are referred to as the **irrotational component** and **solenoidal component** of **F**, respectively. The vector field **F** will be assumed to decrease faster than $1/r$ as $r \to \infty$. The irrotational component can further be written as

$$\mathbf{F}^{\|}(\mathbf{r}) = \nabla \int_{R^3} \nabla' \cdot \left[\frac{\mathbf{F}(\mathbf{r}')}{4\pi R}\right] dV(\mathbf{r}') - \nabla \int_{R^3} \frac{\nabla' \cdot \mathbf{F}(\mathbf{r}')}{4\pi R} dV(\mathbf{r}'). \tag{1.186}$$

Upon using the Gauss's theorem, the first term on the right-hand side approaches to zero so that

$$\mathbf{F}^{\|}(\mathbf{r}) = -\nabla \int_{R^3} \frac{\nabla' \cdot \mathbf{F}(\mathbf{r}')}{4\pi R} dV(\mathbf{r}'). \tag{1.187}$$

Similarly, the solenoidal component can be expressed as

$$\begin{aligned} \mathbf{F}^{\perp}(\mathbf{r}) &= -\nabla \times \int_{R^3} \nabla' \times \left[\frac{1}{4\pi R}\mathbf{F}(\mathbf{r}')\right] dV(\mathbf{r}') + \nabla \times \int_{R^3} \frac{\nabla' \times \mathbf{F}(\mathbf{r}')}{4\pi R} dV(\mathbf{r}') \\ &= \nabla \times \int_{R^3} \frac{\nabla' \times \mathbf{F}(\mathbf{r}')}{4\pi R} dV(\mathbf{r}'), \end{aligned} \tag{1.188}$$

after applying Gauss theorem. Hence, the following theorem is obtained.

Theorem 1.9 Any vector field **F** that decreases faster than $1/r$ as $r \to \infty$ can be expressed by

$$\mathbf{F}(\mathbf{r}) = -\nabla \int_{R^3} \frac{\nabla' \cdot \mathbf{F}(\mathbf{r}')}{4\pi R} dV(\mathbf{r}') + \nabla \times \int_{R^3} \frac{\nabla' \times \mathbf{F}(\mathbf{r}')}{4\pi R} dV(\mathbf{r}'). \tag{1.189}$$

□

This is called **Helmholtz theorem** or **Helmholtz identity**, also known as the **fundamental theorem of vector calculus**. The theorem states that a vector field that decreases rapidly at infinity is uniquely determined by its divergence and curl. An immediate consequence of Helmholtz identity is

$$\int_{R^3} |\mathbf{F}(\mathbf{r})|^2 dV(\mathbf{r}') = \int_{R^3}\int_{R^3} \frac{\nabla \cdot \mathbf{F}(\mathbf{r}) \nabla' \cdot \bar{\mathbf{F}}(\mathbf{r}') + \nabla \times \mathbf{F}(\mathbf{r}) \cdot \nabla' \times \bar{\mathbf{F}}(\mathbf{r}')}{4\pi R} dV(\mathbf{r}) dV(\mathbf{r}'). \tag{1.190}$$

It is noted that the Helmholtz theorem can be generalized to more complicated domains [34].

1.7.2 Helmholtz Theorem for the Field in Finite Region

For the vector field **F** defined in a finite region V bounded by S, it can also be decomposed into the sum of two components

$$\mathbf{F}(\mathbf{r}) = \int_V \mathbf{F}(\mathbf{r}')\delta(\mathbf{r}-\mathbf{r}')dV(\mathbf{r}') = -\nabla^2 \int_V \frac{\mathbf{F}(\mathbf{r}')}{4\pi R} dV(\mathbf{r}') = \mathbf{F}^{\parallel} + \mathbf{F}^{\perp}, \tag{1.191}$$

where

$$\mathbf{F}^{\parallel}(\mathbf{r}) = -\nabla\nabla\cdot \int_V \frac{\mathbf{F}(\mathbf{r}')}{4\pi R} dV(\mathbf{r}'), \mathbf{F}^{\perp}(\mathbf{r}) = \nabla\times\nabla\times \int_V \frac{\mathbf{F}(\mathbf{r}')}{4\pi R} dV(\mathbf{r}'). \tag{1.192}$$

Applying Gauss theorem, the irrotational component can further be written as

$$\begin{aligned}\mathbf{F}^{\parallel}(\mathbf{r}) &= \nabla \int_V \nabla'\cdot\left[\frac{\mathbf{F}(\mathbf{r}')}{4\pi R}\right] dV(\mathbf{r}') - \nabla \int_V \frac{\nabla'\cdot\mathbf{F}(\mathbf{r}')}{4\pi R} dV(\mathbf{r}') \\ &= -\nabla\left[\int_V \frac{\nabla'\cdot\mathbf{F}(\mathbf{r}')}{4\pi R} dV(\mathbf{r}') - \int_S \frac{\mathbf{u}_n(\mathbf{r}')\cdot\mathbf{F}(\mathbf{r}')}{4\pi R} dS(\mathbf{r}')\right],\end{aligned} \tag{1.193}$$

where $\mathbf{u}_n$ is unit outward normal of S. In the same way, the solenoidal component may be expressed by

$$\begin{aligned}\mathbf{F}^{\perp}(\mathbf{r}) &= -\nabla\times \int_V \nabla'\times\left[\frac{1}{4\pi R}\mathbf{F}(\mathbf{r}')\right] dV(\mathbf{r}') + \nabla\times \int_V \frac{\nabla'\times\mathbf{F}(\mathbf{r}')}{4\pi R} dV(\mathbf{r}') \\ &= \nabla\times\left[\int_V \frac{\nabla'\times\mathbf{F}(\mathbf{r}')}{4\pi R} dV(\mathbf{r}') - \int_S \frac{\mathbf{u}_n(\mathbf{r}')\times\mathbf{F}(\mathbf{r}')}{4\pi R} dS(\mathbf{r}')\right].\end{aligned} \tag{1.194}$$

The vector field **F** may then be decomposed according to

$$\begin{aligned}\mathbf{F}(\mathbf{r}) &= -\nabla\left[\int_V \frac{\nabla'\cdot\mathbf{F}(\mathbf{r}')}{4\pi R} dV(\mathbf{r}') - \int_S \frac{\mathbf{u}_n(\mathbf{r}')\cdot\mathbf{F}(\mathbf{r}')}{4\pi R} dS(\mathbf{r}')\right] \\ &\quad + \nabla\times\left[\int_V \frac{\nabla'\times\mathbf{F}(\mathbf{r}')}{4\pi R} dV(\mathbf{r}') - \int_S \frac{\mathbf{u}_n(\mathbf{r}')\times\mathbf{F}(\mathbf{r}')}{4\pi R} dS(\mathbf{r}')\right].\end{aligned} \tag{1.195}$$

Consequently, a vector field defined in a finite region is determined by its divergence and curl, as well as by its values on the boundary.

Remark 1.10 If the vector field **F** is both curl free and divergence free inside *V*, (1.195) reduces to

$$\mathbf{F}(\mathbf{r}) = \nabla \int_S \frac{\mathbf{u}_n(\mathbf{r}') \cdot \mathbf{F}(\mathbf{r}')}{4\pi R} dS(\mathbf{r}') - \nabla \times \int_S \frac{\mathbf{u}_n(\mathbf{r}') \times \mathbf{F}(\mathbf{r}')}{4\pi R} dS(\mathbf{r}'). \tag{1.196}$$

If *S* is a perfect conductor and the vector field **F** stands for the electric field, its tangential component vanishes: $\mathbf{u}_n \times \mathbf{F} = 0$. Let $\mathbf{F} = -\nabla\phi$. Since **F** is harmonic, one may write

$$\begin{cases} \nabla^2 \phi = 0, \mathbf{r} \in V, \\ \mathbf{u}_n \times \nabla\phi = 0, \mathbf{r} \in S. \end{cases} \tag{1.197}$$

The boundary condition in (1.197) implies ϕ is a constant on *S*. As a result,

$$\int_V \nabla \cdot (\phi \nabla \phi) dV = \int_S \phi \mathbf{u}_n \cdot \nabla \phi dS = \text{const} \int_S \mathbf{u}_n \cdot \nabla \phi dS = \text{const} \int_V \nabla \cdot \nabla \phi dV = 0, \tag{1.198}$$

where use has been made of the Gauss theorem and (1.197). The left-hand side of the above equation can be written as

$$\int_V \nabla \cdot (\phi \nabla \phi) dV = \int_V (\nabla \phi \cdot \nabla \phi + \phi \nabla \cdot \nabla \phi) dV = \int_V (\nabla \phi \cdot \nabla \phi) dV.$$

This implies $\mathbf{F} = -\nabla\phi = 0$ inside *V*. It is noted that the derivation of (1.198) is only valid for simply connected region. If *V* is multiply connected, the potential function ϕ can take a different value on each boundary. As a result, a nonzero vector field **F** inside *V* can exist. □

1.7.3 Helmholtz Theorem for Time-Dependent Field

In order to obtain the Helmholtz theorem for a time-dependent vector field $\mathbf{F}(\mathbf{r}, t)$, one may use the retarded Green's function $G(\mathbf{r}, \mathbf{r}'; t, t')$ for the wave equation

$$\begin{cases} \left(\nabla^2 - \dfrac{1}{c^2}\dfrac{\partial^2}{\partial t^2}\right) G(\mathbf{r}, \mathbf{r}'; t, t') = -\delta(\mathbf{r} - \mathbf{r}')\delta(t - t'), \\ G(\mathbf{r}, \mathbf{r}'; t, t') = 0, t < t'. \end{cases} \tag{1.199}$$

It is easy to find the solution of the above equation

$$G(\mathbf{r}, \mathbf{r}'; t, t') = \frac{1}{4\pi R} \delta\left(t - t' - \frac{|\mathbf{r} - \mathbf{r}'|}{c}\right). \tag{1.200}$$

The time-dependent vector field $\mathbf{F}(\mathbf{r}, t)$ can be written as an integration over space and time

$$\begin{aligned}\mathbf{F}(\mathbf{r},t) &= \int_{R^3}\int_{-\infty}^{+\infty} \mathbf{F}(\mathbf{r}',t')\delta(\mathbf{r}-\mathbf{r}')\delta(t-t')dV(\mathbf{r}')dt' \\ &= -\int_{R^3}\int_{-\infty}^{+\infty} \mathbf{F}(\mathbf{r}',t')\left(\nabla^2-\frac{1}{c^2}\frac{\partial^2}{\partial t^2}\right)G(\mathbf{r},\mathbf{r}';t,t')dV(\mathbf{r}')dt'.\end{aligned}$$

Similar to the derivation of (1.189), the above equation can be expressed as

$$\begin{aligned}\mathbf{F}(\mathbf{r},t) &= -\nabla\int_{R^3}\int_{-\infty}^{+\infty} \nabla'\cdot\mathbf{F}(\mathbf{r}',t')G(\mathbf{r},\mathbf{r}';t,t')dV(\mathbf{r}')dt' \\ &+\nabla\times\int_{R^3}\int_{-\infty}^{+\infty} \nabla'\times\mathbf{F}(\mathbf{r}',t')G(\mathbf{r},\mathbf{r}';t,t')dV(\mathbf{r}')dt' \\ &+\frac{1}{c^2}\frac{\partial}{\partial t}\int_{R^3}\int_{-\infty}^{+\infty} \frac{\partial\mathbf{F}(\mathbf{r}',t')}{\partial t'}G(\mathbf{r},\mathbf{r}';t,t')dV(\mathbf{r}')dt'.\end{aligned} \tag{1.201}$$

Substituting (1.200) into (1.201) yields the time-domain Helmholtz theorem

$$\begin{aligned}\mathbf{F}(\mathbf{r},t) &= -\nabla\int_{R^3} \frac{\nabla'\cdot\mathbf{F}(\mathbf{r}',t-|\mathbf{r}-\mathbf{r}'|/c)}{4\pi R}dV(\mathbf{r}') \\ &+\nabla\times\int_{R^3} \frac{\nabla'\times\mathbf{F}(\mathbf{r}',t-|\mathbf{r}-\mathbf{r}'|/c)}{4\pi R}dV(\mathbf{r}') \\ &+\frac{1}{c^2}\frac{\partial}{\partial t}\int_{R^3} \frac{1}{4\pi R}\frac{\partial\mathbf{F}(\mathbf{r}',t-|\mathbf{r}-\mathbf{r}'|/c)}{\partial t'}dV(\mathbf{r}').\end{aligned} \tag{1.202}$$

Note that the time-domain Helmholtz theorem contains a time derivative term that may have both transverse and longitudinal components.

1.8 Curl Operator

Eigenfunctions of the curl operator are useful in expanding solenoidal vector fields and have found applications in some fields of physics. For example, in plasma physics, a magnetic field $\mathbf{B}$ is called a force-free field if $\nabla\times\mathbf{B}=\lambda\mathbf{B}$; in fluid dynamics, a velocity field $\mathbf{v}$ satisfying $\nabla\times\mathbf{v}=\lambda\mathbf{v}$ is called a Beltrami flow [35].

1.8.1 Eigenfunctions of Curl Operator

The eigenvalue problem of the curl operator has interesting applications in electromagnetics [36–38]. The eigenfunction $\mathbf{e}$ of a curl operator is defined by

$$\nabla \times \mathbf{e}(\mathbf{r}) = \lambda \mathbf{e}(\mathbf{r}), \mathbf{r} \in R^3. \tag{1.203}$$

The plane-wave solutions of Maxwell equations are widely used in field analysis for their simplicity [39]. To seek the plane-wave solution of (1.203), one may assume

$$\mathbf{e}(\mathbf{r}) = \mathbf{A}e^{j\mathbf{k}\cdot\mathbf{r}}, \tag{1.204}$$

where $\mathbf{k} = (k_x, k_y, k_z)$, and $\mathbf{A}$ is a constant vector. Introducing (1.204) into (1.203) yields

$$j\mathbf{k} \times \mathbf{A} = \lambda \mathbf{A}.$$

Explicitly this is

$$\begin{bmatrix} -\lambda & -jk_z & jk_y \\ jk_z & -\lambda & -jk_x \\ -jk_y & jk_x & -\lambda \end{bmatrix} \begin{bmatrix} A_x \\ A_y \\ A_z \end{bmatrix} = 0. \tag{1.205}$$

A nonzero solution exists if and only if the determinant of the coefficient matrix is zero

$$\det \begin{bmatrix} -\lambda & -jk_z & jk_y \\ jk_z & -\lambda & -jk_x \\ -jk_y & jk_x & -\lambda \end{bmatrix} = \lambda\left(-\lambda^2 + k^2\right) = 0,$$

where $k = |\mathbf{k}|$. Hence Eq. (1.205) has three eigenvalues $\lambda = nk (n = 0, \pm 1)$. The eigenfunctions corresponding to $\lambda = 0, k, -k$, respectively, satisfy

$$\begin{cases} k_yA_z - k_zA_y = 0 \\ k_zA_x - k_xA_z = 0 \\ k_xA_y - k_yA_x = 0 \end{cases}, \begin{cases} kA_x + jk_zA_y - jk_yA_z = 0 \\ kA_y + jk_xA_z - jk_zA_x = 0 \\ kA_z + jk_yA_x - jk_xA_y = 0 \end{cases}, \begin{cases} kA_x - jk_zA_y + jk_yA_z = 0 \\ kA_y - jk_xA_z + jk_zA_x = 0 \\ kA_z - jk_yA_x + jk_xA_y = 0 \end{cases}.$$

The orthonormal eigenfunctions can be easily found from the above equations as follows:

$$\mathbf{A}_0(\mathbf{k}) = \frac{1}{k}\begin{bmatrix} k_x \\ k_y \\ k_z \end{bmatrix}, \mathbf{A}_n(\mathbf{k}) = \frac{1}{\sqrt{2}k\left(k_x^2 + k_y^2\right)^{1/2}} \begin{bmatrix} jnkk_y - k_xk_z \\ -jnkk_x - k_yk_z \\ k_x^2 + k_y^2 \end{bmatrix}, (n = \pm 1).$$

Let $\mathbf{A}_n(\mathbf{k}) = [A_{nx}(\mathbf{k}), A_{ny}(\mathbf{k}), A_{nz}(\mathbf{k})]^T$. Then, it is readily found that

$$\begin{aligned} &\mathbf{A}_m \cdot \overline{\mathbf{A}}_n = \delta_{mn}(m, n = 0, \pm 1), \\ &\sum_n A_{n\alpha}\overline{A}_{n\beta} = \delta_{\alpha\beta}(\alpha, \beta = x, y, z). \end{aligned} \tag{1.206}$$

The following orthonormal vectors may be introduced:

$$\mathbf{e}_n(\mathbf{r},\mathbf{k}) = \left[e_{nx}(\mathbf{r},\mathbf{k}), e_{ny}(\mathbf{r},\mathbf{k}), e_{nz}(\mathbf{r},\mathbf{k})\right]^T = \frac{1}{(2\pi)^{3/2}}\mathbf{A}_n(\mathbf{k})e^{j\mathbf{k}\cdot\mathbf{r}},$$

which satisfy the orthonormal conditions

$$\begin{aligned} &\int_{R^3} \mathbf{e}_m(\mathbf{r},\mathbf{k}) \cdot \overline{\mathbf{e}}_n(\mathbf{r},\mathbf{k}')dxdydz = \delta_{mn}\delta(\mathbf{k}-\mathbf{k}'), (m, n = 0, \pm 1), \\ &\sum_n \int_{R^3} e_{n\alpha}(\mathbf{r},\mathbf{k})\overline{e}_{n\beta}(\mathbf{r}',\mathbf{k})dk_xdk_ydk_z = \delta_{\alpha\beta}\delta(\mathbf{r}-\mathbf{r}'), (\alpha, \beta = x, y, z), \end{aligned} \tag{1.207}$$

and

$$\begin{aligned} &\nabla \times \mathbf{e}_n = nk\mathbf{e}_n, \\ &\nabla \cdot \mathbf{e}_n = 0\ (n = \pm 1), \\ &\nabla \cdot \mathbf{e}_0 = \frac{jk}{(2\pi)^{3/2}}e^{j\mathbf{k}\cdot\mathbf{r}}. \end{aligned} \tag{1.208}$$

The second equation of (1.207) can be expressed in a dyadic form as

$$\sum_n \int_{R^3} \mathbf{e}_n(\mathbf{r},\mathbf{k})\overline{\mathbf{e}}_n(\mathbf{r}',\mathbf{k})dk_xdk_ydk_z = \overleftrightarrow{\mathbf{I}}\delta(\mathbf{r}-\mathbf{r}'). \tag{1.209}$$

In terms of (1.209), an arbitrary vector $\mathbf{F}$ can then be expanded as follows:

$$\mathbf{F}(\mathbf{r}) = \sum_n \int_{R^3} \mathbf{e}_n(\mathbf{r},\mathbf{k})f_n(\mathbf{k})dk_xdk_ydk_z = \sum_n \mathbf{F}_n(\mathbf{r}), \tag{1.210}$$

where

$$\begin{aligned} &\mathbf{F}_n(\mathbf{r}) = \int_{R^3} \mathbf{e}_n(\mathbf{r},\mathbf{k})f_n(\mathbf{k})dk_xdk_ydk_z, \\ &f_n(\mathbf{k}) = \int_{R^3} \overline{\mathbf{e}}_n(\mathbf{r},\mathbf{k}) \cdot \mathbf{F}(\mathbf{r})dxdydz. \end{aligned} \tag{1.211}$$

Since $\nabla \times \mathbf{F}_0 = 0$ and $\nabla \cdot \mathbf{F}_n = 0$ $(n = \pm 1)$, an arbitrary vector $\mathbf{F}$ may be decomposed into three components: one is irrotational and the other two are solenoidal. This result may be regarded as the **generalized Helmholtz theorem**. By means of (1.208), one may write

$$\mathbf{F}_0(\mathbf{r}) = \nabla\phi(\mathbf{r}), \quad \mathbf{F}_n(\mathbf{r}) = \nabla \times \mathbf{A}_n(\mathbf{r}), n = \pm 1,$$

where

$$\phi(\mathbf{r}) = \frac{-j}{(2\pi)^{3/2}} \int_{R^3} \frac{1}{k} e^{j\mathbf{k}\cdot\mathbf{r}} f_0(\mathbf{k}) dk_x dk_y dk_z,$$

$$\mathbf{A}_n(\mathbf{r}) = \frac{1}{n} \int_{R^3} \frac{1}{k} \mathbf{e}_n(\mathbf{r}, \mathbf{k}) f_n(\mathbf{k}) dk_x dk_y dk_z.$$

As a result, (1.210) can be expressed by

$$\mathbf{F} = \nabla\phi + \nabla \times \mathbf{A}_{-1} + \nabla \times \mathbf{A}_1. \tag{1.212}$$

1.8.2 Plane-Wave Expansions for the Fields and Dyadic Green's Functions

Consider the solution of the generalized Maxwell equations

$$\begin{aligned} \nabla \times \mathbf{H}(\mathbf{r}) &= \mathbf{J}(\mathbf{r}) + j\omega\varepsilon\mathbf{E}(\mathbf{r}), \\ \nabla \times \mathbf{E}(\mathbf{r}) &= -j\omega\mu\mathbf{H}(\mathbf{r}) - \mathbf{J}_m(\mathbf{r}). \end{aligned} \tag{1.213}$$

In terms of (1.210), the fields and the sources have the following expansions:

$$\begin{aligned} \mathbf{E}(\mathbf{r}) &= \sum_n \int_{R^3} \mathbf{e}_n(\mathbf{r}, \mathbf{k}) e_n(\mathbf{k}) dk_x dk_y dk_z, \\ \mathbf{H}(\mathbf{r}) &= \sum_n \int_{R^3} \mathbf{e}_n(\mathbf{r}, \mathbf{k}) h_n(\mathbf{k}) dk_x dk_y dk_z, \\ \mathbf{J}(\mathbf{r}) &= \sum_n \int_{R^3} \mathbf{e}_n(\mathbf{r}, \mathbf{k}) j_n(\mathbf{k}) dk_x dk_y dk_z, \\ \mathbf{J}_m(\mathbf{r}) &= \sum_n \int_{R^3} \mathbf{e}_n(\mathbf{r}, \mathbf{k}) j_{m,n}(\mathbf{k}) dk_x dk_y dk_z. \end{aligned} \tag{1.214}$$

Upon substitution of the field and source expansions into (1.213), the expansion coefficients for the fields may be found as follows:

$$
\begin{aligned}
e_n(\mathbf{k}) &= -\frac{j\omega\mu j_n(\mathbf{k}) + nkj_{m,n}(\mathbf{k})}{n^2k^2 - k_0^2},\\
h_n(\mathbf{k}) &= \frac{nkj_n(\mathbf{k}) - j\omega\varepsilon j_{m,n}(\mathbf{k})}{n^2k^2 - k_0^2},
\end{aligned}
\tag{1.215}
$$

where $k_0 = \omega\sqrt{\mu\varepsilon}$. The field expansions in (1.214) can be written in a compact form as

$$
\begin{aligned}
\mathbf{E}(\mathbf{r}) &= -j\omega\mu\int_{R^3}\overleftrightarrow{\mathbf{G}}_e(\mathbf{r},\mathbf{r}')\cdot\mathbf{J}(\mathbf{r}')dx'dy'dz' - \int_{R^3}\overleftrightarrow{\mathbf{G}}_m(\mathbf{r},\mathbf{r}')\cdot\mathbf{J}_m(\mathbf{r}')dx'dy'dz',\\
\mathbf{H}(\mathbf{r}) &= -j\omega\varepsilon\int_{R^3}\overleftrightarrow{\mathbf{G}}_e(\mathbf{r},\mathbf{r}')\cdot\mathbf{J}_m(\mathbf{r}')dx'dy'dz' + \int_{R^3}\overleftrightarrow{\mathbf{G}}_m(\mathbf{r},\mathbf{r}')\cdot\mathbf{J}(\mathbf{r}')dx'dy'dz',
\end{aligned}
\tag{1.216}
$$

where $\overleftrightarrow{\mathbf{G}}_e$ and $\overleftrightarrow{\mathbf{G}}_e$ are the electric and magnetic dyadic Green's functions, defined by

$$
\begin{aligned}
\overleftrightarrow{\mathbf{G}}_e(\mathbf{r},\mathbf{r}') &= \sum_n\int_{R^3}\frac{\mathbf{e}_n(\mathbf{r},\mathbf{k})\bar{\mathbf{e}}_n(\mathbf{r}',\mathbf{k})}{n^2k^2 - k_0^2}dk_xdk_ydk_z,\\
\overleftrightarrow{\mathbf{G}}_m(\mathbf{r},\mathbf{r}') &= \sum_n\int_{R^3}\frac{nk\mathbf{e}_n(\mathbf{r},\mathbf{k})\bar{\mathbf{e}}_n(\mathbf{r}',\mathbf{k})}{n^2k^2 - k_0^2}dk_xdk_ydk_z.
\end{aligned}
\tag{1.217}
$$

Ignoring the tedious process, the electric and magnetic dyadic Green's functions can be rewritten as

$$
\begin{aligned}
\overleftrightarrow{\mathbf{G}}_e(\mathbf{r},\mathbf{r}') &= \frac{1}{(2\pi)^3}\int_{R^3}\frac{k_0^2\overleftrightarrow{\mathbf{I}} - \mathbf{k}\mathbf{k}}{k_0^2(k^2 - k_0^2)}e^{j\mathbf{k}\cdot(\mathbf{r}-\mathbf{r}')}dk_xdk_ydk_z,\\
\overleftrightarrow{\mathbf{G}}_m(\mathbf{r},\mathbf{r}') &= \frac{1}{(2\pi)^3}\int_{R^3}\frac{\mathbf{k}\times\overleftrightarrow{\mathbf{I}}}{k^2 - k_0^2}je^{j\mathbf{k}\cdot(\mathbf{r}-\mathbf{r}')}dk_xdk_ydk_z.
\end{aligned}
\tag{1.218}
$$

These are the **plane-wave expansions** for the dyadic Green's functions. They can also be derived from the Fourier transform with respect to the position vector **r**. It is noted that the expressions in (1.218) should be taken as a symbolic equality and are meaningful only when these expressions are used as the kernel of an integral operator because they may contain generalized functions [40]. Apparently,

$$
\nabla\times\overleftrightarrow{\mathbf{G}}_e(\mathbf{r},\mathbf{r}') = \overleftrightarrow{\mathbf{G}}_m(\mathbf{r},\mathbf{r}'). \tag{1.219}
$$

It follows from (1.7), (1.216), and (1.219) that the electric and magnetic dyadic Green's functions, respectively, satisfy

$$\begin{aligned} &\nabla\times\nabla\times\overleftrightarrow{\mathbf{G}}_e(\mathbf{r},\mathbf{r}')-k^2\overleftrightarrow{\mathbf{G}}_e(\mathbf{r},\mathbf{r}')=\overleftrightarrow{\mathbf{I}}\delta(\mathbf{r}-\mathbf{r}'),\\ &\nabla\times\nabla\times\overleftrightarrow{\mathbf{G}}_m(\mathbf{r},\mathbf{r}')-k^2\overleftrightarrow{\mathbf{G}}_m(\mathbf{r},\mathbf{r}')=\nabla\times\left[\overleftrightarrow{\mathbf{I}}\delta(\mathbf{r}-\mathbf{r}')\right]. \end{aligned} \tag{1.220}$$

The growing importance of eigenvalue theory in pure and applied mathematics, and in physics and chemistry, has drawn attention to various methods for approximate calculation of eigenvalues. Clearly it is important to develop these methods in a general and theoretical manner, if only because opportunities for particular application may otherwise be inadvertently missed.

-Sydney Henry Gould (Mathematician, 1909–1986)

References

1 Baum, C. E., "On the singularity expansion method for the solution of electromagnetic interaction problems", *AFWL*, Interaction Notes, Vol. 88, 1971.

2 Baum, C. E., "The singularity expansion method", in *Transient Electromagnetic Fields*, edited by L. B. Felsen, New York, NY, USA, Springer-Verlag, pp. 129–179, 1976.

3 Ramm, A. G., "Mathematical foundations of the singularity and eigenmode expansion methods (SEM and EEM)", *J. Math. Anal. Appl.*, Vol. 86, pp. 562–591, 1982.

4 Marks, R. B., "The singular function expansion in time-dependent scattering", *IEEE Trans. Antennas Propagat.*, Vol. 37, pp. 1559–1565, 1989.

5 Cochran, J. A., *The Analysis of Linear Integral Equations*, New York, McGraw-Hill, 1972.

6 Inagaki, N., "Eigenfunctions of composite Hermitian operators with application to discrete and continuous radiating systems", *IEEE Trans. Antennas Propag.*, Vol. 30, No. 7, pp. 571–575, 1982.

7 Pozar, D. M., "Antenna synthesis and optimization using weighted Inagaki modes", *IEEE Trans. Antennas Propagat.*, Vol. 32, pp. 159–165, 1984.

8 Garbacz, R. J., "Modal expansions for resonance scattering phenomena", *Proc. IEEE*, Vol. 53, No. 8, pp. 856–864, 1965.

9 Garbacz, R. J. and R. Turpin, "A generalized expansion for radiated and scattered fields", *IEEE Trans. Antennas Propag.*, Vol. 19, No. 3, pp. 348–358, 1971.

10 Harrington, R. F. and J. R. Mautz, "Theory of characteristic modes for conducting bodies", *IEEE Trans. Antennas Propag.*, Vol. 19, No. 5, pp. 622–628, 1971.

11 Harrington, R. F., J. R. Mautz, and Y. Chang, “Characteristic modes for dielectric and magnetic bodies”, *IEEE Trans. Antennas Propag.*, Vol. 20, No. 2, pp. 194–198, 1972.

12 Gould, S. H., *Variational Methods for Eigenvalue Problems: An Introduction to the Methods of Rayleigh, Ritz, Weinstein, and Aronszajn*, Incorporated, Dover Publications, 1995.

13 Popvić, B. D., “Electromagnetic field theorems”, *IEE Proc.*, Vol. 128, pp. 47–63, 1981.

14 Harrington, R. F., *Time-Harmonic Electromagnetic Fields*, McGraw-Hill Book Company, Inc., 1961.

15 Balanis, C. A., *Advanced Engineering Electromagnetics*, 2nd Ed., John Wiley & Sons, 2012.

16 Geyi, W., *Foundations of Applied Electrodynamics*, New York, Wiley, 2010.

17 Geyi, W., “Stored electromagnetic field energies in general materials”, *J. Opt. Soc. Am. B*, Vol. 36, No. 4, pp. 917–925, 2019.

18 Geyi, W., “A method for the evaluation of small antenna Q”, *IEEE Trans. Antennas and Propagat.*, Vol. 51, pp. 2124–2129, 2003.

19 Courant, R. and D. Hilbert, *Methods of Mathematical Physics*, Vol. 1–2, John Wiley & Sons, 1953.

20 Morse, P. M. and H. Feshbach, *Methods of Theoretical Physics*, McGraw-Hill, 1953.

21 Gustafson, K. E., *Partial Differential Equations and Hilbert Space Methods*, John Wiley & Sons, 1987.

22 Eisenhart, L. P., “Separable systems of Stäckel”, *Ann. Math.*, Vol. 35, No. 2, pp. 284–305, 1934.

23 Levine, H. and J. Schwinger, “On the theory of electromagnetic wave diffraction by an aperture in an infinite plane conducting screen”, *Commun. Pure Appl. Math.*, Vol. 3, No. 4, pp. 355–391, 1950.

24 Tai, C.-T., *Dyadic Green Functions in Electromagnetic Theory*, IEEE Press, 1994.

25 Bladel, J. V., “Some remarks on Green’s functions for infinite spaces”, *IRE Trans. Antennas Propagat*, Vol. 9, pp. 563–566, 1961.

26 Yaghjian, A. D., “Electric dyadic Green’s functions in the source region”, *Proc. IEEE*, Vol. 68, No. 2, pp. 248–263, 1980.

27 Chew, W. C., *Waves and Fields in Inhomogeneous Media*, Van Nostrand Reinhold, 1990.

28 Horn, R. A. and C. R. Johnson, *Matrix Analysis*, 2nd Ed., Cambridge University Press, 2013.

29 Strauss, W. A., *Partial Differential Equations-An Introduction*, John Wiley & Sons, Inc., 1992.

30 Friedman, B., *Principles and Techniques of Applied Mathematics*, John Wiley & Sons, Inc., 1956.

31 Zeidler, E., *Applied Functional Analysis-Applications to Mathematical Physics*, Springer-Verlag, 1995.

32 Kurokawa, K., *An Introduction to Microwave Circuits*, New York, Academic Press, 1969.

33 Mikhlin, S. G., *Variational Methods in Mathematical Physics*, Oxford, Pergamon Press, 1964.

34 Bladel, J. V., "On Helmholtz's theorem in multiply-bounded and multiply-connected regions", *J. Frankl. Inst.*, Vol. 269, No. 6, pp. 445–462, 1960.

35 Yoshida, Z., "Eigenfunction expansions associated with the curl derivatives in cylindrical geometries: completeness of Chandrasekhar-Kendall eigenfunctions", *J. Math. Phys.*, Vol. 33, No. 4, pp. 1252–1256, 1992.

36 Good, R. H., "Particle aspect of the electromagnetic field equations", *Phys. Rev.*, Vol. 105, No. 6, pp. 1914–1919, 1957.

37 Moses, H. E., "Solutions of Maxwell equations in terms of a spinor notation: the direct and inverse problem", *Phys. Rev.*, Vol. 113, No. 6, pp. 1670–1679, 1959.

38 Moses, H. E. and R. T. Prosser, "Initial conditions sources, and currents for prescribed time-dependent acoustic and electromagnetic fields in three dimensions, part 1: the inverse initial value problem. Acoustic and electromagnetic bullets, expanding waves, and imploding waves", *IEEE Trans. Antennas and Propagat.*, Vol. 34, pp. 188–196, 1986.

39 Clemmow, P. C., *The Plane Wave Spectrum Representation of Electromagnetic Fields*, IEEE Press, 1996.

40 Chew, W. C., "Some observations on the spatial and eigenfunction representations of dyadic Green's functions", *IEEE Trans. Antennas Propag.*, Vol. 37, pp. 1322–1327, 1989.

2

Radiation in Waveguide

We do not really deal with mathematical physics, but with physical mathematics; not with the mathematical formulation of physical facts, but with the physical motivation of mathematical methods.

– Arnold Sommerfeld (German physicist, 1868–1951)

In the low-frequency ranges, electric power is often transmitted by a two-wire line. As frequency increases, the two-wire line is no longer suitable for power transmission due to the radiation loss, and must be replaced by a waveguide. Waveguides are the basic building blocks of microwave circuits and their counterparts are connecting wires in low-frequency circuits. Some historical events of waveguides are summarized in Table 2.1 [1, 2]. Most important results on waveguide theory obtained in the first half of last century have been summarized in [3]. The fundamental part of the waveguide theory is to solve an eigenvalue problem with the cutoff wavenumbers being the eigenvalues and eigenfunctions being the corresponding vector modal functions. The cutoff wavenumbers can be expressed as a Rayleigh quotient and the corresponding vector modal functions are the extremal functions that minimize the Rayleigh quotient.

The vector modal functions in a waveguide constitute a complete set and can be used to expand an arbitrary field in the waveguide and solve various problems encountered in waveguide theory. The waveguide problems can be classified into three different categories: the excitation of waveguide, the obstacles in waveguide, and the coupling between waveguides. The excitation of waveguide studies the radiation fields generated by an antenna in the waveguide, which is fed by another waveguide. The waveguide discontinuities are often introduced as passive circuits or elements to achieve various purposes. When a propagating mode is incident upon a waveguide discontinuity, reflected waves and higher-order modes will be excited, and the latter are evanescent and only exist in the vicinity of the

Foundations of Antenna Radiation Theory: Eigenmode Analysis, First Edition. Wen Geyi.

Published 2023 by John Wiley & Sons, Inc.

Table 2.1 Historical events of waveguides.

Year	Event
1893	English mathematician and physicist Oliver Heaviside (1850–1925) theoretically considered various possibilities for waves along wire lines. British physicist Joseph John Thomson (1856–1940) derived the electromagnetic modes in a cylindrical metal cavity.
1884	British physicist Oliver Joseph Lodge (1851–1940) experimentally verified the propagation of EM waves in a metal waveguide.
1897	British scientist Lord Rayleigh (1942–1919) found that the guided waves existed only in a set of normal modes, and to support the modes in the hollow cylinder, the operating frequency must exceed the cutoff frequencies of the corresponding modes.
1899	German theoretical physicist Arnold Sommerfeld (1868–1951) first studied the theory of dielectric waveguide.
1909	Greek physicist Demetrius Hondros (1882–1962) extended Sommerfeld's study on dielectric waveguide.
1930s	American radio engineers George Clark Southworth (1890–1972) studied wave propagation in dielectric rods in 1931, and observed wave propagation in a water-filled copper pipe in 1932, and transmitted waves through air-filled copper pipes up to 20 ft in length in 1933.

discontinuity in the form of stored reactive energy. In order to couple the electromagnetic (EM) field energy from one waveguide to another, one can use antennas or cut one or more small apertures located in the common wall of two waveguides. The three different waveguide problems may be considered as the radiation problems in waveguide by means of equivalence principle.

This chapter features an updated treatment of various dyadic Green's functions for the waveguide, in terms of which the incompleteness problem of the eigenfunctions for expanding a dyadic point source in the conventional studies can be avoided. The dyadic Green's functions are directly constructed from the field expansions in vector modal functions, and are then applied to the study of the radiation problems in waveguide. A time-domain theory for the waveguide is also included for its importance in the study of transient responses in high-speed circuits and ultra-wideband systems.

2.1 Vector Modal Functions for Waveguide

Throughout this chapter, the waveguide wall will be assumed to be perfectly conducting. In a homogeneously filled waveguide, there exist three different types of field modes. The first type of mode has no longitudinal components of both the electric and magnetic field, and is called **transverse EM** (TEM) mode. The second type of mode, called **transverse electric** (TE) mode, has a longitudinal magnetic field component but no longitudinal electric field. The third type of mode does not have a longitudinal magnetic field component but has a longitudinal electric field, and is called **transverse magnetic** (TM) mode. For TE and TM modes, the transverse fields can be represented by the corresponding longitudinal components. For this reason, the theory of a simply connected waveguide can either be approached by the longitudinal components of the fields or by the transverse components. For generality, the transverse fields will be used to study the wave propagations in waveguide in this chapter.

2.1.1 Classification of Vector Modal Functions

The cross section of an arbitrary metal waveguide is shown in Figure 2.1, and is denoted by Ω, and its boundary by Γ. In general, the cross section can be multiply connected. Let $\mathbf{r} = (\boldsymbol{\rho}, z)$, where $\boldsymbol{\rho} = (x, y)$ is the position vector in the waveguide cross section Ω and z denotes the waveguide axis. The waveguide is assumed to be uniform along z direction and is filled with homogeneous medium with medium parameters μ, ε, and σ. The time-harmonic fields travelling along $+z$ direction with propagation constant $\gamma = j\beta$ can be decomposed into the transverse and longitudinal parts as follows:

$$\begin{aligned}\mathbf{E}(\mathbf{r}) &= [\mathbf{e}(\boldsymbol{\rho}) + \mathbf{u}_z e_z(\boldsymbol{\rho})]e^{-\gamma z},\\ \mathbf{H}(\mathbf{r}) &= [\mathbf{h}(\boldsymbol{\rho}) + \mathbf{u}_z h_z(\boldsymbol{\rho})]e^{-\gamma z}.\end{aligned} \tag{2.1}$$

Substituting these into Maxwell equations in source-free region

$$\begin{aligned}\nabla \times \mathbf{H} &= j\omega\varepsilon_e \mathbf{E},\\ \nabla \times \mathbf{E} &= -j\omega\mu \mathbf{H},\\ \nabla \cdot \mathbf{E} &= 0,\\ \nabla \cdot \mathbf{H} &= 0,\end{aligned}$$

where $\varepsilon_e = \varepsilon\left(1 - j\dfrac{\sigma}{\omega\varepsilon}\right)$, the relations between the transverse and longitudinal fields can be obtained

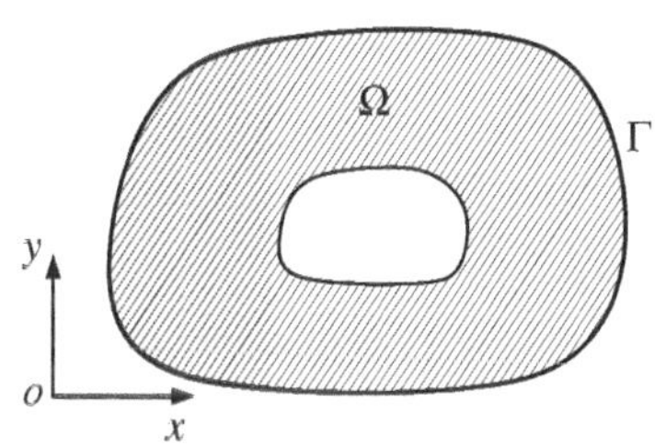

Figure 2.1 An arbitrary waveguide.

$$
\begin{aligned}
&\nabla \times \mathbf{h} = j\omega\varepsilon_e e_z \mathbf{u}_z,\\
&\quad j\beta\mathbf{u}_z \times \mathbf{h} + \mathbf{u}_z \times \nabla h_z = -j\omega\varepsilon_e \mathbf{e},\\
&\nabla \times \mathbf{e} = -j\omega\mu h_z \mathbf{u}_z,\\
&\quad j\beta\mathbf{u}_z \times \mathbf{e} + \mathbf{u}_z \times \nabla e_z = j\omega\mu \mathbf{h},\\
&\nabla \cdot \mathbf{e} = j\beta e_z,\\
&\nabla \cdot \mathbf{h} = j\beta h_z.
\end{aligned}
\tag{2.2}
$$

From (2.2), the transverse electric field is readily found to satisfy the vector Helmholtz equation

$$
\begin{cases}
-\nabla^2 \mathbf{e}(\boldsymbol{\rho}) = k_c^2 \mathbf{e}(\boldsymbol{\rho}), \boldsymbol{\rho} \in \Omega,\\
\mathbf{u}_n \times \mathbf{e}(\boldsymbol{\rho}) = 0, \nabla \cdot \mathbf{e}(\boldsymbol{\rho}) = 0, \boldsymbol{\rho} \in \Gamma,
\end{cases}
\tag{2.3}
$$

where $\nabla^2 = -\nabla\times\nabla\times\mathbf{e} + \nabla\nabla\cdot\mathbf{e}$ and $k_c^2 = \omega^2\mu\varepsilon_e + \gamma^2$. It is known from Section 1.5 that a complete set of orthogonal eigenfunctions $\{\mathbf{e}_1, \mathbf{e}_2, ...\}$ can be constructed from (2.3), and the corresponding eigenvalues satisfy $0 \le k_{c1}^2 \le k_{c2}^2 \le ....$ The eigenfunction $\mathbf{e}_n$ is called nth **vector modal function**, and the corresponding eigenvalue k_{cn} is called **cutoff wavenumber** of the nth vector modal function. From now on, it will be assumed that all vector modal functions are orthonormal

$$
\int_\Omega \mathbf{e}_m \cdot \mathbf{e}_n d\Omega = \delta_{mn}.
\tag{2.4}
$$

An arbitrary vector function $\mathbf{f}$ can then be expressed as a linear combination of the vector modal functions

$$
\mathbf{f} = \sum_{n=1}^{\infty} a_n \mathbf{e}_n
$$

with $a_n = \int_\Omega \mathbf{f} \cdot \mathbf{e}_n d\Omega$. The vector modal function $\mathbf{e}_n$ belongs to one of the following three types:

1. $\nabla \times \mathbf{e}_n = 0,\ \nabla \cdot \mathbf{e}_n = 0,$
2. $\nabla \times \mathbf{e}_n \neq 0,\ \nabla \cdot \mathbf{e}_n = 0,$
3. $\nabla \times \mathbf{e}_n = 0,\ \nabla \cdot \mathbf{e}_n \neq 0.$

The vector modal functions belonging to the first type are called TEM **modes**. For the TEM modes, a scalar potential function ϕ may be introduced such that $\mathbf{e}_n = -\nabla\phi$ and

$$
\begin{cases}
\nabla \cdot \nabla\phi(\boldsymbol{\rho}) = 0, & \boldsymbol{\rho} \in \Omega,\\
\mathbf{u}_n \times \nabla\phi(\boldsymbol{\rho}) = 0, & \boldsymbol{\rho} \in \Gamma.
\end{cases}
\tag{2.5}
$$

The second equation implies that the potential function ϕ is constant along the boundary Γ. If Ω is simply connected, the above equations imply $\mathbf{e}_n = 0$ and a hollow waveguide does not support a TEM mode. If Ω is a multiply connected region (such as a coaxial cable), ϕ may take different values on different conductors. In this case, the waveguide can support a TEM mode. If $\mathbf{e}_n$ is a TEM mode, then $k_{cn}^2 = 0$.

The vector modal functions belonging to second type are called TE **modes** or **H modes**. For $\nabla \times \mathbf{e}_n$ is longitudinal

$$\mathbf{u}_z \times \nabla \times \mathbf{e}_n = \nabla(\mathbf{u}_z \cdot \mathbf{e}_n) - (\mathbf{u}_z \cdot \nabla)\mathbf{e}_n = 0,$$

one may introduce a new function h_{zn} such that

$$\nabla \times \mathbf{e}_n = \mathbf{u}_z k_{cn} h_{zn}. \tag{2.6}$$

The new function h_{zn} is proportional to the longitudinal magnetic field. It follows from (2.3) and (2.6) that

$$\begin{cases} \nabla^2 h_{zn}(\boldsymbol{\rho}) + k_{cn}^2 h_{zn}(\boldsymbol{\rho}) = 0, \boldsymbol{\rho} \in \Omega, \\ \mathbf{u}_n \cdot \nabla h_{zn}(\boldsymbol{\rho}) = 0, \boldsymbol{\rho} \in \Gamma. \end{cases} \tag{2.7}$$

As discussed in Section 1.4, the eigenfunctions of (2.7) also form a complete set. By the definition (2.6), the eigenfunction of (2.7) satisfy

$$\int_\Omega h_{zm} h_{zn} d\Omega = \frac{1}{k_{cm} k_{cn}} \int_\Omega \nabla \times \mathbf{e}_m \cdot \nabla \times \mathbf{e}_n d\Omega = \frac{k_{cm}}{k_{cn}} \int_\Omega \mathbf{e}_m \cdot \mathbf{e}_n d\Omega.$$

This indicates that the set $\{h_{zn}\}$ is orthonormal if the set $\{\mathbf{e}_n\}$ is. On the contrary, if h_{zn} is an eigenfunction of (2.7), one may let

$$\mathbf{e}_n = -\frac{1}{k_{cn}} \mathbf{u}_z \times \nabla h_{zn}. \tag{2.8}$$

It is easy to verify that $\mathbf{e}_n$ satisfies (2.3) and $\nabla \cdot \mathbf{e}_n = 0$. Thus, $\mathbf{e}_n$ is a TE mode. In addition,

$$\int_\Omega \mathbf{e}_m \cdot \mathbf{e}_n d\Omega = \frac{1}{k_{cm} k_{cn}} \int_\Omega \nabla h_{zm} \cdot \nabla h_{zn} d\Omega = \frac{k_{cm}}{k_{cn}} \int_\Omega h_{zm} \cdot h_{zn} d\Omega.$$

Consequently, if the set $\{h_{zn}\}$ is orthonormal, so is the set $\{\mathbf{e}_n\}$. The foregoing analysis shows that there is a one-one correspondence between the set of TE modes and the set of eigenfunctions $\{h_{zn}\}$.

The vector modal functions belonging to the third type are called TM **modes** or **E modes**. One may introduce a new function e_{zn} such that

$$\nabla \cdot \mathbf{e}_n = k_{cn} e_{zn}. \tag{2.9}$$

The new function e_{zn} is proportional to the longitudinal electric field. It follows from (2.3) and (2.9) that

$$\begin{cases} \nabla^2 e_{zn}(\boldsymbol{\rho}) + k_{cn}^2 e_{zn}(\boldsymbol{\rho}) = 0, \ \boldsymbol{\rho} \in \Omega, \\ e_{zn} = 0, \ \boldsymbol{\rho} \in \Gamma. \end{cases} \tag{2.10}$$

As shown from Section 1.4, the eigenfunctions of (2.10) form a complete set. Moreover,

$$\int_\Omega e_{zm} e_{zn} d\Omega = \frac{1}{k_{cm}k_{cn}} \int_\Omega \nabla \cdot \mathbf{e}_m \cdot \nabla \cdot \mathbf{e}_n d\Omega = \frac{k_{cm}}{k_{cn}} \int_\Omega \mathbf{e}_m \cdot \mathbf{e}_n d\Omega.$$

Thus, if the set $\{\mathbf{e}_n\}$ is orthonormal, so is the set $\{e_{zn}\}$. Conversely, the TM modes can be derived from the eigenfunctions e_{zn} of (2.10) through

$$\mathbf{e}_n = -\frac{1}{k_{cn}} \nabla e_{zn}. \tag{2.11}$$

It is easy to show that $\mathbf{e}_n$ defined by (2.11) satisfies (2.3) and is a TM mode. Furthermore,

$$\int_\Omega \mathbf{e}_m \cdot \mathbf{e}_n d\Omega = \frac{1}{k_{cm}k_{cn}} \int_\Omega \nabla e_{zm} \cdot \nabla e_{zn} d\Omega = \frac{k_{cm}}{k_{cn}} \int_\Omega e_{zm} \cdot e_{zn} d\Omega.$$

Therefore, if the set $\{e_{zn}\}$ is orthonormal so is the set $\{\mathbf{e}_n\}$. A one-one correspondence between the set of TM modes and the set of eigenfunctions $\{e_{zn}\}$ is thus established.

In summary, three orthonormal sets can be constructed from the orthonormal set $\{\mathbf{e}_n\}$:

1. $\{\mathbf{u}_z \times \mathbf{e}_n | \mathbf{u}_n \cdot \mathbf{u}_z \times \mathbf{e}_n = 0, \nabla \cdot \mathbf{e}_n = 0, \boldsymbol{\rho} \in \Gamma\}$.
2. $\left\{ \frac{\nabla \cdot \mathbf{e}_n}{k_{cn}} \middle| \nabla \cdot \mathbf{e}_n = 0, \boldsymbol{\rho} \in \Gamma \right\}$ or $\{e_{zn}\}$.
3. $\left\{ \frac{\nabla \times \mathbf{e}_n}{k_{cn}}, \tilde{c} \middle| \mathbf{u}_n \cdot \nabla \left[\frac{\mathbf{u}_z \cdot \nabla \times \mathbf{e}_n}{k_{cn}} \right] = 0, \boldsymbol{\rho} \in \Gamma \right\}$ or $\{h_{zn}\}$.

In the above, $\tilde{c}$ is a constant. According to the boundary conditions, the set $\{\mathbf{e}_n\}$ is most suitable for the expansion of transverse electric field, the set $\{\mathbf{u}_z \times \mathbf{e}_n\}$ is best suited to the expansion of the transverse magnetic field, the set $\{\nabla \cdot \mathbf{e}_n / k_{cn}\}$ is most appropriate for the expansion of longitudinal electric field, and the set $\{\nabla \times \mathbf{e}_n / k_{cn}, \tilde{c}\}$ is most proper for the expansion of longitudinal magnetic field [4, 5].

2.1.2 Vector Modal Functions for Typical Waveguides

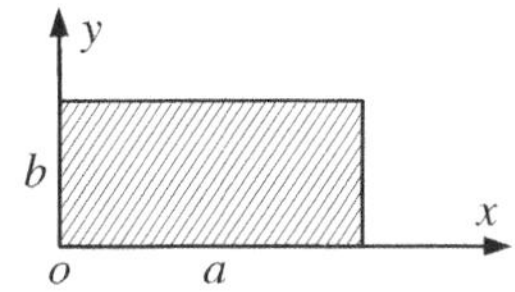

Figure 2.2 Rectangular waveguide.

For a waveguide filled with homogeneous medium, the transverse component for the TE or TM mode can be derived from the longitudinal component. For the TEM mode in a waveguide, one can use the potential function defined by (2.5) to determine the transverse components. The method of separation of variables can be used to solve the Helmholtz equation or Laplace equation [3–15] for waveguides of simple geometry. The vector modal functions for some typical geometries will be summarized in this section. In practice, only the dominant mode is propagating in the waveguide.

2.1.2.1 Rectangular Waveguide

A homogeneous rectangular waveguide shown in Figure 2.2 only supports TE or TM modes. By the method of separation of variables in the rectangular coordinate system, the normalized modal solutions of (2.7) and (2.10) can be easily found as

$$\begin{aligned} e_{zn} &= \sqrt{\frac{4}{ab}}\sin\frac{p\pi}{a}x\sin\frac{q\pi}{b}y, \\ h_{zn} &= \sqrt{\frac{\varepsilon_p\varepsilon_q}{ab}}\cos\frac{p\pi}{a}x\cos\frac{q\pi}{b}y, \end{aligned} \tag{2.12}$$

where $\varepsilon_m = \begin{cases} 1, & m = 0 \\ 2, & m \geq 1 \end{cases}$. Here, the subscript n represents the double index (p, q).

The cutoff wavenumbers for TM and TE modes are, respectively, given by

$$\begin{aligned} k_{cn}^{TM} &= \sqrt{\left(\frac{p\pi}{a}\right)^2 + \left(\frac{q\pi}{b}\right)^2}, \quad p, q = 1, 2, \ldots, \\ k_{cn}^{TE} &= \sqrt{\left(\frac{p\pi}{a}\right)^2 + \left(\frac{q\pi}{b}\right)^2}, \quad p, q = 0, 1, 2, \ldots. \end{aligned} \tag{2.13}$$

The normalized vector modal functions for the TM and TE modes can be obtained from (2.8) and (2.11)

$$\mathbf{e}_n^{TM} = \mathbf{u}_x \frac{1}{k_{cn}^{TM}}\frac{p\pi}{a}\sqrt{\frac{4}{ab}}\cos\frac{p\pi}{a}x\sin\frac{q\pi}{b}y + \mathbf{u}_y\frac{1}{k_{cn}^{TM}}\frac{q\pi}{b}\sqrt{\frac{4}{ab}}\sin\frac{p\pi}{a}x\cos\frac{q\pi}{b}y, \tag{2.14}$$

$$\mathbf{e}_n^{TE} = \mathbf{u}_x \frac{1}{k_{cn}^{TE}}\frac{q\pi}{b}\sqrt{\frac{\varepsilon_p\varepsilon_q}{ab}}\cos\frac{p\pi}{a}x\sin\frac{q\pi}{b}y - \mathbf{u}_y\frac{1}{k_{cn}^{TE}}\frac{p\pi}{a}\sqrt{\frac{\varepsilon_p\varepsilon_q}{ab}}\sin\frac{p\pi}{a}x\cos\frac{q\pi}{b}y, \tag{2.15}$$

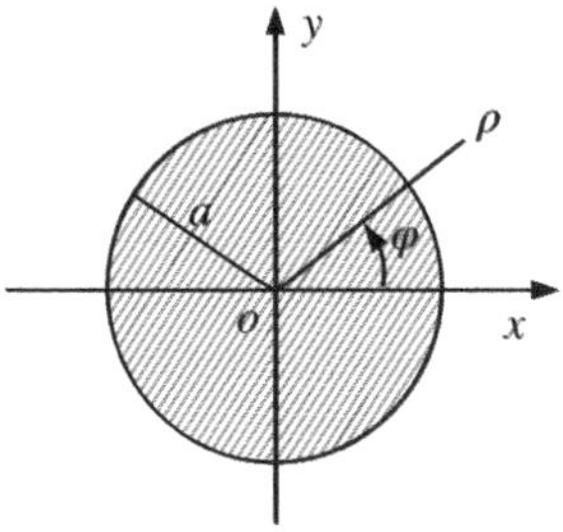

Figure 2.3 Circular waveguide.

respectively. The dominant mode (the mode with the lowest cutoff wavenumber) in the rectangular waveguide is TE_{10} mode.

2.1.2.2 Circular Waveguide

A uniform waveguide of circular cross section of radius a is shown in Figure 2.3, which is not as widely used as rectangular waveguides. The circular waveguide is best described by the cylindrical coordinate system (ρ, φ, z). The normalized modal solutions of (2.7) and (2.10) are readily constructed by the method of separation of variables

$$
\begin{aligned}
e_{zn} &= \sqrt{\frac{\varepsilon_q}{\pi}} \frac{J_q\left(\chi_{qp}\frac{\rho}{a}\right)}{\chi_{qp} J_{q+1}\left(\chi_{qp}\right)} \begin{pmatrix} \cos q\varphi \\ \sin q\varphi \end{pmatrix}, \\
h_{zn} &= \sqrt{\frac{\varepsilon_q}{\pi}} \frac{1}{\sqrt{\chi'^2_{qp} - q^2}} \frac{J_q\left(\chi'_{qp}\frac{\rho}{a}\right)}{J_q\left(\chi'_{qp}\right)} \begin{pmatrix} \cos q\varphi \\ \sin q\varphi \end{pmatrix},
\end{aligned}
\tag{2.16}
$$

where χ_{qp} and χ'_{qp} are, respectively, the pth nonvanishing roots of the equations

$$
J_q\left(\chi_{qp}\right) = 0, \; J'_q\left(\chi'_{qp}\right) = 0. \tag{2.17}
$$

Some lower-order roots of the above equations are listed in Tables 2.2 and 2.3. The cutoff wavenumbers for the TM and TE modes are, respectively, given by

$$
k_{cn}^{TM} = \frac{\chi_{qp}}{a}, \; k_{cn}^{TE} = \frac{\chi'_{qp}}{a}. \tag{2.18}
$$

From (2.8) and (2.11), the normalized vector modal functions for the TM and TE modes can be readily found as follows:

Table 2.2 Roots of $J_q(\chi_{qp}) = 0$.

	$q = 0$	$q = 1$	$q = 2$	$q = 3$
$p = 1$	2.405	3.832	5.136	6.380
$p = 2$	5.520	7.016	8.417	9.761
$p = 3$	8.654	10.173	11.620	13.015

Table 2.3 Roots of $J'_q(\chi'_{qp}) = 0$.

	$q = 0$	$q = 1$	$q = 2$	$q = 3$
$p = 1$	3.832	1.841	3.054	4.201
$p = 2$	7.016	5.331	6.706	8.015
$p = 3$	10.173	8.536	9.969	11.346

$$\mathbf{e}_n^{TM} = -\mathbf{u}_\rho \sqrt{\frac{\varepsilon_q}{\pi}} \frac{J'_q\left(\chi_{qp}\frac{\rho}{a}\right)}{aJ_{q+1}\left(\chi_{qp}\right)} \begin{pmatrix} \cos q\varphi \\ \sin q\varphi \end{pmatrix} \pm \mathbf{u}_\varphi \sqrt{\frac{\varepsilon_q}{\pi}} \frac{q}{\chi_{qp}} \frac{J_q\left(\chi_{qp}\frac{\rho}{a}\right)}{\rho J_{q+1}\left(\chi_{qp}\right)} \begin{pmatrix} \sin q\varphi \\ \cos q\varphi \end{pmatrix}, \tag{2.19}$$

$$\begin{aligned} \mathbf{e}_n^{TE} = &\pm \mathbf{u}_\rho \sqrt{\frac{\varepsilon_q}{\pi}} \frac{q}{\sqrt{\chi'^2_{qp} - q^2}} \frac{J_q\left(\chi'_{qp}\frac{\rho}{a}\right)}{\rho J_q\left(\chi'_{qp}\right)} \begin{pmatrix} \sin q\varphi \\ \cos q\varphi \end{pmatrix} \\ &+ \mathbf{u}_\varphi \sqrt{\frac{\varepsilon_q}{\pi}} \frac{\chi'_{qp}}{\sqrt{\chi'^2_{qp} - q^2}} \frac{J'_q\left(\chi'_{qp}\frac{\rho}{a}\right)}{aJ_q\left(\chi'_{qp}\right)} \begin{pmatrix} \cos q\varphi \\ \sin q\varphi \end{pmatrix}, \end{aligned} \tag{2.20}$$

where $\mathbf{u}_\rho$ and $\mathbf{u}_\varphi$ are unit vectors along ρ and φ direction, respectively. The dominant mode in the circular waveguide is the TE_{11} mode.

2.1.2.3 Coaxial Waveguide

A coaxial waveguide is shown in Figure 2.4. The dominant mode for coaxial waveguide is the TEM mode. The normalized potential function ϕ for the TEM mode can be determined from the Laplace equation (2.5)

$$\phi = \frac{\ln \rho}{\sqrt{2\pi \ln \frac{b}{a}}}.$$

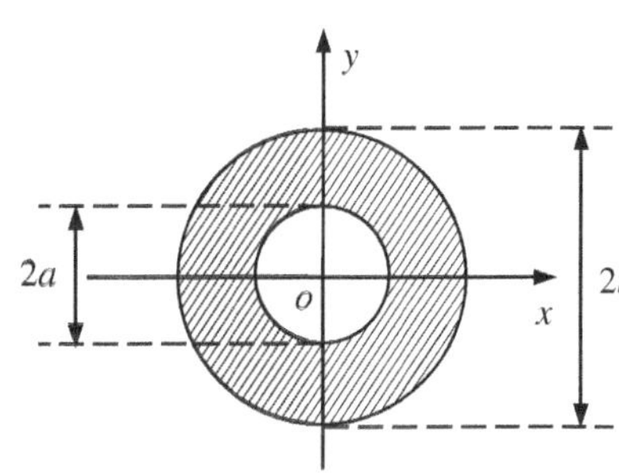

Figure 2.4 Coaxial waveguide.

Other normalized modal solutions for the longitudinal fields can be determined from (2.7) and (2.10):

$$e_{zn} = \sqrt{\frac{\varepsilon_q}{2\pi}} e\left(\chi_{qp}\frac{\rho}{a}\right)\begin{pmatrix}\cos q\varphi \\ \sin q\varphi\end{pmatrix},$$
$$h_{zn} = \sqrt{\frac{\varepsilon_q}{2\pi}} h\left(\chi'_{qp}\frac{\rho}{a}\right)\begin{pmatrix}\cos q\varphi \\ \sin q\varphi\end{pmatrix}, \tag{2.21}$$

where

$$e\left(\chi_{qp}\frac{\rho}{a}\right) = \frac{\pi}{\sqrt{2}} \frac{J_q\left(\chi_{qp}\frac{\rho}{a}\right)N_q\left(\chi_{qp}\right) - N_q\left(\chi_{qp}\frac{\rho}{a}\right)J_q\left(\chi_{qp}\right)}{\sqrt{\frac{J_q^2\left(\chi_{qp}\right)}{J_q^2\left(c\chi_{qp}\right)} - 1}}, q = 0, 1, 2, \ldots,$$

$$h\left(\chi'_{qp}\frac{\rho}{a}\right) = \frac{\pi}{\sqrt{2}} \frac{J_q\left(\chi'_{qp}\frac{\rho}{a}\right)N'_q\left(\chi_{qp}\right) - N_q\left(\chi'_{qp}\frac{\rho}{a}\right)J'_q\left(\chi'_{qp}\right)}{\sqrt{\frac{J_q^2\left(\chi'_{qp}\right)}{J_q^2\left(\chi'_{qp}\frac{b}{a}\right)}\left[1 - \left(\frac{q}{\chi'_{qp}\frac{b}{a}}\right)^2\right] - \left[1 - \left(\frac{q}{\chi'_{qp}}\right)^2\right]}}, q = 0, 1, 2, \ldots,$$

and χ_{qp} and χ'_{qp} are, respectively, the pth nonvanishing roots of the equations

$$J_q\left(\chi_{qp}\frac{b}{a}\right)N_q\left(\chi_{qp}\right) - N_q\left(\chi_{qp}\frac{b}{a}\right)J_q\left(\chi_{qp}\right) = 0,$$
$$J'_q\left(\chi'_{qp}\frac{b}{a}\right)N'_q\left(\chi'_{qp}\right) - N'_q\left(\chi'_{qp}\frac{b}{a}\right)J'_q\left(\chi'_{qp}\right) = 0. \tag{2.22}$$

The cutoff wavenumbers for TM modes and TE modes are, respectively, given by

$$k_{cn}^{TM} = \frac{\chi_{qp}}{a} = \frac{(b-a)\chi_{qp}}{a(b-a)} \approx \frac{\pi p}{b-a}, \ p = 1, 2, \ldots,$$
$$k_{cn}^{TE} = \frac{\chi'_{q1}}{a} = \frac{(b+a)\chi'_{q1}}{a(b+a)} \approx \frac{2q}{b+a}, \ q = 1, 2, \ldots,$$
$$k_{cn}^{TE} = \frac{\chi'_{qp}}{a} = \frac{(b-a)\chi'_{qp}}{a(b-a)} \approx \frac{(p-1)\pi}{b-a}, \ p = 2, 3, \ldots.$$

The normalized vector modal function for the TEM mode is

$$\mathbf{e}_n^{TEM} = \nabla\phi = \mathbf{u}_\rho \frac{1}{\sqrt{2\pi \ln\left(\frac{b}{a}\right)}} \frac{1}{\rho}. \tag{2.23}$$

The normalized vector modal functions for the TM and TE modes are, respectively, given by

$$\mathbf{e}_n^{TM} = -\mathbf{u}_\rho \frac{\chi_{qp}}{a} e'\left(\chi_{qp}\frac{\rho}{a}\right)\sqrt{\frac{\varepsilon_q}{2\pi}}\begin{pmatrix}\cos q\varphi \\ \sin q\varphi\end{pmatrix} \pm \mathbf{u}_\varphi \frac{q}{\rho} e\left(\chi_{qp}\frac{\rho}{a}\right)\sqrt{\frac{\varepsilon_q}{2\pi}}\begin{pmatrix}\sin q\varphi \\ \cos q\varphi\end{pmatrix}, \tag{2.24}$$

$$\mathbf{e}_n^{TE} = \pm \mathbf{u}_\rho \frac{q}{\rho} h\left(\chi'_{qp}\frac{\rho}{a}\right)\sqrt{\frac{\varepsilon_q}{2\pi}}\begin{pmatrix}\sin q\varphi \\ \cos q\varphi\end{pmatrix} + \mathbf{u}_\varphi \frac{\chi'_{qp}}{a} h'\left(\chi'_{qp}\frac{\rho}{a}\right)\sqrt{\frac{\varepsilon_q}{2\pi}}\begin{pmatrix}\cos q\varphi \\ \sin q\varphi\end{pmatrix}. \tag{2.25}$$

Note that e' in (2.24) denotes the derivative with respect to its argument. The dominant TE mode is the $TE_{11}(q = 1, p = 1)$ mode.

2.2 Radiated Fields in Waveguide

The fields in the waveguide can be expressed as a linear combination of the vector modal functions. Instead of point-wise or uniform convergence, the series converges in the mean to the fields, which implies almost everywhere convergence. Great care must be exercised in performing various operations such as term-by-term differential and integral operations of the series. The best practice is to try to avoid such operations.

2.2.1 Modal Expansions for the Fields and Dyadic Green's Functions

Consider a waveguide with a perfect electric wall shown in Figure 2.1. Since the curl of the electric field $\nabla \times \mathbf{E}$ behaves like a magnetic field and $\nabla \times \mathbf{H}$ like an electric field, $\{\mathbf{e}_n\}$ and $\{\nabla \cdot \mathbf{e}_n / k_{cn}\}$ can be used to expand the transverse and longitudinal components of both $\mathbf{E}$ and $\nabla \times \mathbf{H}$, respectively, while $\{\mathbf{u}_z \times \mathbf{e}_n\}$ and $\{\nabla \times \mathbf{e}_n / k_{cn}, \tilde{c}\}$ can be used to expand the transverse and longitudinal components of both $\mathbf{H}$ and $\nabla \times \mathbf{E}$, respectively. The fields excited by the electric current source $\mathbf{J}$ and magnetic current $\mathbf{J}_m$ in the waveguide can thus be expanded in terms of the transverse vector modal functions $\mathbf{e}_n$ as follows:

$$\begin{aligned} \mathbf{E} &= \sum_n V_n \mathbf{e}_n + \sum_n e_n \mathbf{u}_z \frac{\nabla \cdot \mathbf{e}_n}{k_{cn}}, \\ \mathbf{H} &= \sum_n I_n \mathbf{u}_z \times \mathbf{e}_n + \mathbf{u}_z \frac{1}{\sqrt{\Omega}} \int_\Omega \frac{\mathbf{u}_z \cdot \mathbf{H}}{\sqrt{\Omega}} d\Omega + \sum_n h_n \frac{\nabla \times \mathbf{e}_n}{k_{cn}}, \end{aligned} \tag{2.26}$$

$$\nabla \times \mathbf{E} = \sum_{n=1}^{\infty} \mathbf{u}_z \times \mathbf{e}_n \int_{\Omega} \nabla \times \mathbf{E} \cdot \mathbf{u}_z \times \mathbf{e}_n d\Omega + \mathbf{u}_z \frac{1}{\sqrt{\Omega}} \int_{\Omega} \mathbf{u}_z \cdot \frac{\nabla \times \mathbf{E}}{\sqrt{\Omega}} d\Omega$$
$$+ \sum_{n=1}^{\infty} \frac{\nabla \times \mathbf{e}_n}{k_{cn}} \int_{\Omega} \nabla \times \mathbf{E} \cdot \frac{\nabla \times \mathbf{e}_n}{k_{cn}} d\Omega,$$
$$\nabla \times \mathbf{H} = \sum_{n=1}^{\infty} \mathbf{e}_n \int_{\Omega} \nabla \times \mathbf{H} \cdot \mathbf{e}_n d\Omega + \sum_{n=1}^{\infty} \frac{\mathbf{u}_z \nabla \cdot \mathbf{e}_n}{k_{cn}} \int_{\Omega} \nabla \times \mathbf{H} \cdot \frac{\mathbf{u}_z \nabla \cdot \mathbf{e}_n}{k_{cn}} d\Omega, \tag{2.27}$$

where the expansion coefficients V_n and I_n are called **modal voltage** and **modal current**

$$V_n = \int_{\Omega} \mathbf{E} \cdot \mathbf{e}_n d\Omega, \quad I_n = \int_{\Omega} \mathbf{H} \cdot \mathbf{u}_z \times \mathbf{e}_n d\Omega, \tag{2.28}$$

and

$$e_n = \int_{\Omega} \mathbf{u}_z \cdot \mathbf{E} \frac{\nabla \cdot \mathbf{e}_n}{k_{cn}} d\Omega, \quad h_n = \int_{\Omega} \mathbf{H} \cdot \frac{\nabla \times \mathbf{e}_n}{k_{cn}} d\Omega. \tag{2.29}$$

A simple manipulation leads to

$$\nabla \times \mathbf{E} = \sum_n \left(\frac{\partial V_n}{\partial z} + k_{cn} e_n \right) \mathbf{u}_z \times \mathbf{e}_n + \sum_n k_{cn} V_n \frac{\nabla \times \mathbf{e}_n}{k_{cn}},$$
$$\nabla \times \mathbf{H} = \sum_n \left(-\frac{\partial I_n}{\partial z} + k_{cn} h_n \right) \mathbf{e}_n + \mathbf{u}_z \sum_n k_{cn} I_n \frac{\nabla \cdot \mathbf{e}_n}{k_{cn}}. \tag{2.30}$$

Substituting (2.26) and (2.30) into the generalized Maxwell equations

$$\nabla \times \mathbf{E} = -j\omega\mu \mathbf{H} - \mathbf{J}_m,$$
$$\nabla \times \mathbf{H} = (\sigma + j\omega\varepsilon)\mathbf{E} + \mathbf{J}, \tag{2.31}$$

and comparing the transverse and longitudinal components, one may find the equations for the modal voltage and current

$$-\frac{\partial I_n}{\partial z} + k_{cn} h_n = (\sigma + j\omega\varepsilon) V_n + \int_{\Omega} \mathbf{J} \cdot \mathbf{e}_n d\Omega, \tag{2.32}$$

$$k_{cn} I_n = (\sigma + j\omega\varepsilon) e_n + \int_{\Omega} \mathbf{u}_z \cdot \mathbf{J} \frac{\nabla \cdot \mathbf{e}_n}{k_{cn}} d\Omega, \text{for TM modes only}, \tag{2.33}$$

$$\frac{\partial V_n}{\partial z} + k_{cn} e_n = -j\omega\mu I_n - \int_{\Omega} \mathbf{J}_m \cdot \mathbf{u}_z \times \mathbf{e}_n d\Omega, \tag{2.34}$$

$$k_{cn}V_n = -j\omega\mu h_n - \int_\Omega \mathbf{J}_m \cdot \frac{\nabla \times \mathbf{e}_n}{k_{cn}} d\Omega, \text{for TE modes only,} \tag{2.35}$$

$$-j\omega\mu \int_\Omega \frac{\mathbf{H}\cdot\mathbf{u}_z}{\sqrt{\Omega}} d\Omega = \int_\Omega \frac{\mathbf{u}_z \cdot \mathbf{J}_m}{\sqrt{\Omega}} d\Omega, \text{for TE modes only.} \tag{2.36}$$

For the TEM mode, the modal voltage and current satisfy

$$\begin{aligned}\frac{\partial V_n^{TEM}}{\partial z} &= -j\omega\mu I_n^{TEM} - \int_\Omega \mathbf{J}_m \cdot \mathbf{u}_z \times \mathbf{e}_n^{TEM} d\Omega,\\ \frac{\partial I_n^{TEM}}{\partial z} &= -(\sigma + j\omega\varepsilon)V_n^{TEM} - \int_\Omega \mathbf{J}\cdot\mathbf{e}_n^{TEM} d\Omega,\end{aligned} \tag{2.37}$$

from (2.32) and (2.34). From now on, the superscripts TEM, TE, and TM will be used to designate the field quantities related to TEM, TE, and TM modes. By eliminating the modal current in (2.37), the modal voltage for the TEM mode satisfies the inhomogeneous Helmholtz equation

$$\frac{\partial^2 V_n^{TEM}}{\partial z^2} + k_e^2 V_n^{TEM} = j\omega\mu \int_\Omega \mathbf{J}\cdot\mathbf{e}_n^{TEM} d\Omega - \frac{\partial}{\partial z}\int_\Omega \mathbf{J}_m \cdot \mathbf{u}_z \times \mathbf{e}_n^{TEM} d\Omega, \tag{2.38}$$

where

$$k_e^2 = k^2 - j\sigma k\eta = \omega^2\mu\varepsilon_e,\ \ \varepsilon_e = \varepsilon\left(1 - j\frac{\sigma}{\omega\varepsilon}\right). \tag{2.39}$$

Once the modal voltage V_n^{TEM} is determined, the modal current I_n^{TEM} can be found from the first equation of (2.37).

For the TE mode, the modal voltage and current are determined from (2.32), (2.34)–(2.36):

$$\begin{aligned}\frac{\partial V_n^{TE}}{\partial z} &= -j\omega\mu I_n^{TE} - \int_\Omega \mathbf{J}_m \cdot \mathbf{u}_z \times \mathbf{e}_n^{TE} d\Omega,\\ \frac{\partial I_n^{TE}}{\partial z} - k_{cn}^{TE} h_n^{TE} &= -(\sigma + j\omega\varepsilon)V_n^{TE} - \int_\Omega \mathbf{J}\cdot\mathbf{e}_n^{TE} d\Omega,\\ j\omega\mu h_n^{TE} &= -k_{cn}^{TE} V_n^{TE} - \int_\Omega \mathbf{J}_m \cdot \frac{\nabla \times \mathbf{e}_n^{TE}}{k_{cn}^{TE}} d\Omega,\\ -j\omega\mu \int_\Omega \frac{\mathbf{u}_z\cdot\mathbf{H}}{\sqrt{\Omega}} d\Omega &= \int_\Omega \frac{\mathbf{u}_z \cdot \mathbf{J}_m}{\sqrt{\Omega}} d\Omega.\end{aligned} \tag{2.40}$$

It is easy to see that the voltage V_n^{TE} satisfies the inhomogeneous Helmholtz equation

$$\frac{\partial^2 V_n^{TE}}{\partial z^2} + \left[k_e^2 - (k_{cn}^{TE})^2\right] V_n^{TE} = j\omega\mu \int_\Omega \mathbf{J} \cdot \mathbf{e}_n^{TE} d\Omega - \frac{\partial}{\partial z} \int_\Omega \mathbf{J}_m \cdot \mathbf{u}_z \times \mathbf{e}_n^{TE} d\Omega$$
$$+ k_{cn}^{TE} \int_\Omega \mathbf{J}_m \cdot \frac{\nabla \times \mathbf{e}_n^{TE}}{k_{cn}^{TE}} d\Omega. \tag{2.41}$$

Once the modal voltage V_n^{TE} is determined, the modal current I_n^{TE} can be solved from the first equation of (2.40).

For the TM mode, the modal voltage and current are determined from (2.32)–(2.34):

$$\begin{aligned}
\frac{\partial V_n^{TM}}{\partial z} + k_{cn}^{TM} e_n^{TM} &= -j\omega\mu I_n^{TM} - \int_\Omega \mathbf{J}_m \cdot \mathbf{u}_z \times \mathbf{e}_n^{TM} d\Omega, \\
\frac{\partial I_n^{TM}}{\partial z} &= -(\sigma + j\omega\varepsilon) V_n^{TM} - \int_\Omega \mathbf{J} \cdot \mathbf{e}_n^{TM} d\Omega, \\
(\sigma + j\omega\varepsilon) e_n^{TM} &= k_{cn}^{TM} I_n^{TM} - \int_\Omega \mathbf{u}_z \cdot \mathbf{J} \frac{\nabla \cdot \mathbf{e}_n^{TM}}{k_{cn}^{TM}} d\Omega.
\end{aligned} \tag{2.42}$$

The modal current I_n^{TM} satisfies

$$\frac{\partial^2 I_n^{TM}}{\partial z^2} + \left[k_e^2 - (k_{cn}^{TM})^2\right] I_n^{TM} = -\frac{\partial}{\partial z} \int_\Omega \mathbf{J} \cdot \mathbf{e}_n^{TM} d\Omega - k_{cn} \int_\Omega \mathbf{u}_z \cdot \mathbf{J} \frac{\nabla \cdot \mathbf{e}_n^{TM}}{k_{cn}^{TM}} d\Omega$$
$$+ (\sigma + j\omega\varepsilon) \int_\Omega \mathbf{J}_m \cdot \mathbf{u}_z \times \mathbf{e}_n^{TM} d\Omega. \tag{2.43}$$

Once the modal current I_n^{TM} is determined, the modal voltage V_n^{TM} can then be solved from the second equation of (2.42).

Remark 2.1 In a source-free region, the modal voltage and modal current satisfy the transmission line equation

$$\begin{aligned}
\frac{dV_n}{dz} &= -j\beta_n Z_{wn} I_n(z), \\
\frac{dI_n}{dz} &= -j\beta_n Z_{wn}^{-1} V_n(z),
\end{aligned} \tag{2.44}$$

where Z_{wn} is called the **wave impedance** of the nth mode

$$Z_{wn} = \begin{cases} \eta, \text{TEM} \\ \dfrac{\eta k}{\beta_n}, \text{TE} \\ \dfrac{\eta \beta_n}{k}, \text{TM} \end{cases}, \quad \beta_n = \begin{cases} k, \text{TEM} \\ \sqrt{k^2 - k_{cn}^2}, \text{TE or TM} \end{cases}, \tag{2.45}$$

with $k = \omega\sqrt{\mu\varepsilon_e}$, $\eta = \sqrt{\mu/\varepsilon_e}$. If $\beta_n \neq 0$, the solutions of (2.44) can be expressed as

$$\begin{aligned} V_n(z) &= V_n^+(z) + V_n^-(z) = A_n e^{-j\beta_n z} + B_n e^{j\beta_n z}, \\ I_n(z) &= I_n^+(z) - I_n^-(z) = \left(A_n e^{-j\beta_n z} - B_n e^{j\beta_n z}\right) Z_{wn}^{-1}, \end{aligned} \tag{2.46}$$

where the superscript + and − represent wave propagating in $+z$ and $-z$ direction, respectively,

$$\begin{aligned} V_n^+(z) &= A_n e^{-j\beta_n z}, \qquad V_n^-(z) = B_n e^{j\beta_n z}, \\ I_n^+(z) &= A_n Z_{wn}^{-1} e^{-j\beta_n z}, \quad I_n^-(z) = B_n Z_{wn}^{-1} e^{j\beta_n z}. \end{aligned}$$

The **characteristic impedance** for the nth mode is defined by

$$Z_{0n} = \frac{V_n^+}{I_n^+} = \frac{V_n^-}{I_n^-} = Z_{wn}.$$

The **guide wavelength** for the nth mode is defined by

$$\lambda_n = \frac{2\pi}{\beta_n}.$$

□

The method of Green's function will now be applied to solve the one-dimensional inhomogeneous Helmholtz equation. Consider the Green's function $G_n(z, z')$ defined by

$$\frac{\partial^2 G_n(z,z')}{\partial z^2} + \left(k_e^2 - k_{cn}^2\right) G_n(z,z') = -\delta(z - z'), \tag{2.47}$$

and the solution of the inhomogeneous Helmholtz equation

$$\frac{\partial^2 u_n(z)}{\partial z^2} + \left(k_e^2 - k_{cn}^2\right) u_n(z) = f(z), \tag{2.48}$$

where $f(z)$ stands for the source function. Multiplying (2.47) by $u_n(z)$ and (2.48) by $G_n(z, z')$ and subtracting the resultants yield

$$u_n(z)\delta(z - z') = -f(z)G_n(z,z') + G_n(z,z')\frac{\partial^2 u_n(z)}{\partial z^2} - u_n(z)\frac{\partial^2 G_n(z,z')}{\partial z^2}. \tag{2.49}$$

If the radiation conditions at $z = \pm\infty$ are taken into account, the solution of (2.48) can be obtained from (2.49) by taking the integration over $(-\infty, \infty)$:

$$u_n(z) = -\int_{-\infty}^{\infty} f(z')G_n(z,z')dz'. \tag{2.50}$$

Consequently, the expansion coefficients for the three types of modes can be obtained from (2.50) as follows:

$$\begin{aligned} V_n^{TEM}(z) = &-j\omega\mu\int_{\Omega}\int_{-\infty}^{\infty} \mathbf{J}(\mathbf{r}')\cdot\mathbf{e}_n^{TEM}(\boldsymbol{\rho}')G_n^{TEM}(z,z')d\Omega'dz', \\ &-\int_{\Omega}\int_{-\infty}^{\infty} \mathbf{J}_m(\mathbf{r}')\cdot\mathbf{u}_z\times\mathbf{e}_n^{TEM}(\boldsymbol{\rho}')\frac{\partial G_n^{TEM}(z,z')}{\partial z'}d\Omega'dz', \end{aligned} \tag{2.51}$$

$$\begin{aligned} I_n^{TEM}(z) = &\int_{\Omega}\int_{-\infty}^{\infty} \mathbf{J}(\mathbf{r}')\cdot\mathbf{e}_n^{TEM}(\boldsymbol{\rho}')\frac{\partial G_n^{TEM}(z,z')}{\partial z}d\Omega'dz', \\ &+\frac{1}{j\omega\mu}\int_{\Omega}\int_{-\infty}^{\infty} \mathbf{J}_m(\mathbf{r}')\cdot\mathbf{u}_z\times\mathbf{e}_n^{TEM}(\boldsymbol{\rho}')\frac{\partial^2 G_n^{TEM}(z,z')}{\partial z\partial z'}d\Omega'dz' \\ &-\frac{1}{j\omega\mu}\int_{\Omega}\mathbf{J}_m(\mathbf{r})\cdot\mathbf{u}_z\times\mathbf{e}_n^{TEM}(\boldsymbol{\rho})d\Omega, \end{aligned} \tag{2.52}$$

$$\begin{aligned} V_n^{TE}(z) = &-j\omega\mu\int_{\Omega}\int_{-\infty}^{\infty} \mathbf{J}(\mathbf{r}')\cdot\mathbf{e}_n^{TE}(\boldsymbol{\rho}')G_n^{TE}(z,z')d\Omega'dz' \\ &-\int_{\Omega}\int_{-\infty}^{\infty} \mathbf{J}_m(\mathbf{r}')\cdot\mathbf{u}_z\times\mathbf{e}_n^{TE}(\boldsymbol{\rho}')\frac{\partial G_n^{TE}(z,z')}{\partial z'}d\Omega'dz' \\ &-k_{cn}^{TE}\int_{\Omega}\int_{-\infty}^{\infty} \mathbf{J}_m(\mathbf{r}')\cdot\frac{\nabla\times\mathbf{e}_n^{TE}(\boldsymbol{\rho}')}{k_{cn}^{TE}}G_n^{TE}(z,z')d\Omega'dz', \end{aligned} \tag{2.53}$$

$$
\begin{aligned}
I_n^{TE}(z) = & \int_{\Omega} \int_{-\infty}^{\infty} \mathbf{J}(\mathbf{r}') \cdot \mathbf{e}_n^{TE}(\boldsymbol{\rho}') \frac{\partial G_n^{TE}(z, z')}{\partial z} d\Omega' dz' \\
& + \frac{1}{j\omega\mu} \int_{\Omega} \int_{-\infty}^{\infty} \mathbf{J}_m(\mathbf{r}') \cdot \mathbf{u}_z \times \mathbf{e}_n^{TE}(\boldsymbol{\rho}') \frac{\partial^2 G_n^{TE}(z, z')}{\partial z \partial z'} d\Omega' dz' \\
& + \frac{k_{cn}^{TE}}{j\omega\mu} \int_{\Omega} \int_{-\infty}^{\infty} \mathbf{J}_m(\mathbf{r}') \cdot \frac{\nabla \times \mathbf{e}_n^{TE}(\boldsymbol{\rho}')}{k_{cn}^{TE}} \frac{\partial G_n^{TE}(z, z')}{\partial z} d\Omega' dz' \\
& - \frac{1}{j\omega\mu} \int_{\Omega} \mathbf{J}_m(\mathbf{r}) \cdot \mathbf{u}_z \times \mathbf{e}_n^{TE}(\boldsymbol{\rho}) d\Omega,
\end{aligned}
\tag{2.54}
$$

$$
\begin{aligned}
h_n^{TE}(z) = & k_{cn}^{TE} \int_{\Omega} \int_{-\infty}^{\infty} \mathbf{J}(\mathbf{r}') \cdot \mathbf{e}_n^{TE}(\boldsymbol{\rho}') G_n^{TE}(z, z') d\Omega' dz' \\
& + \frac{k_{cn}^{TE}}{j\omega\mu} \int_{\Omega} \int_{-\infty}^{\infty} \mathbf{J}_m(\mathbf{r}') \cdot \mathbf{u}_z \times \mathbf{e}_n^{TE}(\boldsymbol{\rho}') \frac{\partial G_n^{TE}(z, z')}{\partial z'} d\Omega' dz' \\
& + \frac{\left(k_{cn}^{TE}\right)^2}{j\omega\mu} \int_{\Omega} \int_{-\infty}^{\infty} \mathbf{J}_m(\mathbf{r}') \cdot \frac{\nabla \times \mathbf{e}_n^{TE}(\boldsymbol{\rho}')}{k_{cn}^{TE}} G_n^{TE}(z, z') d\Omega' dz' \\
& - \frac{1}{j\omega\mu} \int_{\Omega} \mathbf{J}_m(\mathbf{r}) \cdot \frac{\nabla \times \mathbf{e}_n^{TE}(\boldsymbol{\rho})}{k_{cn}^{TE}} d\Omega,
\end{aligned}
\tag{2.55}
$$

$$
\begin{aligned}
V_n^{TM}(z) = & \frac{1}{\sigma + j\omega\varepsilon} \int_{\Omega} \int_{-\infty}^{\infty} \mathbf{J}(\mathbf{r}') \cdot \mathbf{e}_n^{TM}(\boldsymbol{\rho}') \frac{\partial^2 G_n^{TM}(z, z')}{\partial z \partial z'} d\Omega' dz' \\
& - \frac{1}{\sigma + j\omega\varepsilon} k_{cn}^{TM} \int_{\Omega} \int_{-\infty}^{\infty} \mathbf{u}_z \cdot \mathbf{J}(\mathbf{r}') \frac{\nabla \cdot \mathbf{e}_n^{TM}(\boldsymbol{\rho}')}{k_{cn}^{TM}} \frac{\partial G_n^{TM}(z, z')}{\partial z} d\Omega' dz' \\
& + \int_{\Omega} \int_{-\infty}^{\infty} \mathbf{J}_m(\mathbf{r}') \cdot \mathbf{u}_z \times \mathbf{e}_n^{TM}(\boldsymbol{\rho}') \frac{\partial G_n^{TM}(z, z')}{\partial z} d\Omega' dz' \\
& - \frac{1}{\sigma + j\omega\varepsilon} \int_{\Omega} \mathbf{J}(\mathbf{r}) \cdot \mathbf{e}_n^{TM}(\boldsymbol{\rho}) d\Omega,
\end{aligned}
\tag{2.56}
$$

$$
\begin{aligned}
I_n^{TM}(z) = &- \int_{\Omega}\int_{-\infty}^{\infty} \mathbf{J}(\mathbf{r}') \cdot \mathbf{e}_n^{TM}(\boldsymbol{\rho}') \frac{\partial G_n^{TM}(z,z')}{\partial z'} d\Omega' dz' \\
&+ k_{cn}^{TM} \int_{\Omega}\int_{-\infty}^{\infty} \mathbf{u}_z \cdot \mathbf{J}(\mathbf{r}') \frac{\nabla \cdot \mathbf{e}_n^{TM}(\boldsymbol{\rho}')}{k_{cn}^{TM}} G_n^{TM}(z,z') d\Omega' dz' \\
&- (\sigma + j\omega\varepsilon) \int_{\Omega}\int_{-\infty}^{\infty} \mathbf{J}_m(\mathbf{r}') \cdot \mathbf{u}_z \times \mathbf{e}_n^{TM}(\boldsymbol{\rho}') G_n^{TM}(z,z') d\Omega' dz',
\end{aligned}
\tag{2.57}
$$

$$
\begin{aligned}
e_n^{TM}(z) = &- \frac{k_{cn}^{TM}}{\sigma + j\omega\varepsilon} \int_{\Omega}\int_{-\infty}^{\infty} \mathbf{J}(\mathbf{r}') \cdot \mathbf{e}_n^{TM}(\boldsymbol{\rho}') \frac{\partial G_n^{TM}(z,z')}{\partial z'} d\Omega' dz' \\
&+ \frac{(k_{cn}^{TM})^2}{\sigma + j\omega\varepsilon} \int_{\Omega}\int_{-\infty}^{\infty} \mathbf{u}_z \cdot \mathbf{J}(\mathbf{r}') \frac{\nabla \cdot \mathbf{e}_n^{TM}(\boldsymbol{\rho}')}{k_{cn}^{TM}} G_n^{TM}(z,z') d\Omega' dz' \\
&- \int_{\Omega}\int_{-\infty}^{\infty} \mathbf{J}_m(\mathbf{r}') \cdot \mathbf{u}_z \times \mathbf{e}_n^{TM}(\boldsymbol{\rho}') G_n^{TM}(z,z') d\Omega' dz' \\
&- \frac{1}{\sigma + j\omega\varepsilon} \int_{\Omega} \mathbf{u}_z \cdot \mathbf{J}(\mathbf{r}) \frac{\nabla \cdot \mathbf{e}_n^{TM}(\boldsymbol{\rho})}{k_{cn}^{TM}} d\Omega.
\end{aligned}
\tag{2.58}
$$

In the above, the Green's functions G_n^{TEM}, G_n^{TE}, and G_n^{TM} for the three different modes are, respectively, defined by

$$
\begin{aligned}
&\frac{\partial^2 G_n^{TEM}(z,z')}{\partial z^2} + k_e^2 G_n^{TEM}(z,z') = -\delta(z-z'), \\
&\frac{\partial^2 G_n^{TE}(z,z')}{\partial z^2} + \left[k_e^2 - (k_{cn}^{TE})^2\right] G_n^{TE}(z,z') = -\delta(z-z'), \\
&\frac{\partial^2 G_n^{TM}(z,z')}{\partial z^2} + \left[k_e^2 - (k_{cn}^{TM})^2\right] G_n^{TM}(z,z') = -\delta(z-z').
\end{aligned}
\tag{2.59}
$$

The solutions of these equations are [see (1.68)]

$$
G_n^{TEM,TE,TM}(z,z') = \frac{1}{2\gamma_n^{TEM,TE,TM}} e^{-\gamma_n^{TEM,TE,TM}|z-z'|},
\tag{2.60}
$$

where

$$
\gamma_n^{TEM,TE,TM} = \begin{cases} j\sqrt{k_e^2 - (k_{cn}^{TEM,TE,TM})^2}, k > k_{cn}^{TEM,TE,TM} \\ \sqrt{(k_{cn}^{TEM,TE,TM})^2 - k_e^2}, k < k_{cn}^{TEM,TE,TM} \end{cases},
\tag{2.61}
$$

are the propagation constants for the modes. The first derivative of the Green's function is discontinuous at $z = z'$, and its second derivative is thus singular at $z = z'$:

$$\begin{aligned}\frac{\partial G_n^{TEM,TE,TM}(z,z')}{\partial z} &= -\frac{1}{2}e^{-\gamma_n^{TEM,TE,TM}|z-z'|}\operatorname{sgn}(z-z'),\\ \frac{\partial^2 G_n^{TEM,TE,TM}(z,z')}{\partial z\partial z'} &= \delta(z-z') - \left(\gamma_{cn}^{TEM,TE,TM}\right)^2 G_n^{TEM,TE,TM}(z,z'),\end{aligned} \tag{2.62}$$

where $\operatorname{sgn} x = \begin{cases} 1, & x > 1 \\ -1, & x < 1 \end{cases}$.

The most general field expressions can now be obtained by substituting (2.51)–(2.58) into (2.26). Ignoring the tedious process, the field expansions may be written in a compact form as

$$\mathbf{E}(\mathbf{r}) = -j\omega\mu\int_\Omega\int_{-\infty}^{\infty}\overleftrightarrow{\mathbf{G}}_e(\mathbf{r},\mathbf{r}')\cdot\mathbf{J}(\mathbf{r}')dV(\mathbf{r}') - \int_\Omega\int_{-\infty}^{\infty}\overleftrightarrow{\mathbf{G}}_{em}(\mathbf{r},\mathbf{r}')\cdot\mathbf{J}_m(\mathbf{r}')dV(\mathbf{r}'), \tag{2.63}$$

$$\mathbf{H}(\mathbf{r}) = -j\omega\varepsilon_e\int_\Omega\int_{-\infty}^{\infty}\overleftrightarrow{\mathbf{G}}_m(\mathbf{r},\mathbf{r}')\cdot\mathbf{J}_m(\mathbf{r}')dV(\mathbf{r}') + \int_\Omega\int_{-\infty}^{\infty}\overleftrightarrow{\mathbf{G}}_{me}(\mathbf{r},\mathbf{r}')\cdot\mathbf{J}(\mathbf{r}')dV(\mathbf{r}'), \tag{2.64}$$

where $dV = d\Omega dz$ denotes the differential volume element and the dyadic Green's functions are defined by

$$\begin{aligned}\overleftrightarrow{\mathbf{G}}_e(\mathbf{r},\mathbf{r}') = {} & \frac{1}{2}\sum_n \frac{Z_{wn}^{TEM}}{j\omega\mu}\mathbf{E}_{n\pm}^{TEM}(\mathbf{r})\mathbf{E}_{n\mp}^{TEM}(\mathbf{r}') + \frac{1}{2}\sum_n \frac{Z_{wn}^{TE}}{j\omega\mu}\mathbf{E}_{n\pm}^{TE}(\mathbf{r})\mathbf{E}_{n\mp}^{TE}(\mathbf{r}')\\ & + \frac{1}{2}\sum_n \frac{Z_{wn}^{TM}}{j\omega\mu}\mathbf{E}_{n\pm}^{TM}(\mathbf{r})\mathbf{E}_{n\mp}^{TM}(\mathbf{r}') - \mathbf{u}_z\mathbf{u}_{z'}\frac{1}{k_e^2}\delta(\mathbf{r}-\mathbf{r}'), \quad \begin{matrix} z > z' \\ z < z' \end{matrix},\end{aligned} \tag{2.65}$$

$$\begin{aligned}\overleftrightarrow{\mathbf{G}}_{em}(\mathbf{r},\mathbf{r}') = {} & -\frac{1}{2}\sum_n Z_{wn}^{TEM}\mathbf{E}_{n\pm}^{TEM}(\mathbf{r})\mathbf{H}_{n\mp}^{TEM}(\mathbf{r}') - \frac{1}{2}\sum_n Z_{wn}^{TE}\mathbf{E}_{n\pm}^{TE}(\mathbf{r})\mathbf{H}_{n\mp}^{TE}(\mathbf{r}')\\ & - \frac{1}{2}\sum_n Z_{wn}^{TM}\mathbf{E}_{n\pm}^{TM}(\mathbf{r})\mathbf{H}_{n\mp}^{TM}(\mathbf{r}'), \quad \begin{matrix} z > z' \\ z < z' \end{matrix},\end{aligned} \tag{2.66}$$

$$\begin{aligned}\overleftrightarrow{\mathbf{G}}_m(\mathbf{r},\mathbf{r}') = {} & -\frac{1}{2}\sum_n \frac{Z_{wn}^{TEM}}{j\omega\varepsilon_e}\mathbf{H}_{n\pm}^{TEM}(\mathbf{r})\mathbf{H}_{n\mp}^{TEM}(\mathbf{r}') - \frac{1}{2}\sum_n \frac{Z_{wn}^{TE}}{j\omega\varepsilon_e}\mathbf{H}_{n\pm}^{TE}(\mathbf{r})\mathbf{H}_{n\mp}^{TE}(\mathbf{r}')\\ & - \frac{1}{2}\sum_n \frac{Z_{wn}^{TM}}{j\omega\varepsilon_e}\mathbf{H}_{n\pm}^{TM}(\mathbf{r})\mathbf{H}_{n\mp}^{TM}(\mathbf{r}') - \mathbf{u}_z\mathbf{u}_{z'}\frac{1}{k_e^2}\delta(\mathbf{r}-\mathbf{r}'), \quad \begin{matrix} z > z' \\ z < z' \end{matrix},\end{aligned} \tag{2.67}$$

$$\overleftrightarrow{\mathbf{G}}_{me}(\mathbf{r},\mathbf{r}') = -\frac{1}{2}\sum_n Z_{wn}^{TEM}\mathbf{H}_{n\pm}^{TEM}(\mathbf{r})\mathbf{E}_{n\mp}^{TEM}(\mathbf{r}') - \frac{1}{2}\sum_n Z_{wn}^{TE}\mathbf{H}_{n\pm}^{TE}(\mathbf{r})\mathbf{E}_{n\mp}^{TE}(\mathbf{r}') - \frac{1}{2}\sum_n Z_{wn}^{TM}\mathbf{H}_{n\pm}^{TM}(\mathbf{r})\mathbf{E}_{n\mp}^{TM}(\mathbf{r}'), \quad \begin{matrix} z > z' \\ z < z' \end{matrix}. \tag{2.68}$$

In the above, the **fundamental field patterns**, propagating in $+z$ and $-z$ directions in the infinite waveguide, have been introduced

$$\begin{aligned} \mathbf{E}_{n\pm}^{TEM}(\mathbf{r}) &= \mathbf{e}_n^{TEM}(\boldsymbol{\rho})e^{\mp\gamma_n^{TEM}z}, \\ \mathbf{E}_{n\pm}^{TE}(\mathbf{r}) &= \mathbf{e}_n^{TE}(\boldsymbol{\rho})e^{\mp\gamma_n^{TE}z}, \\ \mathbf{E}_{n\pm}^{TM}(\mathbf{r}) &= \mathbf{e}_n^{TM}(\boldsymbol{\rho})e^{\mp\gamma_n^{TM}z} \pm \mathbf{u}_z\nabla\cdot\mathbf{e}_n^{TM}(\boldsymbol{\rho})\frac{e^{\mp\gamma_n^{TM}z}}{\gamma_n^{TM}}, \end{aligned} \tag{2.69}$$

$$\begin{aligned} \mathbf{H}_{n\pm}^{TEM}(\mathbf{r}) &= \pm\frac{1}{Z_{wn}^{TEM}}\mathbf{u}_z\times\mathbf{e}_n^{TEM}(\boldsymbol{\rho})e^{\mp\gamma_n^{TEM}z}, \\ \mathbf{H}_{n\pm}^{TE}(\mathbf{r}) &= \pm\frac{1}{Z_{wn}^{TE}}\mathbf{u}_z\times\mathbf{e}_n^{TE}(\boldsymbol{\rho})e^{\mp\gamma_n^{TE}z} - \frac{1}{Z_{wn}^{TE}}\nabla\times\mathbf{e}_n^{TE}(\boldsymbol{\rho})\frac{e^{\mp\gamma_n^{TE}z}}{\gamma_n^{TE}}, \\ \mathbf{H}_{n\pm}^{TM}(\mathbf{r}) &= \pm\frac{1}{Z_{wn}^{TM}}\mathbf{u}_z\times\mathbf{e}_n^{TM}(\boldsymbol{\rho})e^{\mp\gamma_n^{TM}z}, \end{aligned} \tag{2.70}$$

where

$$Z_{wn}^{TEM} = \sqrt{\frac{\mu}{\varepsilon_e}}, Z_{wn}^{TE} = \frac{j\omega\mu}{\gamma_n^{TE}}, Z_{wn}^{TM} = \frac{\gamma_n^{TM}}{j\omega\varepsilon_e} \tag{2.71}$$

are the wave impedances of the nth modes. For brevity, the dyadic Green's functions (2.65)–(2.68) will be written as

$$\overleftrightarrow{\mathbf{G}}_e(\mathbf{r},\mathbf{r}') = \frac{1}{2}\sum_n \frac{Z_{wn}}{j\omega\mu}\mathbf{E}_{n\pm}(\mathbf{r})\mathbf{E}_{n\mp}(\mathbf{r}') - \mathbf{u}_z\mathbf{u}_{z'}\frac{1}{k_e^2}\delta(\mathbf{r}-\mathbf{r}'), \quad \begin{matrix} z > z' \\ z < z' \end{matrix}, \tag{2.72}$$

$$\overleftrightarrow{\mathbf{G}}_{em}(\mathbf{r},\mathbf{r}') = -\frac{1}{2}\sum_n Z_{wn}\mathbf{E}_{n\pm}(\mathbf{r})\mathbf{H}_{n\mp}(\mathbf{r}'), \quad \begin{matrix} z > z' \\ z < z' \end{matrix}, \tag{2.73}$$

$$\overleftrightarrow{\mathbf{G}}_m(\mathbf{r},\mathbf{r}') = -\frac{1}{2}\sum_n \frac{Z_{wn}}{j\omega\varepsilon_e}\mathbf{H}_{n\pm}(\mathbf{r})\mathbf{H}_{n\mp}(\mathbf{r}') - \mathbf{u}_z\mathbf{u}_{z'}\frac{1}{k_e^2}\delta(\mathbf{r}-\mathbf{r}'), \quad \begin{matrix} z > z' \\ z < z' \end{matrix}, \tag{2.74}$$

$$\overleftrightarrow{\mathbf{G}}_{me}(\mathbf{r},\mathbf{r}') = -\frac{1}{2}\sum_n Z_{wn}\mathbf{H}_{n\pm}(\mathbf{r})\mathbf{E}_{n\mp}(\mathbf{r}'), \quad \begin{matrix} z > z' \\ z < z' \end{matrix}, \tag{2.75}$$

where the summation is over all the TEM, TE, and TM modes and

$$Z_{wn} = \begin{cases} Z_{wn}^{TEM}, \text{ for TEM modes} \\ Z_{wn}^{TE}, \text{ for TE modes} \\ Z_{wn}^{TM}, \text{ for TM modes} \end{cases}. \tag{2.76}$$

It is noted that the Dirac delta function singularities in (2.72) and (2.74) come from the expansions in terms of the longitudinal field components

$$\begin{aligned} \delta(\boldsymbol{\rho}-\boldsymbol{\rho}') &= \sum_n \frac{\nabla\cdot\mathbf{e}_n^{TM}(\boldsymbol{\rho})}{k_{cn}^{TM}}\frac{\nabla\cdot\mathbf{e}_n^{TM}(\boldsymbol{\rho}')}{k_{cn}^{TM}}, \\ \mathbf{u}_z\mathbf{u}_{z'}\delta(\boldsymbol{\rho}-\boldsymbol{\rho}') &= \mathbf{u}_z\mathbf{u}_{z'}\frac{1}{\Omega} + \sum_n \frac{\nabla\times\mathbf{e}_n^{TE}(\boldsymbol{\rho})}{k_{cn}^{TE}}\frac{\nabla\times\mathbf{e}_n^{TE}(\boldsymbol{\rho}')}{k_{cn}^{TE}}. \end{aligned} \tag{2.77}$$

Remark 2.2 In the treatment of eigenfunction expansions of dyadic Green's functions in waveguides, the dyadic delta source $\overleftrightarrow{\mathbf{I}}\delta(\mathbf{r}-\mathbf{r}')$ must be expanded in terms of both solenoidal and irrotational vector modal functions. Otherwise, erroneous results may be obtained [16, 17]. In our approach, the dyadic Green's functions are derived directly from the field expansions, which avoids the possible errors caused by the incompleteness of the eigenfunctions selected to expand a dyadic point source. □

It is easy to verify that the fundamental field patterns are solutions of Maxwell equations in source-free region

$$\begin{aligned} \nabla\times\mathbf{E}_{n\pm}^{TEM,TE,TM} &= -j\omega\mu\mathbf{H}_{n\pm}^{TEM,TE,TM}, \\ \nabla\times\mathbf{H}_{n\pm}^{TEM,TE,TM} &= j\omega\varepsilon_e\mathbf{E}_{n\pm}^{TEM,TE,TM}, \end{aligned} \tag{2.78}$$

and satisfy the boundary conditions on the waveguide walls

$$\begin{aligned} &\mathbf{u}_n\times\mathbf{E}_{n\pm}^{TEM,TE,TM} = 0, \quad \mathbf{u}_n\times\nabla\times\mathbf{H}_{n\pm}^{TEM,TE,TM} = 0, \\ &\nabla\cdot\mathbf{E}_{n\pm}^{TEM,TE,TM} = 0, \quad \mathbf{u}_n\cdot\mathbf{H}_{n\pm}^{TEM,TE,TM} = 0. \end{aligned} \tag{2.79}$$

Apparently, the fundamental field patterns satisfy the homogeneous vector Helmholtz equation

$$\begin{aligned} \nabla\times\nabla\times\mathbf{E}_{n\pm}^{TEM,TE,TM} - k_e^2\mathbf{E}_{n\pm}^{TEM,TE,TM} &= 0, \\ \nabla\times\nabla\times\mathbf{H}_{n\pm}^{TEM,TE,TM} - k_e^2\mathbf{H}_{n\pm}^{TEM,TE,TM} &= 0. \end{aligned} \tag{2.80}$$

Consider a uniform waveguide excited by the electric current source $\mathbf{J}$ and the magnetic current $\mathbf{J}_m$ confined the in the region V_0 between z_1 and z_2, as shown in

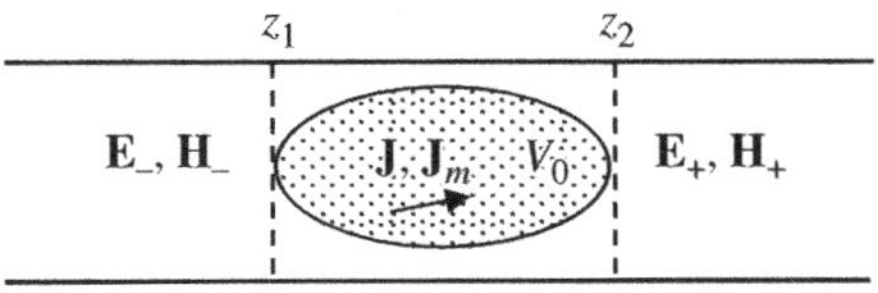

Figure 2.5 Mode excitation.

Figure 2.5. According to (2.63), (2.64) and (2.72)–(2.75), the fields in the region $z \geq z_2$ and $z \leq z_1$ may be expressed as

$$\begin{aligned}
\mathbf{E}_\pm(\mathbf{r}) &= -\frac{1}{2}\sum_n Z_{wn}\mathbf{E}_{n\pm}(\mathbf{r})\int_{V_0}\mathbf{J}(\mathbf{r}')\cdot\mathbf{E}_{n\mp}(\mathbf{r}')dV(\mathbf{r}') \\
&+\frac{1}{2}\sum_n Z_{wn}\mathbf{E}_{n\pm}(\mathbf{r})\int_{V_0}\mathbf{J}_m(\mathbf{r}')\cdot\mathbf{H}_{n\mp}(\mathbf{r}')dV(\mathbf{r}'), \\
\mathbf{H}_\pm(\mathbf{r}) &= -\frac{1}{2}\sum_n Z_{wn}\mathbf{H}_{n\pm}(\mathbf{r})\int_{V_0}\mathbf{J}(\mathbf{r}')\cdot\mathbf{E}_{n\mp}(\mathbf{r}')dV(\mathbf{r}') \\
&+\frac{1}{2}\sum_n Z_{wn}\mathbf{H}_{n\pm}(\mathbf{r})\int_{V_0}\mathbf{J}_m(\mathbf{r}')\cdot\mathbf{H}_{n\mp}(\mathbf{r}')dV(\mathbf{r}'),
\end{aligned} \tag{2.81}$$

where the summation is over all the TEM, TE, and TM modes. Comparing the field expansions in (2.26) with those in (2.81), the transverse fields in the waveguide can be written as

$$\begin{aligned}
\mathbf{E}_{t\pm}(\mathbf{r}) &= \sum_n V_{n\pm}(z)\mathbf{e}_n(\boldsymbol{\rho}), \\
\mathbf{H}_{t\pm}(\mathbf{r}) &= \sum_n I_{n\pm}(z)\mathbf{u}_z\times\mathbf{e}_n(\boldsymbol{\rho}),
\end{aligned} \tag{2.82}$$

where the subscript t is used to signify the transverse components and

$$\begin{aligned}
V_{n\pm}(z) &= -\frac{1}{2}e^{\mp\gamma_n z}Z_{wn}\int_{V_0}\mathbf{J}(\mathbf{r}')\cdot\mathbf{E}_{n\mp}(\mathbf{r}')dV(\mathbf{r}') \\
&+\frac{1}{2}e^{\mp\gamma_n z}Z_{wn}\int_{V_0}\mathbf{J}_m(\mathbf{r}')\cdot\mathbf{H}_{n\mp}(\mathbf{r}')dV(\mathbf{r}'), \\
I_{n\pm}(z) &= \mp\frac{1}{2}e^{\mp\gamma_n z}\int_{V_0}\mathbf{J}(\mathbf{r}')\cdot\mathbf{E}_{n\mp}(\mathbf{r}')dV(\mathbf{r}') \\
&\pm\frac{1}{2}e^{\mp\gamma_n z}\int_{V_0}\mathbf{J}_m(\mathbf{r}')\cdot\mathbf{H}_{n\mp}(\mathbf{r}')dV(\mathbf{r}'),
\end{aligned} \tag{2.83}$$

are the modal voltages and currents.

Example 2.1 Consider a rectangular waveguide excited by a line current extending across the waveguide located at ($x = a/2$, $z = 0$) as illustrated in Figure 2.6. The current density is given by

$$\mathbf{J}(\mathbf{r}) = -\mathbf{u}_y j\delta\left(x - \frac{a}{2}\right)\delta(z).$$

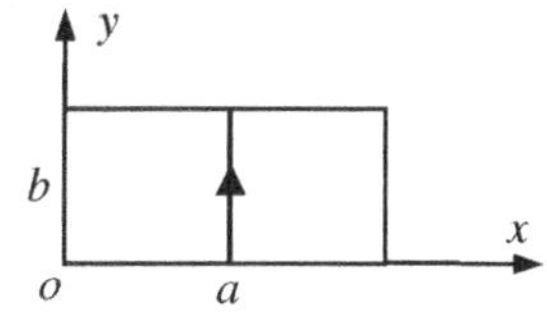

Figure 2.6 Excitation of rectangular waveguide.

Since the line current is uniform in y direction, the fields excited by the current are independent of y. Consequently, only TE_{n0} modes are excited. From (2.15) and (2.65), one may find

$$\overleftrightarrow{\mathbf{G}}_e(\mathbf{r},\mathbf{r}') = \frac{1}{2}\sum_n \frac{1}{\gamma_n^{TE}} \mathbf{e}_n^{TE}(\boldsymbol{\rho}) e^{\mp\gamma_n^{TE} z} \mathbf{e}_n^{TE}(\boldsymbol{\rho}') e^{\pm\gamma_n^{TE} z'}$$

$$= \frac{1}{ab}\sum_{n=1}^{\infty} \frac{1}{\gamma_n^{TE}} \mathbf{u}_y \mathbf{u}_{y'} \sin\frac{n\pi}{a}x \sin\frac{n\pi}{a}x' e^{\mp\gamma_n^{TE} z} e^{\pm\gamma_n^{TE} z'}, \quad \begin{matrix} z > z' \\ z < z' \end{matrix}, \tag{2.84}$$

where

$$\gamma_n^{TE} = \begin{cases} j\sqrt{k_e^2 - \left(\frac{n\pi}{a}\right)^2}, k > \frac{n\pi}{a} \\ \sqrt{\left(\frac{n\pi}{a}\right)^2 - k_e^2}, \quad k < \frac{n\pi}{a} \end{cases}. \tag{2.85}$$

The electric field (2.81) reduces to

$$\mathbf{E}(\mathbf{r}) = -\mathbf{u}_y \frac{\omega\mu}{a}\sum_{n=1}^{\infty} \frac{1}{\gamma_n^{TE}} e^{-\gamma_n^{TE}|z|} \sin\frac{n\pi}{2} \sin\frac{n\pi}{a}x. \tag{2.86}$$

□

2.2.2 Dyadic Green's Functions for Semi-infinite Waveguide

A semi-infinite waveguide is shown in Figure 2.7, where a short-circuit position is introduced at $z = 0$. The dyadic Green's function for the semi-infinite waveguide is given by

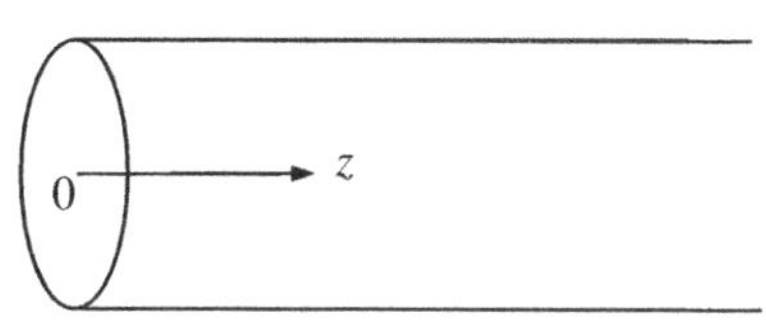

Figure 2.7 A semi-infinite waveguide.

$$\overleftrightarrow{\mathbf{G}}_e(\mathbf{r},\mathbf{r}') = \frac{1}{2}\sum_n \frac{Z_{wn}}{j\omega\mu}[\mathbf{E}_{n\pm}(\mathbf{r})\mathbf{E}_{n\mp}(\mathbf{r}') - \mathbf{E}_{n+}(\mathbf{r})\mathbf{E}_{n+}(\mathbf{r}')] \\ - \mathbf{u}_z\mathbf{u}_{z'}\frac{1}{k_e^2}\delta(\mathbf{r}-\mathbf{r}'), \quad \begin{matrix} z > z' \\ z < z' \end{matrix}. \tag{2.87}$$

The derivation of (2.87) will be discussed in Chapter 3.

2.3 Waveguide Discontinuities

The typical boundary value problems in waveguide theory include the excitation of waveguide, obstacles in the waveguide, and the coupling between waveguides. They all belong to waveguide discontinuities, in which a uniform waveguide is deformed or discontinued. The waveguide discontinuities are often used to achieve various purposes such as exciting a desired mode, matching a given load, obtaining a phase shift in the transmitted wave, changing the propagation direction of the transmitted wave, or coupling an incident field to another waveguide. The waveguide discontinuities will distort the fields in the original uniform waveguide. One of the important tasks of microwave field theory is to establish the circuit or network parameters for various waveguide discontinuities. In most applications, the waveguide supports a single dominant mode. When a discontinuity exists, such as the discontinuity in cross-sectional shape or an obstacle in the waveguide, an infinite number of non-propagating modes will be excited in the vicinity of the discontinuity by the incident propagating mode. Due to the vector nature of the EM fields, the analysis of waveguide discontinuity is usually an extremely complicated process. From a practical point of view, only the circuit parameters are of interest. A traditional technique is to express the circuit parameter in terms of the field distribution in certain area such as in an aperture. The expression for a circuit parameter is said to be **variational** if it is stationary with respect to an arbitrary small variation of the field about its true value. By properly choosing a trial field, the variational expression may yield a good approximation to the circuit parameter. The variational method was introduced by Schwinger during the period from 1940 to 1945 and many useful results have been summarized in [18]. The general field expansions (2.63) and (2.64) can be used as the starting point to solve a number of waveguide discontinuity problems.

2.3.1 Excitation of Waveguide

A waveguide can be excited by a conducting probe antenna connected to a transmission line, as illustrated in Figure 2.8a. According to Schelkunoff–Love equivalence principle, the field inside the waveguide is generated by the equivalent

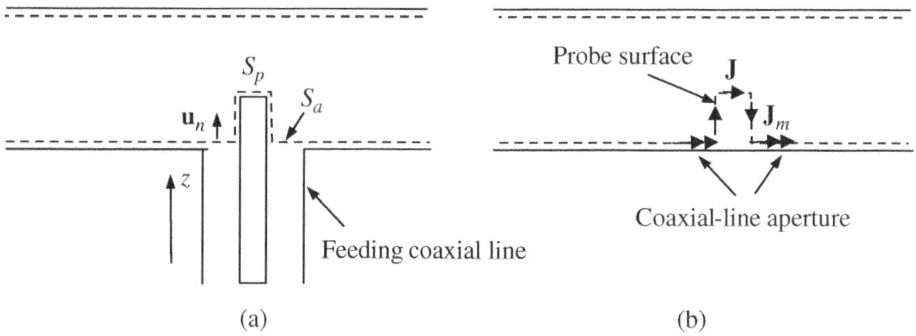

Figure 2.8 Excitation of waveguide. (a) A probe in waveguide. (b) Equivalent problem.

current sources $\mathbf{J} = \mathbf{u}_n \times \mathbf{H}$ and $\mathbf{J}_m = -\mathbf{u}_n \times \mathbf{E}$, with the probe removed and the feeding aperture covered by a perfect conductor, as illustrated in Figure 2.8b. The equivalent electric current $\mathbf{J}$ is distributed on the probe surface denoted by S_p while the equivalent magnetic current $\mathbf{J}_m$ is on the feeding aperture (the coaxial-line aperture) denoted by S_a. The electric field generated by the equivalent sources are given by (2.63):

$$\mathbf{E}(\mathbf{r}) = \mathbf{E}^{in}(\mathbf{r}) + \mathbf{E}^{s}(\mathbf{r}), \tag{2.88}$$

where

$$\begin{aligned} \mathbf{E}^{in}(\mathbf{r}) &= -\int_{S_a} \overleftrightarrow{\mathbf{G}}_{em}(\mathbf{r}, \mathbf{r}') \cdot \mathbf{J}_m(\mathbf{r}') dS(\mathbf{r}'), \\ \mathbf{E}^{s}(\mathbf{r}) &= -j\omega\mu \int_{S_p} \overleftrightarrow{\mathbf{G}}_{e}(\mathbf{r}, \mathbf{r}') \cdot \mathbf{J}(\mathbf{r}') dS(\mathbf{r}'). \end{aligned} \tag{2.89}$$

Here, the superscripts in and s, respectively, designate the incident field applied to the probe and scattered field generated by the induced current on the probe. By setting the observation point $\mathbf{r}$ on the probe surface and applying the boundary condition $\mathbf{u}_n(\mathbf{r}) \times \mathbf{E}(\mathbf{r}) = 0$, the following integral equation for the probe current $\mathbf{J}$ can be obtained:

$$-j\omega\mu \mathbf{u}_n(\mathbf{r}) \times \int_{S_p} \overleftrightarrow{\mathbf{G}}_{e}(\mathbf{r}, \mathbf{r}') \cdot \mathbf{J}(\mathbf{r}') dS(\mathbf{r}') = -\mathbf{u}_n(\mathbf{r}) \times \mathbf{E}^{in}(\mathbf{r}). \tag{2.90}$$

To determine the equivalent magnetic current on the feeding aperture, the excitation conditions must be used. The transverse fields on the feeding aperture may be expressed as

$$\begin{aligned} -\mathbf{u}_n(\mathbf{r}) \times \mathbf{E}(\mathbf{r}) &= -\sum_{n=1}^{\infty} V_n(z) \mathbf{u}_n \times \mathbf{e}_n^{(g)}(\boldsymbol{\rho}), \\ \mathbf{u}_n(\mathbf{r}) \times \mathbf{H}(\mathbf{r}) &= -\sum_{n=1}^{\infty} I_n(z) \mathbf{e}_n^{(g)}(\boldsymbol{\rho}), \end{aligned} \tag{2.91}$$

where $\mathbf{e}_n^{(g)}$ $(n = 1, 2, ...)$ are the normalized vector modal functions in the feeding waveguide, and

$$V_n(z) = A_n e^{-j\beta_n^{(g)} z} + B_n e^{j\beta_n^{(g)} z},$$
$$I_n(z) = \frac{1}{Z_{wn}^{(g)}}\left[A_n e^{-j\beta_n^{(g)} z} - B_n e^{j\beta_n^{(g)} z}\right],$$

are, respectively, the modal voltage and current for the nth modes with

$$Z_{wn}^{(g)} = \begin{cases} \eta, \text{ TEM mode} \\ \dfrac{\eta k}{\beta_n^{(g)}}, \text{ TE mode} \\ \dfrac{\eta \beta_n^{(g)}}{k}, \text{ TM mode} \end{cases}, \quad \beta_n^{(g)} = \begin{cases} k, \text{TEM mode} \\ \sqrt{k^2 - \left(k_{cn}^{(g)}\right)^2}, \text{ TE or TM mode} \end{cases}.$$

In the above, the superscript (g) is used to designate the quantities in the feeding waveguide. Suppose that the feeding waveguide is in a single-mode operation and the antenna is excited by the dominant mode of unit amplitude. The modal voltages and currents can then be written as

$$V_1(z) = e^{-j\beta_1^{(g)} z} + B_1 e^{j\beta_1^{(g)} z}, \qquad V_n(z) = B_n e^{j\beta_n^{(g)} z} (n \geq 2),$$
$$I_1(z) = \frac{1}{Z_{w1}^{(g)}}\left[e^{-j\beta_1^{(g)} z} - B_1 e^{j\beta_1^{(g)} z}\right], \qquad I_n(z) = -\frac{1}{Z_{wn}^{(g)}} B_n e^{j\beta_n^{(g)} z} (n \geq 2).$$

Equation (2.91) can be used to determine the equivalent currents on the feeding aperture $(z = 0)$:

$$\begin{aligned} \mathbf{J}_m &= -(1 + B_1)\mathbf{u}_n \times \mathbf{e}_1^{(g)} - \sum_{n=2}^{\infty} B_n \mathbf{u}_n \times \mathbf{e}_n^{(g)}, \\ \mathbf{J} &= -\frac{1}{Z_{w1}^{(g)}}(1 - B_1)\mathbf{e}_1^{(g)} + \sum_{n=2}^{\infty} \frac{1}{Z_{wn}^{(g)}} B_n \mathbf{e}_n^{(g)}. \end{aligned} \tag{2.92}$$

The expansion coefficients can be determined by the second equation

$$B_1 = 1 + Z_{w1}^{(g)} \int_{S_a} \mathbf{J} \cdot \mathbf{e}_1^{(g)} dS, \; B_n = Z_{wn}^{(g)} \int_{S_a} \mathbf{J} \cdot \mathbf{e}_n^{(g)} dS.$$

Therefore, the equivalent magnetic current $\mathbf{J}_m$ on the feeding aperture can be represented by the equivalent electric current on the same feeding aperture

$$\mathbf{J}_m(\mathbf{r}) = -2\mathbf{u}_n(\mathbf{r}) \times \mathbf{e}_1^{(g)}(\boldsymbol{\rho}) - \sum_{n=1}^{\infty} \mathbf{u}_n(\mathbf{r}) \times \mathbf{e}_n^{(g)}(\boldsymbol{\rho}) Z_{wn}^{(g)} \int_{S_a} \mathbf{J} \cdot \mathbf{e}_n dS. \tag{2.93}$$

Substituting (2.93) into the expressions of the incident fields, one may express the incident fields in terms of the equivalent electric current on the feeding aperture as follows:

$$\mathbf{E}^{in}(\mathbf{r}) = 2\int_{S_a} \overleftrightarrow{\mathbf{G}}_{em}(\mathbf{r},\mathbf{r}')\cdot\left[\mathbf{u}_n(\mathbf{r}')\times\mathbf{e}_1^{(g)}(\mathbf{\rho}')\right]dS(\mathbf{r}')$$
$$+\sum_{n=1}^{\infty}\int_{S_a}\mathbf{J}\cdot\mathbf{e}_n^{(g)}dS\left\{\int_{S_a}\overleftrightarrow{\mathbf{G}}_{em}(\mathbf{r},\mathbf{r}')\cdot\left[\mathbf{u}_n(\mathbf{r}')\times\mathbf{e}_n^{(g)}(\mathbf{\rho}')\right]Z_{wn}^{(g)}dS(\mathbf{r}')\right\}. \tag{2.94}$$

Combination of (2.94) and (2.90) yields the integral equation

$$-j\omega\mu\mathbf{u}_n(\mathbf{r})\times\int_{S_p}\overleftrightarrow{\mathbf{G}}_{e}(\mathbf{r},\mathbf{r}')\cdot\mathbf{J}(\mathbf{r}')dS(\mathbf{r}')$$
$$+\sum_{n=1}^{\infty}\int_{S_a}\mathbf{J}\cdot\mathbf{e}_n^{(g)}dS\left\{\mathbf{u}_n(\mathbf{r})\times\int_{S_a}\overleftrightarrow{\mathbf{G}}_{em}(\mathbf{r},\mathbf{r}')\cdot\left[\mathbf{u}_n(\mathbf{r}')\times\mathbf{e}_n^{(g)}(\mathbf{\rho}')\right]Z_{wn}^{(g)}dS(\mathbf{r}')\right\}$$
$$= -2\mathbf{u}_n(\mathbf{r})\times\int_{S_a}\overleftrightarrow{\mathbf{G}}_{em}(\mathbf{r},\mathbf{r}')\cdot\left[\mathbf{u}_n(\mathbf{r}')\times\mathbf{e}_1^{(g)}(\mathbf{\rho}')\right]dS(\mathbf{r}'). \tag{2.95}$$

The equivalent electric current $\mathbf{J}$ on the boundary $S_p + S_a$ can then be determined by numerically solving the integral Eq. (2.95). Note that the integral Eq. (2.95) is valid for any feeding waveguide and any metal antenna (such as a loop).

2.3.2 Conducting Obstacles in Waveguide

Consider a perfectly conducting obstacle placed in an infinitely long waveguide, as shown in Figure 2.9. The dominant mode, denoted by $(\mathbf{E}^{in}, \mathbf{H}^{in})$, is incident upon the obstacle bounded by S from $z = -\infty$, which induces current $\mathbf{J}$ on the conducting obstacle. The induced current on the conducting obstacle generates the scattered field $(\mathbf{E}^{s}, \mathbf{H}^{s})$, and the latter can be determined from (2.63) and (2.64).

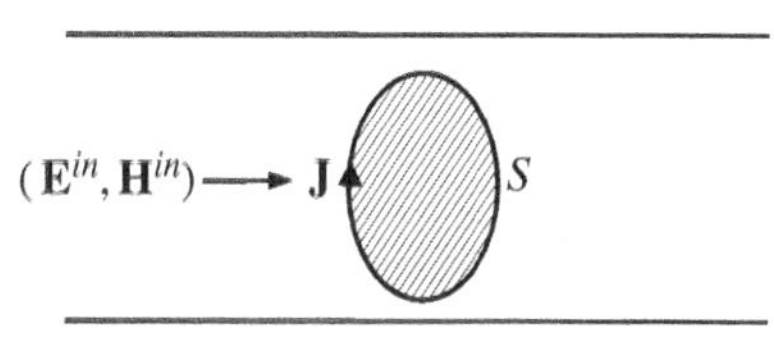

Figure 2.9 Scattering by a conducting obstacle in a waveguide.

The total electric field outside the obstacle is the superposition of the incident field and the scattered field

$$\mathbf{E}(\mathbf{r}) = \mathbf{E}^{in}(\mathbf{r}) + \mathbf{E}^{s}(\mathbf{r}) = \mathbf{E}^{in}(\mathbf{r}) - j\omega\mu \int_S \overleftrightarrow{\mathbf{G}}_e(\mathbf{r}, \mathbf{r}') \cdot \mathbf{J}(\mathbf{r}')dS(\mathbf{r}'). \tag{2.96}$$

On the surface of the scatterer, the tangential component of the electric field must vanish, which leads to

$$j\omega\mu\mathbf{u}_n \times \int_S \overleftrightarrow{\mathbf{G}}_e(\mathbf{r}, \mathbf{r}') \cdot \mathbf{J}(\mathbf{r}')dS(\mathbf{r}') = \mathbf{u}_n \times \mathbf{E}^{in}(\mathbf{r}). \tag{2.97}$$

This is the integral equation for the current $\mathbf{J}$ on the conducting obstacle.

Example 2.2 Figure 2.10 shows a circular conducting post across the narrow side of a rectangular waveguide. The dominant TE_{10} mode is incident upon the post and induces current on the post, which generates higher order modes around the post. Since the electric field of the TE_{10} mode has a y component only and is independent of the y coordinate, and the whole structure is uniform in the y direction, the excited higher order modes must be independent of y, and thus are TE_{n0} modes. For TE_{n0} modes, the magnetic field energy is higher than electric field energy, and the post is thus equivalent to an inductor. The dyadic Green's function in (2.97) is then given by

$$\begin{aligned}\overleftrightarrow{\mathbf{G}}_e(\mathbf{r}, \mathbf{r}') = {} & \frac{1}{ab}\sum_{n=1}^{\infty} \frac{1}{\gamma_n^{TE}} \mathbf{u}_y \mathbf{u}_{y'} \sin\frac{n\pi}{a}x \sin\frac{n\pi}{a}x' e^{\mp\gamma_n^{TE}z} e^{\pm\gamma_n^{TE}z'} \quad \begin{matrix} z > z' \\ z < z' \end{matrix} \\ & - \mathbf{u}_z \mathbf{u}_{z'} \frac{1}{k_e^2}\delta(\mathbf{r} - \mathbf{r}'), \end{aligned} \tag{2.98}$$

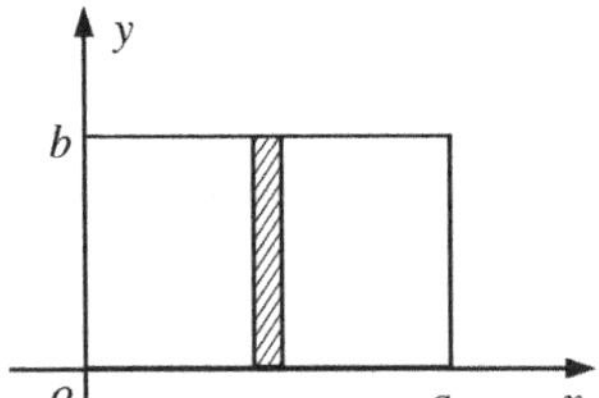

Figure 2.10 Inductive post in rectangular waveguide.

Suppose that the conducting post is very thin and the induced current on the conducting post has only a y component and can be written as $\mathbf{J} = I\delta(x - x_0)\delta(z)\mathbf{u}_y$. The scattered field by the conducting post can then be expressed by

$$\begin{aligned}\mathbf{E}^{s}(\mathbf{r}) &= -j\omega\mu \int_S \overleftrightarrow{\mathbf{G}}_e(\mathbf{r}, \mathbf{r}') \cdot \mathbf{J}(\mathbf{r}')dS(\mathbf{r}') \\ &= -\mathbf{u}_y \frac{j\omega\mu I}{a} \sum_{n=1}^{\infty} \frac{1}{\gamma_n^{TE}} \sin\frac{n\pi}{a}x \sin\frac{n\pi}{a}x_0 e^{-\gamma_n^{TE}|z|}.\end{aligned}$$

Let the incident TE_{10} mode be $\mathbf{E}^{in} = \mathbf{u}_y \sin\frac{\pi}{a}x e^{-\gamma_1^{TE}z}$. Equation (2.97) reduces to

$$\sin\frac{\pi}{a}x e^{-\gamma_1^{TE}z} - \frac{j\omega\mu I}{a}\sum_{n=1}^{\infty}\frac{1}{\gamma_n^{TE}}\sin\frac{n\pi}{a}x \sin\frac{n\pi}{a}x_0 e^{-\gamma_n^{TE}|z|} = 0. \tag{2.99}$$

The above equation is valid on the surface of the conducting post and can be used to determine the induced current I on the inductive post. □

2.3.3 Coupling by Small Aperture

Consider a system of two waveguides coupled by a small aperture S_a bounded by Γ, as shown in Figure 2.11a. The impressed electric current $\mathbf{J}_{imp}$ and magnetic current $\mathbf{J}_{m,\,imp}$ are assumed to be located in waveguide 1 only and there are no impressed sources in waveguide 2. By the Schelkunoff–Love equivalence principle, the original problem can be separated into two equivalent problems as shown in Figure 2.11b,c. In waveguide 1, the fields are produced by the impressed sources $\mathbf{J}_{imp}$, $\mathbf{J}_{m,\,imp}$ and the equivalent magnetic current $\mathbf{J}_{ms} = \mathbf{u}_n \times \mathbf{E}$ over the aperture region S_a with the aperture covered by an electric conductor. In waveguide 2, the fields are produced by the equivalent magnetic current $-\mathbf{J}_{ms}$ with the aperture covered by an electric conductor. When the aperture is absent (i.e. closed by a perfect conductor), the incident fields generated by the impressed sources in waveguide

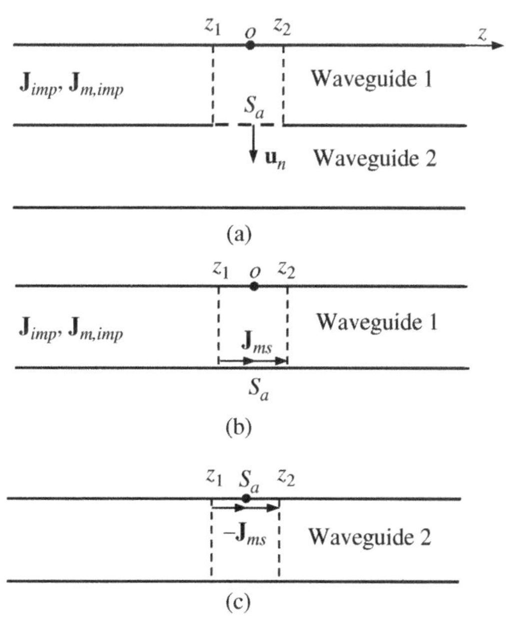

Figure 2.11 Two waveguides coupled by a small aperture. (a) Original problem. (b) Equivalent problem for waveguide 1. (c) Equivalent problem for waveguide 2.

1 are denoted by $\mathbf{E}_{(1)}^{in}, \mathbf{H}_{(1)}^{in}$ (the subscripts or superscripts 1 and 2 will be used to designate waveguide 1 and 2, respectively), and they satisfy the boundary conditions

$$\mathbf{u}_n \times \mathbf{E}_{(1)}^{in} = 0, \ \ \mathbf{u}_n \cdot \mathbf{H}_{(1)}^{in} = 0.$$

The total fields in waveguide 1 can be expressed as the sum of the incident and the scattered fields

$$\mathbf{E}_{(1)} = \mathbf{E}_{(1)}^{in} + \mathbf{E}_{(1)}^{s},$$
$$\mathbf{H}_{(1)} = \mathbf{H}_{(1)}^{in} + \mathbf{H}_{(1)}^{s}.$$

Here, the scattered fields $\mathbf{E}_{(1)}^{s}$ and $\mathbf{H}_{(1)}^{s}$ are produced by the magnetic current $\mathbf{J}_{ms}$ and satisfy

$$\nabla \times \mathbf{H}_{(1)}^{s} = j\omega\varepsilon_0 \mathbf{E}_{(1)}^{s},$$
$$\nabla \times \mathbf{E}_{(1)}^{s} = -j\omega\mu_0 \mathbf{H}_{(1)}^{s} - \mathbf{J}_{ms}.$$

Assume that the magnetic current element $\mathbf{J}_{ms}$ is located between z_1 and z_2. According to (2.81), the scattered fields for $z \geq z_2$ and $z \leq z_1$ in waveguide 1 may be, respectively, expanded in terms of the vector modal functions as follows:

$$\begin{aligned}
\mathbf{E}_{(1)}^{s} &= \sum_{n=1}^{\infty} A_{n+}^{(1)} \mathbf{E}_{n+}^{(1)}, \quad \mathbf{H}_{(1)}^{s} = \sum_{n=1}^{\infty} A_{n+}^{(1)} \mathbf{H}_{n+}^{(1)}, \quad z \geq z_2, \\
\mathbf{E}_{(1)}^{s} &= \sum_{n=1}^{\infty} A_{n-}^{(1)} \mathbf{E}_{n-}^{(1)}, \quad \mathbf{H}_{(1)}^{s} = \sum_{n=1}^{\infty} A_{n-}^{(1)} \mathbf{H}_{n-}^{(1)}, \quad z \leq z_1.
\end{aligned} \tag{2.100}$$

and the expansion coefficients are

$$A_{n+}^{(1)} = \frac{1}{2} Z_{wn}^{(1)} \int_{S_a} \mathbf{J}_{ms} \cdot \mathbf{H}_{n-}^{(1)} dS, \quad A_{n-}^{(1)} = \frac{1}{2} Z_{wn}^{(1)} \int_{S_a} \mathbf{J}_{ms} \cdot \mathbf{H}_{n+}^{(1)} dS. \tag{2.101}$$

A local coordinate system (ξ, ζ) with the origin at the center of the aperture may be introduced as illustrated in Figure 2.12. For a small aperture, the field $\mathbf{H}_{n\pm}^{(1)}$ may be expanded into a Taylor series about the origin

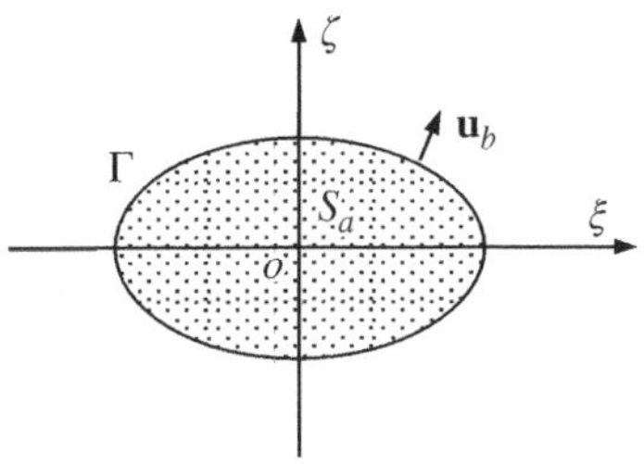

Figure 2.12 Aperture coordinates.

$$\mathbf{H}_{n\pm}^{(1)}(\xi,\zeta) = \mathbf{H}_{n\pm}^{(1)}(0,0) + \xi\frac{\partial \mathbf{H}_{n\pm}^{(1)}(0,0)}{\partial \xi} + \zeta\frac{\partial \mathbf{H}_{n\pm}^{(1)}(0,0)}{\partial \zeta}$$
$$= \mathbf{H}_{n\pm}^{(1)}(0,0) + \mathbf{r}_a \cdot \nabla \mathbf{H}_{n\pm}^{(1)}(0,0),$$

with $\mathbf{r}_a = \xi\mathbf{u}_\xi + \zeta\mathbf{u}_\zeta$. Equation (2.101) can be written as

$$A_{n+}^{(1)} = \frac{1}{2}Z_{wn}^{(1)}\mathbf{H}_{n-}^{(1)}(0,0)\cdot\int_{S_a}\mathbf{J}_{ms}dS + \frac{1}{2}Z_{wn}^{(1)}\int_{S_a}\mathbf{r}_a\cdot\nabla\mathbf{H}_{n-}^{(1)}(0,0)\cdot\mathbf{J}_{ms}dS,$$
$$A_{n-}^{(1)} = \frac{1}{2}Z_{wn}^{(1)}\mathbf{H}_{n+}^{(1)}(0,0)\cdot\int_{S_a}\mathbf{J}_{ms}dS + \frac{1}{2}Z_{wn}^{(1)}\int_{S_a}\mathbf{r}_a\cdot\nabla\mathbf{H}_{n+}^{(1)}(0,0)\cdot\mathbf{J}_{ms}dS. \tag{2.102}$$

Since the magnetic current is confined in the aperture, for an arbitrary function ϕ, one may write

$$\int_{S_a}\nabla\cdot(\phi\mathbf{J}_{ms})dS = \int_{S_a}(\phi\nabla\cdot\mathbf{J}_{ms} + \mathbf{J}_{ms}\cdot\mathbf{u}_\xi)dS = \int_{\partial S_a}(\phi\mathbf{J}_{ms})\cdot\mathbf{u}_b d\Gamma = 0, \tag{2.103}$$

with $\mathbf{u}_b$ being the unit outward normal to the aperture boundary Γ. Making use of (2.103), the first integral on the right-hand side of (2.102) can be written as

$$\int_{S_a}\mathbf{J}_{ms}dS = -\int_{S_a}\mathbf{r}_a\nabla\cdot\mathbf{J}_{ms}dS = j\omega\int_{S_a}\mathbf{r}_a\rho_{ms}dS = j\omega\mu_0\mathbf{m},$$

where

$$\mathbf{m} = \frac{1}{\mu_0}\int_{S_a}\mathbf{r}_a\rho_{ms}dS = \frac{1}{j\omega\mu_0}\int_{S_a}\mathbf{J}_{ms}dS = \frac{1}{j\omega\mu_0}\int_{S_a}(\mathbf{u}_n\times\mathbf{E})dS \tag{2.104}$$

is the magnetic dipole moment. Note that

$$\mathbf{r}_a\cdot\nabla\mathbf{H}_{n\pm}^{(1)}(0,0)\cdot\mathbf{J}_{ms} = \xi J_{ms\xi}\frac{\partial H_{n\pm\xi}^{(1)}(0,0)}{\partial\xi} + \xi J_{ms\zeta}\frac{\partial H_{n\pm\zeta}^{(1)}(0,0)}{\partial\xi}$$
$$+ \zeta J_{ms\xi}\frac{\partial H_{n\pm\xi}^{(1)}(0,0)}{\partial\zeta} + \zeta J_{ms\zeta}\frac{\partial H_{n\pm\zeta}^{(1)}(0,0)}{\partial\zeta},$$

where $\mathbf{H}_{n\pm}^{(1)}(0,0) = H_{n\pm\xi}^{(1)}(0,0)\mathbf{u}_\xi + H_{n\pm\zeta}^{(1)}(0,0)\mathbf{u}_\zeta$. Subtracting and adding similar terms yield

$$
\begin{aligned}
\mathbf{r}_a \cdot \nabla \mathbf{H}_{n\pm}^{(1)}(0,0)\cdot \mathbf{J}_{ms} &= \frac{1}{2}(\xi J_{ms\zeta} - \zeta J_{ms\xi})\left(\frac{\partial H_{n\pm\zeta}^{(1)}(0,0)}{\partial \xi} - \frac{\partial H_{n\pm\xi}^{(1)}(0,0)}{\partial \zeta}\right)\\
&+ \frac{\xi}{2}J_{ms\zeta}\frac{\partial H_{n\pm\xi}^{(1)}(0,0)}{\partial \zeta} + \frac{\zeta}{2}J_{ms\xi}\frac{\partial H_{n\pm\zeta}^{(1)}(0,0)}{\partial \xi} + \xi J_{ms\xi}\frac{\partial H_{n\pm\xi}^{(1)}(0,0)}{\partial \xi}\\
&+ \zeta J_{ms\zeta}\frac{\partial H_{n\pm\zeta}^{(1)}(0,0)}{\partial \zeta} + \frac{\xi}{2}J_{ms\zeta}\frac{\partial H_{n\pm\zeta}^{(1)}(0,0)}{\partial \xi} + \frac{\zeta}{2}J_{ms\xi}\frac{\partial H_{n\pm\xi}^{(1)}(0,0)}{\partial \zeta}.
\end{aligned}
\tag{2.105}
$$

According to (2.78), the first term on the right-hand side can be written as

$$
\frac{1}{2}(\xi J_{ms\zeta} - \zeta J_{ms\xi})\left(\frac{\partial H_{n\pm\zeta}^{(1)}(0,0)}{\partial \xi} - \frac{\partial H_{n\pm\xi}^{(1)}(0,0)}{\partial \zeta}\right) = j\omega\varepsilon_0 \frac{1}{2}\mathbf{E}_{n\pm}^{(1)}\cdot(\mathbf{r}_a \times \mathbf{J}_{ms}).
$$

This gives

$$
j\omega\varepsilon_0 \mathbf{E}_{n\pm}^{(1)}(0,0)\cdot \int_{S_a} \frac{\mathbf{r}_a \times \mathbf{J}_{ms}}{2} dS = -j\omega \mathbf{E}_{n\pm}^{(1)}(0,0)\cdot \mathbf{p}, \tag{2.106}
$$

where

$$
\mathbf{p} = \varepsilon_0 \frac{1}{2}\int_{S_a} (\mathbf{J}_{ms} \times \mathbf{r}_a) dS \tag{2.107}
$$

is the equivalent electric dipole moment of the magnetic current. Setting $\phi = \xi^2/2$, $\zeta^2/2$, and $\xi\zeta$ in (2.103), one may, respectively, find

$$
\begin{aligned}
\int_{S_a} \xi J_{ms\xi} dS &= \frac{j\omega}{2}\int_{S_a} \xi^2 \rho_{ms} dS,\\
\int_{S_a} \zeta J_{ms\zeta} dS &= \frac{j\omega}{2}\int_{S_a} \zeta^2 \rho_{ms} dS,\\
\int_{S_a} (\xi J_{ms\zeta} + \zeta J_{ms\xi}) dS &= j\omega \int_{S_a} \xi\zeta \rho_{ms} dS.
\end{aligned}
$$

Introducing the dyadic magnetic quadrupole $\overleftrightarrow{\mathbf{Q}}^m$ defined by

$$
\begin{aligned}
Q_{\xi\xi}^m &= \frac{1}{\mu_0}\int_{S_a} \xi^2 \rho_{ms} dS,\\
Q_{\zeta\zeta}^m &= \frac{1}{\mu_0}\int_{S_a} \zeta^2 \rho_{ms} dS,\\
Q_{\xi\zeta}^m = Q_{\zeta\xi}^m &= \frac{1}{\mu_0}\int_{S_a} \xi\zeta \rho_{ms} dS,
\end{aligned}
$$

(2.105) can be expressed as

$$\int_{S_a} \mathbf{r}_a \cdot \nabla \mathbf{H}_{n\pm}^{(1)}(0,0) \cdot \mathbf{J}_{ms} dS = j\omega \mathbf{E}_{n\pm}^{(1)}(0,0) \cdot \mathbf{p} + \frac{j\omega\mu_0}{2} \nabla \mathbf{H}_{n\pm}^{(1)}(0,0) : \overleftrightarrow{\mathbf{Q}}^m, \tag{2.108}$$

where the double dot denotes the double product of two dyads. The expansion coefficients in (2.102) are then given by

$$\begin{aligned} A_{n+}^{(1)} &= \frac{1}{2} Z_{wn}^{(1)} \left[-\mathbf{E}_{n-}^{(1)}(0,0) \cdot j\omega \mathbf{p} + \mathbf{H}_{n-}^{(1)}(0,0) \cdot j\omega\mu_0 \mathbf{m} + \frac{\mu_0}{2} \nabla \mathbf{H}_{n-}^{(1)}(0,0) : \overleftrightarrow{\mathbf{Q}}^m \right], \\ A_{n-}^{(1)} &= \frac{1}{2} Z_{wn}^{(1)} \left[-\mathbf{E}_{n+}^{(1)}(0,0) \cdot j\omega \mathbf{p} + \mathbf{H}_{n+}^{(1)}(0,0) \cdot j\omega\mu_0 \mathbf{m} + \frac{\mu_0}{2} \nabla \mathbf{H}_{n+}^{(1)}(0,0) : \overleftrightarrow{\mathbf{Q}}^m \right]. \end{aligned} \tag{2.109}$$

The fields in waveguide 2 are generated by the equivalent magnetic current $-\mathbf{J}_{ms}$. The fields in the regions $z \geq z_2$ and $z \leq z_1$ in waveguide 2 may be, respectively, expanded in terms of the vector modal functions as follows:

$$\mathbf{E}_{(2)}^s = \sum_{n=1}^{\infty} A_{n+}^{(2)} \mathbf{E}_{n+}^{(2)}, \quad \mathbf{H}_{(2)}^s = \sum_{n=1}^{\infty} A_{n+}^{(2)} \mathbf{H}_{n+}^{(2)}, \quad z \geq z_2, \tag{2.110}$$

$$\mathbf{E}_{(2)}^s = \sum_{n=1}^{\infty} A_{n-}^{(2)} \mathbf{E}_{n-}^{(2)}, \quad \mathbf{H}_{(2)}^s = \sum_{n=1}^{\infty} A_{n-}^{(2)} \mathbf{H}_{n-}^{(2)}, \quad z \leq z_1. \tag{2.111}$$

Similarly, the expansion coefficients are given by

$$\begin{aligned} A_{n+}^{(2)} &= -\frac{1}{2} Z_{wn}^{(2)} \left[-\mathbf{E}_{n-}^{(2)}(0,0) \cdot j\omega \mathbf{p} + \mathbf{H}_{n-}^{(2)}(0,0) \cdot j\omega\mu_0 \mathbf{m} + \frac{\mu_0}{2} \nabla \mathbf{H}_{n-}^{(2)}(0,0) : \overleftrightarrow{\mathbf{Q}}^m \right], \\ A_{n-}^{(2)} &= -\frac{1}{2} Z_{wn}^{(2)} \left[-\mathbf{E}_{n+}^{(2)}(0,0) \cdot j\omega \mathbf{p} + \mathbf{H}_{n+}^{(2)}(0,0) \cdot j\omega\mu_0 \mathbf{m} + \frac{\mu_0}{2} \nabla \mathbf{H}_{n+}^{(2)}(0,0) : \overleftrightarrow{\mathbf{Q}}^m \right]. \end{aligned} \tag{2.112}$$

Here, $\mathbf{p}$, $\mathbf{m}$, and $\overleftrightarrow{\mathbf{Q}}^m$ are same as defined before. In most applications, the quadrupole terms in (2.112) can be ignored. The magnetic current $\mathbf{J}_{ms}$ and the charge ρ_{ms} may be determined by numerical methods. Cohn proposed an electrolytic-tank method for determining the aperture parameters of arbitrary shape [19]. For very small apertures, the static field solution for the dipole moments can be readily found [20, 21]. For a small circular aperture of radius a, the dipole moments can be obtained by solving integral equations and they are

$$\mathbf{m} = \frac{8}{3} a^3{}_0 \mathbf{H}_{in}^{(1)}(0), \quad \frac{\mathbf{p}}{\varepsilon_0} = -\frac{4a^3}{3} \mathbf{E}_{in}^{(1)}(0). \tag{2.113}$$

2.4 Transient Fields in Waveguide

Understanding the propagation of time-domain signal in waveguide is required in many situations, such as in high-speed circuits and in the susceptibility studies for electronic systems. In the time domain, one needs to solve the Maxwell equations subject to initial conditions, boundary conditions, excitation conditions, and causality. Consider an arbitrary metal waveguide shown in Figure 2.1. Since the vector modal functions $\{\mathbf{e}_n | n = 1, 2, \ldots\}$ are independent of frequency, they can be used to expand the fields in both frequency and time domain. Similar to (2.26) and (2.27), the transient EM fields in the waveguide can be expressed as

$$\begin{aligned}
\mathbf{E}(\mathbf{r}, t) &= \sum_{n=1}^{\infty} v_n(z, t)\mathbf{e}_n(\boldsymbol{\rho}) + \mathbf{u}_z \sum_{n=1}^{\infty} e_n(z, t)\frac{\nabla \cdot \mathbf{e}_n(\boldsymbol{\rho})}{k_{cn}}, \\
\mathbf{H}(\mathbf{r}, t) &= \sum_{n=1}^{\infty} i_n(z, t)\mathbf{u}_z \times \mathbf{e}_n(\boldsymbol{\rho}) + \mathbf{u}_z \frac{1}{\sqrt{\Omega}} \int_{\Omega} \frac{\mathbf{u}_z \cdot \mathbf{H}(\mathbf{r}, t)}{\sqrt{\Omega}} d\Omega + \sum_{n=1}^{\infty} h_n(z, t)\frac{\nabla \times \mathbf{e}_n(\boldsymbol{\rho})}{k_{cn}},
\end{aligned} \tag{2.114}$$

$$\begin{aligned}
\nabla \times \mathbf{E}(\mathbf{r}, t) &= \sum_{n=1}^{\infty} \left[\frac{\partial v_n(z, t)}{\partial z} + k_{cn} e_n(z, t)\right] \mathbf{u}_z \times \mathbf{e}_n(\boldsymbol{\rho}) + \sum_{n=1}^{\infty} k_{cn} v_n(z, t)\frac{\nabla \times \mathbf{e}_n(\boldsymbol{\rho})}{k_{cn}}, \\
\nabla \times \mathbf{H}(\mathbf{r}, t) &= \sum_{n=1}^{\infty} \left[-\frac{\partial i_n(z, t)}{\partial z} + k_{cn} h_n(z, t)\right] \mathbf{e}_n(\boldsymbol{\rho}) + \mathbf{u}_z \sum_{n=1}^{\infty} k_{cn} i_n(z, t)\frac{\nabla \cdot \mathbf{e}_n(\boldsymbol{\rho})}{k_{cn}},
\end{aligned} \tag{2.115}$$

where v_n and i_n are **time-domain modal voltage** and **time-domain modal current** defined by

$$\begin{aligned}
v_n(z, t) &= \int_{\Omega} \mathbf{E}(\mathbf{r}, t) \cdot \mathbf{e}_n(\boldsymbol{\rho}) d\Omega, \\
i_n(z, t) &= \int_{\Omega} \mathbf{H}(\mathbf{r}, t) \cdot \mathbf{u}_z \times \mathbf{e}_n(\boldsymbol{\rho}) d\Omega,
\end{aligned} \tag{2.116}$$

and the longitudinal components e_n and h_n are defined by

$$\begin{aligned}
e_n(z, t) &= \int_{\Omega} \mathbf{u}_z \cdot \mathbf{E}(\mathbf{r}, t) \frac{\nabla \cdot \mathbf{e}_n(\boldsymbol{\rho})}{k_{cn}} d\Omega, \\
h_n(z, t) &= \int_{\Omega} \mathbf{H}(\mathbf{r}, t) \cdot \frac{\nabla \times \mathbf{e}_n(\boldsymbol{\rho})}{k_{cn}} d\Omega.
\end{aligned}$$

Substituting (2.114) and (2.115) into the generalized Maxwell equations

$$\nabla \times \mathbf{E}(\mathbf{r},t) = -\mu\frac{\partial \mathbf{H}(\mathbf{r},t)}{\partial t} - \mathbf{J}_m(\mathbf{r},t),$$
$$\nabla \times \mathbf{H}(\mathbf{r},t) = \varepsilon\frac{\partial \mathbf{E}(\mathbf{r},t)}{\partial t} + \mathbf{J}(\mathbf{r},t) + \sigma\mathbf{E}(\mathbf{r},t),$$

and comparing the transverse and longitudinal components, one may find the equations for the modal voltages and currents

$$-\frac{\partial i_n}{\partial z} + k_{cn}h_n = \varepsilon\frac{\partial v_n}{\partial t} + \sigma v_n + \int_\Omega \mathbf{J}\cdot\mathbf{e}_n d\Omega, \tag{2.117}$$

$$k_{cn}i_n = \varepsilon\frac{\partial e_n}{\partial t} + \sigma e_n + \int_\Omega \mathbf{u}_z\cdot\mathbf{J}\frac{\nabla\cdot\mathbf{e}_n}{k_{cn}}d\Omega, \text{ for TM modes only,} \tag{2.118}$$

$$\frac{\partial v_n}{\partial z} + e_n k_{cn} = -\mu\frac{\partial i_n}{\partial t} - \int_\Omega \mathbf{J}_m\cdot\mathbf{u}_z\times\mathbf{e}_n d\Omega, \tag{2.119}$$

$$k_{cn}v_n = -\mu\frac{\partial h_n}{\partial t} - \int_\Omega \mathbf{u}_z\cdot\mathbf{J}_m\frac{\mathbf{u}_z\cdot\nabla\times\mathbf{e}_n}{k_{cn}}d\Omega, \text{ for TE modes only,} \tag{2.120}$$

$$-\mu\frac{\partial}{\partial t}\int_\Omega \frac{\mathbf{H}\cdot\mathbf{u}_z}{\sqrt{\Omega}}d\Omega = \int_\Omega \frac{\mathbf{u}_z\cdot\mathbf{J}_m}{\sqrt{\Omega}}d\Omega, \text{ for TE modes only.} \tag{2.121}$$

For the TEM mode, the modal voltage and modal current satisfy

$$\begin{aligned}\frac{\partial v_n^{TEM}}{\partial z} &= -\mu\frac{\partial i_n^{TEM}}{\partial t} - \int_\Omega \mathbf{J}_m\cdot\mathbf{u}_z\times\mathbf{e}_n d\Omega,\\ \frac{\partial i_n^{TEM}}{\partial z} &= -\varepsilon\frac{\partial v_n^{TEM}}{\partial t} - \sigma v_n^{TEM} - \int_\Omega \mathbf{J}\cdot\mathbf{e}_n d\Omega,\end{aligned} \tag{2.122}$$

from (2.117) and (2.119). After eliminating the modal current, the modal voltage for the TEM modes satisfies the wave equation

$$\frac{\partial^2 v_n^{TEM}}{\partial z^2} - \frac{1}{v^2}\frac{\partial^2 v_n^{TEM}}{\partial t^2} - \sigma\frac{\eta}{v}\frac{\partial v_n^{TEM}}{\partial t} = \mu\frac{\partial}{\partial t}\int_\Omega \mathbf{J}\cdot\mathbf{e}_n d\Omega - \frac{\partial}{\partial z}\int_\Omega \mathbf{J}_m\cdot\mathbf{u}_z\times\mathbf{e}_n d\Omega, \tag{2.123}$$

where $v = 1/\sqrt{\mu\varepsilon}$. The modal current i_n^{TEM} can be determined by the time integration of v_n^{TEM}.

For the TE mode, one may find the equations for the modal voltage and current

$$\begin{aligned}
\frac{\partial v_n^{TE}}{\partial z} &= -\mu\frac{\partial i_n^{TE}}{\partial t} - \int_\Omega \mathbf{J}_m \cdot \mathbf{u}_z \times \mathbf{e}_n d\Omega,\\
\frac{\partial i_n^{TE}}{\partial z} - k_{cn}h_n &= -\varepsilon\frac{\partial v_n^{TE}}{\partial t} - \sigma v_n^{TE} - \int_\Omega \mathbf{J}\cdot\mathbf{e}_n d\Omega,\\
\mu\frac{\partial h_n}{\partial t} &= -k_{cn}v_n^{TE} - \int_\Omega \mathbf{u}_z\cdot\mathbf{J}_m\frac{\mathbf{u}_z\cdot\nabla\times\mathbf{e}_n}{k_{cn}}d\Omega,
\end{aligned} \tag{2.124}$$

from (2.117), (2.119), and (2.120). The modal voltage v_n^{TE} satisfies the modified Klein–Gordon equation

$$\begin{aligned}
&\frac{\partial^2 v_n^{TE}}{\partial z^2} - \frac{1}{v^2}\frac{\partial^2 v_n^{TE}}{\partial t^2} - \sigma\frac{\eta}{v}\frac{\partial v_n^{TE}}{\partial t} - k_{cn}^2 v_n^{TE}\\
&= \mu\frac{\partial}{\partial t}\int_\Omega \mathbf{J}\cdot\mathbf{e}_n d\Omega - \frac{\partial}{\partial z}\int_\Omega \mathbf{J}_m\cdot\mathbf{u}_z\times\mathbf{e}_n d\Omega + k_{cn}\int_\Omega \mathbf{u}_z\cdot\mathbf{J}_m\frac{\mathbf{u}_z\cdot\nabla\times\mathbf{e}_n}{k_{cn}}d\Omega.
\end{aligned} \tag{2.125}$$

The modal currents i_n^{TE} can then be determined by the time integration of $\partial v_n^{TE}/\partial z$.

For the TM mode, the equations for the modal voltage and current are found as follows:

$$\begin{aligned}
\frac{\partial v_n^{TM}}{\partial z} + k_{cn}e_n &= -\mu\frac{\partial i_n^{TM}}{\partial t} - \int_\Omega \mathbf{J}_m\cdot\mathbf{u}_z\times\mathbf{e}_n d\Omega,\\
\frac{\partial i_n^{TM}}{\partial z} &= -\varepsilon\frac{\partial v_n^{TM}}{\partial t} - \sigma v_n^{TM} - \int_\Omega \mathbf{J}\cdot\mathbf{e}_n d\Omega,\\
\varepsilon\frac{\partial e_n}{\partial t} &= k_{cn}i_n^{TM} - \sigma e_n - \int_\Omega \mathbf{u}_z\cdot\mathbf{J}\frac{\nabla\cdot\mathbf{e}_n}{k_{cn}}d\Omega,
\end{aligned} \tag{2.126}$$

from (2.117)–(2.119). The modal current i_n^{TM} also satisfies the modified Klein–Gordon equation

$$
\begin{aligned}
&\frac{\partial^2 i_n^{TM}}{\partial z^2} - \frac{1}{v^2}\frac{\partial^2 i_n^{TM}}{\partial t^2} - \sigma\frac{\eta}{v}\frac{\partial i_n^{TM}}{\partial t} - k_{cn}^2 i_n^{TM} \\
&= -\frac{\partial}{\partial z}\int_{\Omega}\mathbf{J}\cdot\mathbf{e}_n d\Omega - k_{cn}\int_{\Omega}\mathbf{u}_z\cdot\mathbf{J}\frac{\nabla\cdot\mathbf{e}_n}{k_{cn}}d\Omega \\
&\quad + \varepsilon\frac{\partial}{\partial t}\int_{\Omega}\mathbf{J}_m\cdot\mathbf{u}_z\times\mathbf{e}_n d\Omega + \sigma\int_{\Omega}\mathbf{J}_m\cdot\mathbf{u}_z\times\mathbf{e}_n d\Omega.
\end{aligned}
\tag{2.127}
$$

The modal voltages v_n^{TM} can then be determined by a time integration of $\partial i_n^{TM}/\partial z$. To find the complete solution of the transient fields in the waveguide, one needs to solve the modified Klein–Gordon equation, which can be done by using the retarded Green's function of the modified Klein–Gordon equation, defined by

$$
\begin{cases}
\left(\dfrac{\partial^2}{\partial z^2} - \dfrac{1}{v^2}\dfrac{\partial^2}{\partial t^2} - \sigma\dfrac{\eta}{v}\dfrac{\partial}{\partial t} - k_{cn}^2\right)G_n(z,t;z',t') = -\delta(z-z')\delta(t-t'), \\
G_n(z,t;z',t')\big|_{t<t'} = 0.
\end{cases}
\tag{2.128}
$$

The solution of (2.128) is [22]

$$
\begin{aligned}
G_n(z,t;z',t') &= \frac{v}{2}e^{-\gamma(t-t')}H\left[(t-t') - \frac{|z-z'|}{v}\right] \\
&\cdot J_0\left[\left(k_{cn}^2v^2-\gamma^2\right)^{1/2}\sqrt{(t-t')^2 - \frac{|z-z'|^2}{v^2}}\right],
\end{aligned}
\tag{2.129}
$$

where $J_0(x)$ is the Bessel function of first kind and $H(x)$ is the unit step function. The retarded Green's function can now be used to solve the modified Klein–Gordon equation with the known source function $f(z, t)$:

$$
\left(\frac{\partial^2}{\partial z^2} - \frac{1}{v^2}\frac{\partial^2}{\partial t^2} - \sigma\frac{\eta}{v}\frac{\partial}{\partial t} - k_{cn}^2\right)u_n(z,t) = f(z,t),
$$

and the solution is

$$
u_n(z,t) = -\int_{-\infty}^{\infty}\int_{-\infty}^{\infty} f(z',t')G_n(z,t;z',t')dt'dz', \quad z\in(-\infty,\infty). \tag{2.130}
$$

For generality, our treatment for the expansions of the transient fields in waveguide has been based on the transverse field patterns (the vector modal functions). The transient fields in waveguide can also expanded in terms of the longitudinal field patterns for simply connected waveguide [23].

Example 2.3 Consider a rectangular waveguide of width a and height b. The waveguide is excited by a line current extending across the waveguide located at $x = a/2,\ z = 0$, and the current density is given by

$$\mathbf{J}(\mathbf{r}, t) = \mathbf{u}_y \delta\left(x - \frac{a}{2}\right)\delta(z) f(t).$$

Since the line current is uniform in y direction, the fields excited by the current are independent of y. As a consequence, only TE_{n0} modes are excited

$$\mathbf{e}_n(x, y) = \mathbf{e}_{n0}^{TE}(x, y) = -\mathbf{u}_y \left(\frac{2}{ab}\right)^{1/2} \sin\frac{n\pi}{a}x, n = 1, 2, 3, \ldots.$$

From (2.130), the modal voltage can be expressed by

$$v_n^{TE}(z, t) = \frac{b\eta}{2}\left(\frac{2}{ab}\right)^{1/2} \sin\frac{n\pi}{2} \int_{-\infty}^{t-|z|/v} \frac{df(t')}{dt'} J_0\left[k_{cn} v \sqrt{(t-t')^2 - \frac{|z|^2}{v^2}}\right] dt', \tag{2.131}$$

where $\eta = \sqrt{\mu/\varepsilon}$. Evidently, the time-domain voltages v_n^{TE} for $n = 2, 4, 6, \ldots$ vanish. The total electric field in the waveguide is given by the first equation of (2.114):

$$\mathbf{E} = -\mathbf{u}_y \left(\frac{2}{ab}\right)^{1/2} \sum_{n=1}^{\infty} v_n^{TE}(z, t) \sin\frac{n\pi}{a}x. \tag{2.132}$$

For a sinusoidal source turned on at $t = 0$, one may expect that the time-domain response approaches the well-known steady-state response as time goes to infinity. Let $f(t) = H(t)\sin\omega t$ in (2.131). The modal voltage (2.131) may be expressed as the sum of two parts

$$v_n^{TE}(z, t) = v_n^{TE}(z, t)|_{\text{steady}} + v_n^{TE}(z, t)|_{\text{transient}}, \quad t > \frac{|z|}{v},$$

where $v_n^{TE}(z, t)|_{\text{steady}}$ and $v_n^{TE}(z, t)|_{\text{transient}}$ represent the steady-state part and the transient part, respectively,

$$\begin{aligned}
v_n^{TE}(z, t)|_{\text{steady}} &= \frac{b\eta}{2}\left(\frac{2}{ab}\right)^{1/2} ka \sin\frac{n\pi}{2} \\
&\quad \times \int_{|z|/a}^{\infty} \cos ka\left(\frac{vt}{a} - u\right) J_0\left[k_{cn} a \sqrt{u^2 - \frac{|z|^2}{a^2}}\right] du, \\
v_n^{TE}(z, t)|_{\text{transient}} &= -\frac{b\eta}{2}\left(\frac{2}{ab}\right)^{1/2} ka \sin\frac{n\pi}{2} \\
&\quad \times \int_{vt/a}^{\infty} \cos ka\left(\frac{vt}{a} - u\right) J_0\left[k_{cn} a \sqrt{u^2 - \frac{|z|^2}{a^2}}\right] du.
\end{aligned}$$

The transient part approaches to zero as $t \to \infty$. The steady-state response is found to be

$$v_n^{TE}(z,t)|_{\text{steady}} = \frac{b\eta}{2}\left(\frac{2}{ab}\right)^{1/2} \frac{k \sin\frac{n\pi}{2}}{\sqrt{\left|k^2 - k_{cn}^2\right|}} \begin{cases} \sin\left(\omega t - |z|\sqrt{k^2 - k_{cn}^2}\right), k > k_{cn} \\ \text{Re}\, e^{j\omega t - |z|\sqrt{k_{cn}^2 - k^2}}, k < k_{cn} \end{cases}$$
$$= \frac{b\eta}{2}\left(\frac{2}{ab}\right)^{1/2} k \sin\frac{n\pi}{2} \,\text{Re}\, e^{j\omega t} \frac{1}{\gamma_n^{TE}} e^{-\gamma_n^{TE}|z|},$$

where γ_n^{TE} is given by (2.85). Thus, as $|z|$ increases, the modal voltages decrease rapidly when $k < k_{cn}$. In other words, only those modes satisfying $k > k_{cn}$ propagate in the steady state. When $v_n^{TE}|_{steady}$ is inserted into (2.132), one may find the steady-state response of the electric field

$$\mathbf{E} = -\mathbf{u}_y \frac{\omega\mu}{a} \sum_{n=1}^{\infty} \frac{1}{\gamma_n^{TE}} e^{-\gamma_n^{TE}|z|} \sin\frac{n\pi}{2} \sin\frac{n\pi}{a}x. \tag{2.133}$$

This agrees with the time-harmonic theory (2.86). □

> The interaction between physics and mathematics has always played an important role in the development of both sciences. The physicist who does not have the latest mathematical knowledge available to him is at a distinct disadvantage. The mathematician who shies away from physical applications will most likely miss important insights and motivations.
>
> *Martin Schechter (American mathematician, 1930–2021)*

References

1 Packard, K. S., "The origin of waveguides: a case of multiple rediscovery", *IEEE Trans. Microw. Theory Tech.*, Vol. 32, pp. 961–969, 1984.

2 Oliner, A., "Historical perspectives on microwave field theory", *IEEE Trans. Microw. Theory Tech.*, Vol. 32, pp. 1022–1045, 1984.

3 Marcuvitz, N., *Waveguide Handbook*, McGraw-Hill Book Company, Inc., 1951.

4 Kurokawa, K., *An Introduction to Microwave Circuits*, New York, Academic Press, 1969.

5 Jones, D. S., *The Theory of Electromagnetism*, Pergamon Press, 1964.

6 Lewin, L., *Advanced Theory of Waveguides*, London, Iliffe and Sons, Limited, 1951.

7 Harrington, R. F., *Time-Harmonic Electromagnetic Fields*, McGraw-Hill Book Company, Inc., 1961.

8 Balanis, C. A., *Advanced Engineering Electromagnetics*, 2nd Ed., John Wiley & Sons, 2012.

9 Montgomery, C. G., R. H. Dicke, and E. M. Purcell, *Principles of Microwave Circuits*, McGraw-Hill, 1948.

10 Mittra, R. and W. W. Lee, *Analytical Techniques in the Theory of Guided Waves*, New York, Macmillan, 1971.

11 Felsen, L. B. and N. Marcuvitz, *Radiation and Scattering of Waves*, IEEE Press, 1994.

12 Pozar, D. M., *Microwave Engineering*, John Wiley & Sons, 1998.

13 Collin, R. E., *Field Theory of Guided Waves*, 2nd Ed., IEEE Press, 1991.

14 Collin, R. E., *Foundations for Microwave Engineering*, 2nd Ed., IEEE Press, 2001.

15 Geyi, W., *Foundations for Radio Frequency Engineering*, World Scientific, 2015.

16 Tai, C.-T., "On the eigenfunction expansions of dyadic Green's functions", *Proc. IEEE*, Vol. 61, No. 4, pp. 480–481, 1973.

17 Tai, C.-T., *Dyadic Green Functions in Electromagnetic Theory*, IEEE Press, 1994.

18 Schwinger, J. and D. S. Saxon, *Discontinuities in Waveguides*, New York, Gordon and Breach, 1968.

19 Cohn, S. B., "Determination of aperture parameters by electrolytic tank measurements", *Proc. IRE*, Vol. 39, pp. 1416–1421, 1951.

20 Bethe, H. A., "Theory of diffraction by small holes", *Phys. Rev.*, Vol. 66, No. 7–8, pp. 163–182, 1944.

21 Stratton, J. A., *Electromagnetic Theory*, New York, McGraw-Hill, 1941.

22 Geyi, W., "A time-domain theory of waveguide", *Prog. Electromagn. Res.*, Vol. 59, pp. 267–297, 2006.

23 Kristensson, G., "Transient electromagnetic wave propagation in waveguides", *J. Electromagn. Waves Appl.*, Vol. 9, pp. 645–671, 1995.

3

Radiation in Cavity Resonator

The career of a young theoretical physicist consists of treating the harmonic oscillator in ever-increasing levels of abstraction.
– Sidney Coleman (American physicist, 1937–2007)

A resonator is a device that oscillates with greater amplitude at its resonant frequencies. A metal cavity resonator is a hollow metal box and has an infinite number of resonant frequencies, and the corresponding resonant modes form a complete set and may be used to expand the electromagnetic (EM) fields inside the cavity. If the set of the modes is ordered according to increasing resonant frequencies, there is always a lowest resonant frequency but no highest one. Hermann Weyl (German mathematician and theoretical physicist, 1885–1955) first investigated the cavity resonator problem by using the theory of linear integral equation and studied the asymptotic distribution of resonant frequencies [1]. The theory of the cavity resonator was then reexamined by a number of authors [e.g. 2–8]. Many books on EM theory and engineering cover the basic theory of cavity resonators [7, 9–23].

In a cavity resonator bounded by a perfect conductor, the vector modal functions can be classified into three types and they are generally necessary for the expansion of the fields. The vector modal functions of the first type behave like the TEM modes in waveguide, both their curl and divergence being zero, and they only exist in a multiply connected region, and are called **static vector modal functions**. The vector modal functions of the second type are similar to the TE modes in a waveguide, with zero divergence and nonzero curl, called **divergenceless vector modal functions** or **natural resonance modes**. The vector modal functions of the third type are similar to the TM modes in a waveguide and have zero curl and nonzero divergence, known as **irrotational vector modal functions**.

Foundations of Antenna Radiation Theory: Eigenmode Analysis, First Edition. Wen Geyi.

Published 2023 by John Wiley & Sons, Inc.

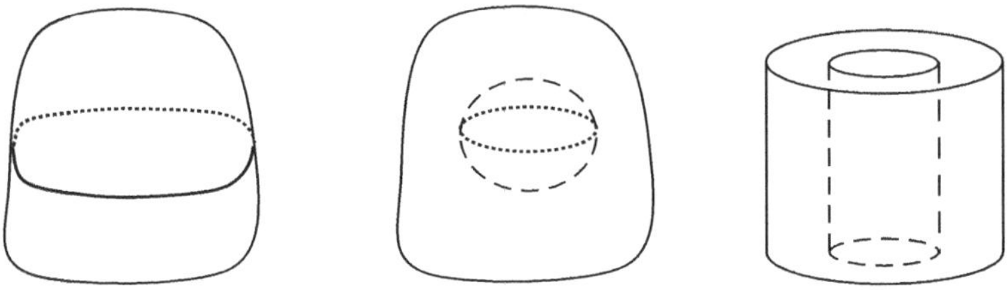

Figure 3.1 Three typical cavity regions. (a) Type 1: simply connected with a simple boundary. (b) Type 2: simply connected with multiple boundaries. (c) Type 3: multiply connected with a simple boundary.

The region occupied by a cavity resonator can be one of the three types illustrated in Figure 3.1. The first type is a simply connected region with a simple boundary such as the one shown in Figure 3.1a, and a sphere belongs to this type. The second type is a simply connected region with multiple boundaries as shown in Figure 3.1b, and the region bounded by two concentric sphere belongs to this type. The third type is a multiply connected region with a simple boundary as shown in Figure 3.1c, and the region bounded by two coaxial cylinders closed at both ends belongs to this type. In type 1, only divergenceless and irrotational vector modal functions can exist; in type 2, static vector modal functions are needed for the expansion of electric field; and in type 3, static vector modal functions are needed for the expansion of magnetic field [2, 5].

All the cavity-related problems can be treated as a radiation problem, and can be studied by using the dyadic Green's functions. This chapter features the derivation of the vector modal functions for the waveguide cavity resonator and the derivation of the modal representations of the dyadic Green's functions. The transient fields generated by the source in the waveguide cavity resonator are examined by the field expansions in terms of the vector modal functions in the corresponding waveguide, and their analytical expressions are obtained.

3.1 Radiated Fields in Cavity Resonator

The EM fields in a cavity resonator are often excited by an external power source, which is coupled to the cavity by a small aperture, a small probe, or a loop. The cavity excitation problems can be reduced to the radiation problems in the cavity resonator.

3.1.1 Classification of Vector Modal Functions for Cavity Resonator

Assume that the wall of the cavity is perfectly conducting and the cavity is filled with homogeneous medium with medium parameters μ and ε, and occupies a region V bounded by S, as illustrated in Figure 3.2. The fields inside a cavity free of source satisfy the Maxwell equations

$$\begin{aligned} &\nabla \times \mathbf{H}(\mathbf{r}) = j\omega\varepsilon\mathbf{E}(\mathbf{r}),\\ &\nabla \times \mathbf{E}(\mathbf{r}) = -j\omega\mu\mathbf{H}(\mathbf{r}),\\ &\nabla \cdot \mathbf{E}(\mathbf{r}) = 0,\\ &\nabla \cdot \mathbf{H}(\mathbf{r}) = 0, \end{aligned} \tag{3.1}$$

and the boundary conditions $\mathbf{u}_n \times \mathbf{E} = 0$ and $\mathbf{u}_n \cdot \mathbf{H} = 0$ on S, where $\mathbf{u}_n$ is the unit outward normal of S. It follows from (3.1) and the boundary conditions that the EM fields inside the cavity satisfy the wave equations

$$\begin{cases} \nabla \times \nabla \times \mathbf{E}(\mathbf{r}) - k^2\mathbf{E}(\mathbf{r}) = 0, & \mathbf{r} \in V,\\ \mathbf{u}_n \times \mathbf{E}(\mathbf{r}) = 0, & \mathbf{r} \in S, \end{cases} \tag{3.2}$$

$$\begin{cases} \nabla \times \nabla \times \mathbf{H}(\mathbf{r}) - k^2\mathbf{H}(\mathbf{r}) = 0, & \mathbf{r} \in V,\\ \mathbf{u}_n \cdot \mathbf{H}(\mathbf{r}) = 0, \mathbf{u}_n \times \nabla \times \mathbf{H}(\mathbf{r}) = 0, & \mathbf{r} \in S. \end{cases} \tag{3.3}$$

Similar to the waveguide theory, one may introduce the eigenvalue problems

$$\begin{cases} -\nabla^2\mathbf{e}(\mathbf{r}) = -k^2\mathbf{e}(\mathbf{r}), & \mathbf{r} \in V,\\ \mathbf{u}_n \times \mathbf{e}(\mathbf{r}) = 0, \nabla \cdot \mathbf{e}(\mathbf{r}) = 0, & \mathbf{r} \in S, \end{cases} \tag{3.4}$$

$$\begin{cases} -\nabla^2\mathbf{h}(\mathbf{r}) = -k^2\mathbf{h}(\mathbf{r}), & \mathbf{r} \in V,\\ \mathbf{u}_n \cdot \mathbf{h}(\mathbf{r}) = 0, \mathbf{u}_n \times \nabla \times \mathbf{h}(\mathbf{r}) = 0, & \mathbf{r} \in S, \end{cases} \tag{3.5}$$

where $-\nabla^2\mathbf{e} = \nabla \times \nabla \times \mathbf{e} - \nabla\nabla \cdot \mathbf{e}$ and k^2 is the eigenvalue to be determined. From Section 1.5, a set of orthogonal eigenfunctions $\{\mathbf{e}_1, \mathbf{e}_2, ...\}$ can be obtained from (3.4), and the corresponding eigenvalues satisfy $0 \le k_{e1}^2 \le k_{e2}^2 \le \cdots$.

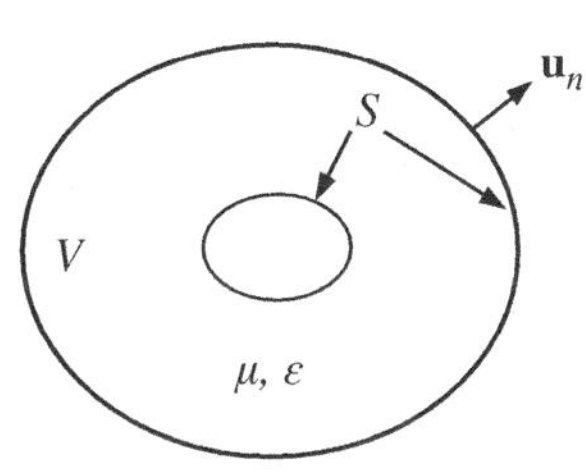

Figure 3.2 An arbitrary metal cavity resonator.

The eigenfunction $\mathbf{e}_n$ is called *n*th **electric vector modal function**. From now on, all the vector modal functions will be assumed to be orthonormal

$$\int_V \mathbf{e}_m \cdot \mathbf{e}_n dV = \delta_{mn}. \tag{3.6}$$

The electric vector modal functions fall into the following three types:

$$\nabla \times \mathbf{e}_n = 0, \quad \nabla \cdot \mathbf{e}_n = 0.$$

$$\nabla \times \mathbf{e}_n \neq 0, \quad \nabla \cdot \mathbf{e}_n = 0.$$

$$\nabla \times \mathbf{e}_n = 0, \quad \nabla \cdot \mathbf{e}_n \neq 0.$$

The electric vector modal functions of the first type only exist in the simply connected region with multiple boundaries shown in Figure 3.1b. For the modes in the first type, a scalar potential function φ_n can be introduced such that $\mathbf{e}_n = \nabla \varphi_n$. Thus,

$$\begin{cases} \nabla^2 \varphi_n(\mathbf{r}) = 0, & \mathbf{r} \in V, \\ \varphi_n(\mathbf{r}) = \text{const.}, & \mathbf{r} \in S. \end{cases} \tag{3.7}$$

For the modal functions in the third type, a scalar potential function ϕ_n may also be introduced such that $\mathbf{e}_n = \nabla \phi_n$. It is readily shown that the scalar functions ϕ_n satisfies

$$\begin{cases} \nabla^2 \phi_n(\mathbf{r}) + k_{en}^2 \phi_n(\mathbf{r}) = 0, & \mathbf{r} \in V, \\ \phi_n(\mathbf{r}) = 0, & \mathbf{r} \in S, \end{cases} \tag{3.8}$$

and

$$\int_V |\mathbf{e}_n,|^2 dV = k_{en}^2 \int_V \phi_n^2 dV. \tag{3.9}$$

Similarly, a complete set of orthonormal **magnetic vector modal functions** $\{\mathbf{h}_1, \mathbf{h}_2, \ldots\}$ can be constructed from (3.5) and they fall into the following three types:

$$\nabla \times \mathbf{h}_n = 0, \quad \nabla \cdot \mathbf{h}_n = 0.$$

$$\nabla \times \mathbf{h}_n \neq 0, \quad \nabla \cdot \mathbf{h}_n = 0.$$

$$\nabla \times \mathbf{h}_n = 0, \quad \nabla \cdot \mathbf{h}_n \neq 0.$$

The magnetic vector modal functions of the first type only exist in the multiply connected region with a simple boundary shown in Figure 3.1c. Physically these modes represent the magnetic field generated by a direct current flowing through the circuit consisting of the center conductor, the short ends, and the out conductor. For the magnetic vector modal functions of the first and third type, one may introduce a scalar potential function ψ_n such that $\mathbf{h}_n = \nabla \psi_n$. Then,

$$\begin{cases} \nabla^2\psi_n(\mathbf{r}) + k_{hn}^2\psi_n(\mathbf{r}) = 0, & \mathbf{r} \in V \\ \dfrac{\partial\psi_n(\mathbf{r})}{\partial n} = 0, & \mathbf{r} \in S, \end{cases} \tag{3.10}$$

and

$$\int_V |\mathbf{h}_n|^2 dV = k_{hn}^2 \int_V \psi_n^2 dV. \tag{3.11}$$

The vector modal functions of the second type in the sets $\{\mathbf{e}_n\}$ and $\{\mathbf{h}_n\}$ are related to each other. In fact, if $\mathbf{e}_n$ is in the second type, a function $\mathbf{h}_n$ can be introduced through

$$\nabla \times \mathbf{e}_n = k_{en}\mathbf{h}_n, \tag{3.12}$$

and it is easy to verify that $\mathbf{h}_n$ belongs to the second type and satisfies

$$\begin{cases} \nabla \times \nabla \times \mathbf{h}_n(\mathbf{r}) - k_{en}^2\mathbf{h}_n(\mathbf{r}) = 0, & \mathbf{r} \in V, \\ \mathbf{u}_n \times \nabla \times \mathbf{h}_n(\mathbf{r}) = 0, & \mathbf{r} \in S. \end{cases}$$

To show that the normal component of $\mathbf{h}_n$ is zero, let us consider the integration of $\mathbf{u}_n \cdot \mathbf{h}_n$ over an arbitrary part of S, denoted ΔS:

$$\int_{\Delta S} \mathbf{u}_n \cdot \mathbf{h}_n dS = k_{en}^{-1} \int_{\Delta S} \mathbf{u}_n \cdot \nabla \times \mathbf{e}_n dS = k_{en}^{-1} \int_{\Delta\Gamma} \mathbf{e}_n \cdot \mathbf{u}_\Gamma d\Gamma, \tag{3.13}$$

where $\Delta\Gamma$ is the closed contour around ΔS and $\mathbf{u}_\Gamma$ is the unit tangent vector along the contour; dS and $d\Gamma$, respectively, denote the differential surface element on S and the differential line element along $\Delta\Gamma$. The right-hand side of (3.13) vanishes for the tangential component of $\mathbf{e}_n$ must be zero. Hence, $\mathbf{u}_n \cdot \mathbf{h}_n = 0$ for ΔS is arbitrary. Therefore, $\mathbf{h}_n$ satisfies (3.5) and the corresponding eigenvalue is k_{en}^2. If $\mathbf{h}_m$ is another vector modal function corresponding to $\mathbf{e}_m$ belonging to the second type, the inner product between $\mathbf{h}_n$ and $\mathbf{h}_m$ is

$$\begin{aligned} \int_V \mathbf{h}_m \cdot \mathbf{h}_n dV &= (k_{em}k_{en})^{-1} \int_V \nabla \times \mathbf{e}_m \cdot \nabla \times \mathbf{e}_n dV \\ &= (k_{em}k_{en})^{-1} \int_S \mathbf{u}_n \times \mathbf{e}_m \cdot \nabla \times \mathbf{e}_n dS \\ &\quad + \left(\frac{k_{en}}{k_{em}}\right) \int_V \mathbf{e}_m \cdot \mathbf{e}_n dV = \delta_{mn}. \end{aligned}$$

Consequently, the vector modal functions $\mathbf{h}_n$ in the second type can be derived from the vector modal functions $\mathbf{e}_n$ in the second type, and they are orthonormal. Conversely if $\mathbf{h}_n$ is in the second type, one can define $\mathbf{e}_n$ through

$$\nabla \times \mathbf{h}_n = k_{hn}\mathbf{e}_n. \tag{3.14}$$

A similar discussion shows that $\mathbf{e}_n$ is an eigenfunction of (3.4) with k_{hn} being the eigenvalue. So the completeness of the two sets is still guaranteed if the vector modal functions belonging to the second type in $\{\mathbf{e}_n\}$ and $\{\mathbf{h}_n\}$ are related through either (3.12) or (3.14). Hereafter, (3.12) and (3.14) will be assumed and $k_{e,n} = k_{h,n}$ will be denoted by k_n. Note that the complete set $\{\mathbf{e}_n\}$ is most appropriate for the expansion of electric field, and the complete set $\{\mathbf{h}_n\}$ is most appropriate for the expansion of the magnetic field.

3.1.2 Modal Expansions for the Fields and Dyadic Green's Functions

If the cavity contains an impressed electric current $\mathbf{J}$, a magnetic current $\mathbf{J}_m$, and a lossy medium with $\varepsilon_e = \varepsilon(1 - j(\sigma/\omega\varepsilon))$, the EM fields in the cavity satisfy the Maxwell equations

$$\begin{aligned} \nabla \times \mathbf{H} &= j\omega\varepsilon_e\mathbf{E} + \mathbf{J}, \\ \nabla \times \mathbf{E} &= -j\omega\mu\mathbf{H} - \mathbf{J}_m, \end{aligned} \tag{3.15}$$

with the boundary conditions $\mathbf{u}_n \times \mathbf{E} = 0$ and $\mathbf{u}_n \cdot \mathbf{H} = 0$ on the boundary S. Since $\nabla \times \mathbf{E}$ behaves like a magnetic field and $\nabla \times \mathbf{H}$ behaves like an electric field, the fields and their curls can be expanded in terms of the vector modal functions as follows:

$$\begin{aligned} \mathbf{E} &= \sum_n V_n\mathbf{e}_n + \sum_\nu V_\nu\mathbf{e}_\nu, \\ \mathbf{H} &= \sum_n I_n\mathbf{h}_n + \sum_\tau I_\tau\mathbf{h}_\tau, \end{aligned} \tag{3.16}$$

$$\begin{aligned} \nabla \times \mathbf{E} &= \sum_n \mathbf{h}_n \int_V \nabla \times \mathbf{E} \cdot \mathbf{h}_n dV + \sum_\tau \mathbf{h}_\tau \int_V \nabla \times \mathbf{E} \cdot \mathbf{h}_\tau dV, \\ \nabla \times \mathbf{H} &= \sum_n \mathbf{e}_n \int_V \nabla \times \mathbf{H} \cdot \mathbf{e}_n dV + \sum_\nu \mathbf{e}_\nu \int_V \nabla \times \mathbf{H} \cdot \mathbf{e}_\nu dV, \end{aligned} \tag{3.17}$$

where the subscript n denotes the vector modal functions of the second type, and the Greek subscripts ν and τ for the vector modal functions of the first or third type

$$V_{n(\nu)} = \int_V \mathbf{E} \cdot \mathbf{e}_{n(\nu)} dV, \; I_{n(\tau)} = \int_V \mathbf{H} \cdot \mathbf{h}_{n(\tau)} dV. \tag{3.18}$$

The expansion coefficients in (3.17) can be determined through integration by parts

$$\int_V \nabla \times \mathbf{E} \cdot \mathbf{h}_n dV = \int_V \mathbf{E} \cdot \nabla \times \mathbf{h}_n dV + \int_S (\mathbf{E} \times \mathbf{h}_n) \cdot \mathbf{u}_n dS = k_n V_n,$$

$$\int_V \nabla \times \mathbf{E} \cdot \mathbf{h}_\tau dV = \int_V \mathbf{E} \cdot \nabla \times \mathbf{h}_\tau dV + \int_S (\mathbf{E} \times \mathbf{h}_\tau) \cdot \mathbf{u}_n dS = 0,$$

$$\int_V \nabla \times \mathbf{H} \cdot \mathbf{e}_n dS = \int_V \mathbf{H} \cdot \nabla \times \mathbf{e}_n dV + \int_S (\mathbf{H} \times \mathbf{e}_n) \cdot \mathbf{u}_n dS = k_n I_n,$$

$$\int_V \nabla \times \mathbf{H} \cdot \mathbf{e}_\nu dS = \int_V \mathbf{H} \cdot \nabla \times \mathbf{e}_\nu dV + \int_S (\mathbf{H} \times \mathbf{e}_\nu) \cdot \mathbf{u}_n dS = 0,$$

where the boundary conditions on a perfect conductor have been used. Thus, (3.17) can be written as

$$\nabla \times \mathbf{E} = \sum_n k_n V_n \mathbf{h}_n, \quad \nabla \times \mathbf{H} = \sum_n k_n I_n \mathbf{e}_n. \tag{3.19}$$

Substituting (3.16) and (3.19) into (3.15) and equating the expansion coefficients of the vector modal functions, one may find the equations relating the expansion coefficients

$$\begin{aligned} j\omega\varepsilon_e V_n - k_n I_n &= -\int_V \mathbf{J} \cdot \mathbf{e}_n dV, \\ j\omega\varepsilon_e V_\nu &= -\int_V \mathbf{J} \cdot \mathbf{e}_\nu dV, \\ j\omega\mu I_n + k_n V_n &= -\int_V \mathbf{J}_m \cdot \mathbf{h}_n dV, \\ j\omega\mu I_\tau &= -\int_V \mathbf{J}_m \cdot \mathbf{h}_\tau dV. \end{aligned} \tag{3.20}$$

Solving these equations, the field expansion coefficients can be found as follows:

$$\begin{aligned} V_n &= -\frac{j\omega\mu}{k_n^2 - k_e^2} \int_V \mathbf{J} \cdot \mathbf{e}_n dV - \frac{k_n}{k_n^2 - k_e^2} \int_V \mathbf{J}_m \cdot \mathbf{h}_n dV, \\ I_n &= \frac{k_n}{k_n^2 - k_e^2} \int_V \mathbf{J} \cdot \mathbf{e}_n dV + \frac{1}{j\omega\mu} \frac{k_e^2}{k_n^2 - k_e^2} \int_V \mathbf{J}_m \cdot \mathbf{h}_n dV, \end{aligned} \tag{3.21}$$

$$
\begin{aligned}
V_\nu &= -\frac{1}{j\omega\varepsilon_e}\int_V \mathbf{J}\cdot\mathbf{e}_\nu dV,\\
I_\tau &= -\frac{1}{j\omega\mu}\int_V \mathbf{J}_m\cdot\mathbf{h}_\tau dV,
\end{aligned}
\tag{3.22}
$$

where $k_e^2 = \omega^2\mu\varepsilon_e$. Insertion of (3.21) and (3.22) into (3.16) gives the field expressions

$$
\begin{aligned}
\mathbf{E} &= -\sum_n\left(\frac{j\omega\mu}{k_n^2-k_e^2}\int_V \mathbf{J}\cdot\mathbf{e}_n dV + \frac{k_n}{k_n^2-k_e^2}\int_V \mathbf{J}_m\cdot\mathbf{h}_n dV\right)\mathbf{e}_n\\
&\quad -\sum_\nu\left(\frac{1}{j\omega\varepsilon_e}\int_V \mathbf{J}\cdot\mathbf{e}_\nu dV\right)\mathbf{e}_\nu,\\
\mathbf{H} &= \sum_n\left(\frac{k_n}{k_n^2-k_e^2}\int_V \mathbf{J}\cdot\mathbf{e}_n dV + \frac{1}{j\omega\mu}\frac{k_e^2}{k_n^2-k_e^2}\int_V \mathbf{J}_m\cdot\mathbf{h}_n dV\right)\mathbf{h}_n\\
&\quad -\sum_\tau\left(\frac{1}{j\omega\mu}\int_V \mathbf{J}_m\cdot\mathbf{h}_\tau dV\right)\mathbf{h}_\tau.
\end{aligned}
\tag{3.23}
$$

It can be seen that the expansion coefficients before the modes of the second type contain a factor of the form $1/\left(k_n^2 - k_e^2\right)$. If the medium filled in the cavity is lossless, the fields become infinite at $k = k_n$, a resonance phenomenon well known in a resonant circuit. It is noted that the expansion coefficients before the modes of the first and third types become unbounded as the operating frequency ω approach to zero. The field expressions in (3.23) can be written in dyadic form as

$$
\begin{aligned}
\mathbf{E}(\mathbf{r}) &= -j\omega\mu\int_V \overleftrightarrow{\mathbf{G}}_e(\mathbf{r},\mathbf{r}')\cdot\mathbf{J}(\mathbf{r}')dV - \int_V \overleftrightarrow{\mathbf{G}}_{em}(\mathbf{r},\mathbf{r}')\cdot\mathbf{J}_m(\mathbf{r}')d\mathbf{r}',\\
\mathbf{H}(\mathbf{r}) &= -j\omega\varepsilon_e\int_V \overleftrightarrow{\mathbf{G}}_m(\mathbf{r},\mathbf{r}')\cdot\mathbf{J}_m(\mathbf{r}')dV + \int_V \overleftrightarrow{\mathbf{G}}_{me}(\mathbf{r},\mathbf{r}')\cdot\mathbf{J}(\mathbf{r}')dV,
\end{aligned}
\tag{3.24}
$$

where the dyadic Green's functions are defined by

$$
\overleftrightarrow{\mathbf{G}}_e(\mathbf{r},\mathbf{r}') = \sum_n \frac{1}{k_n^2-k_e^2}\mathbf{e}_n(\mathbf{r})\mathbf{e}_n(\mathbf{r}') - \sum_\nu \frac{1}{k_e^2}\mathbf{e}_\nu(\mathbf{r})\mathbf{e}_\nu(\mathbf{r}'),
\tag{3.25}
$$

$$
\overleftrightarrow{\mathbf{G}}_{em}(\mathbf{r},\mathbf{r}') = \sum_n \frac{k_n}{k_n^2-k_e^2}\mathbf{e}_n(\mathbf{r})\mathbf{h}_n(\mathbf{r}'),
\tag{3.26}
$$

$$\overleftrightarrow{\mathbf{G}}_m(\mathbf{r},\mathbf{r}') = \sum_n \frac{1}{k_n^2 - k_e^2}\mathbf{h}_n(\mathbf{r})\mathbf{h}_n(\mathbf{r}') - \frac{1}{k_e^2}\sum_\tau \mathbf{h}_\tau(\mathbf{r})\mathbf{h}_\tau(\mathbf{r}'), \tag{3.27}$$

$$\overleftrightarrow{\mathbf{G}}_{me}(\mathbf{r},\mathbf{r}') = \sum_n \frac{k_n}{k_n^2 - k_e^2}\mathbf{h}_n(\mathbf{r})\mathbf{e}_n(\mathbf{r}'). \tag{3.28}$$

3.2 Cavity with Openings

In previous studies, it is assumed that the cavity is closed by perfect conducting walls. In practical situations, cavities often have apertures through which the cavity exchanges energy with the outside. A general aperture coupling problem between the cavity region and the outside is shown in Figure 3.3a. The impressed currents $\mathbf{J}$ and $\mathbf{J}_m$ are confined in the cavity region. The impressed currents outside the cavity are denoted by $\mathbf{J}'$ and $\mathbf{J}'_m$. By Schelkunoff–Love equivalent principle, the original problem can be separated into two equivalent problems as shown in Figure 3.3b,c. Inside the cavity, the fields are produced by the impressed sources $\mathbf{J}$ and $\mathbf{J}_m$, and the equivalent magnetic current $\mathbf{J}_{ms} = \mathbf{u}_n \times \mathbf{E}$ over the aperture region S_a, with the aperture covered by an electric conductor. Outside the cavity, the fields are produced by the impressed sources $\mathbf{J}'$ and $\mathbf{J}'_m$, and the equivalent magnetic current $-\mathbf{J}_{ms}$ (the minus sign ensures that the tangential electrical field is continuous across the aperture). The fields inside the cavity can be easily obtained from (3.23). The fields outside the cavity resonator can be expressed in terms of the impressed sources $\mathbf{J}'$ and $\mathbf{J}'_m$, and the equivalent magnetic current $-\mathbf{J}_{ms}$ through

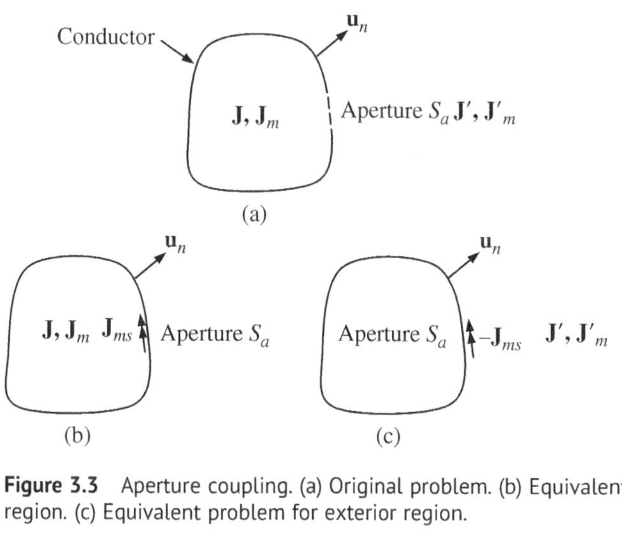

Figure 3.3 Aperture coupling. (a) Original problem. (b) Equivalent problem for cavity region. (c) Equivalent problem for exterior region.

the integral representation or the modal expansions of the fields discussed in Chapter 4. The condition that the tangential magnetic field must be continuous across the aperture may be used to determine the equivalent magnetic current $\mathbf{J}_{ms}$. In the following, the above idea will be applied to the study on the coupling between waveguides and cavity resonator.

3.2.1 Cavity with One Port

The EM field energy may be coupled to cavity resonator by a waveguide through an aperture. Let Ω be the cross section of the waveguide at $z = 0$, where the reference plane T (input terminal) intersects with the waveguide, as shown in Figure 3.4. The cross section Ω and metallic wall S' form the cavity resonator region. Suppose that the waveguide only supports the dominant mode and the waveguide extends to infinity in $-z$ direction. The dominant mode of unit amplitude is incident from $z = -\infty$, and excites higher order modes in the neighborhood of the reference plane. The transverse fields in the waveguide region $z < 0$ can be expanded in terms of the vector modal functions in the waveguide as follows:

$$
\begin{aligned}
-\mathbf{u}_z \times \mathbf{E} &= -(e^{-\gamma_1 z} + \Gamma e^{\gamma_1 z})\mathbf{u}_z \times \mathbf{e}_1^{(g)} - \sum_{m=2}^{\infty} V_m^{(g)} e^{\gamma_m z}\mathbf{u}_z \times \mathbf{e}_m^{(g)}, \\
-\mathbf{u}_z \times \mathbf{H} &= (e^{-\gamma_1 z} - \Gamma e^{\gamma_1 z})Z_{w1}^{-1}\mathbf{e}_1^{(g)} - \sum_{m=2}^{\infty} V_m^{(g)} Z_{wm}^{-1} e^{\gamma_m z}\mathbf{e}_m^{(g)},
\end{aligned}
\tag{3.29}
$$

where the superscript (g) signifies the quantities related to waveguide, $V_m^{(g)}$ are the modal voltages, and Γ is the refection coefficient of the dominant mode at $z = 0$. Assume that the cavity does not contain any impressed sources. The fields in the cavity region $z > 0$ can then be obtained from (3.23):

$$
\begin{aligned}
\mathbf{E} &= \sum_n \frac{k_n}{k_n^2 - k_e^2}\mathbf{e}_n^{(r)} \int_\Omega \mathbf{J}_{ms} \cdot \mathbf{h}_n^{(r)} d\Omega, \\
\mathbf{H} &= \sum_n \frac{1}{j\omega\mu} \frac{k_e^2}{k_n^2 - k_e^2}\mathbf{h}_n^{(r)} \int_\Omega \mathbf{J}_{ms} \cdot \mathbf{h}_n^{(r)} d\Omega - \sum_\tau \frac{1}{j\omega\mu}\mathbf{h}_\tau^{(r)} \int_\Omega \mathbf{J}_{ms} \cdot \mathbf{h}_\tau^{(r)} d\Omega,
\end{aligned}
\tag{3.30}
$$

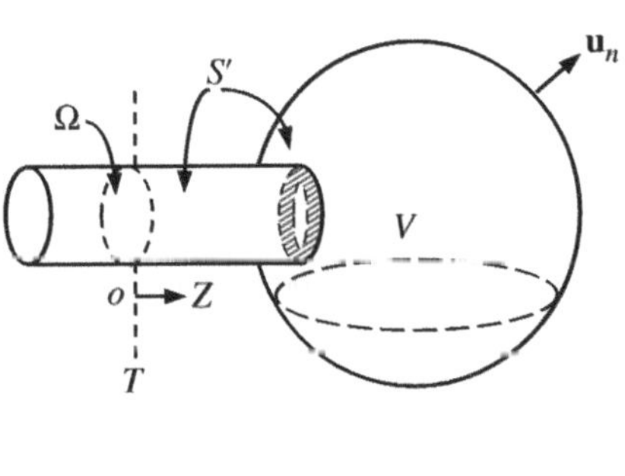

Figure 3.4 Coupling between waveguide and cavity resonator.

where the superscript (r) denotes the resonator modes, and $\mathbf{J}_{ms} = -\mathbf{u}_z \times \mathbf{E}$ is the equivalent magnetic current at the input terminal. Making use of the first equation of (3.29), the integrals over the aperture in (3.30) can be expressed as

$$\int_\Omega \mathbf{J}_{ms} \cdot \mathbf{h}_n^{(r)} d\Omega = -\sum_{m=1}^{\infty} V_m^{(g)} \int_\Omega \mathbf{u}_z \times \mathbf{e}_m^{(g)} \cdot \mathbf{h}_n^{(r)} d\Omega,$$

$$\int_\Omega \mathbf{J}_{ms} \cdot \mathbf{h}_\tau^{(r)} d\Omega = -\sum_{m=1}^{\infty} V_m^{(g)} \int_\Omega \mathbf{u}_z \times \mathbf{e}_m^{(g)} \cdot \mathbf{h}_\tau^{(r)} d\Omega,$$

where $V_1^{(g)} = 1 + \Gamma$ stands for the terminal voltage for the dominant mode in the waveguide. Since the tangential magnetic field must be continuous at $z = 0$, an equation relating the modes in the waveguide and those in the cavity can be obtained from the second equations of (3.29) and (3.30):

$$\begin{aligned} I_1^{(g)} \mathbf{e}_1^{(g)} - \sum_{m=2}^{\infty} V_{m,g} Z_{wm}^{-1} \mathbf{e}_m^{(g)} &= -\sum_n \frac{1}{j\omega\mu} \frac{k_e^2}{k_n^2 - k_e^2} \mathbf{u}_z \times \mathbf{h}_n^{(r)} \int_\Omega \mathbf{J}_{ms} \cdot \mathbf{h}_n^{(r)} d\Omega \\ &\quad + \sum_\tau \frac{1}{j\omega\mu} \mathbf{u}_z \times \mathbf{h}_\tau^{(r)} \int_\Omega \mathbf{J}_{ms} \cdot \mathbf{h}_\tau^{(r)} d\Omega, \end{aligned}$$

where $I_1^{(g)} = (1-\Gamma) Z_{w1}^{-1}$ represents the modal current for the dominant mode in the waveguide. Multiplying both sides by $\mathbf{e}_1^{(g)}$ and taking the integration over the input terminal, the terminal current can be found as follows:

$$\begin{aligned} I_1^{(g)} &= \sum_n \frac{1}{j\omega\mu} \frac{k_e^2}{k_n^2 - k_e^2} \int_\Omega \mathbf{u}_z \times \mathbf{e}_1^{(g)} \cdot \mathbf{h}_n^{(r)} d\Omega \int_\Omega \mathbf{J}_{ms} \cdot \mathbf{h}_n^{(r)} d\Omega \\ &\quad - \sum_\tau \frac{1}{j\omega\mu} \int_\Omega \mathbf{u}_z \times \mathbf{e}_1^{(g)} \cdot \mathbf{h}_\tau^{(r)} d\Omega \int_\Omega \mathbf{J}_{ms} \cdot \mathbf{h}_\tau^{(r)} d\Omega. \end{aligned}$$

The input admittance is then given by the ratio of the terminal current over the terminal voltage

$$\begin{aligned} Y &= \frac{I_1^{(g)}}{V_1^{(g)}} = \frac{1}{Z_{w1}} \frac{1-\Gamma}{1+\Gamma} \\ &= -\sum_n \frac{1}{j\omega\mu} \frac{k_e^2}{k_n^2 - k_e^2} \left(\sum_{m=1}^{\infty} \frac{V_m^{(g)}}{V_1^{(g)}} \int_\Omega \mathbf{u}_z \times \mathbf{e}_m^{(g)} \cdot \mathbf{h}_n^{(r)} d\Omega \right) \int_\Omega \mathbf{u}_z \times \mathbf{e}_1^{(g)} \cdot \mathbf{h}_n^{(r)} d\Omega \\ &\quad + \sum_\tau \frac{1}{j\omega\mu} \left(\sum_{m=1}^{\infty} \frac{V_m^{(g)}}{V_1^{(g)}} \int_\Omega \mathbf{u}_z \times \mathbf{e}_m^{(g)} \cdot \mathbf{h}_\tau^{(r)} d\Omega \right) \int_\Omega \mathbf{u}_z \times \mathbf{e}_1^{(g)} \cdot \mathbf{h}_\tau^{(r)} d\Omega. \end{aligned} \tag{3.31}$$

The above expression for the input admittance can be rewritten as

$$Y=\sum_n \frac{1}{j\omega_n\mu}\frac{j+\dfrac{\sigma}{\omega\varepsilon}}{\dfrac{1}{Q_n}+j\left(\dfrac{\omega}{\omega_n}-\dfrac{\omega_n}{\omega}\right)}\left(\sum_{m=1}^{\infty}\frac{V_m^{(g)}}{V_1^{(g)}}\int_\Omega \mathbf{u}_z\times\mathbf{e}_m^{(g)}\cdot\mathbf{h}_n^{(r)}d\Omega\right)\int_\Omega \mathbf{u}_z\times\mathbf{e}_1^{(g)}\cdot\mathbf{h}_n^{(r)}d\Omega$$
$$+\sum_\tau \frac{1}{j\omega\mu}\left(\sum_{m=1}^{\infty}\frac{V_m^{(g)}}{V_1^{(g)}}\int_\Omega \mathbf{u}_z\times\mathbf{e}_m^{(g)}\cdot\mathbf{h}_\tau^{(r)}d\Omega\right)\int_\Omega \mathbf{u}_z\times\mathbf{e}_1^{(g)}\cdot\mathbf{h}_\tau^{(r)}d\Omega, \tag{3.32}$$

where $Q_n=\omega_n\varepsilon/\sigma$. If only the dominant mode exists at the input terminal, input admittance (3.32) reduces to

$$Y=\sum_n \frac{1}{j\omega_n\mu}\frac{j+\dfrac{\sigma}{\omega\varepsilon}}{\dfrac{1}{Q_n}+j\left(\dfrac{\omega}{\omega_n}-\dfrac{\omega_n}{\omega}\right)}\left(\int_\Omega \mathbf{u}_z\times\mathbf{e}_1^{(g)}\cdot\mathbf{h}_n^{(r)}d\Omega\right)^2$$
$$+\sum_\tau \frac{1}{j\omega\mu}\left(\int_\Omega \mathbf{u}_z\times\mathbf{e}_1^{(g)}\cdot\mathbf{h}_\tau^{(r)}d\Omega\right)^2. \tag{3.33}$$

Remark 3.1 Slater studied the input impedance of a cavity resonator [7]. In his expression for the input impedance, a capacitance term is missing due to the incorrect use of vector modal functions. Teichmann and Wigner investigated the field expansions in a cavity excited through holes [8]. They found that the natural resonance modes (the second type) were incomplete and it was necessary to add irrotational vector modal functions in order to permit expansion of relevant fields. The contribution of this added component to the input impedance is an inductance term (see the second term on the right-hand side of (3.33)), which must be included for the cavity type 3 shown in Figure 3.1c. □

3.2.2 Cavity with Two Ports

Consider now a cavity with two openings, each of which is connected with a waveguide. Let Ω_i and $T_i (i=1,2)$ denote the cross section and input terminal of waveguide i, respectively. The cross sections Ω_1 and Ω_2 at the input terminals, and the metallic wall S' consisting of the wall of the cavity and part of the wall of the waveguide, form the cavity resonator region, as illustrated in Figure 3.5. It will be assumed that both the waveguides only support the dominant mode. The dominant mode of unit amplitude is incident from $z_1=-\infty$ in the waveguide 1, which is coupled to the waveguide 2 through the cavity resonator. The transverse fields in

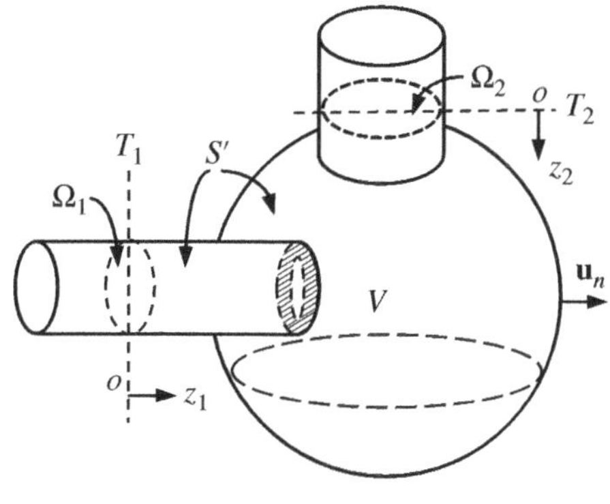

Figure 3.5 Cavity resonator with two ports.

the region $z_1 < 0$ in waveguide 1 can be expanded in the vector modal functions $\left\{\mathbf{e}_n^{(g)}\right\}$ as follows:

$$-\mathbf{u}_{z_1} \times \mathbf{E} = -(e^{-\gamma_1 z_1} + \Gamma e^{\gamma_1 z_1})\mathbf{u}_{z_1} \times \mathbf{e}_1^{(g)} - \sum_{m=2}^{\infty} V_m^{(g)} e^{\gamma_m z_1} \mathbf{u}_{z_1} \times \mathbf{e}_m^{(g)},$$
$$-\mathbf{u}_{z_1} \times \mathbf{H} = (e^{-\gamma_1 z_1} - \Gamma e^{\gamma_1 z_1}) Z_{w1}^{-1} \mathbf{e}_1^{(g)} - \sum_{m=2}^{\infty} V_m^{(g)} Z_{wm}^{-1} e^{\gamma_m z_1} \mathbf{e}_m^{(g)},$$
(3.34)

while the transverse fields in the region $z_2 < 0$ in waveguide 2 can be expressed in the vector modal functions $\left\{\mathbf{e}'^{(g)}_n\right\}$ as

$$-\mathbf{u}_{z_2} \times \mathbf{E}' = -V'^{(g)}_1 e^{\gamma'_1 z_2} \mathbf{u}_{z_2} \times \mathbf{e}'^{(g)}_1 - \sum_{m=2}^{\infty} V'^{(g)}_m e^{\gamma'_m z_2} \mathbf{u}_{z_2} \times \mathbf{e}'^{(g)}_m,$$
$$-\mathbf{u}_{z_2} \times \mathbf{H}' = -V'^{(g)}_1 e^{\gamma'_1 z_2} Z'^{-1}_{w1} \mathbf{e}'^{(g)}_1 - \sum_{m=2}^{\infty} V'^{(g)}_m Z'^{-1}_{wm} e^{\gamma'_m z_2} \mathbf{e}'^{(g)}_m,$$
(3.35)

If the cavity does not contain any impressed sources, the fields in the cavity region can be expressed in terms of the equivalent magnetic currents on the input terminals, and can be obtained from (3.23):

$$\mathbf{E} = \sum_n \frac{k_n}{k_n^2 - k_e^2} \mathbf{e}_n^{(r)} \left(\int_{\Omega_1} \mathbf{J}_{ms} \cdot \mathbf{h}_n^{(r)} d\Omega + \int_{\Omega_2} \mathbf{J}_{ms} \cdot \mathbf{h}_n^{(r)} d\Omega \right),$$
$$\mathbf{H} = \sum_n \frac{1}{j\omega\mu} \frac{k_e^2}{k_n^2 - k_e^2} \mathbf{h}_n^{(r)} \left(\int_{\Omega_1} \mathbf{J}_{ms} \cdot \mathbf{h}_n^{(r)} d\Omega + \int_{\Omega_2} \mathbf{J}_{ms} \cdot \mathbf{h}_n^{(r)} d\Omega \right)$$
$$-\sum_\tau \frac{1}{j\omega\mu} \mathbf{h}_\tau^{(r)} \left(\int_{\Omega_1} \mathbf{J}_{ms} \cdot \mathbf{h}_\tau^{(r)} d\Omega + \int_{\Omega_2} \mathbf{J}_{ms} \cdot \mathbf{h}_\tau^{(r)} d\Omega \right).$$
(3.36)

The integrals containing the equivalent magnetic currents can be calculated by using the first equation of (3.34) and (3.35):

$$\int_{\Omega_1} \mathbf{J}_{ms} \cdot \mathbf{h}_n^{(r)} d\Omega = -\sum_{m=1}^{\infty} V_m^{(g)} \int_{\Omega_1} \mathbf{u}_{z_1} \times \mathbf{e}_m^{(g)} \cdot \mathbf{h}_n^{(r)} d\Omega,$$

$$\int_{\Omega_2} \mathbf{J}_{ms} \cdot \mathbf{h}_n^{(r)} d\Omega = -\sum_{m=1}^{\infty} V'^{(g)}_m \int_{\Omega_2} \mathbf{u}_{z_2} \times \mathbf{e}'^{(g)}_m \cdot \mathbf{h}_n^{(r)} d\Omega,$$

$$\int_{\Omega_1} \mathbf{J}_{ms} \cdot \mathbf{h}_\tau^{(r)} d\Omega = -\sum_{m=1}^{\infty} V_m^{(g)} \int_{\Omega_1} \mathbf{u}_{z_1} \times \mathbf{e}_m^{(g)} \cdot \mathbf{h}_\tau^{(r)} d\Omega,$$

$$\int_{\Omega_2} \mathbf{J}_{ms} \cdot \mathbf{h}_\tau^{(r)} d\Omega = -\sum_{m=1}^{\infty} V'^{(g)}_m \int_{\Omega_2} \mathbf{u}_{z_2} \times \mathbf{e}'^{(g)}_m \cdot \mathbf{h}_\tau^{(r)} d\Omega.$$

Since the tangential magnetic field must be continuous at the input terminals $z_1 = 0$ and $z_2 = 0$, from the second equations of (3.34)–(3.36), one may find

$$\begin{aligned}(1-\Gamma)Z_{w1}^{-1}\mathbf{e}_1^{(g)} - \sum_{m=2}^{\infty} V_{m,g} Z_{wm}^{-1}\mathbf{e}_m^{(g)} &= -\sum_n \frac{1}{j\omega\mu}\frac{k_e^2}{k_n^2-k_e^2}\mathbf{u}_{z_1} \\ &\quad\times \mathbf{h}_n^{(r)}\left(\int_{\Omega_1} \mathbf{J}_{ms}\cdot\mathbf{h}_n^{(r)} d\Omega + \int_{\Omega_2} \mathbf{J}_{ms}\cdot\mathbf{h}_n^{(r)} d\Omega\right) \\ &\quad+ \sum_\tau \frac{1}{j\omega\mu}\mathbf{u}_{z_1}\times\mathbf{h}_\tau^{(r)}\left(\int_{\Omega_1} \mathbf{J}_{ms}\cdot\mathbf{h}_\tau^{(r)} d\Omega + \int_{\Omega_2} \mathbf{J}_{ms}\cdot\mathbf{h}_\tau^{(r)} d\Omega\right),\end{aligned} \tag{3.37}$$

$$\begin{aligned}-V'^{(g)}_1 Z'^{-1}_{w1}\mathbf{e}'^{(g)}_1 - \sum_{m=2}^{\infty} V'^{(g)}_m Z'^{-1}_{wm}\mathbf{e}'^{(g)}_m &= -\sum_n \frac{1}{j\omega\mu}\frac{k_e^2}{k_n^2-k_e^2}\mathbf{u}_{z_2} \\ &\quad\times \mathbf{h}_n^{(r)}\left(\int_{\Omega_1} \mathbf{J}_{ms}\cdot\mathbf{h}_n^{(r)} d\Omega + \int_{\Omega_2} \mathbf{J}_{ms}\cdot\mathbf{h}_n^{(r)} d\Omega\right) \\ &\quad+ \sum_\tau \frac{1}{j\omega\mu}\mathbf{u}_{z_2}\times\mathbf{h}_\tau^{(r)}\left(\int_{\Omega_1} \mathbf{J}_{ms}\cdot\mathbf{h}_\tau^{(r)} d\Omega + \int_{\Omega_2} \mathbf{J}_{ms}\cdot\mathbf{h}_\tau^{(r)} d\Omega\right).\end{aligned} \tag{3.38}$$

Taking the scalar product of $\mathbf{e}_1^{(g)}$ with (3.37) and $\mathbf{e}'^{(g)}_1$ with (3.38), and integrating the resultants over Ω_1 and Ω_2, respectively, yield

$$(1-\Gamma)Z_{w1}^{-1} = \sum_n \frac{k_e^2 \int\limits_{\Omega_1} \mathbf{u}_{z_1} \times \mathbf{e}_1^{(g)} \cdot \mathbf{h}_n^{(r)} d\Omega}{j\omega\mu\left(k_n^2 - k_e^2\right)}$$

$$\left(- \sum_{m=1}^{\infty} V_m^{(g)} \int\limits_{\Omega_1} \mathbf{u}_{z_1} \times \mathbf{e}_m^{(g)} \cdot \mathbf{h}_n^{(r)} d\Omega - \sum_{m=1}^{\infty} V'^{(g)}_m \int\limits_{\Omega_2} \mathbf{u}_{z_2} \times \mathbf{e}'^{(g)}_m \cdot \mathbf{h}_n^{(r)} d\Omega \right)$$

$$- \sum_\tau \frac{\int\limits_{\Omega_1} \mathbf{u}_{z_1} \times \mathbf{e}_1^{(g)} \cdot \mathbf{h}_\tau^{(r)} d\Omega}{j\omega\mu}$$

$$\left(- \sum_{m=1}^{\infty} V_m^{(g)} \int\limits_{\Omega_1} \mathbf{u}_{z_1} \times \mathbf{e}_m^{(g)} \cdot \mathbf{h}_\tau^{(r)} d\Omega - \sum_{m=1}^{\infty} V'^{(g)}_m \int\limits_{\Omega_2} \mathbf{u}_{z_2} \times \mathbf{e}'^{(g)}_m \cdot \mathbf{h}_\tau^{(r)} d\Omega \right), \tag{3.39}$$

$$- V'^{(g)}_1 Z'^{-1}_{w1} = \sum_n \frac{k_e^2 \int\limits_{\Omega_2} \mathbf{u}_{z_2} \times \mathbf{e}'^{(g)}_1 \cdot \mathbf{h}_n^{(r)} d\Omega}{j\omega\mu\left(k_n^2 - k_e^2\right)}$$

$$\left(- \sum_{m=1}^{\infty} V_m^{(g)} \int\limits_{\Omega_1} \mathbf{u}_{z_1} \times \mathbf{e}_m^{(g)} \cdot \mathbf{h}_n^{(r)} d\Omega - \sum_{m=1}^{\infty} V'^{(g)}_m \int\limits_{\Omega_2} \mathbf{u}_{z_2} \times \mathbf{e}'^{(g)}_m \cdot \mathbf{h}_n^{(r)} d\Omega \right)$$

$$- \sum_\tau \frac{\int\limits_{\Omega_2} \mathbf{u}_{z_2} \times \mathbf{e}'^{(g)}_1 \cdot \mathbf{h}_\tau^{(r)} d\Omega}{j\omega\mu}$$

$$\left(- \sum_{m=1}^{\infty} V_m^{(g)} \int\limits_{\Omega_1} \mathbf{u}_{z_1} \times \mathbf{e}_m^{(g)} \cdot \mathbf{h}_\tau^{(r)} d\Omega - \sum_{m=1}^{\infty} V'^{(g)}_m \int\limits_{\Omega_2} \mathbf{u}_{z_2} \times \mathbf{e}'^{(g)}_m \cdot \mathbf{h}_\tau^{(r)} d\Omega \right). \tag{3.40}$$

Using the terminal voltages and currents for the dominant modes in waveguides 1 and 2, defined by

$$V_1 = V_1^{(g)},\ V_2 = V'^{(g)}_1,\ I_1 = (1-\Gamma)Z_{w1}^{-1},\ I_2 = -V'^{(g)}_1 Z'^{-1}_{w1},$$

(3.39) and (3.40) can be rewritten in admittance matrix form as

$$\begin{bmatrix} I_1 \\ I_2 \end{bmatrix} = \begin{bmatrix} Y_{11} & Y_{12} \\ Y_{21} & Y_{22} \end{bmatrix} \begin{bmatrix} V_1 \\ V_2 \end{bmatrix},$$

where the elements of the admittance matrix are given by

$$Y_{11} = -\sum_n \frac{k_e^2 \int\limits_{\Omega_1} \mathbf{u}_{z_1} \times \mathbf{e}_1^{(g)} \cdot \mathbf{h}_n^{(r)} d\Omega}{j\omega\mu\left(k_n^2 - k_e^2\right)} \left(\sum_{m=1}^{\infty} \frac{V_m^{(g)}}{V_1^{(g)}} \int\limits_{\Omega_1} \mathbf{u}_{z_1} \times \mathbf{e}_m^{(g)} \cdot \mathbf{h}_n^{(r)} d\Omega \right)$$

$$+ \sum_\tau \frac{\int\limits_{\Omega_1} \mathbf{u}_{z_1} \times \mathbf{e}_1^{(g)} \cdot \mathbf{h}_\tau^{(r)} d\Omega}{j\omega\mu} \left(\sum_{m=1}^{\infty} \frac{V_m^{(g)}}{V_1^{(g)}} \int\limits_{\Omega_1} \mathbf{u}_{z_1} \times \mathbf{e}_m^{(g)} \cdot \mathbf{h}_\tau^{(r)} d\Omega \right),$$

$$Y_{12} = -\sum_n \frac{k_e^2 \int\limits_{\Omega_1} \mathbf{u}_{z_1} \times \mathbf{e}_1^{(g)} \cdot \mathbf{h}_n^{(r)} d\Omega}{j\omega\mu\left(k_n^2 - k_e^2\right)} \left(\sum_{m=1}^{\infty} \frac{V_m'^{(g)}}{V_1'^{(g)}} \int\limits_{\Omega_2} \mathbf{u}_{z_2} \times \mathbf{e}_m'^{(g)} \cdot \mathbf{h}_n^{(r)} d\Omega \right)$$

$$+ \sum_\tau \frac{\int\limits_{\Omega_1} \mathbf{u}_{z_1} \times \mathbf{e}_1^{(g)} \cdot \mathbf{h}_\tau^{(r)} d\Omega}{j\omega\mu} \left(\sum_{m=1}^{\infty} \frac{V_m'^{(g)}}{V_1'^{(g)}} \int\limits_{\Omega_2} \mathbf{u}_{z_2} \times \mathbf{e}_m'^{(g)} \cdot \mathbf{h}_\tau^{(r)} d\Omega \right),$$

$$Y_{21} = -\sum_n \frac{k_e^2 \int\limits_{\Omega_2} \mathbf{u}_{z_2} \times \mathbf{e}_1'^{(g)} \cdot \mathbf{h}_n^{(r)} d\Omega}{j\omega\mu\left(k_n^2 - k_e^2\right)} \left(\sum_{m=1}^{\infty} \frac{V_m^{(g)}}{V_1^{(g)}} \int\limits_{\Omega_1} \mathbf{u}_{z_1} \times \mathbf{e}_m^{(g)} \cdot \mathbf{h}_n^{(r)} d\Omega \right)$$

$$+ \sum_\tau \frac{\int\limits_{\Omega_2} \mathbf{u}_{z_2} \times \mathbf{e}_1'^{(g)} \cdot \mathbf{h}_\tau^{(r)} d\Omega}{j\omega\mu} \left(\sum_{m=1}^{\infty} \frac{V_m^{(g)}}{V_1^{(g)}} \int\limits_{\Omega_1} \mathbf{u}_{z_1} \times \mathbf{e}_m^{(g)} \cdot \mathbf{h}_\tau^{(r)} d\Omega \right),$$

$$Y_{22} = -\sum_n \frac{k_e^2 \int\limits_{\Omega_2} \mathbf{u}_{z_2} \times \mathbf{e}_1'^{(g)} \cdot \mathbf{h}_n^{(r)} d\Omega}{j\omega\mu\left(k_n^2 - k_e^2\right)} \left(\sum_{m=1}^{\infty} \frac{V_m'^{(g)}}{V_1'^{(g)}} \int\limits_{\Omega_1} \mathbf{u}_{z_1} \times \mathbf{e}_m'^{(g)} \cdot \mathbf{h}_n^{(r)} d\Omega \right)$$

$$+ \sum_\tau \frac{\int\limits_{\Omega_2} \mathbf{u}_{z_2} \times \mathbf{e}_1'^{(g)} \cdot \mathbf{h}_\tau^{(r)} d\Omega}{j\omega\mu} \left(\sum_{m=1}^{\infty} \frac{V_m'^{(g)}}{V_1'^{(g)}} \int\limits_{\Omega_2} \mathbf{u}_{z_2} \times \mathbf{e}_m'^{(g)} \cdot \mathbf{h}_\tau^{(r)} d\Omega \right).$$

3.3 Waveguide Cavity Resonator

The evaluation of the vector modal functions in an arbitrary metal cavity is not easy, and one needs to resort to numerical methods. When the metal cavity consists of a section of a uniform metal waveguide, the analysis of the field distribution in the metal cavity can be carried out by means of the modal theory of waveguide. Notice that only the cavity types 1 and 3 shown in Figure 3.1 are possible for forming a waveguide cavity resonator.

3.3.1 Field Expansions by Vector Modal Functions of Waveguide

Consider a waveguide cavity with a perfect electric wall of length L, as sketched in Figure 3.6. The EM fields and the current sources $\mathbf{J}$ and $\mathbf{J}_m$ inside the cavity can be expanded in terms of the vector modal functions $\{\mathbf{e}_n\}$ in the waveguide

$$\begin{aligned}
\mathbf{E}(\mathbf{r}) &= \sum_n V_n(z)\mathbf{e}_n(\boldsymbol{\rho}) + \sum_n e_n(z)\mathbf{u}_z \frac{\nabla \cdot \mathbf{e}_n(\boldsymbol{\rho})}{k_{cn}}, \\
\mathbf{H}(\mathbf{r}) &= \sum_n I_n(z)\mathbf{u}_z \times \mathbf{e}_n(\boldsymbol{\rho}) + \frac{\mathbf{u}_z}{\sqrt{\Omega}} \int_\Omega \frac{\mathbf{u}_z}{\sqrt{\Omega}} \cdot \mathbf{H}(\boldsymbol{\rho}', z) d\Omega + \sum_n h_n(z) \frac{\nabla \times \mathbf{e}_n(\boldsymbol{\rho})}{k_{cn}},
\end{aligned} \tag{3.41}$$

$$\begin{aligned}
\nabla \times \mathbf{E}(\mathbf{r}) &= \sum_n \left[\frac{\partial V_n(z)}{\partial z} + k_{cn} e_n(z)\right] \mathbf{u}_z \times \mathbf{e}_n(\boldsymbol{\rho}) + \sum_n k_{cn} V_n(z) \frac{\nabla \times \mathbf{e}_n(\boldsymbol{\rho})}{k_{cn}}, \\
\nabla \times \mathbf{H}(\mathbf{r}) &= \sum_n \left[-\frac{\partial I_n(z)}{\partial z} + k_{cn} h_n(z)\right] \mathbf{e}_n(\boldsymbol{\rho}) + \sum_n k_{cn} I_n(z) \mathbf{u}_z \frac{\nabla \cdot \mathbf{e}_n(\boldsymbol{\rho})}{k_{cn}},
\end{aligned} \tag{3.42}$$

$$\begin{aligned}
\mathbf{J}(\mathbf{r}) &= \sum_n j_n(z)\mathbf{e}_n(\boldsymbol{\rho}) + \sum_n j_{zn}(z)\mathbf{u}_z \frac{\nabla \cdot \mathbf{e}_n(\boldsymbol{\rho})}{k_{cn}}, \\
\mathbf{J}_m(\mathbf{r}) &= \sum_n j'_n(z)\mathbf{u}_z \times \mathbf{e}_n(\boldsymbol{\rho}) + \frac{\mathbf{u}_z}{\sqrt{\Omega}} \int_\Omega \frac{\mathbf{u}_z}{\sqrt{\Omega}} \cdot \mathbf{J}_m(\boldsymbol{\rho}', z) d\Omega + \sum_n j'_{zn}(z) \frac{\nabla \times \mathbf{e}_n(\boldsymbol{\rho})}{k_{cn}},
\end{aligned} \tag{3.43}$$

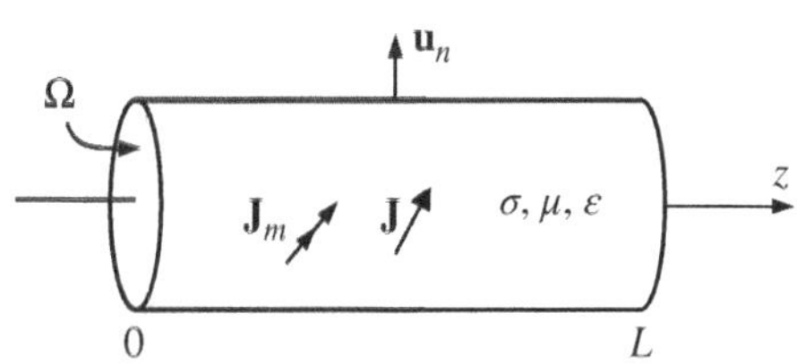

Figure 3.6 A metal cavity formed by a waveguide.

where $\mathbf{r} = (\boldsymbol{\rho}, z)$ and $\boldsymbol{\rho} = (x, y)$, and the expansion coefficients are determined by

$$V_n = \int_\Omega \mathbf{E} \cdot \mathbf{e}_n d\Omega, \qquad I_n = \int_\Omega \mathbf{H} \cdot \mathbf{u}_z \times \mathbf{e}_n d\Omega,$$

$$e_n = \int_\Omega \mathbf{E} \cdot \mathbf{u}_z \frac{\nabla \cdot \mathbf{e}_n}{k_{cn}} d\Omega, \quad h_n = \int_\Omega \mathbf{H} \cdot \frac{\nabla \times \mathbf{e}_n}{k_{cn}} d\Omega,$$

$$j_n = \int_\Omega \mathbf{J} \cdot \mathbf{e}_n d\Omega, \qquad j'_n = \int_\Omega \mathbf{J}_m \cdot \mathbf{u}_z \times \mathbf{e}_n d\Omega,$$

$$j_{zn} = \int_\Omega \mathbf{J} \cdot \mathbf{u}_z \frac{\nabla \cdot \mathbf{e}_n}{k_{cn}} d\Omega, \quad j'_{zn} = \int_\Omega \mathbf{J}_m \cdot \frac{\nabla \times \mathbf{e}_n}{k_{cn}} d\Omega.$$

On substitution of (3.41)–(3.43) into the generalized Maxwell equations

$$\begin{aligned} \nabla \times \mathbf{E} &= -j\omega\mu\mathbf{H} - \mathbf{J}_m, \\ \nabla \times \mathbf{H} &= (\sigma + j\omega\varepsilon)\mathbf{E} + \mathbf{J}, \end{aligned} \tag{3.44}$$

one may find

$$\begin{aligned} &\sum_n \left(\frac{\partial V_n}{\partial z} + k_{cn} e_n\right) \mathbf{u}_z \times \mathbf{e}_n + \sum_n k_{cn} V_n \frac{\nabla \times \mathbf{e}_n}{k_{cn}} \\ &= -j\omega\mu \sum_n I_n \mathbf{u}_z \times \mathbf{e}_n - j\omega\mu \frac{\mathbf{u}_z}{\sqrt{\Omega}} \int_\Omega \frac{\mathbf{u}_z}{\sqrt{\Omega}} \cdot \mathbf{H} d\Omega \\ &\quad - j\omega\mu \sum_n h_n \frac{\nabla \times \mathbf{e}_n}{k_{cn}} - \sum_n j'_n \mathbf{u}_z \times \mathbf{e}_n - \frac{\mathbf{u}_z}{\sqrt{\Omega}} \int_\Omega \frac{\mathbf{u}_z}{\sqrt{\Omega}} \cdot \mathbf{J}_m d\Omega - \sum_n j'_{zn} \frac{\nabla \times \mathbf{e}_n}{k_{cn}}, \end{aligned} \tag{3.45}$$

$$\begin{aligned} \sum_n \left(-\frac{\partial I_n}{\partial z} + k_{cn} h_n\right) \mathbf{e}_n + \sum_n k_{cn} I_n \mathbf{u}_z \frac{\nabla \cdot \mathbf{e}_n}{k_{cn}} &= \sum_n (\sigma + j\omega\varepsilon) V_n \mathbf{e}_n \\ &+ \sum_n (\sigma + j\omega\varepsilon) e_n \mathbf{u}_z \frac{\nabla \cdot \mathbf{e}_n}{k_{cn}} + \sum_n j_n \mathbf{e}_n + \sum_n j_{zn} \mathbf{u}_z \frac{\nabla \cdot \mathbf{e}_n}{k_{cn}}. \end{aligned} \tag{3.46}$$

The equations for the expansion coefficients can be obtained by comparing the transverse and longitudinal components on both sides of Eqs. (3.45) and (3.46):

$$-\frac{\partial I_n}{\partial z} + k_{cn} h_n = (\sigma + j\omega\varepsilon) V_n + j_n, \tag{3.47}$$

$$k_{cn} I_n = (\sigma + j\omega\varepsilon) e_n + j_{zn}, \text{ for TM modes only,} \tag{3.48}$$

$$\frac{\partial V_n}{\partial z} + e_n k_{cn} = -j\omega\mu I_n - j'_n, \tag{3.49}$$

$$k_{cn} V_n = -j\omega\mu h_n - j'_{zn}, \text{ for TE modes only,} \tag{3.50}$$

$$-j\omega\mu \int_\Omega \frac{\mathbf{u}_z}{\sqrt{\Omega}} \cdot \mathbf{H} d\Omega = \int_\Omega \frac{\mathbf{u}_z}{\sqrt{\Omega}} \cdot \mathbf{J}_m d d\Omega, \text{ for TE modes only.} \tag{3.51}$$

For the TEM mode, the modal voltage and current must satisfy

$$\begin{aligned} \frac{\partial V_n^{TEM}}{\partial z} &= -j\omega\mu I_n^{TEM} - j_n'^{\,TEM}, \\ \frac{\partial I_n^{TEM}}{\partial z} &= -(\sigma + j\omega\varepsilon) V_n^{TEM} - j_n^{TEM}, \end{aligned} \tag{3.52}$$

from (3.47) and (3.49). From now on, the superscripts TEM, TE, and TM will be used to signify the quantities related to the TEM, TE, and TM modes in the waveguide. The equation for the modal voltage can be easily found from (3.52) by eliminating the modal current

$$\frac{\partial^2 V_n^{TEM}}{\partial z^2} + k_e^2 V_n^{TEM} = j\omega\mu j_n^{TEM} - \frac{\partial}{\partial z} j_n'^{\,TEM}. \tag{3.53}$$

Once the modal voltage is known, the modal current I_n^{TEM} can then be determined from (3.49).

For the TE modes, the equations for the modal voltage and current are given by

$$\begin{aligned} \frac{\partial V_n^{TE}}{\partial z} &= -j\omega\mu I_n^{TE} - j_n'^{\,TE}, \\ \frac{\partial I_n^{TE}}{\partial z} - k_{cn}^{TE} h_n^{TE} &= -(\sigma + j\omega\varepsilon) V_n^{TE} - j_n^{TE}, \\ j\omega\mu h_n^{TE} &= -k_{cn}^{TE} V_n^{TE} - j'^{\,TE}_{zn}, \\ -j\omega\mu \int_\Omega \frac{\mathbf{u}_z}{\sqrt{\Omega}} \cdot \mathbf{H} d\Omega &= \int_\Omega \frac{\mathbf{u}_z}{\sqrt{\Omega}} \cdot \mathbf{J}_m d\Omega, \end{aligned} \tag{3.54}$$

from (3.47), (3.49)–(3.51). After elimination, the modal voltage V_n^{TE} is found to satisfy the inhomogeneous wave equation

$$\frac{\partial^2 V_n^{TE}}{\partial z^2} + \left[k_e^2 - \left(k_{cn}^{TE}\right)^2\right] V_n^{TE} = j\omega\mu j_n^{TE} - \frac{\partial}{\partial z} j'^{\,TE}_n + k_{cn}^{TE} j'^{\,TE}_{zn}. \tag{3.55}$$

The modal current I_n^{TE} can be determined from the modal voltage V_n^{TE} by using the first equation of (3.54).

For the TM mode, the modal voltage and current satisfy

$$\begin{aligned}
&\frac{\partial V_n^{TM}}{\partial z} + k_{cn}^{TM} e_n^{TM} = -j\omega\mu I_n^{TM} - j'^{TM}_n, \\
&\frac{\partial I_n^{TM}}{\partial z} = -(\sigma + j\omega\varepsilon) V_n^{TM} - j_n^{TM}, \\
&(\sigma + j\omega\varepsilon) e_n^{TM} = k_{cn}^{TM} I_n^{TM} - j_{zn}^{TM},
\end{aligned} \tag{3.56}$$

from (3.47) to (3.49). The modal current I_n^{TM} satisfies the wave equation

$$\frac{\partial^2 I_n^{TM}}{\partial z^2} + \left[k_e^2 - \left(k_{cn}^{TM}\right)^2\right] I_n^{TM} = -\frac{\partial}{\partial z} j_n^{TM} - k_{cn} j_{zn}^{TM} + (\sigma + j\omega\varepsilon) j'^{TM}_n. \tag{3.57}$$

The modal voltage V_n^{TM} can then be determined by the second equation of (3.56).

Since the tangential electric field on the electric conductor must be zero, the modal voltage satisfies the homogeneous Dirichlet boundary conditions at both ends of the cavity resonator

$$V_n\big|_{z=0} = V_n\big|_{z=L} = 0. \tag{3.58}$$

According to (3.47) and that the normal component of the magnetic field on an electric conductor is zero, the modal current must satisfy the homogeneous Neumann boundary conditions at the two ends of the cavity resonator

$$\left.\frac{\partial I_n}{\partial z}\right|_{z=0} = \left.\frac{\partial I_n}{\partial z}\right|_{z=L} = 0. \tag{3.59}$$

On account of the boundary conditions (3.58), the solutions of (3.53) and (3.55) for the modal voltages have the following expansions:

$$V_n^{TEM,TE} = \sum_{l=1}^{\infty} a_{nl}^{TEM,TE} \sqrt{\frac{2}{L}} \sin\frac{l\pi}{L} z. \tag{3.60}$$

When these are introduced into (3.53) and (3.55), one may find

$$\begin{aligned}
&\sum_{l=1}^{\infty} a_{nl}^{TEM,TE} \left[k_e^2 - \left(k_{nl}^{TEM,TE}\right)^2\right] \sqrt{\frac{2}{L}} \sin\frac{l\pi}{L} z \\
&= j\omega\mu j_n^{TEM,TE} - \frac{\partial}{\partial z} j'^{TEM,TE}_n + k_{cn}^{TEM,TE} j'^{TEM,TE}_{zn},
\end{aligned} \tag{3.61}$$

where

$$\left(k_{nl}^{TEM,TE}\right)^2 = \left(k_{cn}^{TEM,TE}\right)^2 + \left(\frac{l\pi}{L}\right)^2. \tag{3.62}$$

Note that $k_{cn}^{TEM} = 0$. Multiplying both sides of (3.61) by $\sqrt{2/L}\sin(l\pi z/L)$ and taking the integration overt the interval [0, l], the expansion coefficients in (3.61) are found to be

$$\begin{aligned} a_{nl}^{TEM,TE} &= -\frac{j\omega\mu}{\left(\Gamma_n^{TEM,TE}\right)^2}\int_0^L j_n^{TEM,TE}(z)\sqrt{\frac{2}{L}}\sin\frac{l\pi}{L}zdz \\ &\quad -\frac{1}{\left(\Gamma_n^{TEM,TE}\right)^2}\int_0^L j'^{TEM,TE}_n(z)\frac{l\pi}{L}\sqrt{\frac{2}{L}}\cos\frac{l\pi}{L}zdz \\ &\quad -\frac{k_{cn}^{TEM,TE}}{\left(\Gamma_n^{TEM,TE}\right)^2}\int_0^L j'^{TEM,TE}_{zn}(z)\sqrt{\frac{2}{L}}\sin\frac{l\pi}{L}zdz, \end{aligned} \tag{3.63}$$

where

$$\left(\Gamma_n^{TEM,TE}\right)^2 = \left(k_{nl}^{TEM,TE}\right)^2 - k_e^2. \tag{3.64}$$

The modal currents for TEM and TE modes are obtained from (3.49) while the longitudinal components of magnetic field are determined by the last two equations of (3.54). Thus,

$$\begin{aligned} I_n^{TEM,TE} &= -\frac{1}{j\omega\mu}\frac{\partial V_n^{TEM,TE}}{\partial z} - \frac{1}{j\omega\mu}j'^{TEM,TE}_n \\ &= -\frac{1}{j\omega\mu}\sum_{l=1}^{\infty} a_{nl}^{TEM,TE}\frac{l\pi}{L}\sqrt{\frac{2}{L}}\cos\frac{l\pi}{L}z - \frac{1}{j\omega\mu}j'^{TEM,TE}_n, \\ h_n^{TE} &= -\frac{1}{j\omega\mu}k_{cn}^{TE}V_n^{TE} - \frac{1}{j\omega\mu}j'^{TEM,TE}_{zn} \\ &= -\frac{1}{j\omega\mu}k_{cn}^{TE}\sum_{l=1}^{\infty} a_{nl}^{TE}\sqrt{\frac{2}{L}}\sin\frac{l\pi}{L}z - \frac{1}{j\omega\mu}j'^{TEM,TE}_{zn}, \\ -j\omega\mu\int_\Omega &\frac{\mathbf{u}_z}{\sqrt{\Omega}}\cdot\mathbf{H}d\Omega = \int_\Omega \frac{\mathbf{u}_z}{\sqrt{\Omega}}\cdot\mathbf{J}_m d\Omega. \end{aligned}$$

From the boundary condition (3.59), the solution of (3.57) for modal current can be expressed as

$$I_n^{TM} = \sum_{l=0}^{\infty} a_{nl}^{TM} \sqrt{\frac{\varepsilon_l}{L}} \cos\frac{l\pi}{L} z. \tag{3.65}$$

Substituting the above expansion into (3.57), one may find

$$\begin{aligned}\sum_{l=0}^{\infty} a_{nl}^{TM}\left[k_e^2 - \left(k_{nl}^{TM}\right)^2\right]\sqrt{\frac{\varepsilon_l}{L}} \cos\frac{l\pi}{L} z \\ = -\frac{\partial}{\partial z} j_n^{TM} - k_{cn}^{TM} j_{zn}^{TM} + (\sigma + j\omega\varepsilon) j'^{TM}_n,\end{aligned} \tag{3.66}$$

where

$$\left(k_{nl}^{TM}\right)^2 = \left(k_{cn}^{TM}\right)^2 + \left(\frac{l\pi}{L}\right)^2. \tag{3.67}$$

Similarly, the expansion coefficients a_{nl}^{TM} in (3.66) can be determined by using the orthonormality of the functions $\sqrt{\varepsilon_l/L}\cos(l\pi z/L)$:

$$\begin{aligned}a_{nl}^{TM} &= \frac{1}{\left(\Gamma_n^{TM}\right)^2}\frac{l\pi}{L}\int_0^L j_n^{TM}\sqrt{\frac{\varepsilon_l}{L}}\sin\frac{l\pi}{L} z dz + \frac{k_{cn}^{TM}}{\left(\Gamma_n^{TM}\right)^2}\int_0^L j_{zn}^{TM}\sqrt{\frac{\varepsilon_l}{L}}\cos\frac{l\pi}{L} z dz \\ &\quad - \frac{\sigma + j\omega\varepsilon}{\left(\Gamma_n^{TM}\right)^2}\int_0^L j'^{TM}_{zn}\sqrt{\frac{\varepsilon_l}{L}}\cos\frac{l\pi}{L} z dz,\end{aligned} \tag{3.68}$$

where

$$\left(\Gamma_n^{TM}\right)^2 = \left(k_{nl}^{TM}\right)^2 - k_e^2. \tag{3.69}$$

In deriving (3.68), the electric current has been assumed to vanish at the two ends of the cavity resonator. The modal voltage and longitudinal component of the electric field are then, respectively, determined from the last two equations of (3.56):

$$\begin{aligned}V_n^{TM} &= \frac{1}{\sigma + j\omega\varepsilon}\sum_{l=0}^{\infty} a_{nl}^{TM}\frac{l\pi}{L}\sqrt{\frac{\varepsilon_l}{L}}\sin\frac{l\pi}{L} z - \frac{1}{\sigma + j\omega\varepsilon} j_n^{TM}, \\ e_n^{TM} &= \frac{k_{cn}^{TM}}{\sigma + j\omega\varepsilon}\sum_{l=0}^{\infty} a_{nl}^{TM}\sqrt{\frac{\varepsilon_l}{L}}\cos\frac{l\pi}{L} z - \frac{1}{\sigma + j\omega\varepsilon} j_{zn}^{TM}.\end{aligned}$$

Substituting the expansions for the modal voltages and currents into (3.41) and (3.42), one obtains the field expansions

$$\begin{aligned}
\mathbf{E} &= \sum_n \sum_{l=0}^{\infty} a_{nl}^{TEM} \mathbf{e}_{nl}^{TEM} + \sum_n \sum_{l=0}^{\infty} a_{nl}^{TE} \mathbf{e}_{nl}^{TE} \\
&+ \frac{1}{\sigma + j\omega\varepsilon} \sum_n \sum_{l=0}^{\infty} a_{nl}^{TM} k_{nl}^{TM} \mathbf{e}_{nl}^{TM} - \frac{1}{\sigma + j\omega\varepsilon} \mathbf{J}^{TM}, \\
\mathbf{H} &= -\frac{1}{j\omega\mu} \sum_n \sum_{l=0}^{\infty} a_{nl}^{TEM} k_{nl}^{TEM} \mathbf{h}_{nl}^{TEM} - \frac{1}{j\omega\mu} \sum_n \sum_{l=0}^{\infty} a_{nl}^{TE} k_{nl}^{TE} \mathbf{h}_{nl}^{TE} \\
&+ \sum_n \sum_{l=0}^{\infty} a_{nl}^{TM} \mathbf{h}_{nl}^{TM} - \frac{1}{j\omega\mu} (\mathbf{J}_m^{TEM} + \mathbf{J}_m^{TE}),
\end{aligned} \tag{3.70}$$

$$\begin{aligned}
\nabla \times \mathbf{E} &= \sum_n \sum_{l=0}^{\infty} a_{nl}^{TEM} k_{cn}^{TEM} \mathbf{h}_{nl}^{TEM} + \sum_n \sum_{l=0}^{\infty} a_{nl}^{TE} k_{cn}^{TE} \mathbf{h}_{nl}^{TE} \\
&- j\omega\mu \sum_n \sum_{l=0}^{\infty} a_{nl}^{TM} \mathbf{h}_{nl}^{TM} - \mathbf{J}_m^{TM}, \\
\nabla \times \mathbf{H} &= (\sigma + j\omega\varepsilon) \sum_n \sum_{l=0}^{\infty} a_{nl}^{TEM} \mathbf{e}_{nl}^{TEM} + (\sigma + j\omega\varepsilon) \sum_n \sum_{l=0}^{\infty} a_{nl}^{TE} \mathbf{e}_{nl}^{TE} \\
&+ \sum_n \sum_{l=0}^{\infty} a_{nl}^{TM} k_{cn}^{TM} \mathbf{e}_{nl}^{TM} + \mathbf{J}^{TEM} + \mathbf{J}^{TE},
\end{aligned} \tag{3.71}$$

where the normalized electric vector modal functions $\mathbf{e}_{nl}$ and the normalized magnetic vector modal functions $\mathbf{h}_{nl}$ for the waveguide cavity resonator have been introduced, and they are defined by

$$\begin{aligned}
\mathbf{e}_{nl}^{TEM} &= \mathbf{e}_n^{TEM} \sqrt{\frac{2}{L}} \sin \frac{l\pi}{L} z, \\
\mathbf{h}_{nl}^{TEM} &= \mathbf{u}_z \times \mathbf{e}_n^{TEM} \sqrt{\frac{\varepsilon_l}{L}} \cos \frac{l\pi}{L} z,
\end{aligned} \tag{3.72}$$

$$\begin{aligned}
\mathbf{e}_{nl}^{TE} &= \mathbf{e}_n^{TE} \sqrt{\frac{2}{L}} \sin \frac{l\pi}{L} z, \\
\mathbf{h}_{nl}^{TE} &= \frac{1}{k_{nl}^{TE}} \left[\mathbf{u}_z \times \mathbf{e}_n^{TE} \frac{l\pi}{L} \sqrt{\frac{2}{L}} \cos \frac{l\pi}{L} z + k_{cn}^{TE} \frac{\nabla \times \mathbf{e}_n^{TE}}{k_{cn}^{TE}} \sqrt{\frac{2}{L}} \sin \frac{l\pi}{L} z \right],
\end{aligned} \tag{3.73}$$

$$\begin{aligned}
\mathbf{e}_{nl}^{TM} &= \frac{1}{k_{nl}^{TM}} \left[\frac{l\pi}{L} \mathbf{e}_n^{TM} \sqrt{\frac{\varepsilon_l}{L}} \sin \frac{l\pi}{L} z + \mathbf{u}_z k_{cn}^{TM} \frac{\nabla \cdot \mathbf{e}_n^{TM}}{k_{cn}^{TM}} \sqrt{\frac{\varepsilon_l}{L}} \cos \frac{l\pi}{L} z \right], \\
\mathbf{h}_{nl}^{TM} &= \mathbf{u}_z \times \mathbf{e}_n^{TM} \sqrt{\frac{\varepsilon_l}{L}} \cos \frac{l\pi}{L} z.
\end{aligned} \tag{3.74}$$

Other source terms in (3.70) and (3.71) are given by

$$\mathbf{J}^{TEM} = \sum_n j_n^{TEM} \mathbf{e}_n^{TEM},$$

$$\mathbf{J}^{TE} = \sum_n j_n^{TE} \mathbf{e}_n^{TE},$$

$$\mathbf{J}^{TM} = \sum_{n=1}^{\infty} j_n^{TM} \mathbf{e}_n^{TM} + \sum_{n=1}^{\infty} j_{zn}^{TM} \mathbf{u}_z,$$

$$\mathbf{J}_m^{TEM} = \sum_n j_n'^{TEM} \mathbf{u}_z \times \mathbf{e}_n^{TEM},$$

$$\mathbf{J}_m^{TE} = \sum_n j_n'^{TE} \mathbf{u}_z \times \mathbf{e}_n^{TE} + \sum_n j_{zn}'^{TE} \frac{\nabla \times \mathbf{e}_n^{TE}}{k_{cn}^{TE}} + \mathbf{u}_z \frac{1}{\sqrt{\Omega}} \int_\Omega \frac{\mathbf{u}_z \cdot \mathbf{J}_m}{\sqrt{\Omega}} d\Omega,$$

$$\mathbf{J}_m^{TM} = \sum_n j_n'^{TM} \mathbf{u}_z \times \mathbf{e}_n^{TM},$$

and they denote the expansions of the current distributions in terms of the vector modal functions of various types. In terms of the normalized vector modal functions, the expansion coefficients (3.63) and (3.68) can be expressed by

$$a_{nl}^{TEM,TE} = -\frac{j\omega\mu}{\left(\Gamma_n^{TEM,TE}\right)^2} \int_\Omega \int_0^L \mathbf{J} \cdot \mathbf{e}_{nl}^{TEM,TE} dV - \frac{k_{nl}^{TEM,TE}}{\left(\Gamma_n^{TEM,TE}\right)^2} \int_\Omega \int_0^L \mathbf{J}_m \cdot \mathbf{h}_{nl}^{TEM,TE} dV, \tag{3.75}$$

$$a_{nl}^{TM} = \frac{k_{nl}^{TM}}{\left(\Gamma_n^{TM}\right)^2} \int_\Omega \int_0^L \mathbf{J} \cdot \mathbf{e}_{nl}^{TM} dV - \frac{\sigma + j\omega\varepsilon}{\left(\Gamma_n^{TM}\right)^2} \int_\Omega \int_0^L \mathbf{J}_m \cdot \mathbf{h}_{nl}^{TM} dV, \tag{3.76}$$

where $dV = d\Omega dz$ denotes the differential volume element. The electric field in (3.70) can be rewritten as

$$\begin{aligned}
\mathbf{E} = &-\sum_n \sum_{l=0}^{\infty} \mathbf{e}_{nl}^{TEM} \frac{j\omega\mu}{\left(\Gamma_n^{TEM}\right)^2} \int_\Omega \int_0^L \mathbf{J} \cdot \mathbf{e}_{nl}^{TEM} dV - \sum_n \sum_{l=0}^{\infty} \mathbf{e}_{nl}^{TE} \frac{j\omega\mu}{\left(\Gamma_n^{TE}\right)^2} \int_\Omega \int_0^L \mathbf{J} \cdot \mathbf{e}_{nl}^{TE} dV \\
&-\sum_n \sum_{l=0}^{\infty} \mathbf{e}_{nl}^{TM} \frac{j\omega\mu}{\left(\Gamma_n^{TM}\right)^2} \int_\Omega \int_0^L \mathbf{J} \cdot \mathbf{e}_{nl}^{TM} dV - \sum_n \sum_{l=0}^{\infty} \mathbf{e}_{nl}^{TEM} \frac{1}{\left(\Gamma_n^{TEM}\right)^2} \frac{l\pi}{L} \int_\Omega \int_0^L \mathbf{J}_m \cdot \mathbf{h}_{nl}^{TEM} dV \\
&-\sum_n \sum_{l=0}^{\infty} \mathbf{e}_{nl}^{TE} \frac{k_{nl}^{TE}}{\left(\Gamma_n^{TE}\right)^2} \int_\Omega \int_0^L \mathbf{J}_m \cdot \mathbf{h}_{nl}^{TE} dV - \sum_n \sum_{l=0}^{\infty} \mathbf{e}_{nl}^{TM} \frac{k_{nl}^{TM}}{\left(\Gamma_n^{TM}\right)^2} \int_\Omega \int_0^L \mathbf{J}_m \cdot \mathbf{h}_{nl}^{TM} dV \\
&-\frac{j\omega\mu}{k_e^2} \sum_n \sum_{l=0}^{\infty} \mathbf{e}_{nl}^{TM} \int_\Omega \int_0^L \mathbf{J} \cdot \mathbf{e}_{nl}^{TM} dV + \frac{j\omega\mu}{k_e^2} \sum_n j_n^{TM} \mathbf{e}_n^{TM} + \frac{j\omega\mu}{k_e^2} \sum_n j_{zn}^{TM} \mathbf{u}_z \frac{\nabla \cdot \mathbf{e}_n^{TM}}{k_{cn}^{TM}}.
\end{aligned} \tag{3.77}$$

3.3.2 Modal Representations of Dyadic Green's Functions

The electric current **J** may be considered as a vector of the electric type. For this reason, $j_n(z)$ can be expanded in terms of the functions $\sqrt{2/L}\sin(l\pi z/L)$ while $j_{zn}(z)$ can be expanded in terms of the function $\sqrt{\varepsilon_l/L}\cos(l\pi z/L)$. The last three terms on the right-hand side of (3.77) can be written as

$$-\frac{j\omega\mu}{k_e^2}\sum_n\sum_{l=0}^{\infty}\mathbf{e}_{nl}^{TM}\int_{\Omega}\int_0^L\mathbf{J}\cdot\mathbf{e}_{nl}^{TM}dV+\frac{j\omega\mu}{k_e^2}\sum_n j_n^{TM}\mathbf{e}_n^{TM}+\frac{j\omega\mu}{k_e^2}\sum_n j_{zn}^{TM}\mathbf{u}_z\frac{\nabla\cdot\mathbf{e}_n^{TM}}{k_{cn}^{TM}}$$

$$=\frac{j\omega\mu}{k_e^2}\sum_n\sum_{l=0}^{\infty}\tilde{\mathbf{e}}_{nl}^{TM}\int_{\Omega}\int_0^L\mathbf{J}\cdot\tilde{\mathbf{e}}_{nl}^{TM}dV,$$

where $\tilde{\mathbf{e}}_{nl}^{TM}$ is the normalized vector modal function defined by

$$\tilde{\mathbf{e}}_{nl}^{TM}=\frac{k_{cn}^{TM}}{k_{nl}^{TM}}\mathbf{e}_n^{TM}\sqrt{\frac{\varepsilon_l}{L}}\sin\frac{l\pi}{L}z-\frac{1}{k_{nl}^{TM}}\frac{l\pi}{L}\mathbf{u}_z\frac{\nabla\cdot\mathbf{e}_n^{TM}}{k_{cn}^{TM}}\sqrt{\frac{\varepsilon_l}{L}}\cos\frac{l\pi}{L}z. \tag{3.78}$$

The electric field (3.77) can be expressed in a compact form by introducing dyadic notations

$$\mathbf{E}(\mathbf{r})=-j\omega\mu\int_{\Omega}\int_0^L\overleftrightarrow{\mathbf{G}}_e(\mathbf{r},\mathbf{r}')\cdot\mathbf{J}(\mathbf{r}')dV-\int_{\Omega}\int_0^L\overleftrightarrow{\mathbf{G}}_{em}(\mathbf{r},\mathbf{r}')\cdot\mathbf{J}_m(\mathbf{r}')dV, \tag{3.79}$$

where the dyadic Green's functions are defined by

$$\begin{aligned}\overleftrightarrow{\mathbf{G}}_e(\mathbf{r},\mathbf{r}')&=\sum_n\sum_{l=0}^{\infty}\frac{1}{\left(\Gamma_n^{TEM}\right)^2}\mathbf{e}_{nl}^{TEM}(\mathbf{r})\mathbf{e}_{nl}^{TEM}(\mathbf{r}')+\sum_n\sum_{l=0}^{\infty}\frac{1}{\left(\Gamma_n^{TE}\right)^2}\mathbf{e}_{nl}^{TE}(\mathbf{r})\mathbf{e}_{nl}^{TE}(\mathbf{r}')\\&\quad+\sum_n\sum_{l=0}^{\infty}\frac{1}{\left(\Gamma_n^{TM}\right)^2}\mathbf{e}_{nl}^{TM}(\mathbf{r})\mathbf{e}_{nl}^{TM}(\mathbf{r}')-\sum_n\sum_{l=0}^{\infty}\frac{1}{k_e^2}\tilde{\mathbf{e}}_{nl}^{TM}(\mathbf{r})\tilde{\mathbf{e}}_{nl}^{TM}(\mathbf{r}'),\end{aligned} \tag{3.80}$$

$$\begin{aligned}\overleftrightarrow{\mathbf{G}}_{em}(\mathbf{r},\mathbf{r}')&=\sum_n\sum_{l=0}^{\infty}\frac{k_{nl}^{TEM}}{\left(\Gamma_n^{TEM}\right)^2}\mathbf{e}_{nl}^{TEM}(\mathbf{r})\mathbf{h}_{nl}^{TEM}(\mathbf{r}')+\sum_n\sum_{l=0}^{\infty}\frac{k_{nl}^{TE}}{\left(\Gamma_n^{TE}\right)^2}\mathbf{e}_{nl}^{TE}(\mathbf{r})\mathbf{h}_{nl}^{TE}(\mathbf{r}')\\&\quad+\sum_n\sum_{l=0}^{\infty}\frac{k_{nl}^{TM}}{\left(\Gamma_n^{TM}\right)^2}\mathbf{e}_{nl}^{TM}(\mathbf{r})\mathbf{h}_{nl}^{TM}(\mathbf{r}').\end{aligned} \tag{3.81}$$

Similarly, the magnetic current $\mathbf{J}_m$ may be considered as a vector of the type of magnetic field. Therefore, $j_n'(z)$ can be expanded in terms of the functions

$\sqrt{\varepsilon_l/L}\cos(l\pi z/L)$ while $j'_{zn}(z)$ can be expanded in terms of the function $\sqrt{2/L}\sin(l\pi z/L)$. The magnetic field in (3.70) can be written as

$$\mathbf{H}(\mathbf{r}) = -j\omega\varepsilon_e \int\limits_{\Omega}\int\limits_{0}^{L} \overleftrightarrow{\mathbf{G}}_m(\mathbf{r},\mathbf{r}')\cdot\mathbf{J}_m(\mathbf{r}')dV + \int\limits_{\Omega}\int\limits_{0}^{L} \overleftrightarrow{\mathbf{G}}_{me}(\mathbf{r},\mathbf{r}')\cdot\mathbf{J}(\mathbf{r}')dV, \tag{3.82}$$

where the dyadic Green's functions are defined by

$$\begin{aligned}\overleftrightarrow{\mathbf{G}}_m(\mathbf{r},\mathbf{r}') &= \sum_n\sum_{l=0}^{\infty}\frac{1}{\left(\Gamma_n^{TEM}\right)^2}\mathbf{h}_{nl}^{TEM}(\mathbf{r})\mathbf{h}_{nl}^{TEM}(\mathbf{r}') + \sum_n\sum_{l=0}^{\infty}\frac{1}{\left(\Gamma_n^{TE}\right)^2}\mathbf{h}_{nl}^{TE}(\mathbf{r})\mathbf{h}_{nl}^{TE}(\mathbf{r}')\\ &+ \sum_n\sum_{l=0}^{\infty}\frac{1}{\left(\Gamma_n^{TM}\right)^2}\mathbf{h}_{nl}^{TM}(\mathbf{r})\mathbf{h}_{nl}^{TM}(\mathbf{r}')\\ &- \sum_n\sum_{l=0}^{\infty}\frac{1}{k_e^2}\tilde{\mathbf{h}}_{nl}^{TE}(\mathbf{r})\tilde{\mathbf{h}}_{nl}^{TE}(\mathbf{r}') - \frac{1}{k_e^2}\frac{\mathbf{u}_z\mathbf{u}_{z'}}{\Omega}\delta(z-z'),\end{aligned} \tag{3.83}$$

$$\begin{aligned}\overleftrightarrow{\mathbf{G}}_{me}(\mathbf{r},\mathbf{r}') &= \sum_n\sum_{l=0}^{\infty}\frac{k_{nl}^{TEM}}{\left(\Gamma_n^{TEM}\right)^2}\mathbf{h}_{nl}^{TEM}(\mathbf{r})\mathbf{e}_{nl}^{TEM}(\mathbf{r}') + \sum_n\sum_{l=0}^{\infty}\frac{k_{nl}^{TE}}{\left(\Gamma_n^{TE}\right)^2}\mathbf{h}_{nl}^{TE}(\mathbf{r})\mathbf{e}_{nl}^{TE}(\mathbf{r}')\\ &+ \sum_n\sum_{l=0}^{\infty}\frac{k_{nl}^{TM}}{\left(\Gamma_n^{TM}\right)^2}\mathbf{h}_{nl}^{TM}(\mathbf{r})\mathbf{e}_{nl}^{TM}(\mathbf{r}').\end{aligned} \tag{3.84}$$

In (3.83), the vector modal function $\tilde{\mathbf{h}}_{nl}^{TE}$ is defined by

$$\tilde{\mathbf{h}}_{nl}^{TE}(\mathbf{r}) = \frac{k_{cn}^{TE}}{k_{nl}^{TE}}\mathbf{u}_z\times\mathbf{e}_n^{TE}(\boldsymbol{\rho})\sqrt{\frac{2}{L}}\cos\frac{l\pi}{L}z - \frac{1}{k_{nl}^{TE}}\frac{l\pi}{L}\frac{\nabla\times\mathbf{e}_n^{TE}(\boldsymbol{\rho})}{k_{cn}^{TE}}\sqrt{\frac{2}{L}}\sin\frac{l\pi}{L}z. \tag{3.85}$$

It is easy to verify that the normalized vector modal functions for the waveguide cavity resonator satisfy the following relations:

$$\nabla\times\mathbf{h}_{n0}^{TEM} = 0,\quad \nabla\cdot\mathbf{h}_{n0}^{TEM} = 0, \tag{3.86}$$

$$\nabla\times\tilde{\mathbf{e}}_{nl}^{TM} = 0, \tag{3.87}$$

$$\nabla\times\tilde{\mathbf{h}}_{nl}^{TE} = 0, \tag{3.88}$$

$$\nabla\times\mathbf{e}_{nl}^{TEM} = k_{nl}^{TEM}\mathbf{h}_{nl}^{TEM},\quad \nabla\times\mathbf{h}_{nl}^{TEM} = k_{nl}^{TEM}\mathbf{e}_{nl}^{TEM}, \tag{3.89}$$

$$\nabla \times \mathbf{e}_{nl}^{TE} = k_{nl}^{TE}\mathbf{h}_{nl}^{TE}, \quad \nabla \times \mathbf{h}_{nl}^{TE} = k_{nl}^{TE}\mathbf{e}_{nl}^{TE}, \tag{3.90}$$

$$\nabla \times \mathbf{e}_{nl}^{TM} = k_{nl}^{TM}\mathbf{h}_{nl}^{TM}, \quad \nabla \times \mathbf{h}_{nl}^{TM} = k_{nl}^{TM}\mathbf{e}_{nl}^{TM}. \tag{3.91}$$

All the vector modal functions satisfy the vector wave equations (3.4) and (3.5). Furthermore, the irrotational vector modal functions $\tilde{\mathbf{e}}_{nl}^{TM}$ can be expressed as the gradient of a scalar function

$$\tilde{\mathbf{e}}_{nl}^{TM} = \nabla \varphi_{nl}^{TM}, \tag{3.92}$$

where

$$\varphi_{nl}^{TM} = -\frac{1}{k_{nl}^{TM}}\frac{\nabla \cdot \mathbf{e}_{n}^{TM}}{k_{cn}^{TM}}\sqrt{\frac{\varepsilon_l}{L}}\sin\frac{l\pi}{L}z \tag{3.93}$$

satisfies

$$\begin{cases} \nabla^2 \varphi_{nl}^{TM}(\mathbf{r}) + \left(k_{nl}^{TM}\right)^2 \varphi_{nl}^{TM}(\mathbf{r}) = 0, \\ \varphi_{nl}^{TM}(\mathbf{r})\Big|_{\Gamma} = 0. \end{cases} \tag{3.94}$$

Similarly, the irrotational vector modal function $\tilde{\mathbf{h}}_{nl}^{TE}$ can be expressed as the gradient of a scalar function

$$\tilde{\mathbf{h}}_{nl}^{TE} = \nabla \psi_{nl}^{TE}, \tag{3.95}$$

where

$$\psi_{nl}^{TE} = \frac{1}{k_{nl}^{TE}}\mathbf{u}_z \cdot \frac{\nabla \times \mathbf{e}_{n}^{TE}}{k_{cn}^{TE}}\sqrt{\frac{2}{L}}\cos\frac{l\pi}{L}z$$

satisfies

$$\begin{cases} \nabla^2 \psi_{nl}^{TE}(\mathbf{r}) + \left(k_{nl}^{TE}\right)^2 \psi_{nl}^{TE}(\mathbf{r}) = 0, \\ \dfrac{\partial \psi_{nl}^{TE}(\mathbf{r})}{\partial n}\Bigg|_{\Gamma} = 0. \end{cases} \tag{3.96}$$

In summary, the normalized electric and magnetic vector modal functions used to expand the EM fields in a waveguide cavity may be divided into three types:

$$\begin{aligned} &\text{Electric}: \begin{cases} \text{Type 1}: & \{0\}, \\ \text{Type 2}: & \{\mathbf{e}_{nl}^{TEM}, \mathbf{e}_{nl}^{TE}, \mathbf{e}_{nl}^{TM}\}, \\ \text{Type 3}: & \{\tilde{\mathbf{e}}_{nl}^{TM}\}. \end{cases} \\ &\text{Magnetic}: \begin{cases} \text{Type 1}: & \{\mathbf{h}_{n0}^{TEM}\}, \\ \text{Type 2}: & \{\mathbf{h}_{nl}^{TEM}, \mathbf{h}_{nl}^{TE}, \mathbf{h}_{nl}^{TM}\}, \\ \text{Type 3}: & \{\tilde{\mathbf{h}}_{nl}^{TE}\}. \end{cases} \end{aligned} \tag{3.97}$$

The vector modal functions of type 1 are both irrotational and divergenceless. Such vector modal functions are needed for the expansion of the magnetic field in the waveguide cavity resonators consisting of multiple conductor transmission lines such as a coaxial cable, and they correspond to the field generated by the direct current flowing along the multiple conductors through the short ends. It is also noted that the electric vector modal function of type 1 does not exist even in the waveguide cavity resonator consisting of multiple conductor transmission lines due to the existence of the short ends. The vector modal functions of type 2 are divergenceless while those of type 3 are irrotational.

3.4 Vector Modal Functions for Typical Waveguide Cavity Resonators

In this section, the normalized vector modal functions for typical waveguide cavities will be summarized. They can be obtained directly from the vector modal functions of the corresponding waveguide; their derivations are fairly straightforward, and only the results will be given.

3.4.1 Rectangular Waveguide Cavity

For a rectangular waveguide cavity resonator shown in Figure 3.7, the normalized vector modal functions actually represent the possible standing-wave patterns that can exist in the rectangular waveguide, and can be obtained from the vector modal functions of the rectangular waveguide

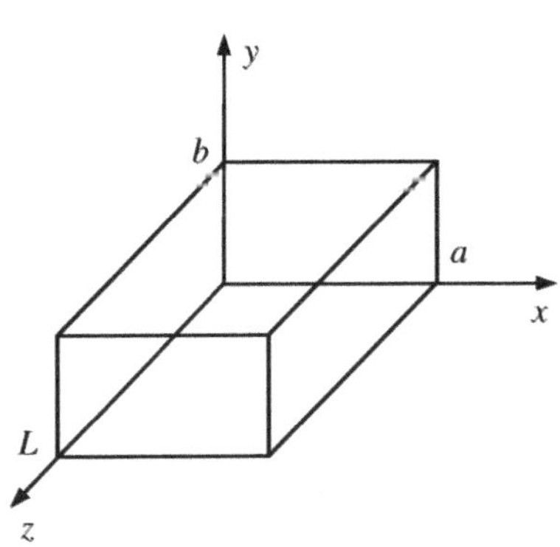

Figure 3.7 Rectangular waveguide cavity.

$$
\begin{aligned}
\mathbf{e}_{nl}^{TE} &= \mathbf{u}_x \frac{1}{k_{cn}^{TE}} \frac{q\pi}{b} \sqrt{\frac{2\varepsilon_p \varepsilon_q}{Lab}} \cos\frac{p\pi}{a}x \sin\frac{q\pi}{b}y \sin\frac{l\pi}{L}z \\
&- \mathbf{u}_y \frac{1}{k_{cn}^{TE}} \frac{p\pi}{a} \sqrt{\frac{2\varepsilon_p \varepsilon_q}{Lab}} \sin\frac{p\pi}{a}x \cos\frac{q\pi}{b}y \sin\frac{l\pi}{L}z,
\end{aligned} \tag{3.98}
$$

$$
\begin{aligned}
\mathbf{e}_{nl}^{TM} &= \mathbf{u}_x \frac{1}{k_{nl}^{TM} k_{cn}^{TM}} \frac{l\pi}{L} \frac{p\pi}{a} \sqrt{\frac{4\varepsilon_l}{Lab}} \cos\frac{p\pi}{a}x \sin\frac{q\pi}{b}y \sin\frac{l\pi}{L}z \\
&+ \mathbf{u}_y \frac{1}{k_{nl}^{TM} k_{cn}^{TM}} \frac{l\pi}{L} \frac{q\pi}{b} \sqrt{\frac{4\varepsilon_l}{Lab}} \sin\frac{p\pi}{a}x \cos\frac{q\pi}{b}y \sin\frac{l\pi}{L}z \\
&- \mathbf{u}_z \frac{k_{cn}^{TM}}{k_{nl}^{TM}} \sqrt{\frac{4\varepsilon_l}{Lab}} \sin\frac{p\pi}{a}x \sin\frac{q\pi}{b}y \cos\frac{l\pi}{L}z,
\end{aligned} \tag{3.99}
$$

$$
\begin{aligned}
\tilde{\mathbf{e}}_{nl}^{TM} &= \mathbf{u}_x \frac{1}{k_{nl}^{TM}} \frac{p\pi}{a} \sqrt{\frac{4\varepsilon_l}{Lab}} \cos\frac{p\pi}{a}x \sin\frac{q\pi}{b}y \sin\frac{l\pi}{L}z \\
&+ \mathbf{u}_y \frac{1}{k_{nl}^{TM}} \frac{q\pi}{b} \sqrt{\frac{4\varepsilon_l}{Lab}} \sin\frac{p\pi}{a}x \cos\frac{q\pi}{b}y \sin\frac{l\pi}{L}z \\
&+ \mathbf{u}_z \frac{1}{k_{nl}^{TM}} \frac{l\pi}{L} \sqrt{\frac{4\varepsilon_l}{Lab}} \sin\frac{p\pi}{a}x \sin\frac{q\pi}{b}y \cos\frac{l\pi}{L}z,
\end{aligned} \tag{3.100}
$$

$$
\begin{aligned}
\tilde{\mathbf{h}}_{nl}^{TE} &= \mathbf{u}_x \frac{1}{k_{nl}^{TE}} \frac{p\pi}{a} \sqrt{\frac{2\varepsilon_p \varepsilon_q}{Lab}} \sin\frac{p\pi}{a}x \cos\frac{q\pi}{b}y \cos\frac{l\pi}{L}z \\
&+ \mathbf{u}_y \frac{1}{k_{nl}^{TE}} \frac{q\pi}{b} \sqrt{\frac{2\varepsilon_p \varepsilon_q}{Lab}} \cos\frac{p\pi}{a}x \sin\frac{q\pi}{b}y \cos\frac{l\pi}{L}z \\
&+ \mathbf{u}_z \frac{1}{k_{nl}^{TE}} \frac{l\pi}{L} \sqrt{\frac{2\varepsilon_p \varepsilon_q}{Lab}} \cos\frac{p\pi}{a}x \cos\frac{q\pi}{b}y \sin\frac{l\pi}{L}z,
\end{aligned} \tag{3.101}
$$

where k_{cn}^{TE} and k_{cn}^{TM} are, respectively, the cutoff wavenumbers for TE and TM modes in the rectangular waveguide.

3.4.2 Circular Waveguide Cavity

The circular waveguide cavity has a high quality factor (QF) and wide operating frequency range, and is often used for wavemeters to measure frequency [12]. The vector modal functions for circular waveguide cavity resonator (Figure 3.8) are given by

$$
\begin{aligned}
\mathbf{e}_{nl}^{TE} &= \pm \mathbf{u}_\rho \sqrt{\frac{2\varepsilon_q}{L\pi}} \frac{q}{\sqrt{\chi_{qp}'^2 - q^2}} \frac{1}{\rho} \frac{J_q\left(\chi_{qp}' \frac{\rho}{a}\right)}{J_q\left(\chi_{qp}'\right)} \begin{pmatrix} \sin q\varphi \\ \cos q\varphi \end{pmatrix} \sin\frac{l\pi}{L}z \\
&+ \mathbf{u}_\varphi \sqrt{\frac{2\varepsilon_q}{L\pi}} \frac{\chi_{qp}'}{\sqrt{\chi_{qp}'^2 - q^2}} \frac{1}{a} \frac{J_q'\left(\chi_{qp}' \frac{\rho}{a}\right)}{J_q\left(\chi_{qp}'\right)} \begin{pmatrix} \cos q\varphi \\ \sin q\varphi \end{pmatrix} \sin\frac{l\pi}{L}z,
\end{aligned} \tag{3.102}
$$

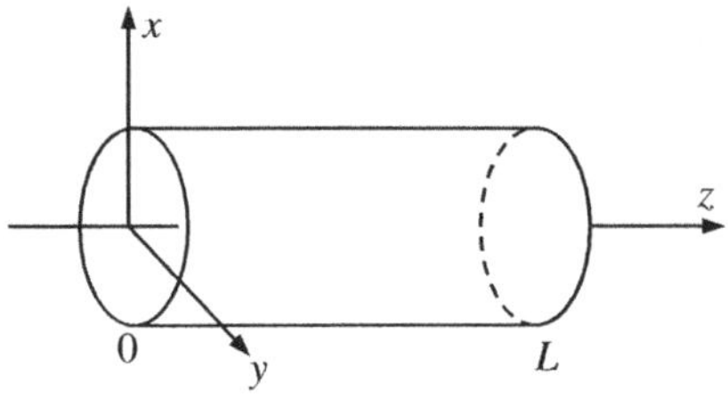

Figure 3.8 Circular waveguide cavity resonator.

$$
\begin{aligned}
\mathbf{e}_{nl}^{TM} = &-\mathbf{u}_{\rho}\sqrt{\frac{\varepsilon_q\varepsilon_l}{\pi L}}\frac{1}{k_{nl}^{TM}}\frac{l\pi}{L}\frac{1}{a}\frac{J_q'\left(\chi_{qp}\frac{\rho}{a}\right)}{J_{q+1}\left(\chi_{qp}\right)}\begin{pmatrix}\cos q\varphi\\ \sin q\varphi\end{pmatrix}\sin\frac{l\pi}{L}z\\
&\pm\mathbf{u}_{\varphi}\sqrt{\frac{\varepsilon_q\varepsilon_l}{\pi L}}\frac{1}{k_{nl}^{TM}}\frac{l\pi}{L}\frac{q}{\chi_{qp}\rho}\frac{1}{J_{q+1}\left(\chi_{qp}\right)}J_q\left(\chi_{qp}\frac{\rho}{a}\right)\begin{pmatrix}\sin q\varphi\\ \cos q\varphi\end{pmatrix}\sin\frac{l\pi}{L}z\\
&+\mathbf{u}_{z}\sqrt{\frac{\varepsilon_q\varepsilon_l}{\pi L}}\frac{1}{k_{nl}^{TM}}\frac{\chi_{qp}}{a}\frac{1}{a}\frac{J_q\left(\chi_{qp}\frac{\rho}{a}\right)}{J_{q+1}\left(\chi_{qp}\right)}\begin{pmatrix}\cos q\varphi\\ \sin q\varphi\end{pmatrix}\cos\frac{l\pi}{L}z,
\end{aligned}
\tag{3.103}
$$

$$
\begin{aligned}
\tilde{\mathbf{e}}_{nl}^{TM} = &-\mathbf{u}_{\rho}\sqrt{\frac{\varepsilon_q\varepsilon_l}{\pi L}}\frac{k_{cn}^{TM}}{k_{nl}^{TM}}\frac{J_q'\left(\chi_{qp}\frac{\rho}{a}\right)}{aJ_{q+1}\left(\chi_{qp}\right)}\begin{pmatrix}\cos q\varphi\\ \sin q\varphi\end{pmatrix}\sin\frac{l\pi}{L}z\\
&\pm\mathbf{u}_{\varphi}\sqrt{\frac{\varepsilon_q\varepsilon_l}{\pi L}}\frac{k_{cn}^{TM}}{k_{nl}^{TM}}\frac{q}{\chi_{qp}\rho J_{q+1}\left(\chi_{qp}\right)}J_q\left(\chi_{qp}\frac{\rho}{a}\right)\begin{pmatrix}\sin q\varphi\\ \cos q\varphi\end{pmatrix}\sin\frac{l\pi}{L}z\\
&-\mathbf{u}_{z}\sqrt{\frac{\varepsilon_q\varepsilon_l}{\pi L}}\frac{1}{k_{nl}^{TM}}\frac{l\pi}{L}\frac{J_q\left(\chi_{qp}\frac{\rho}{a}\right)}{aJ_{q+1}\left(\chi_{qp}\right)}\begin{pmatrix}\cos q\varphi\\ \sin q\varphi\end{pmatrix}\cos\frac{l\pi}{L}z,
\end{aligned}
\tag{3.104}
$$

$$
\begin{aligned}
\tilde{\mathbf{h}}_{nl}^{TE} = &\pm\mathbf{u}_{\varphi}\sqrt{\frac{2\varepsilon_q}{\pi L}}\frac{k_{cn}^{TE}}{k_{nl}^{TE}}\frac{q}{\sqrt{\chi'^2_{qp}-q^2}}\frac{J_q\left(\chi'_{qp}\frac{\rho}{a}\right)}{\rho J_q\left(\chi'_{qp}\right)}\begin{pmatrix}\sin q\varphi\\ \cos q\varphi\end{pmatrix}\cos\frac{l\pi}{L}z\\
&-\mathbf{u}_{\rho}\sqrt{\frac{2\varepsilon_q}{\pi L}}\frac{k_{cn}^{TE}}{k_{nl}^{TE}}\frac{\chi'_{qp}}{\sqrt{\chi'^2_{qp}-q^2}}\frac{J_q'\left(\chi'_{qp}\frac{\rho}{a}\right)}{aJ_q\left(\chi'_{qp}\right)1}\begin{pmatrix}\cos q\varphi\\ \sin q\varphi\end{pmatrix}\cos\frac{l\pi}{L}z\\
&+\mathbf{u}_{z}\sqrt{\frac{2\varepsilon_q}{\pi L}}\frac{1}{k_{nl}^{TE}k_{cn}^{TE}}\frac{l\pi}{L}\left(\frac{\chi'_{qp}}{a}\right)^2\frac{1}{J_q\left(\chi'_{qp}\right)\sqrt{\chi'^2_{qp}-q^2}}\begin{pmatrix}\cos q\varphi\\ \sin q\varphi\end{pmatrix}\sin\frac{l\pi}{L}z.
\end{aligned}
\tag{3.105}
$$

The notations used in the above expressions are defined in Section 2.1.

3.4.3 Coaxial Waveguide Cavity

The vector modal functions for a coaxial waveguide cavity resonator (Figure 3.9) are

$$\mathbf{e}_{nl}^{TEM} = \mathbf{u}_\rho \frac{1}{\sqrt{2\pi \ln c_1}} \frac{1}{\rho} \sqrt{\frac{2}{L}} \sin \frac{l\pi}{L} z,$$

$$\mathbf{e}_{nl}^{TE} = \pm \mathbf{u}_\rho \sqrt{\frac{\varepsilon_q}{\pi L}} \frac{q}{\rho} h\left(\chi'_{qp} \frac{\rho}{a}\right) \begin{pmatrix} \sin q\varphi \\ \cos q\varphi \end{pmatrix} \sin \frac{l\pi}{L} z$$
$$+ \mathbf{u}_\varphi \sqrt{\frac{\varepsilon_q}{\pi L}} \frac{\chi'_{qp}}{a} h'\left(\chi'_{qp} \frac{\rho}{a}\right) \begin{pmatrix} \cos q\varphi \\ \sin q\varphi \end{pmatrix} \sin \frac{l\pi}{L} z,$$

$$\mathbf{e}_{nl}^{TM} = -\mathbf{u}_\rho \sqrt{\frac{\varepsilon_q \varepsilon_l}{2\pi L}} \frac{1}{k_{nl}^{TM}} \frac{l\pi}{L} \frac{\chi_{qp}}{a} e'\left(\chi_{qp} \frac{\rho}{a}\right) \begin{pmatrix} \cos q\varphi \\ \sin q\varphi \end{pmatrix} \sin \frac{l\pi}{L} z$$
$$\pm \mathbf{u}_\varphi \sqrt{\frac{\varepsilon_q \varepsilon_l}{2\pi L}} \frac{1}{k_{nl}^{TM}} \frac{l\pi}{L} \frac{q}{\rho} e\left(\chi_{qp} \frac{\rho}{a}\right) \begin{pmatrix} \sin q\varphi \\ \cos q\varphi \end{pmatrix} \sin \frac{l\pi}{L} z$$
$$+ \mathbf{u}_z \sqrt{\frac{\varepsilon_q \varepsilon_l}{2\pi L}} \frac{1}{k_{nl}^{TM}} \left(\frac{\chi_{qp}}{a}\right)^2 e\left(\chi_{qp} \frac{\rho}{a}\right) \begin{pmatrix} \cos q\varphi \\ \sin q\varphi \end{pmatrix} \cos \frac{l\pi}{L} z,$$

$$\tilde{\mathbf{e}}_{nl}^{TM} = -\mathbf{u}_\rho \sqrt{\frac{\varepsilon_q \varepsilon_l}{2\pi L}} \frac{k_{cn}^{TM}}{k_{nl}^{TM}} \frac{\chi_{qp}}{a} e'\left(\chi_{qp} \frac{\rho}{a}\right) \begin{pmatrix} \cos q\varphi \\ \sin q\varphi \end{pmatrix} \sin \frac{l\pi}{L} z$$
$$\pm \mathbf{u}_\varphi \sqrt{\frac{\varepsilon_q \varepsilon_l}{2\pi L}} \frac{q}{\rho} e\left(\chi_{qp} \frac{\rho}{a}\right) \begin{pmatrix} \sin q\varphi \\ \cos q\varphi \end{pmatrix} \frac{k_{cn}^{TM}}{k_{nl}^{TM}} \sin \frac{l\pi}{L} z$$
$$- \mathbf{u}_z \sqrt{\frac{\varepsilon_q \varepsilon_l}{2\pi L}} \frac{1}{k_{nl}^{TM} k_{cn}^{TM}} \frac{l\pi}{L} \left(\frac{\chi_{qp}}{a}\right)^2 e\left(\chi_{qp} \frac{\rho}{a}\right) \begin{pmatrix} \cos q\varphi \\ \sin q\varphi \end{pmatrix} \cos \frac{l\pi}{L} z,$$

$$\tilde{\mathbf{h}}_{nl}^{TE} = \pm \mathbf{u}_\varphi \sqrt{\frac{\varepsilon_q}{\pi L}} \frac{k_{cn}^{TE}}{k_{nl}^{TE}} \frac{q}{\rho} h\left(\chi'_{qp} \frac{\rho}{a}\right) \begin{pmatrix} \sin q\varphi \\ \cos q\varphi \end{pmatrix} \cos \frac{l\pi}{L} z$$
$$- \mathbf{u}_\rho \sqrt{\frac{\varepsilon_q}{\pi L}} \frac{k_{cn}^{TE}}{k_{nl}^{TE}} \frac{\chi'_{qp}}{a} h'\left(\chi'_{qp} \frac{\rho}{a}\right) \begin{pmatrix} \cos q\varphi \\ \sin q\varphi \end{pmatrix} \cos \frac{l\pi}{L} z$$
$$+ \mathbf{u}_z \frac{1}{k_{nl}^{TE} k_{cn}^{TE}} \frac{l\pi}{L} \left(\frac{\chi'_{qp}}{a}\right)^2 h\left(\chi'_{qp} \frac{\rho}{a}\right) \sqrt{\frac{\varepsilon_q}{2\pi}} \begin{pmatrix} \cos q\varphi \\ \sin q\varphi \end{pmatrix} \sqrt{\frac{2}{L}} \sin \frac{l\pi}{L} z.$$

For the notations used above, please refer to Section 2.1.

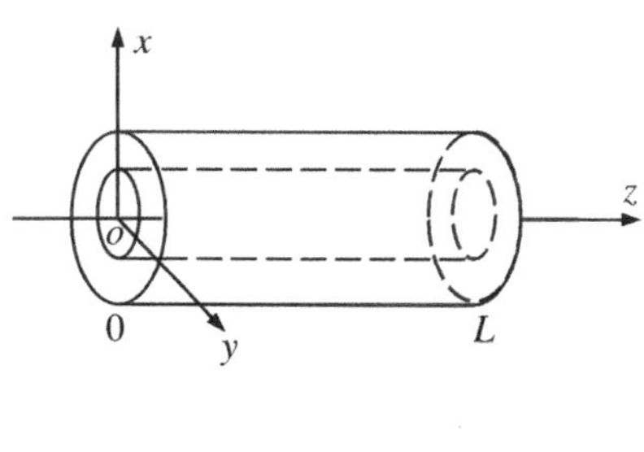

Figure 3.9 Coaxial waveguide cavity resonator.

3.5 Radiation in Waveguide Revisited

An infinite waveguide can be formed by extending the two ends of the waveguide cavity to infinity. For this purpose, one can make use of the coordinate transform $z \to z - L/2$ to shift the origin of the original coordinate system to the center of the cavity. The waveguide cavity becomes an infinite waveguide as the length L tends to infinity, and the dyadic Green's functions for the waveguide cavity can be shown to approach correspondingly to those for the waveguide discussed in Chapter 2. By the coordinate transform $z \to z - L/2$, the discrete eigenvalues $2l\pi/L$ or $(2l-1)\pi/L$ in the dyadic Green's functions for the waveguide cavity become closer together as $L \to \infty$, and eventually they merge in the limit into the positive axis. As a result, the infinite sums become integrals. After expanding the electric dyadic Green's function (3.80), one may find that a number of summations must be replaced by integrals as $L \to \infty$. They are summarized below.

1) $$\lim_{L\to\infty}\sum_{l=0}^{\infty}\frac{1}{\left(\Gamma_n^{TEM,TE}\right)^2}\frac{2}{L}\sin\frac{l\pi}{L}z_s'\sin\frac{l\pi}{L}z_s = \frac{1}{2\gamma_n^{TEM,TE}}e^{-\gamma_n^{TEM,TE}|z-z'|},$$

2) $$\lim_{L\to\infty}\sum_{l=0}^{\infty}\frac{1}{\left(k_{nl}^{TM}\Gamma_n^{TM}\right)^2}\left(\frac{l\pi}{L}\right)^2\frac{\varepsilon_l}{L}\sin\frac{l\pi}{L}z_s'\sin\frac{l\pi}{L}z_s$$
$$= -\frac{1}{2k_e^2}\left(\gamma_n^{TM}e^{-\gamma_n^{TM}|z-z'|} - k_{cn}^{TM}e^{-k_{cn}^{TM}|z-z'|}\right),$$

3) $$\lim_{L\to\infty}\sum_{l=0}^{\infty}\frac{1}{\left(k_{nl}^{TM}\Gamma_n^{TM}\right)^2}\frac{\varepsilon_l}{L}\cos\frac{l\pi}{L}z_s'\cos\frac{l\pi}{L}z_s$$
$$= \frac{1}{2k_{cn}^{TM}\gamma_n^{TM}k_e^2}\left(k_{cn}^{TM}e^{-\gamma_n^{TM}|z-z'|} - \gamma_n^{TM}e^{-k_{cn}^{TM}|z-z'|}\right),$$

4) $$\lim_{L\to\infty}\sum_{l=0}^{\infty}\frac{1}{\left(k_{nl}^{TM}\Gamma_n^{TM}\right)^2}\frac{l\pi}{L}\frac{\varepsilon_l}{L}\cos\frac{l\pi}{L}z_s'\sin\frac{l\pi}{L}z_s$$
$$= \frac{1}{2k_e^2}\left(e^{-\gamma_n^{TM}|z-z'|} - e^{-k_{cn}^{TM}|z-z'|}\right)\operatorname{sgn}(z-z'),$$

5) $$\lim_{L\to\infty}\sum_{l=0}^{\infty}\left(\frac{k_{cn}^{TM}}{k_{nl}^{TM}k_e}\right)^2\frac{\varepsilon_l}{L}\sin\frac{l\pi}{L}z_s'\sin\frac{l\pi}{L}z_s = \frac{e^{-k_{cn}^{TM}|z-z'|}}{2k_{cn}^{TM}},$$

6) $$\lim_{L\to\infty}\sum_{l=0}^{\infty}\left(\frac{1}{k_{cn}^{TM}k_{nl}^{TM}k_e}\frac{l\pi}{L}\right)^2\frac{\varepsilon_l}{L}\cos\frac{l\pi}{L}z'\cos\frac{l\pi}{L}z$$
$$= \frac{1}{\left(k_e k_{cn}^{TM}\right)^2}\left[\delta(z-z') + \frac{k_{cn}^{TM}e^{-k_{cn}^{TM}|z-z'|}}{2}\right],$$

7) $$\lim_{L\to\infty}\sum_{l=0}^{\infty}\frac{1}{\left(k_{nl}^{TM}k_e\right)^2}\frac{l\pi}{L}\frac{\varepsilon_l}{L}\sin\frac{l\pi}{L}z_s'\cos\frac{l\pi}{L}z_s = -\frac{1}{2k_e^2}e^{-k_{cn}^{TM}|z-z'|}\operatorname{sgn}(z-z'),$$

8) $\lim_{L\to\infty}\sum_{l=0}^{\infty}\frac{k_{cn}^{TM}}{\left(k_{nl}^{TM}k_e\right)^2}\frac{l\pi}{L}\frac{\varepsilon_l}{L}\cos\frac{l\pi}{L}z'\sin\frac{l\pi}{L}z=\frac{k_{cn}^{TM}}{2k_e^2}e^{-k_{cn}^{TM}|z-z'|}\operatorname{sgn}(z-z')$.

In the above, $z_s = z-(L/2), z_s' = z'-(L/2)$, and

$$\gamma_n^{TEM,TE,TM} = \begin{cases} j\sqrt{k_e^2-\left(k_{cn}^{TEM,TE,TM}\right)^2}, k > k_{cn}^{TEM,TE,TM} \\ \sqrt{\left(k_{cn}^{TEM,TE,TM}\right)^2-k_e^2}, k < k_{cn}^{TEM,TE,TM} \end{cases}$$

are the propagation constants of the modes in waveguide. After the limiting process, the electric dyadic Green's function (3.80) can be written as

$$\begin{aligned}\overleftrightarrow{\mathbf{G}}_e(\mathbf{r},\mathbf{r}') &= \sum_n \frac{1}{2\gamma_n^{TEM}}\mathbf{E}_{n\pm}^{TEM}(\mathbf{r})\mathbf{E}_{n\mp}^{TEM}(\mathbf{r}') \\ &+\sum_n \frac{1}{2\gamma_n^{TE}}\mathbf{E}_{n\pm}^{TE}(\mathbf{r})\mathbf{E}_{n\mp}^{TE}(\mathbf{r}') - \sum_n \frac{\gamma_n^{TM}}{2k_e^2}\mathbf{E}_{n\pm}^{TM}(\mathbf{r})\mathbf{E}_{n\mp}^{TM}(\mathbf{r}') \\ &-\sum_n \mathbf{u}_z\mathbf{u}_{z'}\frac{\nabla\cdot\mathbf{e}_n^{TM}(\boldsymbol{\rho})}{k_{cn}^{TM}}\frac{\nabla\cdot\mathbf{e}_n^{TM}(\boldsymbol{\rho}')}{k_{cn}^{TM}}\frac{1}{k_e^2}\delta(z-z'), \quad \begin{matrix} z>z' \\ z<z' \end{matrix},\end{aligned} \tag{3.106}$$

where the fundamental field patterns (2.69) in the waveguide have been used. Equation (3.106) agrees with (2.65) for the waveguide. It can be shown that other dyadic Green's functions also approach to their counterparts in the waveguide as $L\to\infty$.

A semi-infinite waveguide shorted at $z=0$ can be formed from a waveguide cavity by letting its length L approach to infinity. The dyadic Green's function for the semi-infinite waveguide can be found as follows:

$$\begin{aligned}\overleftrightarrow{\mathbf{G}}_e(\mathbf{r},\mathbf{r}') &= \frac{1}{2}\sum_n \frac{Z_{wn}}{j\omega\mu}\left[\mathbf{E}_{n\pm}(\mathbf{r})\mathbf{E}_{n\mp}(\mathbf{r}') - \mathbf{E}_{n+}(\mathbf{r})\mathbf{E}_{n+}(\mathbf{r}')\right] \\ &-\mathbf{u}_z\mathbf{u}_{z'}\frac{1}{k_e^2}\delta(\mathbf{r}-\mathbf{r}'), \quad \begin{matrix} z>z' \\ z<z' \end{matrix},\end{aligned} \tag{3.107}$$

where the summation is over all types of modes.

3.6 Transient Fields in Cavity Resonator

The time-domain response inside a metal cavity resonator is uniquely determined by the boundary conditions, initial conditions, and source conditions. A transient source in cavity will excite an infinite number of modes and the total fields in the cavity are the linear combination of these modes. In the following, our study will be confined to the time-domain response of waveguide cavity resonator. Consider a waveguide cavity with a perfect electric wall of length L, as shown in Figure 3.6. The transient EM fields inside the waveguide cavity, excited by current sources $\mathbf{J}$ and $\mathbf{J}_m$, can be expanded in terms of the transverse vector modal functions $\mathbf{e}_n$ in the waveguide

$$\mathbf{E}(\mathbf{r},t) = \sum_{n=1}^{\infty} v_n(z,t)\mathbf{e}_n(\boldsymbol{\rho}) + \mathbf{u}_z \sum_{n=1}^{\infty} \frac{\nabla \cdot \mathbf{e}_n(\boldsymbol{\rho})}{k_{cn}} e_n(z,t),$$
$$\mathbf{H}(\mathbf{r},t) = \sum_{n=1}^{\infty} i_n(z,t)\mathbf{u}_z \times \mathbf{e}_n(\boldsymbol{\rho}) + \mathbf{u}_z \frac{1}{\sqrt{\Omega}} \int_\Omega \frac{\mathbf{u}_z \cdot \mathbf{H}}{\sqrt{\Omega}} d\Omega + \sum_{n=1}^{\infty} \frac{\nabla \times \mathbf{e}_n(\boldsymbol{\rho})}{k_{cn}} h_n(z,t), \tag{3.108}$$

where

$$v_n = \int_\Omega \mathbf{E} \cdot \mathbf{e}_n d\Omega, \qquad i_n = \int_\Omega \mathbf{H} \cdot \mathbf{u}_z \times \mathbf{e}_n d\Omega,$$
$$e_n = \int_\Omega \mathbf{u}_z \cdot \mathbf{E}\left(\frac{\nabla \cdot \mathbf{e}_{tn}}{k_{cn}}\right) d\Omega, \quad h_n = \int_\Omega \mathbf{H} \cdot \left(\frac{\nabla \times \mathbf{e}_{tn}}{k_{cn}}\right) d\Omega.$$

Similar to the study of the transient fields in waveguide, the modal voltage v_n^{TEM} and the modal current i_n^{TEM} for the TEM mode, respectively, satisfy the wave equations

$$\frac{\partial^2 v_n^{TEM}}{\partial z^2} - \frac{1}{v^2}\frac{\partial^2 v_n^{TEM}}{\partial t^2} - \sigma \frac{\eta}{v}\frac{\partial v_n^{TEM}}{\partial t} = \frac{\eta}{v}\frac{\partial}{\partial t}\int_\Omega \mathbf{J} \cdot \mathbf{e}_n d\Omega - \frac{\partial}{\partial z}\int_\Omega \mathbf{J}_m \cdot \mathbf{u}_z \times \mathbf{e}_n d\Omega,$$
$$\frac{\partial^2 i_n^{TEM}}{\partial z^2} - \frac{1}{v^2}\frac{\partial^2 i_n^{TEM}}{\partial t^2} - \sigma \frac{\eta}{v}\frac{\partial i_n^{TEM}}{\partial t} = \sigma \int_\Omega \mathbf{J}_m \cdot \mathbf{u}_z \times \mathbf{e}_n d\Omega$$
$$- \frac{\partial}{\partial z}\int_\Omega \mathbf{J} \cdot \mathbf{e}_n d\Omega + \frac{1}{\eta v}\frac{\partial}{\partial t}\int_\Omega \mathbf{J}_m \cdot \mathbf{u}_z \times \mathbf{e}_n d\Omega. \tag{3.109}$$

Once v_n^{TEM} (or i_n^{TEM}) are known, i_n^{TEM} (or v_n^{TEM}) can be determined by the time integration of v_n^{TEM} (or i_n^{TEM}). The modal voltage v_n^{TE} for the TE mode satisfies the modified Klein–Gordon equation

$$\frac{\partial^2 v_n^{TE}}{\partial z^2} - \frac{1}{v^2}\frac{\partial^2 v_n^{TE}}{\partial t^2} - \sigma \frac{\eta}{v}\frac{\partial v_n^{TE}}{\partial t} - k_{cn}^2 v_n^{TE} = \frac{\eta}{v}\frac{\partial}{\partial t}\int_\Omega \mathbf{J} \cdot \mathbf{e}_n d\Omega - \frac{\partial}{\partial z}\int_\Omega \mathbf{J}_m \cdot \mathbf{u}_z \times \mathbf{e}_n d\Omega$$
$$+ k_{cn}\int_\Omega (\mathbf{u}_z \cdot \mathbf{J}_m)\left(\frac{\mathbf{u}_z \cdot \nabla \times \mathbf{e}_n}{k_{cn}}\right) d\Omega. \tag{3.110}$$

The modal current i_n^{TE} for the TE mode can be determined by the time integration of $\partial v_n^{TE}/\partial z$:

$$\begin{aligned} i_n^{TE}(z,t) &= -\frac{\eta}{v}\int_{-\infty}^{t}\frac{\partial v_n^{TE}(z,t')}{\partial z}dt' \\ &-\frac{\eta}{v}\int_{-\infty}^{t}\left\{\int_{\Omega}\mathbf{J}_m(\mathbf{r},t')\cdot[\mathbf{u}_z\times\mathbf{e}_n(\boldsymbol{\rho})]d\Omega(\boldsymbol{\rho})\right\}dt'. \end{aligned} \tag{3.111}$$

The modal current i_n^{TM} for the TM mode also satisfies the modified Klein–Gordon equation

$$\begin{aligned} \frac{\partial^2 i_n^{TM}}{\partial z^2}-\frac{1}{v^2}\frac{\partial^2 i_n^{TM}}{\partial t^2}-\sigma\frac{\eta}{v}\frac{\partial i_n^{TM}}{\partial t}-k_{cn}^2 i_n^{TM} &= \sigma\int_{\Omega}\mathbf{J}_m\cdot\mathbf{u}_z\times\mathbf{e}_n d\Omega-\frac{\partial}{\partial z}\int_{\Omega}\mathbf{J}\cdot\mathbf{e}_n d\Omega \\ &+\frac{1}{\eta v}\frac{\partial}{\partial t}\int_{\Omega}\mathbf{J}_m\cdot\mathbf{u}_z\times\mathbf{e}_n d\Omega-k_{cn}\int_{\Omega}\mathbf{u}_z\cdot\mathbf{J}\left(\frac{\nabla\cdot\mathbf{e}_n}{k_{cn}}\right)d\Omega. \end{aligned} \tag{3.112}$$

The modal voltage v_n^{TM} can then be determined by the time integration of $\partial i_n^{TM}/\partial z$:

$$v_n^{TM}(z,t) = -\eta v\int_{-\infty}^{t}\frac{\partial i_n^{TM}(z,t')}{\partial z}dt'-\eta v\int_{-\infty}^{t}\left[\int_{\Omega}\mathbf{J}(\mathbf{r},t')\cdot\mathbf{e}_n(\boldsymbol{\rho})d\Omega(\boldsymbol{\rho})\right]dt'. \tag{3.113}$$

Similar to (3.58) and (3.59), the time-domain modal voltage must satisfy the homogeneous Dirichlet boundary conditions

$$v_n(z,t)\Big|_{z=0}=v_n(z,t)\Big|_{z=L}=0, \tag{3.114}$$

and the time-domain current must satisfy the homogeneous Neumann boundary conditions

$$\frac{\partial i_n(z,t)}{\partial z}\bigg|_{z=0}=\frac{\partial i_n(z,t)}{\partial z}\bigg|_{z=L}=0. \tag{3.115}$$

In order to solve (3.109), (3.110), and (3.112) subject to the boundary conditions (3.114) and (3.115), one may introduce the retarded Green's function

$$\begin{cases} \left(\dfrac{\partial^2}{\partial z^2}-\dfrac{1}{v^2}\dfrac{\partial^2}{\partial t^2}-\sigma\dfrac{\eta}{v}\dfrac{\partial}{\partial t}-k_{cn}^2\right)G_n^v(z,t;z',t')=-\delta(z-z')\delta(t-t'), \\ G_n^v(z,t;z',t')\Big|_{t<t'}=0, \\ G_n^v(z,t;z',t')\Big|_{z=z_1}=G_n^v(z,t;z',t')\Big|_{z=z_2}=0, \end{cases} \tag{3.116}$$

for the modal voltage, and the retarded Green's function

$$\begin{cases} \left(\dfrac{\partial^2}{\partial z^2} - \dfrac{1}{v^2}\dfrac{\partial^2}{\partial t^2} - \sigma\dfrac{\eta}{v}\dfrac{\partial}{\partial t} - k_{cn}^2\right) G_n^i(z,t;\, z',t') = -\delta(z-z')\delta(t-t'), \\ G_n^i(z,t;\, z',t')\Big|_{t<t'} = 0, \\ \left.\dfrac{\partial G_n^i(z,t;\, z',t')}{\partial z}\right|_{z=z_1} = \left.\dfrac{\partial G_n^i(z,t;\, z',t')}{\partial z}\right|_{z=z_2} = 0, \end{cases} \tag{3.117}$$

for the modal current. The solutions of these equations are [24]

$$\begin{aligned} G_n^v(z,t;z',t') = {} & \sum_{m=1}^{\infty} \frac{2v}{L} \sin\frac{m\pi}{L}(z-z_1) \sin\frac{m\pi}{L}(z'-z_1) \\ & \cdot e^{-\gamma(t-t')} \frac{\sin\left[v(t-t')\sqrt{k_{cn}^2 + \left(\frac{m\pi}{L}\right)^2 - \left(\frac{\gamma}{v}\right)^2}\right]}{\sqrt{k_{cn}^2 + \left(\frac{m\pi}{L}\right)^2 - \left(\frac{\gamma}{v}\right)^2}} H(t-t'), \end{aligned} \tag{3.118}$$

$$\begin{aligned} G_n^i(z,t;z',t') = {} & \sum_{m=0}^{\infty} \frac{\varepsilon_m v}{L} \cos\frac{m\pi}{L}(z-z_1) \cos\frac{m\pi}{L}(z'-z_1) \\ & \cdot e^{-\gamma(t-t')} \frac{\sin\left[v(t-t')\sqrt{k_{cn}^2 + \left(\frac{m\pi}{L}\right)^2 - \left(\frac{\gamma}{v}\right)^2}\right]}{\sqrt{k_{cn}^2 + \left(\frac{m\pi}{L}\right)^2 - \left(\frac{\gamma}{v}\right)^2}} H(t-t'), \end{aligned} \tag{3.119}$$

where $\gamma = \sigma/2\varepsilon$. If one of the ends of the waveguide cavity extends to infinity, say, $L \to \infty$, the discrete values $m\pi/L$ become a continuum. In this case, equations (3.118) and (3.119) become

$$\begin{aligned} G_n^v(z,t;z',t')\Big|_{z_2\to\infty} = {} & -\frac{v}{\pi} e^{-\gamma(t-t')} \\ & \cdot \int_0^{\infty} [\cos k(z+z'-2z_1) - \cos k(z-z')] \frac{\sin\left[v(t-t')\sqrt{k_{cn}^2 + k^2 - \left(\frac{\gamma}{v}\right)^2}\right]}{\sqrt{k_{cn}^2 + k^2 - \left(\frac{\gamma}{v}\right)^2}} dk, \end{aligned}$$

$$\begin{aligned} G_n^i(z,t;z',t')\Big|_{z_2\to\infty} = {} & \frac{v}{\pi} e^{-\gamma(t-t')} \\ & \cdot \int_0^{\infty} [\cos k(z+z'-2z_1) + \cos k(z-z')] \frac{\sin\left[v(t-t')\sqrt{k_{cn}^2 + k^2 - \left(\frac{\gamma}{v}\right)^2}\right]}{\sqrt{k_{cn}^2 + k^2 - \left(\frac{\gamma}{v}\right)^2}} dk. \end{aligned}$$

The above integrations can be carried out, and the retarded Green's functions are found to be

$$e^{\gamma(t-t')}G_n^v(z,t;z',t')|_{L\to\infty}$$
$$= -\frac{v}{2}J_0\left[\left(k_{cn}^2v^2-\gamma^2\right)^{1/2}\sqrt{(t-t')^2-\frac{|z+z'-2z_1|^2}{v^2}}\right]H\left[v(t-t')-|z+z'-2z_1|\right]$$
$$+\frac{v}{2}J_0\left[\left(k_{cn}^2v^2-\gamma^2\right)^{1/2}\sqrt{(t-t')^2-\frac{|z-z'|^2}{v^2}}\right]H\left[v(t-t')-|z-z'|\right], \tag{3.120}$$

$$e^{\gamma(t-t')}G_n^i(z,t;z',t')|_{L\to\infty}$$
$$= \frac{v}{2}J_0\left[\left(k_{cn}^2v^2-\gamma^2\right)^{1/2}\sqrt{(t-t')^2-\frac{|z+z'-2z_1|^2}{v^2}}\right]H\left[v(t-t')-|z+z'-2z_1|\right]$$
$$+\frac{v}{2}J_0\left[\left(k_{cn}^2v^2-\gamma^2\right)^{1/2}\sqrt{(t-t')^2-\frac{|z-z'|^2}{v^2}}\right]H\left[v(t-t')-|z-z'|\right]. \tag{3.121}$$

The retarded Green's functions (3.120) and (3.121) can be used to solve the modified Klein–Gordon equations

$$\left(\frac{\partial^2}{\partial z^2}-\frac{1}{v^2}\frac{\partial^2}{\partial t^2}-\sigma\frac{\eta}{v}\frac{\partial}{\partial t}-k_{cn}^2\right)v_n(z,t)=f(z,t),\ z_1<z<z_2,$$
$$\left(\frac{\partial^2}{\partial z^2}-\frac{1}{v^2}\frac{\partial^2}{\partial t^2}-\sigma\frac{\eta}{v}\frac{\partial}{\partial t}-k_{cn}^2\right)i_n(z,t)=g(z,t),\ z_1<z<z_2,$$

subject to the boundary conditions (3.114) and (3.115), and the solutions are

$$v_n(z,t) = -\int_{z_1}^{z_2}dz'\int_{-\infty}^{\infty}f(z',t')G_n^v(z,t;z',t')dt',\ \ z_1<z<z_2,$$
$$i_n(z,t) = -\int_{z_1}^{z_2}dz'\int_{-\infty}^{\infty}g(z',t')G_n^i(z,t;z',t')dt',\ \ z_1<z<z_2. \tag{3.122}$$

According to (3.122), the solutions of (3.110) and (3.112) can be expressed by

$$v_n^{TE}(z,t) = -\frac{\eta}{v}\int_{z_1}^{z_2}dz'\int_{-\infty}^{\infty}G_n^v(z,t;z',t')dt'\int_{\Omega}\frac{\partial}{\partial t'}\mathbf{J}(\boldsymbol{\rho}',z',t')\cdot\mathbf{e}_n(\boldsymbol{\rho}')d\Omega(\boldsymbol{\rho}')$$
$$-\int_{z_1}^{z_2}dz'\int_{-\infty}^{\infty}\frac{\partial G_n^v(z,t;z',t')}{\partial z'}dt'\int_{\Omega}\mathbf{J}_m(\boldsymbol{\rho}',z',t')\cdot\mathbf{u}_z\times\mathbf{e}_n(\boldsymbol{\rho}')d\Omega(\boldsymbol{\rho}')$$
$$-k_{cn}\int_{z_1}^{z_2}dz'\int_{-\infty}^{\infty}G_n^v(z,t;z',t')dt'\int_{\Omega}\mathbf{u}_z\cdot\mathbf{J}_m(\boldsymbol{\rho}',z',t')\frac{\mathbf{u}_z\cdot\nabla\times\mathbf{e}_n(\boldsymbol{\rho}')}{k_{cn}}d\Omega(\boldsymbol{\rho}'), \tag{3.123}$$

$$i_n^{TM}(z,t) = -\int_{z_1}^{z_2} dz' \int_{-\infty}^{\infty} \frac{\partial G_n^i(z,t;z',t')}{\partial z'} dt' \int_{\Omega} \mathbf{J}(\boldsymbol{\rho}',z',t') \cdot \mathbf{e}_n(\boldsymbol{\rho}') d\Omega(\boldsymbol{\rho}')$$
$$-\frac{1}{\eta v}\int_{z_1}^{z_2} dz' \int_{-\infty}^{\infty} G_n^i(z,t;z',t') dt' \int_{\Omega} \frac{\partial}{\partial t'} \mathbf{J}_m(\boldsymbol{\rho}',z',t') \cdot \mathbf{u}_z \times \mathbf{e}_n(\boldsymbol{\rho}') d\Omega(\boldsymbol{\rho}')$$
$$+ k_{cn}\int_{z_1}^{z_2} dz' \int_{-\infty}^{\infty} G_n^i(z,t;z',t') dt' \int_{\Omega} \mathbf{u}_z \cdot \mathbf{J}(\boldsymbol{\rho}',z',t') \frac{\nabla \cdot \mathbf{e}_n(\boldsymbol{\rho}')}{k_{cn}} d\Omega(\boldsymbol{\rho}'). \tag{3.124}$$

In deriving the above expressions, it has been assumed that all sources are confined inside the cavity. It should be notified that the time-domain voltage and current do not satisfy the homogeneous boundary conditions (3.114) and (3.115) at $z = 0$ or $z = L$ if the magnetic current $\mathbf{J}_m$ is tightly pressed on the electric wall $z = 0$ or $z = L$.

Example 3.1 Consider a shorted rectangular waveguide shown in Figure 3.10. The shorted waveguide is excited by a line current extending across the waveguide centered at ($x = a/2$, $z = z_0$):

$$\mathbf{J}(\mathbf{r},t) = \mathbf{u}_y \delta\left(x - \frac{a}{2}\right)\delta(z - z_0) f(t).$$

By the symmetry of the structure and the excitation condition, only TE$_{n0}$ modes will be excited. If the input signal is of the form $f(t) = H(t)\sin \omega t$ and the heat loss is ignored, the time-domain voltages may be found from (3.120) and (3.123) as follows:

$$\begin{aligned} v_n^{TE}(z,t) &= \frac{b\eta}{2}\left(\frac{2}{ab}\right)^{1/2} ka \sin\frac{n\pi}{2} \\ &\cdot \left\{ \int_{|z-z_0|/a}^{vt/a} \cos ka\left(\frac{vt}{a} - u\right) J_0\left[k_{cn}a\sqrt{u^2 - \frac{|z-z_0|^2}{a^2}}\right] du \right. \\ &\left. - \int_{|z+z_0|/a}^{vt/a} \cos ka\left(\frac{vt}{a} - u\right) J_0\left[k_{cn}a\sqrt{u^2 - \frac{|z+z_0|^2}{a^2}}\right] du \right\}. \end{aligned}$$

Due to the existence of radiation loss in the shorted waveguide, the time-domain responses may be divided into the sum of a steady-state part and a transient part

$$v_n^{TE}(z,t) = v_n^{TE}(z,t)\big|_{steady} + v_n^{TE}(z,t)\big|_{transient},$$

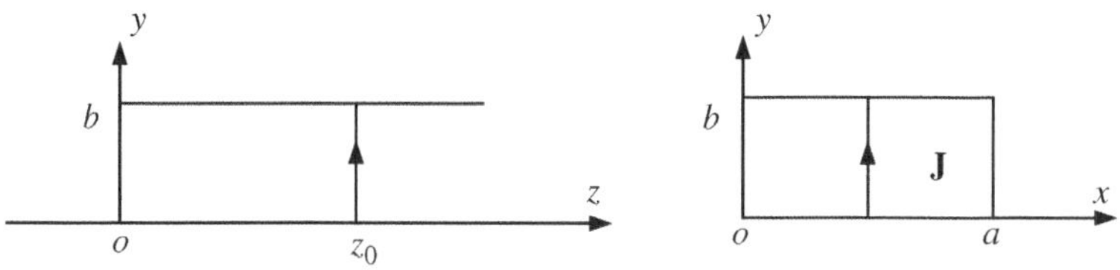

Figure 3.10 A shorted rectangular waveguide excited by a centered current source.

where

$$
\begin{aligned}
v_n^{TE}(z,t)\big|_{steady} &= \frac{b\eta}{2}\left(\frac{2}{ab}\right)^{1/2} ka \sin\frac{n\pi}{2} \\
&\cdot\left\{\int_{|z-z_0|/a}^{\infty} \cos ka\left(\frac{vt}{a-u}\right) J_0\left[k_{cn}a\sqrt{u^2-\frac{|z-z_0|^2}{a^2}}\right] du\right. \\
&\left.-\int_{|z+z_0|/a}^{\infty} \cos ka\left(\frac{vt}{a-u}\right) J_0\left[k_{cn}a\sqrt{u^2-\frac{|z+z_0|^2}{a^2}}\right] du\right\}, \\
v_n^{TE}(z,t)\big|_{transient} &= -\frac{b\eta}{2}\left(\frac{2}{ab}\right)^{1/2} ka \sin\frac{n\pi}{2} \\
&\cdot\left\{\int_{vt/a}^{\infty} \cos ka\left(\frac{vt}{a-u}\right) J_0\left[k_{cn}a\sqrt{u^2-\frac{|z-z_0|^2}{a^2}}\right] du\right. \\
&\left.-\int_{vt/a}^{\infty} \cos ka\left(\frac{vt}{a-u}\right) J_0\left[k_{cn}a\sqrt{u^2-\frac{|z+z_0|^2}{a^2}}\right] du\right\}.
\end{aligned}
$$

The transient part approaches to zero as $t \to \infty$. The integrals in the steady-state part can be carried out

$$
v_n^{TE}(z,t)\big|_{\text{steady}} = \frac{b\eta}{2}\left(\frac{2}{ab}\right)^{1/2} \frac{ka}{\sqrt{\left|(ka)^2-(k_{cn}a)^2\right|}} \sin\frac{n\pi}{2}
$$

$$
\times\begin{cases}
\sin\left(ka\dfrac{vt}{a}-\dfrac{|z-z_0|}{a}\sqrt{(ka)^2-(k_{cn}a)^2}\right) \\
-\sin\left(ka\dfrac{vt}{a}-\dfrac{|z+z_0|}{a}\sqrt{(ka)^2-(k_{cn}a)^2}\right), & k > k_{cn} \\
\cos\left(ka\dfrac{vt}{a}\right)\exp\left[-\dfrac{|z-z_0|}{a}\sqrt{(k_{cn}a)^2-(ka)^2}\right] \\
-\cos\left(ka\dfrac{vt}{a}\right)\exp\left[-\dfrac{|z+z_0|}{a}\sqrt{(k_{cn}a)^2-(ka)^2}\right], & k < k_{cn}
\end{cases}
$$

In the region $0 < z < z_0$, the steady-state response may be rewritten as

$$v_n^{TE}(z,t)\big|_{\text{steady}} = \frac{1}{\sqrt{2}}\left(\frac{b}{a}\right)^{1/2}\frac{\eta k}{\beta_n}\sin\frac{n\pi}{2}\cdot\begin{cases} 2\sin(\beta_n z)\cos(\omega t - \beta_n z_0), & k > k_{cn} \\ \cos(\omega t)\exp[\beta_n(z - z_0)] - \\ \cos(\omega t)\exp[-\beta_n(z + z_0)], & k < k_{cn} \end{cases},$$

where $\beta_n = \left|k^2 - k_{cn}^2\right|^{1/2}$. The time-domain voltage for the TE_{n0} mode in the shorted waveguide is a standing wave if the operating frequency is higher than the cutoff frequency of the TE_{n0} mode. The time-domain currents can be determined by (3.111):

$$\begin{aligned} i_n^{TE}(z,t) &= \left(\frac{2b}{a}\right)^{1/2}\sin\frac{n\pi}{2}\left[-\frac{1}{2}\sin\omega\left(t - \frac{|z+z_0|}{v}\right) - \frac{1}{2}\sin\omega\left(t - \frac{|z+z_0|}{v}\right)\right] \\ &+ \left(\frac{2b}{a}\right)^{1/2}\sin\frac{n\pi}{2}\left\{\frac{k_{cn}(z+z_0)}{2}\int_0^{t-|z+z_0|/v}\frac{J_1\left[k_{cn}v\sqrt{(t-t')^2 - \frac{|z+z_0|^2}{v^2}}\right]}{\sqrt{(t-t')^2 - \frac{|z+z_0|^2}{v^2}}}\sin\omega t'dt'\right. \\ &\left. - \frac{k_{cn}(z-z_0)}{2}\int_0^{t-|z-z_0|/v}\frac{J_1\left[k_{cn}v\sqrt{(t-t')^2 - \frac{|z+z_0|^2}{v^2}}\right]}{\sqrt{(t-t')^2 - \frac{|z+z_0|^2}{v^2}}}\sin\omega t'dt'\right\}. \end{aligned}$$

The steady-state part of $i_n^{TE}(z,t)$ is

$$\begin{aligned} i_n^{TE}(z,t)\big|_{steady} \\ &= \left(\frac{2b}{a}\right)^{1/2}\sin\frac{n\pi}{2}\left[-\frac{1}{2}\sin\omega\left(t - \frac{|z+z_0|}{v}\right) - \frac{1}{2}\sin\omega\left(t - \frac{|z+z_0|}{v}\right)\right] \\ &+ \left(\frac{2b}{a}\right)^{1/2}\sin\frac{n\pi}{2}\left\{\frac{k_{cn}(z+z_0)}{2}\int_{|z+z_0|/v}^{\infty}\frac{J_1\left[k_{cn}v\sqrt{u^2 - \frac{|z+z_0|^2}{v^2}}\right]}{\sqrt{u^2 - \frac{|z+z_0|^2}{v^2}}}\sin\omega(t-u)du\right. \\ &\left. - \frac{k_{cn}(z-z_0)}{2}\int_{|z-z_0|/v}^{\infty}\frac{J_1\left[k_{cn}v\sqrt{u^2 - \frac{|z+z_0|^2}{v^2}}\right]}{\sqrt{u^2 - \frac{|z+z_0|^2}{v^2}}}\sin\omega(t-u)du\right\}. \end{aligned}$$

Assuming $k > k_{cn}$, one may find the steady-state part

$$i_n^{TE}(z,t)\big|_{steady} = -\frac{2}{\sqrt{2}}\left(\frac{b}{a}\right)^{1/2}\sin\frac{n\pi}{2}\cos(\beta_n z)\sin(\omega t - \beta_n z_0)$$

for $0 < z < z_0$. Let $V_n^{TE}(z)$ and $I_n^{TE}(z)$ denote the phasors of the steady responses $v_n^{TE}(z,t)\big|_{\text{steady}}$ and $i_n^{TE}(z,t)\big|_{\text{steady}}$, respectively. Then,

$$V_n^{TE}(z) = \frac{2}{\sqrt{2}}\left(\frac{b}{a}\right)^{1/2}\frac{\eta k}{\beta_n}\sin\frac{n\pi}{2}\sin(\beta_n z)e^{-j\beta_n z_0},$$

$$I_n^{TE}(z) = j\frac{2}{\sqrt{2}}\left(\frac{b}{a}\right)^{1/2}\sin\frac{n\pi}{2}\cos(\beta_n z)e^{-j\beta_n z_0},$$

for $k > k_{cn}$. Since the currents are assumed to be in positive z-direction, the impedances for the TE modes at $z \in (0, z_0)$ are given by

$$Z_n(z) = \frac{V_n^{TE}(z)}{-I_n^{TE}(z)} = j\frac{\eta k}{\beta_n}\tan(\beta_n z), k > k_{cn}.$$

This is a well-known result in time-harmonic EM theory. □

> Not only in geometry, but to a still more astonishing degree in physics, has it become more and more evident that as soon as we have succeeded in unraveling fully the natural laws which govern reality, we find them to be expressible by mathematical relations of surprising simplicity and architectonic perfection. It seems to me to be one of the chief objects of mathematical instruction to develop the faculty of perceiving this simplicity and harmony.
>
> – *Hermann Weyl (German mathematician, theoretical physicist, 1885–1955)*

References

1 Weyl, H., "Uber die randwertaufgabe der strahlungstheorie und asympotische spektralgesetze", *J. Reine. Angew. Math.*, Vol. 143, pp. 177–202, 1913.
2 Bladel, J. V., "On Helmholtz's theorem in multiply-bounded and multiply-connected regions", *J. Frankl. Inst.*, Vol. 269, No. 6, pp. 445–462, 1960.
3 Goubau, G., *Electromagnetische Wellenleiter und Holraume*, Stuttgart, Germany, Wissenschaftliche Verlagsgesellschaft mbH, pp. 80–97, 1955.
4 Jones, D. S., *The Theory of Electromagnetism*, Pergamon Press, 1964.

5 Kurokawa, K., "The Expansions of electromagnetic fields in cavities", *IRE Trans. Microw. Theory Tech.*, Vol. 6, pp. 178–187, 1958.

6 Schelkunoff, S. A., "On representation of electromagnetic fields in cavities in terms of natural modes of oscillation", *J. Appl. Phys.*, Vol. 24, pp. 262–267, 1953.

7 Slater, J. C., *Microwave Electronics*, New York, D. Van Nostrand Co., Inc., pp. 57–83, 1950.

8 Teichmann, T. and E. P. Wigner, "Electromagnetic field expansion in loss-free cavities excited through holes", *J. Appl. Phys.*, Vol. 24, pp. 262–267, 1953.

9 Balanis, C. A., *Advanced Engineering Electromagnetics*, 2nd Ed., John Wiley & Sons, 2012.

10 Bladel, J. V., *Electromagnetic Fields*, IEEE Press, 2007.

11 Collin, R. E., *Field Theory of Guided Waves*, 2nd Ed., IEEE Press, 1991.

12 Collin, R. E., *Foundations for Microwave Engineering*, 2nd Ed., IEEE Press, 2001.

13 Eom, H. J., *Electromagnetic Wave Theory for Boundary-Value Problems: An Advanced Course on Analytical Methods*, Springer, 2004.

14 Greiner, W., *Classical Electrodynamics*, Springer, 1998.

15 Harrington, R. F., *Time-Harmonic Electromagnetic Fields*, McGraw-Hill Book Company, Inc., 1961.

16 Hill, D. A., *Electromagnetic Fields in Cavities: Deterministic and Statistical Theories*, Wiley-IEEE Press, 2009.

17 Ishimaro, A., *Electromagnetic Wave Propagation, Radiation and Scattering*, Prentice Hall, 1991.

18 Jackson, J. D., *Classical Electrodynamics*, 3rd Ed., New York, John Wiley & Sons, 1998.

19 Kurokawa, K., *An Introduction to Microwave Circuits*, New York, Academic Press, 1969.

20 Montgomery, C. G., R. H. Dicke, and E. M. Purcell, *Principles of Microwave Circuits*, McGraw-Hill, 1948.

21 Pozar, D. M., *Microwave Engineering*, John Wiley & Sons, 1998.

22 Ramo, S. and J. R. Whinnery, *Fields and Waves in Modern Radio*, John and Wiley & Sons, 1953.

23 Smythe, W. R., *Static and Dynamic Electricity*, McGraw-Hill, 1950.

24 Geyi, W., "Time-domain theory of metal cavity resonator", *Prog. Electromagn. Res.*, Vol. 78, pp. 219–253, 2008.

4

Radiation in Free Space (I)

Generic Properties

Intuition is more important to discovery than logic.
–Henri Poincare (French mathematician, 1854–1912)

This chapter studies the antenna radiation in free space. Physically an antenna consists of a scatterer and an impressed source. The impressed source generates an incident field, and the latter induces currents (conduction current or displacement current) on the scatterer, which gives off electromagnetic (EM) waves. Antenna theory usually contains three different but related subjects: generic properties of antenna, antenna analysis, and antenna synthesis. The generic properties of antenna are meant to be valid for all antennas, and they are the fundamentals of antenna design. Antenna analysis examines the radiation properties of antenna with a specified current distribution while the antenna synthesis determines the scatterer and its geometry (type of antenna) so that a desired radiation pattern can be achieved. Antennas are used for both transmission and reception. In most applications, the EM field energy is fed to the scatterer by a feeding waveguide. Since the free space may be treated as a spherical waveguide, the antenna can be viewed as a waveguide junction connecting the feeding waveguide and the spherical waveguide, transforming guided modes into spherical modes in transmitting mode, or transforming spherical modes into guided modes in receiving mode.

There are many parameters for characterizing antenna, including gain, bandwidth, efficiency, input impedance, radiation pattern, beamwidth, sidelobes, front-to-back ratio, and polarization. Many books have been focused on the evaluation and optimization of these parameters for various antennas [1–25]. To satisfy one parameter requirement, one may have to sacrifice one or more other parameter levels. A useful performance index for describing antenna is the product of antenna gain and bandwidth since they must be maximized simultaneously for most applications. It can be shown that antenna fractional bandwidth is reciprocal to antenna quality factor (QF). Thus, the product of antenna gain and bandwidth

Foundations of Antenna Radiation Theory: Eigenmode Analysis, First Edition. Wen Geyi.

Published 2023 by John Wiley & Sons, Inc.

can be expressed as the ratio of antenna gain over antenna QF. Most of the antenna parameters are subject to certain limitations that depend on the antenna size, which can be understood by treating the free space as a spherical waveguide. The propagating modes supported by the spherical waveguide depend on the size of the smallest circumscribing sphere of antenna. The bigger the antenna size (the size of the circumscribing sphere), the more propagating modes in the spherical waveguide will be excited. When the antenna size becomes very small, no propagating modes can exist in the spherical waveguide and all the spherical modes are rapidly cut off, which implies a large stored field energy of antenna and small radiation power, and therefore a large QF. For its importance, a detailed theory of spherical waveguide will be presented in this chapter, based on which various generic properties of antenna can be established.

Some highlights of this chapter include a novel treatment of the dyadic Green's functions for the spherical waveguide, a new derivation for the stored field energy in a general medium, a definition of the stored field energy of antenna in an arbitrary medium based on a conservation law for the stored field energy in the medium, the field and circuit methods for evaluating antenna QF, a comprehensive study of the modal quality factors (modal QFs), the upper bounds on the product of gain and bandwidth (PGB) for both directional and omnidirectional antenna with their applications in small antenna designs, and a time-domain theory for the field expansions in the spherical waveguide.

4.1 Antenna Parameters

Antenna parameters are used to evaluate antenna performances and most of them are defined in transmitting mode. Since antenna is a passive linear reciprocal device in many situations, most of these parameters are implicitly valid for antenna in receiving mode. Antenna parameters are often mutually restricted in the sense that one of them is satisfied at the cost of the degradation of the others.

4.1.1 Power, Efficiencies, and Input Impedance

An arbitrary transmitting antenna system and a receiving antenna system are shown in Figure 4.1. The definitions of the power quantities that appear in Figure 4.1 are listed in Table 4.1, and they are defined at the input terminals (i.e. reference planes) of the matching network and antenna and are used to define various efficiencies. The power accepted by the transmitting antenna is given by

$$P^a = \frac{1}{2}\,\mathrm{Re}\,V\bar{I} = P^{in} - P^{ref} - P^{loss}_m,$$

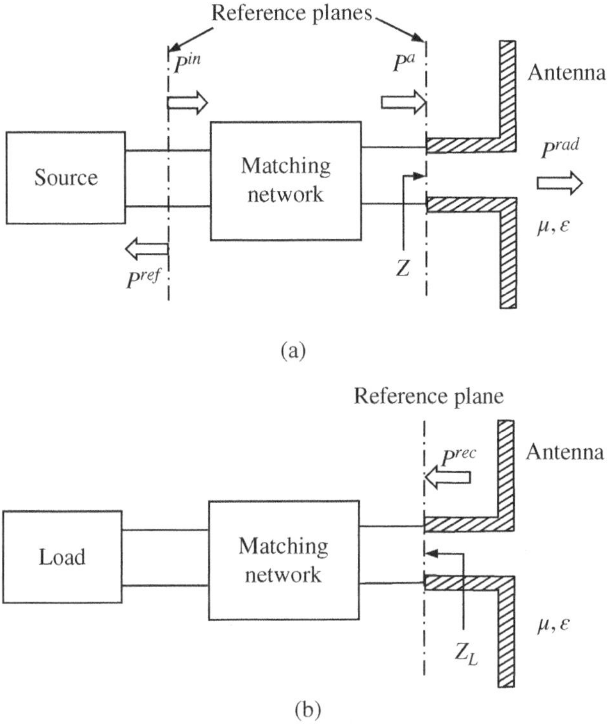

Figure 4.1 (a) Transmitting antenna. (b) Receiving antenna.

Table 4.1 Power quantities.

Parameter	Definition
P^{in}	Power input to the matching network of transmitting antenna.
P^{a}	Power accepted by transmitting antenna.
P^{rad}	Power radiated by transmitting antenna.
P^{ref}	Power reflected back to the source by matching network.
P^{rec}	Power received by the load of receiving antenna.

where V and I are, respectively, the modal voltage and modal current at the antenna input terminal for the dominant mode in the feeding waveguide, and P_m^{loss} is the power loss in the matching network. Let **E** and **H**, respectively, denote the electric field and magnetic field produced by the antenna. The radiated power of the antenna can be represented by

$$P^{rad} = \frac{1}{2}\int_S \text{Re}\left(\mathbf{E} \times \overline{\mathbf{H}}\right) \cdot \mathbf{u}_n dS, \tag{4.1}$$

where S is an arbitrary surface enclosing the antenna. Not all the input power to the antenna will be radiated to free space. The power loss may come from the impedance mismatch that causes portion of the input power reflected back to the transmitter, or from the imperfect conductors and dielectrics that cause portion of the input power to be dissipated into heat. Various efficiencies associated with antenna are summarized in Table 4.2.

The **input impedance** of antenna is defined as the ratio of the voltage to current at the input terminal of the antenna. The **bandwidth** of an antenna is defined as the range of frequencies within which the performance of the antenna, with respect to some characteristics (such as the input impedance, return loss, gain, radiation efficiency, pattern, beamwidth, polarization, sidelobe level, and beam direction), conforms to a specified standard. Antenna bandwidth is an important quantity, which measures the quality of signal transmission such as signal distortion. For broadband antennas, the bandwidth is usually expressed as the ratio of the upper-to-lower frequencies of acceptable operation. For narrow-band (NB) antennas, the bandwidth is expressed as a ratio (percentage) of the frequency difference (upper minus lower) over the center frequency of the bandwidth (fractional bandwidth). The bandwidth can be enhanced by introducing multiple resonant frequencies, losses, parasitic elements, loading or changing matching network.

Table 4.2 Efficiencies.

Parameter	Definition
Radiation efficiency	$e_r = \frac{P^{rad}}{P^a}$: Describe the conduction and dielectric losses of the antenna.
Matching network efficiency	$e_{m1} = \frac{P^{in} - P^{ref}}{P^{in}} = 1 - \lvert\Gamma\rvert^2$ (mismatch efficiency): Describe the mismatch of the matching network. $e_{m2} = \frac{P^a}{P^{in} - P^{ref}}$: Describe the loss in the matching network. $e_m = \frac{P^a}{P^{in}} = \frac{P^{in} - P^{ref}}{P^{in}} \cdot \frac{P^a}{P^{in} - P^{ref}} = e_{m1}e_{m2}$: Describe the mismatch and the loss in the matching network.
Antenna efficiency	$e_t = \frac{P^{rad}}{P^{in}} = \frac{P^{in} - P^{ref}}{P^{in}} \frac{P^a}{P^{in} - P^{ref}} \frac{P^{rad}}{P^a} = e_{m1}e_{m2}e_r$: Describe the mismatch and the losses in the matching network and antenna.

4.1.2 Field Regions, Radiation Pattern, Radiation Intensity, Directivity, and Gain

The space around an EM radiator can be divided into **reactive near-field region**, **radiating near-field region**, and **far-field region**. The reactive near-field and the far-field region are, respectively, defined by $r < R_1 = 0.62\sqrt{D^3/\lambda}$ and $r > R_2 = 2D^2/\lambda$. Here, r is the distance from the radiator, D is the largest dimension of the radiator, and λ is the wavelength. The radiating near-field region is defined by $R_1 < r < R_2$, as illustrated in Figure 4.2.

Antennas are used to communicate wirelessly and they are usually located in the far-field region of each other. Many antenna parameters are determined in the far-field region. In this region, the radiation pattern (the angular field distribution) does not change shape with distance; the electric field **E** and magnetic field **H** are orthogonal to each other and are in phase, and they both fall off with distance as $1/r$; the Poynting vector only has a radial component. The far-field region is also called **Fraunhofer region**, named after Joseph von Fraunhofer (German optician, 1787–1826). In the reactive near-field region, the relationship between the electric field **E** and magnetic field **H** is very complicated and the fields change rapidly with the distance. In this region, the Poynting vector contains both radial component and transverse components. The radial component represents the radiating power and the transverse components are reactive. The radiating near-field region is also called **Fresnel region**, named after French physicist Augustin-Jean Fresnel (1788–1827). It is a transition region where the reactive field becomes smaller than the radiating field.

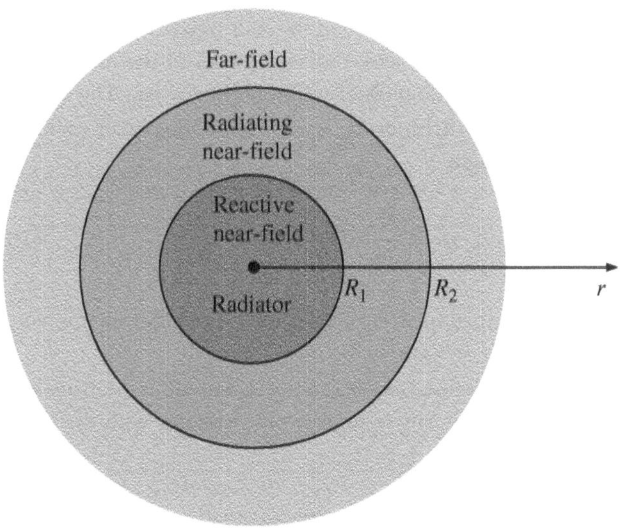

Figure 4.2 Field regions of radiator.

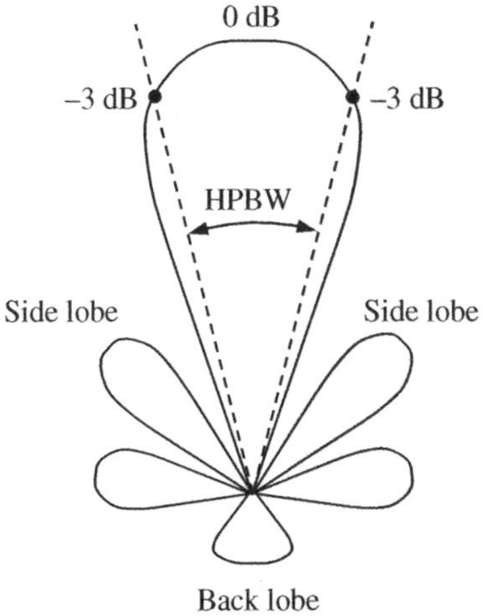

Figure 4.3 Radiation pattern.

The **radiation pattern** of antenna is a mathematical function or a graphical representation of the radiation properties of the antenna as a function of space coordinates. In most cases, the radiation pattern is determined in the far-field region. Radiation properties can be power flux density, radiation intensity, field strength, phase, or polarization. For a linearly polarized antenna, the radiation pattern is usually described by E-plane and H-plane pattern. The **E-plane** is defined as the plane containing the electric field vector and the direction of the maximum radiation and the **H-plane** is defined as the plane containing the magnetic field vector and the direction of maximum radiation. The antenna radiation pattern magnitude must be plotted relative to a recognized standard. The most common standard level is that of a perfect isotropic radiator (antenna), which would radiate energy equally in all directions.

A radiation pattern can be divided into various parts, called lobes, as illustrated in Figure 4.3. A **major lobe** refers to the radiation lobe which contains the direction of maximum radiation. All other lobes are called **minor lobes**. A **side lobe** refers to the minor lobe adjacent to the major lobe. A **back lobe** is a minor lobe which directs energy toward the direction opposite to the major lobe. The **half power beam width** (HPBW) is the angle between the half-power (−3 dB) points of the main lobe, when referenced to the peak radiated power of the main lobe.

Let $\mathbf{u}_r$ be the unit vector along a far-field observation point $\mathbf{r} = r\mathbf{u}_r$. The **radiation intensity** of an antenna in the direction $\mathbf{u}_r$ is defined as the power radiated from the antenna per unit solid angle

$$U(\mathbf{u}_r) = \frac{r^2}{2}\,\mathrm{Re}\left[\mathbf{E}(\mathbf{r}) \times \overline{\mathbf{H}}(\mathbf{r})\right] \cdot \mathbf{u}_r = \frac{r^2}{2\eta}|\mathbf{E}(\mathbf{r})|^2, \tag{4.2}$$

where $\eta = \sqrt{\mu/\varepsilon}$ is the intrinsic impedance of the medium. The radiation intensity for an isotropic radiator is $U(\mathbf{u}_r) = P^{rad}/4\pi$.

The **directivity** of an antenna is defined as the ratio of its radiation intensity in a given direction $\mathbf{u}_r$ to the radiation intensity of an isotropic radiator with the same power radiated by the antenna

$$D(\mathbf{u}_r) = \frac{U(\mathbf{u}_r)}{P^{rad}/4\pi}. \tag{4.3}$$

Theoretically there is no mathematical limit to the directivity that can be obtained from the currents confined in an arbitrarily small volume. However, the high field intensities around a small antenna with a high directivity will produce high energy storage around the antenna, large power dissipation, low radiation efficiency, and narrow bandwidth. For antennas with two orthogonal polarization components, one may introduce the partial directivity of an antenna for a given polarization component. In a spherical coordinate system, one may write

$$U(\mathbf{u}_r) = U_\theta(\mathbf{u}_r) + U_\varphi(\mathbf{u}_r),$$

where

$$U_\theta(\mathbf{u}_r) = \frac{r^2}{2\eta}|E_\theta(\mathbf{r})|^2, \quad U_\varphi(\mathbf{u}_r) = \frac{r^2}{2\eta}|E_\varphi(\mathbf{r})|^2.$$

The directivity can be written as

$$D(\mathbf{u}_r) = D_\theta(\mathbf{u}_r) + D_\varphi(\mathbf{u}_r), \tag{4.4}$$

where

$$D_\theta(\mathbf{u}_r) = \frac{U_\theta(\mathbf{u}_r)}{P^{rad}/4\pi}, \quad D_\varphi(\mathbf{u}_r) = \frac{U_\varphi(\mathbf{u}_r)}{P^{rad}/4\pi}$$

are the partial directivities for θ and φ component, respectively.

The **absolute gain** of an antenna is defined as the ratio of its radiation intensity in a given direction $\mathbf{u}_r$ to the radiation intensity of an isotropic radiator with the same power accepted by the antenna

$$G(\mathbf{u}_r) = \frac{U(\mathbf{u}_r)}{P^a/4\pi} = e_r D(\mathbf{u}_r). \tag{4.5}$$

The old definition of the gain is

$$G_{old}(\mathbf{u}_r) = \frac{U(\mathbf{u}_r)}{P^{in}/4\pi} = e_t D(\mathbf{u}_r). \tag{4.6}$$

This is also called **absolute gain** or **realized gain**, which has included the effects of matching network. The gain of an antenna often refers to the maximum gain and is usually given in decibels, such as dBi (with respect to an isotropic radiator) or dBd (with respect to a half-wave dipole). Similarly, one may introduce partial gains in a spherical coordinate system

$$G(\mathbf{u}_r) = G_\theta(\mathbf{u}_r) + G_\varphi(\mathbf{u}_r). \tag{4.7}$$

4.1.3 Vector Effective Length, Equivalent Area, and Antenna Factor

Let $\mathbf{r} = r\mathbf{u}_r$ be an observation point. The EM fields of antenna in a homogeneous and isotropic medium with permeability μ and permittivity ε can be expressed as an integral over the source region V_0 that contains electric current $\mathbf{J}$ and magnetic current $\mathbf{J}_m$ [26]

$$\begin{aligned}\mathbf{E}(\mathbf{r}) &= -jk\eta\int_{V_0} G(\mathbf{r},\mathbf{r}')\mathbf{J}(\mathbf{r}')dV(\mathbf{r}') - \frac{\eta}{jk}\int_{V_0}\nabla'\cdot\mathbf{J}(\mathbf{r}')\nabla'G(\mathbf{r},\mathbf{r}')dV(\mathbf{r}')\\ &\quad - \int_{V_0}\mathbf{J}_m(\mathbf{r}')\times\nabla'G(\mathbf{r},\mathbf{r}')dV(\mathbf{r}'),\\ \mathbf{H}(\mathbf{r}) &= -j\frac{k}{\eta}\int_{V_0} G(\mathbf{r},\mathbf{r}')\mathbf{J}_m(\mathbf{r}')dV(\mathbf{r}') - \frac{1}{j\eta k}\int_{V_0}\nabla'\cdot\mathbf{J}_m(\mathbf{r}')\nabla'G(\mathbf{r},\mathbf{r}')dV(\mathbf{r}')\\ &\quad + \int_{V_0}\mathbf{J}(\mathbf{r}')\times\nabla'G(\mathbf{r},\mathbf{r}')dV(\mathbf{r}'),\end{aligned} \tag{4.8}$$

where $G(\mathbf{r}, \mathbf{r}') = e^{-jkR}/4\pi R$ with $R = |\mathbf{r} - \mathbf{r}'|$, $k = \omega\sqrt{\mu\varepsilon}$, and $\eta = \sqrt{\mu/\varepsilon}$. By the Gauss theorem, the terms containing the divergence of the currents can be expressed by

$$\begin{aligned}\int_{V_0}\nabla'\cdot\mathbf{J}(\mathbf{r}')\nabla'G(\mathbf{r},\mathbf{r}')dV(\mathbf{r}') &= -\int_{V_0}[\mathbf{J}(\mathbf{r}')\cdot\nabla']\nabla'G(\mathbf{r},\mathbf{r}')dV(\mathbf{r}'),\\ \int_{V_0}\nabla'\cdot\mathbf{J}_m(\mathbf{r}')\nabla'G(\mathbf{r},\mathbf{r}')dV(\mathbf{r}') &= -\int_{V_0}[\mathbf{J}_m(\mathbf{r}')\cdot\nabla']\nabla'G(\mathbf{r},\mathbf{r}')dV(\mathbf{r}'),\end{aligned}$$

and the EM fields in (4.8) can be rewritten as

$$\begin{aligned}\mathbf{E}(\mathbf{r}) &= -jk\eta\int_{V_0}\left(\overleftrightarrow{\mathbf{I}} + \frac{1}{k^2}\nabla\nabla\right)G(\mathbf{r},\mathbf{r}')\cdot\mathbf{J}(\mathbf{r}')dV(\mathbf{r}') - \int_{V_0}\mathbf{J}_m(\mathbf{r}')\times\nabla'G(\mathbf{r},\mathbf{r}')dV(\mathbf{r}'),\\ \mathbf{H}(\mathbf{r}) &= -j\frac{k}{\eta}\int_{V_0}\left(\overleftrightarrow{\mathbf{I}} + \frac{1}{k^2}\nabla\nabla\right)G(\mathbf{r},\mathbf{r}')\cdot\mathbf{J}_m(\mathbf{r}')dV(\mathbf{r}') + \int_{V_0}\mathbf{J}(\mathbf{r}')\times\nabla'G(\mathbf{r},\mathbf{r}'dV(\mathbf{r}'),\end{aligned} \tag{4.9}$$

where $\overleftrightarrow{\mathbf{I}}$ is the identity dyadic. Let $\mathbf{u}_R = (\mathbf{r} - \mathbf{r}')/|\mathbf{r} - \mathbf{r}'|$. The following expressions can be obtained by vector calculus

$$\begin{aligned}&\nabla'G(\mathbf{r},\mathbf{r}') = \left(jk + \frac{1}{R}\right)G(\mathbf{r},\mathbf{r}')\mathbf{u}_R,\\ &[\mathbf{J}(\mathbf{r}')\cdot\nabla']\nabla'G(\mathbf{r},\mathbf{r}') = G(\mathbf{r},\mathbf{r}')\left[-k^2 + \frac{3}{R}\left(jk + \frac{1}{R}\right)\right][\mathbf{J}(\mathbf{r}')\cdot\mathbf{u}_R]\mathbf{u}_R\\ &- G(\mathbf{r},\mathbf{r}')\frac{\mathbf{J}(\mathbf{r}')}{R}\left(jk + \frac{1}{R}\right).\end{aligned}$$

If the distance R is sufficiently large, the terms higher than R^{-1} can be neglected. Thus, one may write

$$[\mathbf{J}(\mathbf{r}')\cdot\nabla']\nabla' G(\mathbf{r},\mathbf{r}') \approx -k^2 G(\mathbf{r},\mathbf{r}')[\mathbf{J}(\mathbf{r}')\cdot\mathbf{u}_R]\mathbf{u}_R. \tag{4.10}$$

In the far-field region defined by $r \gg r'$, $kr \gg 1$, the following approximations can be made:

$$R = |\mathbf{r}-\mathbf{r}'| \approx r - \mathbf{u}_r\cdot\mathbf{r}', \quad \frac{1}{|\mathbf{r}-\mathbf{r}'|} \approx \frac{1}{r}, \quad \frac{e^{-jk|\mathbf{r}-\mathbf{r}'|}}{|\mathbf{r}-\mathbf{r}'|}\mathbf{u}_R \approx \frac{e^{-jkr}}{r} e^{jk\mathbf{u}_r\cdot\mathbf{r}'}\mathbf{u}_r. \tag{4.11}$$

It is readily found from (4.8), (4.10), and (4.11) that the far-zone fields have the asymptotic expressions

$$\mathbf{E}(\mathbf{r}) = \frac{e^{-jkr}}{r}\left[\mathbf{E}_\infty(\mathbf{u}_r) + O\left(\frac{1}{r}\right)\right], \mathbf{H}(\mathbf{r}) = \frac{e^{-jkr}}{r}\left[\mathbf{H}_\infty(\mathbf{u}_r) + O\left(\frac{1}{r}\right)\right]. \tag{4.12}$$

The vector fields $\mathbf{E}_\infty$ and $\mathbf{H}_\infty$ are defined on the unit sphere Ω, and are known as the **electric far-field pattern** and **magnetic far-field pattern**, respectively. They are independent of the distance r and are given by

$$\begin{aligned}
\mathbf{E}_\infty(\mathbf{u}_r) &= -\frac{jk\eta}{4\pi}\int_{V_0}\left[\mathbf{J}-(\mathbf{J}\cdot\mathbf{u}_r)\mathbf{u}_r + \frac{1}{\eta}\mathbf{J}_m\times\mathbf{u}_r\right]e^{jk\mathbf{u}_r\cdot\mathbf{r}'}dV(\mathbf{r}'),\\
\mathbf{H}_\infty(\mathbf{u}_r) &= -\frac{jk}{4\pi\eta}\int_{V_0}[\mathbf{J}_m-(\mathbf{J}_m\cdot\mathbf{u}_r)\mathbf{u}_r - \eta\mathbf{J}\times\mathbf{u}_r]e^{jk\mathbf{u}_r\cdot\mathbf{r}'}dV(\mathbf{r}').
\end{aligned} \tag{4.13}$$

The far-field patterns are transverse and satisfy

$$\begin{aligned}
&\eta\mathbf{H}_\infty(\mathbf{u}_r) = \mathbf{u}_r\times\mathbf{E}_\infty(\mathbf{u}_r),\\
&\mathbf{u}_r\cdot\mathbf{E}_\infty(\mathbf{u}_r) = \mathbf{u}_r\cdot\mathbf{H}_\infty(\mathbf{u}_r) = 0.
\end{aligned} \tag{4.14}$$

The far-zone fields satisfy the **Silver–Müller radiation condition**

$$\lim_{r\to\infty} r(\mathbf{u}_r\times\mathbf{E} - \eta\mathbf{H}) = 0, \tag{4.15}$$

and can also be expressed by

$$\mathbf{E}(\mathbf{r}) = -\frac{jk\eta I}{4\pi r}e^{-jkr}\mathbf{L}(\mathbf{u}_r), \tag{4.16}$$

where I is the input current at the feeding plane and $\mathbf{L}$ is called the **antenna vector effective length**, defined by

$$\mathbf{L}(\mathbf{u}_r) = \frac{1}{I}\int_{V_0}\left[\mathbf{J}-(\mathbf{J}\cdot\mathbf{u}_r)\mathbf{u}_r + \frac{1}{\eta}\mathbf{J}_m\times\mathbf{u}_r\right]e^{jk\mathbf{u}_r\cdot\mathbf{r}'}dV(\mathbf{r}'). \tag{4.17}$$

The open circuit voltage at the antenna feeding plane induced by an incident field $\mathbf{E}^{in}$ from the direction $-\mathbf{u}_r$ is given by [26]

$$V_{oc}(\mathbf{u}_r) = -\frac{1}{I}\int_{V_0} \mathbf{E}^{in}(\mathbf{r}') \cdot \mathbf{J}(\mathbf{r}') dV(\mathbf{r}'), \tag{4.18}$$

which results from the reciprocity of transmitting and receiving antenna. The incident field is given by

$$\mathbf{E}^{in}(\mathbf{r}) = \mathbf{E}^{in}(o)e^{jk\mathbf{u}_r \cdot \mathbf{r}},$$

where $\mathbf{E}^{in}(o)$ is the field strength at the origin (antenna position) and is perpendicular to $\mathbf{u}_r$. The open circuit voltage (4.18) can thus be written as

$$V_{oc}(\mathbf{u}_r) = -\frac{1}{I}\mathbf{E}^{in}(o) \cdot \int_{V_0} \mathbf{J}(\mathbf{r}')e^{jk\mathbf{u}_r \cdot \mathbf{r}'} dV(\mathbf{r}') = -\mathbf{E}^{in}(o) \cdot \mathbf{L}(\mathbf{u}_r).$$

The above relation is often used as the definition of the antenna vector effective length in most literature. According to the equivalent circuit for the receiving antenna shown in Figure 4.4, the received power by the load is

$$p^{rec}(\mathbf{u}_r) = \frac{1}{2}\left|\frac{V_{oc}(\mathbf{u}_r)}{Z + Z_L}\right|^2 \operatorname{Re} Z_L = \frac{1}{2}\left|\frac{\mathbf{E}^{in}(o) \cdot \mathbf{L}(\mathbf{u}_r)}{Z + Z_L}\right|^2 \operatorname{Re} Z_L,$$

where Z is the antenna input impedance.

The **antenna equivalent area** is a transverse area, defined as the ratio of received power to the power flux density of the incident plane wave from the direction $-\mathbf{u}_r$:

$$A_e(\mathbf{u}_r) = \frac{p^{rec}(\mathbf{u}_r)}{\left|\mathbf{E}^{in}(o)\right|^2/2\eta} = \lambda^2\left|\frac{\mathbf{E}^{in}(o)}{Z + Z_L} \cdot \frac{\mathbf{L}(\mathbf{u}_r)}{\lambda}\right|^2 \frac{\eta \operatorname{Re} Z_L}{\left|\mathbf{E}^{in}(o)\right|^2}. \tag{4.19}$$

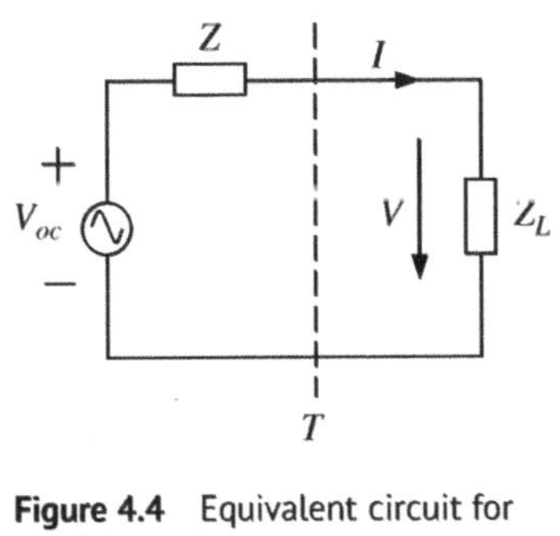

Figure 4.4 Equivalent circuit for receiving antenna.

If the receiving antenna is conjugately matched and there is no polarization loss, the antenna equivalent area can be simplified to

$$A_e(\mathbf{u}_r) = \frac{\lambda^2}{4}\left(\frac{\eta}{\operatorname{Re} Z_L}\right)\left|\frac{\mathbf{L}(\mathbf{u}_r)}{\lambda}\right|^2$$

The **antenna factor** is defined as the ratio of incident electric field strength to the induced terminal voltage

$$AF(\mathbf{u}_r) = \frac{\left|\mathbf{E}^{in}(o)\right|}{\left|V(\mathbf{u}_r)\right|},$$

where $V(\mathbf{u}_r)$ stands for the induced terminal voltage at the reference plane of the receiving antenna due to the incident field from the direction $-\mathbf{u}_r$. From the equivalent circuit of a receiving antenna, one may express the terminal voltage in terms of the open circuit voltage as follows:

$$V(\mathbf{u}_r) = \frac{Z_L}{Z_L + Z} V_{oc}(\mathbf{u}_r).$$

Hence, the relationship between the antenna factor and antenna vector effective length is given by

$$AF(\mathbf{u}_r) = \frac{\left|\mathbf{E}^{in}(o)\right|}{\left|\mathbf{E}^{in}(o) \cdot \mathbf{L}(\mathbf{u}_r)\right|} \left|1 + \frac{Z}{Z_L}\right|. \tag{4.20}$$

If there is no polarization loss, the antenna factor (4.20) reduces to

$$AF(\mathbf{u}_r) = \left|1 + \frac{Z}{Z_L}\right| \frac{1}{\left|\mathbf{L}(\mathbf{u}_r)\right|}.$$

Let S_∞ be a large closed surface enclosing the antenna. The transmitting and receiving properties of antenna can be expressed in the antenna vector effective length and are summarized below.

1) Poynting vector

$$S(\mathbf{r}) = \frac{1}{2\eta} |\mathbf{E}(\mathbf{r})|^2 = \frac{\eta |I|^2}{8r^2} \left|\frac{\mathbf{L}(\mathbf{u}_r)}{\lambda}\right|^2. \tag{4.21}$$

2) Radiation intensity

$$U(\mathbf{u}_r) = \frac{r^2}{2\eta} |\mathbf{E}(\mathbf{r})|^2 = \frac{\eta |I|^2}{8} \left|\frac{\mathbf{L}(\mathbf{u}_r)}{\lambda}\right|^2. \tag{4.22}$$

3) Directivity

$$D(\mathbf{u}_r) = \frac{4\pi U(\mathbf{u}_r)}{P^{rad}} = \frac{\pi\eta}{R^{rad}} \left|\frac{\mathbf{L}(\mathbf{u}_r)}{\lambda}\right|^2. \tag{4.23}$$

4) Gain

$$G(\mathbf{u}_r) = e_r D(\mathbf{u}_r) = \frac{\pi\eta}{R^{rad} + R^{loss}} \left|\frac{\mathbf{L}(\mathbf{u}_r)}{\lambda}\right|^2. \tag{4.24}$$

5) Radiated power

$$P^{rad} = \frac{\eta |I|^2}{8} \int\limits_{S_\infty} \left| \frac{\mathbf{L}(\mathbf{u}_r)}{\lambda} \right|^2 d\Omega. \tag{4.25}$$

6) Radiation resistance

$$R^{rad} = \frac{2P^{rad}}{|I|^2} = \frac{\eta}{4} \int\limits_{S_\infty} \left| \frac{\mathbf{L}(\mathbf{u}_r)}{\lambda} \right|^2 d\Omega. \tag{4.26}$$

In the above, the radiation resistance R^{rad} and the loss resistance R^{loss} are, respectively, defined by

$$R^{rad} = 2P^{rad}/|I|^2, \quad R^{loss} = 2P^{loss}_{ant}/|I|^2$$

with $P^{loss}_{ant} = P^a - P^{rad}$ representing the power loss in the antenna.

Remark 4.1 If there are no magnetic sources, the Poynting vector in the far-field region can be expressed as

$$\mathbf{S} = \frac{1}{2}\,\mathrm{Re}\left(\mathbf{E} \times \overline{\mathbf{H}}\right) = \mathbf{u}_r \frac{k^2\eta}{32\pi^2 r^2} \left| \mathbf{u}_r \times \int\limits_{V_0} \mathbf{J} e^{jk\mathbf{u}_r \cdot \mathbf{r}'} dV(\mathbf{r}') \right|^2. \tag{4.27}$$

The total radiated power by the current $\mathbf{J}$ is

$$P^{rad} = \int\limits_{S_\infty} \mathbf{S} \cdot \mathbf{u}_n dS(\mathbf{r}') = \frac{k^2\eta}{32\pi^2} \int\limits_{S_\infty} \left| \mathbf{u}_r \times \int\limits_{V_0} \mathbf{J} e^{jk\mathbf{u}_r \cdot \mathbf{r}'} dV(\mathbf{r}') \right|^2 d\Omega(\mathbf{r}). \tag{4.28}$$

From the Poynting theorem, the radiated power can also be calculated by the **method of induced electromotive force**

$$\begin{aligned} P^{rad} &= -\frac{1}{2}\,\mathrm{Re} \int\limits_{V_0} \overline{\mathbf{J}}(\mathbf{r}') \cdot \mathbf{E}(\mathbf{r}') dV(\mathbf{r}') \\ &= \frac{k\eta}{8\pi} \int\limits_{V_0} \int\limits_{V_0} \overline{\mathbf{J}}(\mathbf{r}) \cdot \left(\overleftrightarrow{\mathbf{I}} + \frac{1}{k^2} \nabla\nabla \right) \frac{\sin(k|\mathbf{r} - \mathbf{r}'|)}{|\mathbf{r} - \mathbf{r}'|} \cdot \mathbf{J}(\mathbf{r}') dV(\mathbf{r}) dV(\mathbf{r}'). \end{aligned} \tag{4.29}$$

□

4.1.4 Antenna Quality Factor

According to the IEEE Standard Definitions of Terms for Antennas, the **antenna** QF of a resonant antenna is defined as the ratio of 2π times energy stored in the fields excited by the antenna to the energy radiated per cycle

$$Q_I = \frac{\omega \tilde{W}}{P^{rad}} = \frac{\omega\left(\tilde{W}_e + \tilde{W}_m\right)}{P^{rad}}, \tag{4.30}$$

where $\tilde{W} = \tilde{W}_e + \tilde{W}_m$; $\tilde{W}_e$ stands for the stored electric field energy of antenna, $\tilde{W}_m$ for the stored magnetic field energy of antenna (their definitions will be given in Section 4.3.2), ω is the frequency, and P^{rad} is the total radiated power. In most publications, the traditional definition of antenna QF is often used

$$Q_{II} = \begin{cases} \dfrac{2\omega \tilde{W}_m}{P^{rad}}, & \tilde{W}_m > \tilde{W}_e \\ \dfrac{2\omega \tilde{W}_e}{P^{rad}}, & \tilde{W}_e > \tilde{W}_m \end{cases}, \tag{4.31}$$

which has a conditional statement and is more difficult to handle than (4.30) in theoretical study. The antenna QF defined by (4.31) applies to an antenna tuned to resonance only, while the IEEE standard definition (4.30) applies to an antenna under any conditions, at resonance or above resonance. Both definitions give the exact same values when the antenna is tuned to resonance.

4.2 Theory of Spherical Waveguide

The free space may be considered as a spherical waveguide, whose theory can be established in the same way as that of metal waveguide. The vector modal functions for the spherical waveguide can be constructed from the spherical harmonics, and they are all independent of frequency. The fields in the spherical waveguide can be expressed as a series expansion in terms of the vector modal functions and the spherical harmonics, from which the fundamental field pattern or spherical vector wave functions (SVWFs) for the spherical waveguide can be identified. The SVWFs were first reported by Hansen [27], and they are fundamental to the study of radiation theory [28–36].

The radiated fields can also be represented by an integral over the source region as shown in (4.8). The integral representation is often used to study the radiated field and build the integral equation for the source distribution. In order to find the field distribution outside the source region, an integration over the source region must be carried out for each field point. However, in the series expansion of the fields, the expansion coefficients are expressed as the integrals over the source region and the integrations are only performed once. As soon as the coefficients are determined, the evaluation of field distribution only involves the sum of a series, which reduces the computational burden most of the time compared to the integral representation [37–39].

4.2.1 Vector Modal Functions for Spherical Waveguide

The spherical harmonics originate from the solution of Laplace equation in the spherical coordinate system

$$\nabla^2 \phi = 0. \tag{4.32}$$

The operator ∇^2 in the spherical coordinate system can be decomposed into

$$\nabla^2 = \frac{1}{r^2}\frac{\partial}{\partial r}\left(r^2\frac{\partial}{\partial r}\right) + \frac{1}{r^2}\nabla^2_{\theta\varphi}, \tag{4.33}$$

where

$$\nabla^2_{\theta\varphi} = \frac{1}{\sin\theta}\frac{\partial}{\partial\theta}\left(\sin\theta\frac{\partial}{\partial\theta}\right) + \frac{1}{\sin^2\theta}\frac{\partial^2}{\partial\varphi^2} \tag{4.34}$$

is the **surface Laplacian operator** on the unit sphere. By separation of variables, the solution of Laplace equation may be assumed to be of the form

$$\phi(r,\theta,\varphi) = R(r)Y(\theta,\varphi).$$

Once this is inserted into (4.32), two differential equations result

$$\frac{d}{dr}\left(r^2\frac{dR}{dr}\right) - \lambda R = 0, \tag{4.35}$$

$$-\nabla^2_{\theta\varphi}Y(\theta,\varphi) = \lambda Y(\theta,\varphi). \tag{4.36}$$

The eigenvalue problem (4.36) is subject to the boundary conditions that $Y(\theta, \varphi)$ must be finite and periodic in φ with the period 2π [i.e. $Y(\theta, 0) = Y(\theta, 2\pi)$], and may be solved by the method of separation of variables. The eigensolutions are called **spherical harmonics** $Y_{nml}(\theta, \varphi)$. The spherical harmonics form a complete set, and satisfy

$$-\nabla^2_{\theta\varphi}Y_{nml}(\theta,\varphi) = n(n+1)Y_{nml}(\theta,\varphi), \tag{4.37}$$

where

$$\begin{gathered} Y_{nml}(\theta,\varphi) = P_n^m(\cos\theta)f_{ml}(\varphi), \\ f_{ml}(\varphi) = \begin{cases} \cos m\varphi, l = e \\ \sin m\varphi, l = o \end{cases}, \\ n = 0, 1, 2, ..., m = 0, 1, 2, ..., n, \end{gathered} \tag{4.38}$$

and $P_n^m(\cos\theta)$ are the associated Legendre functions. It is easy to verify the following orthogonal relationships for the spherical harmonics:

$$\int_0^{2\pi} d\varphi \int_0^{\pi} Y_{nml} Y_{n'm'l'} \sin\theta d\theta = \begin{cases} 0, & [n,m,l] \neq [n',m',l'] \\ \tilde{N}_{nm}^2, & [n,m,l] = [n',m',l'] \end{cases}, \tag{4.39}$$

$$\begin{aligned} &\int_0^{2\pi} d\varphi \int_0^{\pi} \left(\sin\theta \frac{\partial Y_{nml}}{\partial\theta} \frac{\partial Y_{n'm'l'}}{\partial\theta} + \frac{1}{\sin\theta} \frac{\partial Y_{nml}}{\partial\varphi} \frac{\partial Y_{n'm'l'}}{\partial\varphi} \right) d\theta \\ &\quad = \begin{cases} 0, & [n,m,l] \neq [n',m',l'] \\ N_{nm}^2, & [n,m,l] = [n',m',l'] \end{cases}, \end{aligned} \tag{4.40}$$

where

$$\tilde{N}_{nm}^2 = (1+\delta_{m0}) \frac{2\pi}{2n+1} \frac{(n+m)!}{(n-m)!}, N_{nm}^2 = n(n+1)\tilde{N}_{nm}^2, \quad \delta_{m0} = \begin{cases} 1, m = 0 \\ 0, m \neq 0 \end{cases}.$$

Theorem 4.1 (Completeness of Spherical Harmonics)
Let $f(\theta, \varphi)$ be a continuous function defined on a spherical surface. Then, the following expansion holds

$$f(\theta,\varphi) = \sum_{n=0}^{\infty} \sum_{m \leq n} \sum_{l=e,o} A_{nml} Y_{nml}(\theta,\varphi),$$

where A_{nml} are the expansion coefficients

$$A_{nml} = \frac{1}{\tilde{N}_{nm}^2} \int_0^{2\pi} d\varphi \int_0^{\pi} f(\theta,\varphi) Y_{nml}(\theta,\varphi) \sin\theta d\theta.$$

□

For an arbitrary scalar field ϕ, the symbol $\nabla_{\theta\varphi}$ will be used to denote the **surface gradient operator** defined on the unit sphere

$$\nabla_{\theta\varphi}\phi = \mathbf{u}_\theta \frac{\partial\phi}{\partial\theta} + \mathbf{u}_\varphi \frac{1}{\sin\theta} \frac{\partial\phi}{\partial\varphi}. \tag{4.41}$$

The conventional 3D gradient operator ∇ in the spherical coordinate system is related to the surface gradient operator $\nabla_{\theta\varphi}$ by

$$\nabla\phi = \mathbf{u}_r \frac{\partial\phi}{\partial r} + \frac{1}{r} \nabla_{\theta\varphi}\phi. \tag{4.42}$$

For a vector field $\mathbf{A} = \mathbf{u}_r A_r + \mathbf{u}_\theta A_\theta + \mathbf{u}_\varphi A_\varphi$, one may introduce a **surface divergence operator** $\nabla_{\theta\varphi}\cdot$ defined on the unit sphere:

$$\nabla_{\theta\varphi}\cdot\mathbf{A} = \frac{1}{\sin\theta}\frac{\partial}{\partial\theta}(\sin\theta A_\theta) + \frac{1}{\sin\theta}\frac{\partial A_\varphi}{\partial\varphi}. \tag{4.43}$$

Note that the conventional 3D divergence operator in the spherical coordinate system can be written as

$$\nabla\cdot\mathbf{A} = \frac{1}{r^2}\frac{\partial}{\partial r}(r^2 A_r) + \frac{1}{r}\nabla_{\theta\varphi}\cdot\mathbf{A}. \tag{4.44}$$

Similar to the waveguide theory, it is convenient to introduce the normalized **vector modal functions**

$$\begin{aligned}
\mathbf{e}_{nml}(\theta,\varphi) &= \frac{1}{N_{nm}}\nabla_{\theta\varphi}Y_{nml}(\theta,\varphi)\\
&= -\mathbf{u}_\theta\frac{1}{N_{nm}\sin\theta}[(n+1)\cos\theta P_n^m(\cos\theta) - (n-m+1)P_{n+1}^m(\cos\theta)]f_{ml}(\varphi)\\
&\quad + \mathbf{u}_\varphi\frac{1}{N_{nm}\sin\theta}P_n^m(\cos\theta)f'_{ml}(\varphi),\\
\mathbf{h}_{nml}(\theta,\varphi) &= \mathbf{u}_r\times\mathbf{e}_{nml}(\theta,\varphi) = -\mathbf{u}_\theta\frac{1}{N_{nm}\sin\theta}P_n^m(\cos\theta)f'_{ml}(\varphi)\\
&\quad - \mathbf{u}_\varphi\frac{1}{N_{nm}\sin\theta}[(n+1)\cos\theta P_n^m(\cos\theta) - (n-m+1)P_{n+1}^m(\cos\theta)]f_{ml}(\varphi),
\end{aligned} \tag{4.45}$$

which satisfy the orthonormal conditions:

$$\begin{aligned}
&\int_\Omega \mathbf{e}_{nml}\cdot\mathbf{e}_{n'm'l'}d\Omega = \delta_{mm'}\delta_{nn'}\delta_{ll'},\\
&\int_\Omega \mathbf{h}_{nml}\cdot\mathbf{h}_{n'm'l'}d\Omega = \delta_{mm'}\delta_{nn'}\delta_{ll'},\\
&\int_\Omega \mathbf{e}_{nml}\cdot\mathbf{h}_{n'm'l'}d\Omega = 0,
\end{aligned} \tag{4.46}$$

where Ω is the unit sphere and $d\Omega$ is the differential element of the solid angle. Furthermore,

$$\begin{aligned}
&\nabla_{\theta\varphi}\cdot\mathbf{e}_{nml} = -\frac{1}{N_{nm}}n(n+1)Y_{nml},\\
&\nabla_{\theta\varphi}\cdot\mathbf{h}_{nml} = 0.
\end{aligned} \tag{4.47}$$

Remark 4.2 The vector modal functions in (4.45) contains sinθ in the denominator, which becomes zero as θ approaches zero. To show that the vector modal functions are regular at $\theta = 0$, one may write the associated Legendre function as

$$P_n^m(z) = \left(1 - z^2\right)^{\frac{1}{2}} \tilde{P}_n^m(z),$$

for $m \geq 1$ with

$$\tilde{P}_n^m(z) = \left(1 - z^2\right)^{\frac{m-1}{2}} \frac{d^m P(z)}{dz^m}.$$

The vector modal function can be expressed as

$$\begin{aligned}
\mathbf{e}_{nml} &= -\mathbf{u}_\theta \frac{1}{N_{nm}} \left[(n+1)\cos\theta \tilde{P}_n^m(\cos\theta) - (n-m+1)\tilde{P}_{n+1}^m(\cos\theta)\right] f_{ml}(\varphi) \\
&\quad + \mathbf{u}_\varphi \frac{1}{N_{nm}} \tilde{P}_n^m(\cos\theta) f'_{ml}(\varphi).
\end{aligned} \tag{4.48}$$

For $m = 0$, this becomes

$$\begin{aligned}
\mathbf{e}_{n0l} &= -\mathbf{u}_\theta \frac{n+1}{N_{n0}\sin\theta} [\cos\theta P_n(\cos\theta) - P_{n+1}(\cos\theta)] f_{0l}(\varphi) \\
&= \mathbf{u}_\theta \frac{n+1}{N_{n0}\sin\theta\cos\theta} \left[\sin^2\theta P_n(\cos\theta) - P_n(\cos\theta) + \cos\theta P_{n+1}(\cos\theta)\right] f_{0l}(\varphi) \\
&= \mathbf{u}_\theta \frac{1}{N_{n0}} \tan\theta \left[(n+1)P_n(\cos\theta) - \frac{d}{dz} P_{n+1}(z)\Big|_{z=\cos\theta}\right] f_{0l}(\varphi).
\end{aligned} \tag{4.49}$$

As θ approaches to zero, both (4.48) and (4.49) become regular. □

In summary, two complete sets of modal functions for the spherical waveguide have been constructed:

1) $\{Y_{nml}(\theta, \varphi)\}$.
2) $\{\mathbf{e}_{nml}(\theta, \varphi), \mathbf{h}_{nml}(\theta, \varphi)\}$.

The first set is suitable for the expansion of any scalar fields in the spherical waveguide, while the second set is suitable for the expansion of any transverse vector fields, and the completeness of the latter will be demonstrated in the next section.

4.2.2 Modal Expansions of Fields and Dyadic Green's Functions

Consider the time-harmonic EM fields generated by an electric current source $\mathbf{J}$ and a magnetic current source $\mathbf{J}_m$. The fields satisfy the generalized Maxwell equations

$$\begin{aligned}\nabla \times \mathbf{H} &= j\omega\varepsilon\mathbf{E} + \mathbf{J},\\ \nabla \times \mathbf{E} &= -j\omega\mu\mathbf{H} - \mathbf{J}_m.\end{aligned} \tag{4.50}$$

All the sources are assumed to be confined in a finite region V_0, as illustrated in Figure 4.5. In the spherical coordinate system, the fields and sources can be decomposed into a transverse component and a radial component

$$\begin{aligned}&\mathbf{E} = \mathbf{E}_t + \mathbf{u}_r E_r, \mathbf{H} = \mathbf{H}_t + \mathbf{u}_r H_r,\\ &\mathbf{J} = \mathbf{J}_t + \mathbf{u}_r J_r, \mathbf{J}_m = \mathbf{J}_{mt} + \mathbf{u}_r J_{mr},\end{aligned}$$

where $\mathbf{u}_r$ is the unit vector in the direction of increasing r and the subscript t denotes the transverse component. By taking the vector and scalar product of (4.50) with the radial vector $\mathbf{r}$, one may break the Maxwell equations into

$$\begin{aligned}j\omega\varepsilon(\mathbf{r} \times \mathbf{E}) + \mathbf{r} \times \mathbf{J} &= \nabla(\mathbf{r} \cdot \mathbf{H}) - (\mathbf{r} \cdot \nabla)\mathbf{H} - \mathbf{H},\\ -j\omega\mu(\mathbf{r} \times \mathbf{H}) - \mathbf{r} \times \mathbf{J}_m &= \nabla(\mathbf{r} \cdot \mathbf{E}) - (\mathbf{r} \cdot \nabla)\mathbf{E} - \mathbf{E},\end{aligned} \tag{4.51}$$

$$\begin{aligned}j\omega\varepsilon(\mathbf{r} \cdot \mathbf{E}) &= -\nabla \cdot (\mathbf{r} \times \mathbf{H}_t) - \mathbf{r} \cdot \mathbf{J},\\ j\omega\mu(\mathbf{r} \cdot \mathbf{H}) &= \nabla \cdot (\mathbf{r} \times \mathbf{E}_t) - \mathbf{r} \cdot \mathbf{J}_m.\end{aligned} \tag{4.52}$$

Equation (4.52) shows that the radial components of the EM fields can be represented by the transverse components. The operator $\mathbf{r} \cdot \nabla$ in (4.51) can be explicitly expressed as the sum of two components

$$(\mathbf{r} \cdot \nabla)\mathbf{F} = r\frac{\partial \mathbf{F}}{\partial r} = r\mathbf{u}_r\frac{\partial}{\partial r}F_r + r\frac{\partial}{\partial r}\mathbf{F}_t.$$

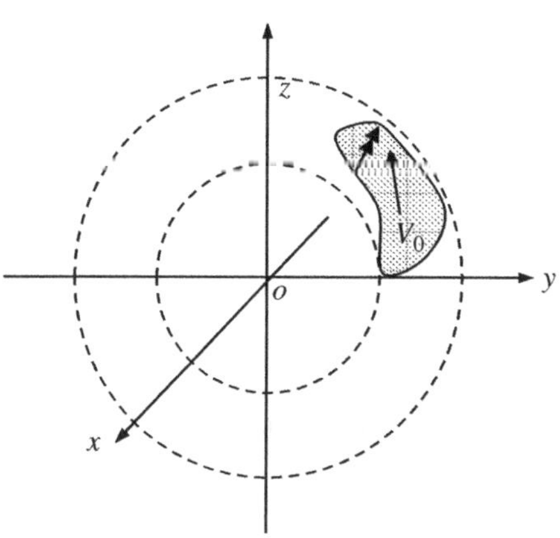

Figure 4.5 Current sources in spherical coordinate system.

By inserting the above expression into (4.51), one may find the relations between the transverse and radial field components of the fields

$$\begin{aligned}
&\frac{1}{r}\nabla_{\theta\varphi}(\mathbf{r}\cdot\mathbf{H}) - r\frac{\partial \mathbf{H}_t}{\partial r} - \mathbf{H}_t = j\omega\varepsilon(\mathbf{r}\times\mathbf{E}_t) + \mathbf{r}\times\mathbf{J}_t,\\
&\frac{1}{r}\nabla_{\theta\varphi}(\mathbf{r}\cdot\mathbf{E}) - r\frac{\partial \mathbf{E}_t}{\partial r} - \mathbf{E}_t = -j\omega\mu(\mathbf{r}\times\mathbf{H}_t) - \mathbf{r}\times\mathbf{J}_{mt}.
\end{aligned} \tag{4.53}$$

The radial components in (4.53) can be eliminated by using (4.52), to get the equations for the transverse fields

$$\begin{aligned}
&-j\omega\mu\frac{\partial}{\partial r}(r\mathbf{H}_t) + \frac{1}{r^2}\nabla_{\theta\varphi}\nabla_{\theta\varphi}\cdot(\mathbf{r}\times\mathbf{E}_t) + k^2(\mathbf{r}\times\mathbf{E}_t) = j\omega\mu(\mathbf{r}\times\mathbf{J}_t) + \frac{1}{r}\nabla_{\theta\varphi}(\mathbf{r}\cdot\mathbf{J}_m),\\
&-j\omega\varepsilon\frac{\partial}{\partial r}(r\mathbf{E}_t) - \frac{1}{r^2}\nabla_{\theta\varphi}\nabla_{\theta\varphi}\cdot(\mathbf{r}\times\mathbf{H}_t) - k^2(\mathbf{r}\times\mathbf{H}_t) = -j\omega\varepsilon(\mathbf{r}\times\mathbf{J}_{mt}) + \frac{1}{r}\nabla_{\theta\varphi}(\mathbf{r}\cdot\mathbf{J}).
\end{aligned} \tag{4.54}$$

To facilitate the expansion of the vector fields, it is convenient to make use of the theorem by Wilcox [34]:

Theorem 4.2 If $\mathbf{A}_t$ is a field of tangent vectors defined on the sphere S, there exist two scalar functions $U(r, \theta, \varphi)$ and $V(r, \theta, \varphi)$ defined on S such that the transverse field $\mathbf{A}_t$ can be expressed by

$$\mathbf{A}_t = \nabla_{\theta\varphi}U + \mathbf{u}_r \times \nabla_{\theta\varphi}V. \tag{4.55}$$

Notice that $\nabla_{\theta\varphi}\cdot(\mathbf{u}_r \times \nabla_{\theta\varphi}V) = 0$. □

Equation (4.55) is the Helmholtz theorem on the sphere. Since the spherical harmonics form a complete set of orthogonal functions, each function defined on a sphere can be expanded as a series in terms of the spherical harmonics

$$\begin{aligned}
U(r,\theta,\varphi) &= \sum_{n,m,l} C_{nml}(r)Y_{nml}(\theta,\varphi),\\
V(r,\theta,\varphi) &= \sum_{n,m,l} D_{nml}(r)Y_{nml}(\theta,\varphi).
\end{aligned}$$

It follows from these expansions and (4.55) that the transverse EM fields may be expressed by

$$\begin{aligned}
\mathbf{E}_t(\mathbf{r}) &= \sum_{n,m,l}\left[V_{nml}^{TM}(r)\mathbf{e}_{nml}(\theta,\varphi) + V_{nml}^{TE}(r)\mathbf{h}_{nml}(\theta,\varphi)\right],\\
\mathbf{H}_t(\mathbf{r}) &= \sum_{n,m,l}\left[I_{nml}^{TM}(r)\mathbf{h}_{nml}(\theta,\varphi) - I_{nml}^{TE}(r)\mathbf{e}_{nml}(\theta,\varphi)\right],
\end{aligned} \tag{4.56}$$

where V_{nml} and I_{nml} are called (spherical) **modal voltages** and (spherical) **modal currents**, respectively, and the superscripts TE and TM are used to designate the TE and TM parts of the fields:

$$\begin{cases} \mathbf{E}_t^{TE} = \sum\limits_{n,m,l} V_{nml}^{TE} \mathbf{h}_{nml}, \\ \mathbf{H}_t^{TE} = -\sum\limits_{n,m,l} I_{nml}^{TE} \mathbf{e}_{nml}, \end{cases} \quad \begin{cases} \mathbf{E}_t^{TM} = \sum\limits_{n,m,l} V_{nml}^{TM} \mathbf{e}_{nml}, \\ \mathbf{H}_t^{TM} = \sum\limits_{n,m,l} I_{nml}^{TM} \mathbf{h}_{nml}. \end{cases} \tag{4.57}$$

For the TM part, it can be shown from (4.52) that the corresponding radial field component H_r is zero in a source-free region. Hence, it is a TM wave. Similarly, the TE part is a TE wave, whose radial field component E_r is zero. Equations in (4.56) indicate that the set $\{\mathbf{e}_{nml}, \mathbf{h}_{nml}\}$ is complete and can be used to expand any transverse vector fields. Taking the cross product of (4.56) with $\mathbf{u}_r$ gives

$$\begin{aligned} \mathbf{u}_r \times \mathbf{E}_t &= \sum_{n,m,l} \left(V_{nml}^{TM} \mathbf{h}_{nml} - V_{nml}^{TE} \mathbf{e}_{nml} \right), \\ \mathbf{u}_r \times \mathbf{H}_t &= \sum_{n,m,l} \left(-I_{nml}^{TM} \mathbf{e}_{nml} - I_{nml}^{TE} \mathbf{h}_{nml} \right). \end{aligned} \tag{4.58}$$

The radial components can be determined from (4.52) and (4.58) as follows:

$$\begin{aligned} j\omega\varepsilon(\mathbf{u}_r \cdot \mathbf{E}) &= -\nabla \cdot (\mathbf{u}_r \times \mathbf{H}_t) - \mathbf{u}_r \cdot \mathbf{J} \\ &= \sum_{n,m,l} \left(I_{nml}^{TM} \nabla \cdot \mathbf{e}_{nml} + I_{nml}^{TE} \nabla \cdot \mathbf{h}_{nml} \right) - \mathbf{u}_r \cdot \mathbf{J}, \\ j\omega\mu(\mathbf{u}_r \cdot \mathbf{H}) &= \nabla \cdot (\mathbf{u}_r \times \mathbf{E}_t) - \mathbf{u}_r \cdot \mathbf{J}_m \\ &= \sum_{n,m,l} \left(V_{nml}^{TM} \nabla \cdot \mathbf{h}_{nml} - V_{nml}^{TE} \nabla \cdot \mathbf{e}_{nml} \right) - \mathbf{u}_r \cdot \mathbf{J}_m. \end{aligned}$$

On account of (4.44) and (4.47), the above expressions can be written as

$$\begin{aligned} E_r &= \sum_{n,m,l} \frac{\eta}{jkr} I_{nml}^{TM} \nabla_{\theta\varphi} \cdot \mathbf{e}_{nml} - \frac{\eta}{jk} J_r, \\ H_r &= -\sum_{n,m,l} \frac{1}{j\eta kr} V_{nml}^{TE} \nabla_{\theta\varphi} \cdot \mathbf{e}_{nml} - \frac{1}{jk\eta} J_{mr}. \end{aligned} \tag{4.59}$$

The total fields can then be determined by combining (4.56) and (4.59):

$$\begin{aligned} \mathbf{E} &= \sum_{n,m,l} \left(V_{nml}^{TM} \mathbf{e}_{nml} + V_{nml}^{TE} \mathbf{h}_{nml} \right) + \sum_{n,m,l} \frac{\eta}{jkr} I_{nml}^{TM} \mathbf{u}_r \nabla_{\theta\varphi} \cdot \mathbf{e}_{nml} - \frac{\eta}{jk} J_r \mathbf{u}_r, \\ \mathbf{H} &= \sum_{n,m,l} \left(I_{nml}^{TM} \mathbf{h}_{nml} - I_{nml}^{TE} \mathbf{e}_{nml} \right) - \sum_{n,m,l} \frac{1}{j\eta kr} V_{nml}^{TE} \mathbf{u}_r \nabla_{\theta\varphi} \cdot \mathbf{e}_{nml} - \frac{1}{jk\eta} J_{mr} \mathbf{u}_r. \end{aligned} \tag{4.60}$$

In order to find the equations for modal voltages and currents, one can use (4.54) for the transverse fields. Making use of the following relations

$$\nabla_{\theta\varphi} \cdot (\mathbf{u}_r \times \mathbf{E}_t) = \sum_{n,m,l} \frac{1}{N_{nm}} n(n+1) V_{nml}^{TE} Y_{nml},$$

$$\nabla_{\theta\varphi} \nabla_{\theta\varphi} \cdot (\mathbf{u}_r \times \mathbf{E}_t) = \sum_{n,m,l} n(n+1) V_{nml}^{TE} \mathbf{e}_{nml},$$

$$\nabla_{\theta\varphi} \cdot (\mathbf{u}_r \times \mathbf{H}_t) = \sum_{n,m,l} \frac{1}{N_{nm}} n(n+1) I_{nml}^{TM} Y_{nml},$$

$$\nabla_{\theta\varphi} \nabla_{\theta\varphi} \cdot (\mathbf{u}_r \times \mathbf{H}_t) = \sum_{n,m,l} n(n+1) I_{nml}^{TM} \mathbf{e}_{nml},$$

and substituting (4.56) and (4.58) into (4.54), one may find

$$\begin{aligned}
&\sum_{n,m,l} \left\{ \left[j\omega\mu \frac{d\left(rI_{nml}^{TE}\right)}{dr} - \beta_n^2 r V_{nml}^{TE} \right] \mathbf{e}_{nml} + \left[-j\omega\mu \frac{d\left(rI_{nml}^{TM}\right)}{dr} + k^2 r V_{nml}^{TM} \right] \mathbf{h}_{nml} \right\} \\
&= j\omega\mu(\mathbf{r} \times \mathbf{J}_t) + \frac{1}{r} \nabla_{\theta\varphi}(\mathbf{r} \cdot \mathbf{J}_m), \\
&\sum_{n,m,l} \left\{ \left[-j\omega\varepsilon \frac{d\left(rV_{nml}^{TM}\right)}{dr} + \beta_n^2 \left(rI_{nml}^{TM}\right) \right] \mathbf{e}_{nml} - \left[j\omega\varepsilon \frac{d\left(rV_{nml}^{TE}\right)}{dr} - k^2 \left(rI_{nml}^{TE}\right) \right] \mathbf{h}_{nml} \right\} \\
&= -j\omega\varepsilon(\mathbf{r} \times \mathbf{J}_{mt}) + \frac{1}{r} \nabla_{\theta\varphi}(\mathbf{r} \cdot \mathbf{J}),
\end{aligned} \tag{4.61}$$

where β_n is similar to the propagation constant in waveguide, and is defined by

$$\beta_n^2 = k^2 - \frac{n(n+1)}{r^2}. \tag{4.62}$$

If the orthonormal properties of the functions $\mathbf{e}_{nml}$ and $\mathbf{h}_{nml}$ are applied to (4.61), the equations relating the modal voltage and current can be obtained as follows:

$$\begin{aligned}
-j\omega\mu \frac{d\left(rI_{nml}^{TE}\right)}{dr} + \beta_n^2 r V_{nml}^{TE} &= -\int_\Omega \left[j\omega\mu(\mathbf{r} \times \mathbf{J}_t) \cdot \mathbf{e}_{nml} + \frac{1}{r} \nabla_{\theta\varphi}(\mathbf{r} \cdot \mathbf{J}_m) \cdot \mathbf{e}_{nml} \right] d\Omega \\
&= -\int_\Omega \left[j\omega\mu(\mathbf{r} \times \mathbf{J}_t) \cdot \mathbf{e}_{nml} - \frac{1}{r} (\mathbf{r} \cdot \mathbf{J}_m) \nabla_{\theta\varphi} \cdot \mathbf{e}_{nml} \right] d\Omega,
\end{aligned} \tag{4.63}$$

$$\begin{aligned}
-j\omega\varepsilon \frac{d\left(rV_{nml}^{TE}\right)}{dr} + k^2 r I_{nml}^{TE} &= \int_\Omega \left[-j\omega\varepsilon(\mathbf{r} \times \mathbf{J}_{mt}) \cdot \mathbf{h}_{nml} + \frac{1}{r} \nabla_{\theta\varphi}(\mathbf{r} \cdot \mathbf{J}) \cdot \mathbf{h}_{nml} \right] d\Omega \\
&= -\int_\Omega j\omega\varepsilon(\mathbf{r} \times \mathbf{J}_{mt}) \cdot \mathbf{h}_{nml} d\Omega,
\end{aligned} \tag{4.64}$$

$$
\begin{aligned}
-j\omega\mu\frac{d\left(rI_{nml}^{TM}\right)}{dr}+k^2rV_{nml}^{TM}&=\iint_{\Omega}\left[j\omega\mu(\mathbf{r}\times\mathbf{J}_t)\cdot\mathbf{h}_{nml}+\frac{1}{r}\nabla_{\theta\varphi}(\mathbf{r}\cdot\mathbf{J}_m)\cdot\mathbf{h}_{nml}\right]d\Omega\\
&=\int_{\Omega}j\omega\mu(\mathbf{r}\times\mathbf{J}_t)\cdot\mathbf{h}_{nml}d\Omega,
\end{aligned}
\tag{4.65}
$$

$$
\begin{aligned}
-j\omega\varepsilon\frac{d\left(rV_{nml}^{TM}\right)}{dr}+\beta_n^2rI_{nml}^{TM}&=\iint_{\Omega}\left[-j\omega\varepsilon(\mathbf{r}\times\mathbf{J}_{mt})\cdot\mathbf{e}_{nml}+\frac{1}{r}\nabla_{\theta\varphi}(\mathbf{r}\cdot\mathbf{J})\cdot\mathbf{e}_{nml}\right]d\Omega\\
&=\iint_{\Omega}\left[-j\omega\varepsilon(\mathbf{r}\times\mathbf{J}_{mt})\cdot\mathbf{e}_{nml}-\frac{1}{r}(\mathbf{r}\cdot\mathbf{J})\nabla_{\theta\varphi}\cdot\mathbf{e}_{nml}\right]d\Omega.
\end{aligned}
\tag{4.66}
$$

After eliminating I_{nml}^{TE} and V_{nml}^{TM} from the above equations, the modal voltage V_{nml}^{TE} and the modal current I_{nml}^{TM} are found to satisfy the inhomogeneous equations for the spherical Bessel functions

$$
\begin{aligned}
\frac{d^2V_{nml}^{TE}}{dr^2}+\frac{2}{r}\frac{dV_{nml}^{TE}}{dr}+\beta_n^2V_{nml}^{TE}&=\frac{1}{r}\frac{d}{dr}\int_{\Omega}r\mathbf{J}_m\cdot\mathbf{e}_{nml}d\Omega\\
&\quad+\frac{1}{r}\iint_{\Omega}\left[j\omega\mu r\mathbf{J}\cdot\mathbf{h}_{nml}+\frac{1}{r}(\mathbf{r}\cdot\mathbf{J}_m)\nabla_{\theta\varphi}\cdot\mathbf{e}_{nml}\right]d\Omega,\\
\frac{d^2I_{nml}^{TM}}{dr^2}+\frac{2}{r}\frac{dI_{nml}^{TM}}{dr}+\beta_n^2I_{nml}^{TM}&=-\frac{1}{r}\frac{d}{dr}\int_{S}r\mathbf{J}\cdot\mathbf{e}_{nml}d\Omega\\
&\quad+\frac{1}{r}\iint_{\Omega}\left[j\omega\varepsilon r\mathbf{J}_m\cdot\mathbf{h}_{nml}-\frac{1}{r}(\mathbf{r}\cdot\mathbf{J})\nabla_{\theta\varphi}\cdot\mathbf{e}_{nml}\right]d\Omega.
\end{aligned}
\tag{4.67}
$$

Once V_{nml}^{TE} and I_{nml}^{TM} are known, I_{nml}^{TE} and V_{nml}^{TM} can then be determined from (4.64) and (4.65). To solve the equations in (4.67), one may use the Green's function defined by

$$
\frac{d^2G_n(r,r')}{dr^2}+\frac{2}{r}\frac{dG_n(r,r')}{dr}+\beta_n^2G_n(r,r')=-\frac{1}{r^2}\delta(r-r').
\tag{4.68}
$$

The corresponding homogeneous equation has two independent solutions $x_1(r)=j_n(kr)$ and $x_2(r)=h_n^{(2)}(kr)$, where $j_n(kr)$ is the spherical Bessel function of first kind and $h_n^{(2)}(kr)$ the spherical Hankel function of second kind. From the theory of differential equations, the solution of (4.68) can be written as

$$G_n(r,r') = ax_1(r) + bx_2(r) + x_2(r)\int_0^r \frac{-\delta(r''-r')x_1(r'')}{r''^2\Delta(r'')}dr'' + x_1(r)\int_r^\infty \frac{-\delta(r''-r')x_2(r'')}{r''^2\Delta(r'')}dr'', \tag{4.69}$$

where $\Delta(r)$ is the Wronskian determinant

$$\Delta(r) = \begin{vmatrix} x_1 & x_2 \\ dx_1/dr & dx_2/dr \end{vmatrix} = -\frac{j}{kr^2}.$$

The Green's function (4.69) can be rewritten as

$$G_n(r,r') = \begin{cases} ax_1(r) + bx_2(r) - jkx_1(r)x_2(r'), & r < r' \\ ax_1(r) + bx_2(r) - jkx_2(r)x_1(r'), & r > r' \end{cases}. \tag{4.70}$$

The constants a and b can be determined from the boundary conditions at $r = 0$ and $r = \infty$. Since the field is finite at $r = 0$, the term $bx_2(r)$ must be excluded from the solution (4.70). On the other hand, the field is outgoing as $r \to \infty$. Consequently, the term $ax_1(r)$ must be excluded from (4.70) as well. As a result

$$G_n(r,r') = \begin{cases} -jkh_n^{(2)}(kr')j_n(kr), & r < r' \\ -jkh_n^{(2)}(kr)j_n(kr'), & r > r' \end{cases}. \tag{4.71}$$

The partial derivatives of the Green's function $G_n(r, r')$ are given by

$$\frac{\partial G_n(r,r')}{\partial kr} = \begin{cases} -jkh_n^{(2)}(kr')\dot{j}_n(kr), & r < r' \\ -jk\dot{h}_n^{(2)}(kr)j_n(kr'), & r > r' \end{cases}, \tag{4.72}$$

$$\frac{\partial G_n(r,r')}{\partial kr'} = \begin{cases} -jk\dot{h}_n^{(2)}(kr')j_n(kr), & r < r' \\ -jkh_n^{(2)}(kr)\dot{j}_n(kr'), & r > r' \end{cases}, \tag{4.73}$$

where $\dot{j}_n(kr) = dj_n(kr)/dkr$, $\dot{h}_n^{(2)}(kr) = dh_n^{(2)}(kr)/dkr$. The partial derivatives of the Green's function are discontinuous at $r = r'$:

$$\left.\frac{\partial G_n(r,r')}{\partial kr'}\right|_{r\to r'_+} - \left.\frac{\partial G_n(r,r')}{\partial kr'}\right|_{r\to r'_-} = \frac{1}{r'^2}.$$

Symbolically, the second partial derivative must have a delta function singularity at $r = r'$:

$$\frac{\partial^2 G_n(r,r')}{\partial kr\partial kr'} = \frac{1}{kr'^2}\delta(r-r') + \begin{cases} -jk\dot{h}_n^{(2)}(kr')\dot{j}_n(kr), & r < r' \\ -jk\dot{h}_n^{(2)}(kr)\dot{j}_n(kr'), & r > r' \end{cases}. \tag{4.74}$$

Let us consider the solution of the inhomogeneous spherical Bessel equation

$$\frac{d^2u_n(r)}{dr^2} + \frac{2}{r}\frac{du_n(r)}{dr} + \beta_n^2 u_n(r) = f_n(r). \tag{4.75}$$

This can be rewritten as

$$\frac{d^2[ru_n(r)]}{dr^2} + \beta_n^2[ru_n(r)] = rf_n(r). \tag{4.76}$$

Similarly, (4.68) can be rewritten as

$$\frac{d^2[rG_n(r,r')]}{dr^2} + \beta_n^2[rG_n(r,r')] = -\frac{1}{r}\delta(r-r'). \tag{4.77}$$

Multiplying (4.76) and (4.77) by rG_n and ru_n, respectively, and subtracting the resultants yield

$$\begin{aligned}&\frac{d^2[ru_n(r)]}{dr^2}rG_n(r,r') - \frac{d^2[rG_n(r,r')]}{dr^2}ru_n(r)\\ &= rG_n(r,r')rf_n(r) + u_n(r)\delta(r-r').\end{aligned} \tag{4.78}$$

The integration of (4.78) with respect to r over the interval $(0, \infty)$ leads to

$$u_n(r') = -\int_0^\infty G_n(r,r')r^2 f_n(r)dr + \left[rG_n(r,r')\frac{\partial[ru_n(r)]}{\partial r} - ru_n(r)\frac{\partial[rG_n(r,r')]}{\partial r}\right]_{r=o}^{\infty}. \tag{4.79}$$

On account of (4.71) and the radiation condition, the term in the square bracket must be zero. Since $G_n(r, r')$ is symmetric function of r and r', the solution of (4.75) can be obtained from (4.79) and can be expressed by

$$u_n(r) = -\int_0^\infty G_n(r,r')f_n(r')r'^2dr'. \tag{4.80}$$

According to (4.80), the solutions of the equations in (4.67) can be found as follows:

$$\begin{aligned}V_{nml}^{TE}(r) = &\int_{V_0}\frac{1}{r'}\frac{d}{dkr'}[G_n(r,r')kr']\mathbf{J}_m(\mathbf{r}')\cdot\mathbf{e}_{nml}(\theta',\varphi')dV(\mathbf{r}')\\ &-jk\eta\int_{V_0}G_n(r,r')\mathbf{J}(\mathbf{r}')\cdot\mathbf{h}_{nml}(\theta',\varphi')dV(\mathbf{r}')\\ &-\int_{V_0}\frac{1}{r'}G_n(r,r')[\mathbf{u}_{r'}\cdot\mathbf{J}_m(\mathbf{r}')]\nabla_{\theta\varphi}\cdot\mathbf{e}_{nml}(\theta',\varphi')dV(\mathbf{r}'),\end{aligned} \tag{4.81}$$

$$
\begin{aligned}
I_{nml}^{TM}(r) &= -\int_{V_0} \frac{1}{r'}\frac{d}{dkr'}[kr'G_n(r,r')]\mathbf{J}(\mathbf{r}')\cdot\mathbf{e}_{nml}(\theta',\varphi')dV(\mathbf{r}') \\
&\quad -j\frac{k}{\eta}\int_{V_0} G_n(r,r')\mathbf{J}_m(\mathbf{r}')\cdot\mathbf{h}_{nml}(\theta',\varphi')dV(\mathbf{r}') \\
&\quad +\int_{V_0} \frac{1}{r'}G_n(r,r')[\mathbf{u}_{r'}\cdot\mathbf{J}(\mathbf{r}')]\nabla_{\theta\varphi}\cdot\mathbf{e}_{nml}(\theta',\varphi')dV(\mathbf{r}').
\end{aligned}
\tag{4.82}
$$

Other modal voltages and currents in (4.60) can be obtained from (4.64) and (4.65):

$$
\begin{aligned}
I_{nml}^{TE}(r) &= \frac{j}{kr\eta}\int_{V_0} \frac{1}{kr'}\frac{d^2}{drdkr'}[krkr'G_n(r,r')]\mathbf{J}_m(\mathbf{r}')\cdot\mathbf{e}_{nml}(\theta',\varphi')dV(\mathbf{r}') \\
&\quad +\frac{1}{r}\int_{V_0} \frac{d}{dkr}krG_n(r,r')\mathbf{J}(\mathbf{r}')\cdot\mathbf{h}_{nml}(\theta',\varphi')dV(\mathbf{r}') \\
&\quad -\frac{j}{kr\eta}\int_{V_0} \frac{1}{kr'}\frac{d}{dr}krG_n(r,r')[\mathbf{u}_{r'}\cdot\mathbf{J}_m(\mathbf{r}')]\nabla_{\theta\varphi}\cdot\mathbf{e}_{nml}(\theta',\varphi')dV(\mathbf{r}') \\
&\quad -\frac{j}{k\eta}\int_{S} \mathbf{J}_m(\mathbf{r}')\cdot\mathbf{e}_{nml}(\theta',\varphi')d\Omega',
\end{aligned}
\tag{4.83}
$$

$$
\begin{aligned}
V_{nml}^{TM}(r) &= -\frac{j\eta}{kr}\int_{V_0} \frac{1}{kr'}\frac{d^2}{drdkr'}[krkr'G_n(r,r')]\mathbf{J}(\mathbf{r}')\cdot\mathbf{e}_{nml}(\theta',\varphi')dV(\mathbf{r}') \\
&\quad +\frac{1}{r}\int_{V_0} \frac{d}{dkr}[krG_n(r,r')]\mathbf{J}_m(\mathbf{r}')\cdot\mathbf{h}_{nml}(\theta',\varphi')dV(\mathbf{r}') \\
&\quad +\frac{j\eta}{kr}\int_{V_0} \frac{1}{kr'}\frac{d}{dr}[krG_n(r,r')][\mathbf{u}_{r'}\cdot\mathbf{J}(\mathbf{r}')]\nabla_{\theta\varphi}\cdot\mathbf{e}_{nml}(\theta',\varphi')dV(\mathbf{r}') \\
&\quad +\frac{j\eta}{k}\int_{S} \mathbf{J}(\mathbf{r}')\cdot\mathbf{e}_{nml}(\theta',\varphi')d\Omega.
\end{aligned}
\tag{4.84}
$$

Substituting (4.81)–(4.84) into (4.60) and neglecting the tedious process, the EM fields may be written in a compact form as

$$
\begin{aligned}
\mathbf{E}(\mathbf{r}) &= -jk\eta\int_{V_0} \overleftrightarrow{\mathbf{G}}_e(\mathbf{r},\mathbf{r}')\cdot\mathbf{J}(\mathbf{r}')dV(\mathbf{r}') - \int_{V_0} \overleftrightarrow{\mathbf{G}}_m(\mathbf{r},\mathbf{r}')\cdot\mathbf{J}_m(\mathbf{r}')dV(\mathbf{r}'), \\
\mathbf{H}(\mathbf{r}) &= -j\frac{k}{\eta}\int_{V_0} \overleftrightarrow{\mathbf{G}}_e(\mathbf{r},\mathbf{r}')\cdot\mathbf{J}_m(\mathbf{r}')dV(\mathbf{r}') + \int_{V_0} \overleftrightarrow{\mathbf{G}}_m(\mathbf{r},\mathbf{r}')\cdot\mathbf{J}(\mathbf{r}')dV(\mathbf{r}'),
\end{aligned}
\tag{4.85}
$$

where the dyadic Green's functions are defined by

$$\overleftrightarrow{\mathbf{G}}_e(\mathbf{r},\mathbf{r}') = \sum_{n,m,l} \frac{-jk}{N_{nm}^2} \begin{cases} \mathbf{M}_{nml}^{(1)}(\mathbf{r})\mathbf{M}_{nml}^{(2)}(\mathbf{r}') + \mathbf{N}_{nml}^{(1)}(\mathbf{r})\mathbf{N}_{nml}^{(2)}(\mathbf{r}'), r < r' \\ \mathbf{M}_{nml}^{(2)}(\mathbf{r})\mathbf{M}_{nml}^{(1)}(\mathbf{r}') + \mathbf{N}_{nml}^{(2)}(\mathbf{r})\mathbf{N}_{nml}^{(1)}(\mathbf{r}'), r > r' \end{cases}$$
$$- \frac{1}{k^2}\delta(\mathbf{r}-\mathbf{r}')\mathbf{u}_r\mathbf{u}_{r'},$$
$$\overleftrightarrow{\mathbf{G}}_m(\mathbf{r},\mathbf{r}') = \sum_{n,m,l} \frac{-jk^2}{N_{nm}^2} \begin{cases} \mathbf{N}_{nml}^{(1)}(\mathbf{r})\mathbf{M}_{nml}^{(2)}(\mathbf{r}') + \mathbf{M}_{nml}^{(1)}(\mathbf{r})\mathbf{N}_{nml}^{(2)}(\mathbf{r}'), r < r' \\ \mathbf{N}_{nml}^{(2)}(\mathbf{r})\mathbf{M}_{nml}^{(1)}(\mathbf{r}') + \mathbf{M}_{nml}^{(2)}(\mathbf{r})\mathbf{N}_{nml}^{(1)}(\mathbf{r}'), r > r' \end{cases}. \tag{4.86}$$

In (4.86), the SVWFs or **fundamental field patterns** $\mathbf{M}_{nml}^{(i)}$ and $\mathbf{N}_{nml}^{(i)}$ in spherical waveguide have been introduced, and they are defined by

$$\begin{aligned} \mathbf{M}_{nml}^{(i)}(\mathbf{r}) &= -\frac{N_{nm}}{kr}\tilde{z}_n^{(i)}(kr)\mathbf{h}_{nml}(\theta,\varphi), \\ \mathbf{N}_{nml}^{(i)}(\mathbf{r}) &= \frac{N_{nm}}{kr}\dot{\tilde{z}}_n^{(i)}(kr)\mathbf{e}_{nml}(\theta,\varphi) - \mathbf{u}_r\frac{N_{nm}}{kr}z_n^{(i)}(kr)\nabla_{\theta\varphi}\cdot\mathbf{e}_{nml}(\theta,\varphi), \\ (i &= 1,2) \end{aligned} \tag{4.87}$$

where

$$z_n^{(1)}(kr) = j_n(kr), \qquad z_n^{(2)}(kr) = h_n^{(2)}(kr),$$
$$\tilde{z}_n^{(i)}(kr) = krz_n^{(i)}(kr), \quad \dot{\tilde{z}}_n^{(i)}(kr) = \frac{d\tilde{z}_n^{(i)}(kr)}{dkr}.$$

Explicitly, the SVWFs are given by

$$\begin{aligned} \mathbf{M}_{nml}^{(i)}(\mathbf{r}) &= z_n^{(i)}(kr)\frac{1}{\sin\theta}\frac{\partial Y_{nml}(\theta,\varphi)}{\partial\varphi}\mathbf{u}_\theta - z_n^{(i)}(kr)\frac{\partial Y_{nml}(\theta,\varphi)}{\partial\theta}\mathbf{u}_\varphi, \\ \mathbf{N}_{nml}^{(i)}(\mathbf{r}) &= \frac{n(n+1)}{kr}z_n^{(i)}(kr)Y_{nml}(\theta,\varphi)\mathbf{u}_r + \frac{1}{kr}\frac{d\left[krz_n^{(i)}(kr)\right]}{dkr}\frac{\partial Y_{nml}(\theta,\varphi)}{\partial\theta}\mathbf{u}_\theta \\ &\quad + \frac{1}{kr}\frac{d\left[krz_n^{(i)}(kr)\right]}{dkr}\frac{1}{\sin\theta}\frac{\partial Y_{nml}(\theta,\varphi)}{\partial\varphi}\mathbf{u}_\varphi, \quad (i = 1,2) \end{aligned} \tag{4.88}$$

Example 4.1 (Field Expansions Outside the Circumscribing Sphere of the Source)

Outside the circumscribing sphere of the source region, the fields in (4.85) are outgoing and reduce to

$$\begin{aligned} \mathbf{E} &= -\sum_{n,m,l}\left(\alpha_{nml}^{(2)}\mathbf{M}_{nml}^{(2)} + \beta_{nml}^{(2)}\mathbf{N}_{nml}^{(2)}\right), \\ \mathbf{H} &= \frac{1}{j\eta}\sum_{n,m,l}\left(\alpha_{nml}^{(2)}\mathbf{N}_{nml}^{(2)} + \beta_{nml}^{(2)}\mathbf{M}_{nml}^{(2)}\right), \end{aligned} \tag{4.89}$$

where $\alpha_{nml}^{(2)}$ and $\beta_{nml}^{(2)}$ are constants determined by the sources

$$
\begin{aligned}
\alpha_{nml}^{(2)} &= \frac{k^2\eta}{N_{nm}^2}\int_{V_0}\mathbf{J}\cdot\mathbf{M}_{nml}^{(1)}dV - \frac{jk^2}{N_{nm}^2}\int_{V_0}\mathbf{J}_m\cdot\mathbf{N}_{nml}^{(1)}dV,\\
\beta_{nml}^{(2)} &= \frac{k^2\eta}{N_{nm}^2}\int_{V_0}\mathbf{J}\cdot\mathbf{N}_{nml}^{(1)}dV - \frac{jk^2}{N_{nm}^2}\int_{V_0}\mathbf{J}_m\cdot\mathbf{M}_{nml}^{(1)}dV.
\end{aligned}
\tag{4.90}
$$

The expansions (4.89) may be divided into the TE and TM parts

$$
\begin{cases}
\mathbf{E}^{TE} = -\sum\limits_{n,m,l}\alpha_{nml}^{(2)}\mathbf{M}_{nml}^{(2)}\\
\mathbf{H}^{TE} = \dfrac{1}{j\eta}\sum\limits_{n,m,l}\alpha_{nml}^{(2)}\mathbf{N}_{nml}^{(2)}
\end{cases},
\quad
\begin{cases}
\mathbf{E}^{TM} = -\sum\limits_{n,m,l}\beta_{nml}^{(2)}\mathbf{N}_{nml}^{(2)}\\
\mathbf{H}^{TM} = \dfrac{1}{j\eta}\sum\limits_{n,m,l}\beta_{nml}^{(2)}\mathbf{M}_{nml}^{(2)}
\end{cases}.
\tag{4.91}
$$

From (4.89), the transverse fields can be expressed in terms of the vector modal functions of the spherical waveguide

$$
\begin{aligned}
kr\mathbf{E}_t &= \sum_{n,m,l}N_{nm}\left(\alpha_{nml}^{(2)}\tilde{h}_n^{(2)}\mathbf{h}_{nml} - \beta_{nml}^{(2)}\dot{\tilde{h}}_n^{(q)}\mathbf{e}_{nml}\right),\\
kr\mathbf{H}_t &= \frac{1}{j\eta}\sum_{n,m,l}N_{nm}\left(\alpha_{nml}^{(2)}\dot{\tilde{h}}_n^{(2)}\mathbf{e}_{nml} - \beta_{nml}^{(2)}\tilde{h}_n^{(2)}\mathbf{h}_{nml}\right).
\end{aligned}
\tag{4.92}
$$

In the far-field region, the SVWFs must be outgoing and their behavior can be determined by the asymptotic formulas

$$
h_n^{(2)}(kr) \to \frac{1}{kr}j^{n+1}e^{-jkr},\ \frac{1}{kr}\dot{\tilde{h}}_n^{(2)}(kr) \to \frac{1}{kr}j^n e^{-jkr},\quad r\to\infty,
\tag{4.93}
$$

and (4.92) becomes

$$
\begin{aligned}
\mathbf{E}_t &= \frac{e^{-jkr}}{kr}\sum_{n,m,l}N_{nm}j^n\left(\alpha_{nml}^{(2)}j\mathbf{h}_{nml} - \beta_{nml}^{(2)}\mathbf{e}_{nml}\right),\\
\mathbf{H}_t &= \frac{1}{j\eta}\frac{e^{-jkr}}{kr}\sum_{n,m,l}N_{nm}j^n\left(\alpha_{nml}^{(2)}\mathbf{e}_{nml} - \beta_{nml}^{(2)}j\mathbf{h}_{nml}\right).
\end{aligned}
\tag{4.94}
$$

□

Example 4.2 (Field Expansions in the Spherical Shell Region Between Sources)

In the spherical shell region V between two source regions as illustrated in Figure 4.6, the fields in (4.85) can be expressed as the sum of the fields generated by the sources outside the shell

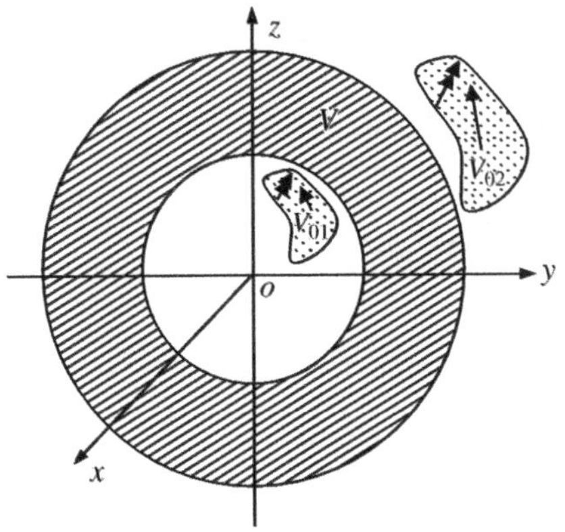

Figure 4.6 Spherical shell region between two sources.

$$
\begin{aligned}
\mathbf{E}(\mathbf{r}) &= -jk\eta \int_{V_{01}} \overleftrightarrow{\mathbf{G}}_e(\mathbf{r},\mathbf{r}') \cdot \mathbf{J}(\mathbf{r}') dV(\mathbf{r}') - \int_{V_{01}} \overleftrightarrow{\mathbf{G}}_m(\mathbf{r},\mathbf{r}') \cdot \mathbf{J}_m(\mathbf{r}') dV(\mathbf{r}') \\
&\quad -jk\eta \int_{V_{02}} \overleftrightarrow{\mathbf{G}}_e(\mathbf{r},\mathbf{r}') \cdot \mathbf{J}(\mathbf{r}') dV(\mathbf{r}') - \int_{V_{02}} \overleftrightarrow{\mathbf{G}}_m(\mathbf{r},\mathbf{r}') \cdot \mathbf{J}_m(\mathbf{r}') dV(\mathbf{r}'), \\
\mathbf{H}(\mathbf{r}) &= -j\frac{k}{\eta} \int_{V_{01}} \overleftrightarrow{\mathbf{G}}_e(\mathbf{r},\mathbf{r}') \cdot \mathbf{J}_m(\mathbf{r}') \mathrm{d}V(\mathbf{r}') + \int_{V_{01}} \overleftrightarrow{\mathbf{G}}_m(\mathbf{r},\mathbf{r}') \cdot \mathbf{J}(\mathbf{r}') dV(\mathbf{r}') \\
&\quad -j\frac{k}{\eta} \int_{V_{02}} \overleftrightarrow{\mathbf{G}}_e(\mathbf{r},\mathbf{r}') \cdot \mathbf{J}_m(\mathbf{r}') \mathrm{d}V(\mathbf{r}') + \int_{V_{02}} \overleftrightarrow{\mathbf{G}}_m(\mathbf{r},\mathbf{r}') \cdot \mathbf{J}(\mathbf{r}') dV(\mathbf{r}').
\end{aligned}
\tag{4.95}
$$

Inserting (4.86) into (4.95), the fields can be expressed in terms of SVWFs as

$$
\begin{aligned}
\mathbf{E} &= -\sum_{n,m,l} \left(\alpha_{nml}^{(1)} \mathbf{M}_{nml}^{(1)} + \beta_{nml}^{(1)} \mathbf{N}_{nml}^{(1)}\right) - \sum_{n,m,l} \left(\alpha_{nml}^{(2)} \mathbf{M}_{nml}^{(2)} + \beta_{nml}^{(2)} \mathbf{N}_{nml}^{(2)}\right), \\
\mathbf{H} &= \frac{1}{j\eta} \sum_{n,m,l} \left(\alpha_{nml}^{(1)} \mathbf{N}_{nml}^{(1)} + \beta_{nml}^{(1)} \mathbf{M}_{nml}^{(1)}\right) + \frac{1}{j\eta} \sum_{n,m,l} \left(\alpha_{nml}^{(2)} \mathbf{N}_{nml}^{(2)} + \beta_{nml}^{(2)} \mathbf{M}_{nml}^{(2)}\right),
\end{aligned}
\tag{4.96}
$$

where the expansion coefficients can be determined by the sources

$$
\begin{aligned}
\alpha_{nml}^{(1)} &= \frac{k^2\eta}{N_{nm}^2} \int_{V_{02}} \mathbf{J} \cdot \mathbf{M}_{nml}^{(2)} dV - \frac{jk^2}{N_{nm}^2} \int_{V_{02}} \mathbf{J}_m \cdot \mathbf{N}_{nml}^{(2)} dV, \\
\beta_{nml}^{(1)} &= \frac{k^2\eta}{N_{nm}^2} \int_{V_{02}} \mathbf{J} \cdot \mathbf{N}_{nml}^{(2)} dV - \frac{jk^2}{N_{nm}^2} \int_{V_{02}} \mathbf{J}_m \cdot \mathbf{M}_{nml}^{(2)} dV,
\end{aligned}
\tag{4.97}
$$

$$
\begin{aligned}
\alpha_{nml}^{(2)} &= \frac{k^2\eta}{N_{nm}^2}\int_{V_{01}} \mathbf{J}\cdot\mathbf{M}_{nml}^{(1)}dV - \frac{jk^2}{N_{nm}^2}\int_{V_{01}} \mathbf{J}_m\cdot\mathbf{N}_{nml}^{(1)}dV,\\
\beta_{nml}^{(2)} &= \frac{k^2\eta}{N_{nm}^2}\int_{V_{01}} \mathbf{J}\cdot\mathbf{N}_{nml}^{(1)}dV - \frac{jk^2}{N_{nm}^2}\int_{V_{01}} \mathbf{J}_m\cdot\mathbf{M}_{nml}^{(1)}dV.
\end{aligned}
\tag{4.98}
$$

Insertion of (4.87) into (4.96) gives the expressions for the transverse fields in terms of the vector modal functions

$$
\begin{aligned}
\mathbf{E}_t &= \sum_{n,m,l}\left(V_{nml}^{TM}\mathbf{e}_{nml} + V_{nml}^{TE}\mathbf{h}_{nml}\right),\\
\mathbf{H}_t &= \sum_{n,m,l}\left(I_{nml}^{TM}\mathbf{h}_{nml} - I_{nml}^{TE}\mathbf{e}_{nml}\right).
\end{aligned}
\tag{4.99}
$$

The modal voltages and currents in (4.99) are the superposition of the outward-going and inward-going waves

$$
\begin{cases} V_{nml}^{TE} = V_{nml}^{TE+} + V_{nml}^{TE-} \\ I_{nml}^{TE} = I_{nml}^{TE+} + I_{nml}^{TE-} \end{cases},\quad
\begin{cases} V_{nml}^{TM} = V_{nml}^{TM+} + V_{nml}^{TM-} \\ I_{nml}^{TM} = I_{nml}^{TM+} + I_{nml}^{TM-} \end{cases},
$$

where the superscripts + and − denote outward-going and inward-going waves, respectively, defined by

$$
\begin{cases} V_{nml}^{TE+}(r) = \dfrac{N_{nm}}{kr}\hat{\alpha}_{nml}^{(2)}\tilde{h}_n^{(2)}(kr) \\ V_{nml}^{TE-}(r) = \dfrac{N_{nm}}{kr}\hat{\alpha}_{nml}^{(1)}\tilde{h}_n^{(1)}(kr) \end{cases},\quad
\begin{cases} V_{nml}^{TM+}(r) = -\dfrac{N_{nm}}{kr}\hat{\beta}_{nml}^{(2)}\dot{\tilde{h}}_n^{(2)}(kr) \\ V_{nml}^{TM-}(r) = -\dfrac{N_{nm}}{kr}\hat{\beta}_{nml}^{(1)}\dot{\tilde{h}}_n^{(1)}(kr) \end{cases},
$$

$$
\begin{cases} I_{nml}^{TE+}(r) = -\dfrac{1}{j\eta}\dfrac{N_{nm}}{kr}\hat{\alpha}_{nml}^{(2)}\dot{\tilde{h}}_n^{(2)}(kr) \\ I_{nml}^{TE-}(r) = -\dfrac{1}{j\eta}\dfrac{N_{nm}}{kr}\hat{\alpha}_{nml}^{(1)}\dot{\tilde{h}}_n^{(1)}(kr) \end{cases},\quad
\begin{cases} I_{nml}^{TM+}(r) = -\dfrac{1}{j\eta}\dfrac{N_{nm}}{kr}\hat{\beta}_{nml}^{(2)}\tilde{h}_n^{(2)}(kr) \\ I_{nml}^{TM-}(r) = -\dfrac{1}{j\eta}\dfrac{N_{nm}}{kr}\hat{\beta}_{nml}^{(1)}\tilde{h}_n^{(1)}(kr) \end{cases}.
$$

The radially directed **wave impedances** for TE modes and TM modes are defined by

$$
\begin{aligned}
Z_n^{TE}(r) &= \frac{V_{nml}^{TE+}(r)}{I_{nml}^{TE+}(r)} = -j\eta\frac{\tilde{h}_n^{(2)}(kr)}{\dot{\tilde{h}}_n^{(2)}(kr)},\\
Z_n^{TM}(r) &= \frac{V_{nml}^{TM+}(r)}{I_{nml}^{TM+}(r)} = j\eta\frac{\dot{\tilde{h}}_n^{(2)}(kr)}{\tilde{h}_n^{(2)}(kr)}.
\end{aligned}
\tag{4.100}
$$

As $r \to \infty$, the wave impedances become real and approach to η. By numerical tabulation, it can be shown that the above wave impedances are predominantly reactive when $kr < n$, and predominantly resistive when $kr > n$. Therefore, the value $kr = n$ is approximately the point of gradual cutoff. If one considers β_n as the propagation constant of the spherical waveguide modes, the above behavior can be easily understood from (4.62). □

4.2.3 Properties of Spherical Vector Wave Functions

The SVWFs defined by (4.87) can be represented in terms of the scalar wave function

$$\psi_{nml}^{(i)}(\mathbf{r}) = z_n^{(i)}(kr)Y_{nml}(\theta, \varphi) \tag{4.101}$$

as

$$\begin{aligned} \mathbf{M}_{nml}^{(i)} &= \nabla \times \left[\mathbf{r}\psi_{nml}^{(i)}\right] = \nabla\psi_{nml}^{(i)} \times \mathbf{r}, \\ \mathbf{N}_{nml}^{(i)} &= \frac{1}{k}\nabla \times \nabla \times \left[\mathbf{r}\psi_{nml}^{(i)}\right]. \end{aligned} \quad (i = 1, 2) \tag{4.102}$$

Evidently the SVWFs satisfy the Maxwell-like equations

$$\begin{aligned} \nabla \times \mathbf{M}_{nml}^{(i)} &= k\mathbf{N}_{nml}^{(i)}, \\ \nabla \times \mathbf{N}_{nml}^{(i)} &= k\mathbf{M}_{nml}^{(i)}, \end{aligned} \tag{4.103}$$

and the vector wave equations

$$\begin{aligned} \nabla \times \nabla \times \mathbf{M}_{nml}^{(i)} - k^2\mathbf{M}_{nml}^{(i)} &= 0, \\ \nabla \times \nabla \times \mathbf{N}_{nml}^{(i)} - k^2\mathbf{N}_{nml}^{(i)} &= 0. \end{aligned} \tag{4.104}$$

The orthogonality properties of SVWFs on a sphere S of radius r can be easily derived from orthogonality properties (4.39) and (4.40) of the spherical harmonics on a unit sphere, and they are summarized below:

1) $$\int_S \mathbf{M}_{nml}^{(i)} \cdot \mathbf{M}_{n'm'l'}^{(i)} dS = \begin{cases} 0, & [n, m, l] \neq [n', m', l'] \\ \left[rz_n^{(i)}(kr)\right]^2 N_{nm}^2, & [n, m, l] = [n', m', l'] \end{cases}.$$

2) $$\int_S \mathbf{M}_{nml}^{(i)} \cdot \overline{\mathbf{M}}_{n'm'l'}^{(i)} dS = \begin{cases} 0, & [n, m, l] \neq [n', m', l'] \\ \left|rz_n^{(i)}(kr)\right|^2 N_{nm}^2, & [n, m, l] = [n', m', l'] \end{cases}.$$

3) $$\int_S \mathbf{N}_{nml}^{(i)} \cdot \mathbf{N}_{n'm'l'}^{(i)} dS = \begin{cases} 0, & [n,m,l] \neq [n',m',l'] \\ \frac{1}{k^2}\left\{n(n+1)\left[z_n^{(i)}(kr)\right]^2 + \left[\dot{z}_n^{(i)}(kr)\right]^2\right\}N_{nm}^2, & [n,m,l] = [n',m',l'] \end{cases}.$$

4) $\int\limits_S \mathbf{N}_{nml}^{(i)} \cdot \overline{\mathbf{N}}_{n'm'l'}^{(i)} dS = \begin{cases} 0, & [n,m,l] \neq [n',m',l'] \\ \frac{1}{k^2}\left\{ n(n+1)\left|z_n^{(i)}(kr)\right|^2 + \left|\dot{\tilde{z}}_n^{(i)}(kr)\right|^2 \right\} N_{nm}^2, & [n,m,l] = [n',m',l'] \end{cases}.$

5) $\int\limits_S \mathbf{M}_{nml}^{(i)} \cdot \mathbf{N}_{n'm'l'}^{(i)} dS = \int\limits_S \mathbf{M}_{nml}^{(i)} \cdot \overline{\mathbf{N}}_{n'm'l'}^{(i)} dS = 0.$

From (4.93), the far-field expressions for the SVWFs can be readily found

$$\mathbf{M}_{nml}^{(2)} \approx -j^{n+1} N_{nm} \frac{e^{-jkr}}{kr} \mathbf{h}_{nml}, \ \mathbf{N}_{nml}^{(2)} \approx j^n N_{nm} \frac{e^{-jkr}}{kr} \mathbf{e}_{nml}. \tag{4.105}$$

Explicitly, these are

$$\begin{aligned} \mathbf{M}_{nml}^{(2)}(\mathbf{r}) &\approx j^{n+1} \frac{e^{-jkr}}{kr} \left[\frac{1}{\sin\theta} \frac{\partial Y_{nml}(\theta,\varphi)}{\partial\varphi} \mathbf{u}_\theta - \frac{\partial Y_{nml}(\theta,\varphi)}{\partial\theta} \mathbf{u}_\varphi \right], \\ \mathbf{N}_{nml}^{(2)}(\mathbf{r}) &\approx j^n \frac{e^{-jkr}}{kr} \left[\frac{\partial Y_{nml}(\theta,\varphi)}{\partial\theta} \mathbf{u}_\theta + \frac{1}{\sin\theta} \frac{\partial Y_{nml}(\theta,\varphi)}{\partial\varphi} \mathbf{u}_\varphi \right]. \end{aligned} \tag{4.106}$$

According to the field expansions (4.91), one may introduce the TE and TM modal fields

$$\begin{cases} \mathbf{E}^{TE_{nml}^{(2)}} = -\mathbf{M}_{nml}^{(2)}, \\ \mathbf{H}^{TE_{nml}^{(2)}} = \frac{1}{j\eta} \mathbf{N}_{nml}^{(2)}, \end{cases} \quad \begin{cases} \mathbf{E}^{TM_{nml}^{(2)}} = -\mathbf{N}_{nml}^{(2)}, \\ \mathbf{H}^{TM_{nml}^{(2)}} = \frac{1}{j\eta} \mathbf{M}_{nml}^{(2)}. \end{cases} \tag{4.107}$$

The radiated powers of the TE and TM modal fields are equal and given by

$$\begin{aligned} P_{TE_{nml}^{(2)}}^{rad} = P_{TM_{nml}^{(2)}}^{rad} &= \frac{1}{2} \int\limits_{S_\infty} \mathrm{Re}\left(\mathbf{E}^{TE_{nml}^{(2)}} \times \overline{\mathbf{H}}^{TE_{nml}^{(2)}} \right) \cdot \mathbf{u}_r dS = \frac{1}{2} \int\limits_{S_\infty} \mathrm{Re}\left(\mathbf{E}^{TM_{nml}^{(2)}} \times \overline{\mathbf{H}}^{TM_{nml}^{(2)}} \right) \cdot \mathbf{u}_r dS \\ &= \frac{1}{2\eta} \int\limits_{S_\infty} \left\{ \frac{1}{(kr)^2} \left[\frac{1}{\sin\theta} \frac{\partial Y_{nml}(\theta,\varphi)}{\partial\varphi} \right]^2 + \left[\frac{\partial Y_{nml}(\theta,\varphi)}{\partial\theta} \right]^2 \right\} dS = \frac{N_{nm}^2}{2k^2\eta}, \end{aligned} \tag{4.108}$$

where use has been made of (4.40).

4.2.4 Far-Zone Fields

The far-fields (4.94) can be rewritten as

$$\mathbf{E}(\mathbf{r}) = \frac{e^{-jkr}}{r} \mathbf{E}_\infty(\mathbf{u}_r), \ \mathbf{H}(\mathbf{r}) = \frac{e^{-jkr}}{r} \mathbf{H}_\infty(\mathbf{u}_r), \tag{4.109}$$

where $\mathbf{E}_\infty$ and $\mathbf{H}_\infty$ are the electric and magnetic far-field patterns

$$\begin{aligned}\mathbf{E}_\infty(\mathbf{u}_r) &= \frac{1}{k}\sum_{n,m,l} j^n N_{nm}\left(j\alpha_{nml}^{(2)}\mathbf{h}_{nml} - \beta_{nml}^{(2)}\mathbf{e}_{nml}\right),\\ \mathbf{H}_\infty(\mathbf{u}_r) &= \frac{1}{k\eta}\sum_{n,m,l} j^n N_{nm}\left(-j\alpha_{nml}^{(2)}\mathbf{e}_{nml} - \beta_{nml}^{(2)}\mathbf{h}_{nml}\right),\end{aligned} \tag{4.110}$$

and satisfy (4.14). It is easy to verify that the far-fields (4.109) satisfy the Silver–Müller radiation condition (4.15). The radiation intensity in the direction $\mathbf{u}_r$ is given by

$$\begin{aligned}U(\mathbf{u}_r) &= \frac{1}{2}r^2\,\mathrm{Re}\left(\mathbf{E}\times\overline{\mathbf{H}}\right)\cdot\mathbf{u}_r = \frac{r^2}{2\eta}|\mathbf{E}|^2\\ &= \frac{1}{2\eta}\left|\sum_{n,m,l}\frac{j^n}{k}N_{nm}\left(j\alpha_{nml}^{(2)}\mathbf{h}_{nml} - \beta_{nml}^{(2)}\mathbf{e}_{nml}\right)\right|^2.\end{aligned} \tag{4.111}$$

The total radiated power of the source is

$$P^{rad} = \frac{1}{2}\,\mathrm{Re}\int_{S_\infty}\mathbf{E}\times\overline{\mathbf{H}}\cdot\mathbf{u}_r dS = \frac{1}{2k^2\eta}\sum_{n,m,l}N_{nm}^2\left(\left|\alpha_{nml}^{(2)}\right|^2 + \left|\beta_{nml}^{(2)}\right|^2\right). \tag{4.112}$$

Remark 4.3 Once the electric or magnetic radiation pattern is known, one can use the expansions for the far-field patterns in (4.110) to determine the expansion coefficients $\alpha_{nml}^{(2)}$ and $\beta_{nml}^{(2)}$, based on which the fields in the near-field region can be determined from (4.89). Similarly, one can determine the far-fields from the near-fields. The spherical modal analysis affords an easy way to transform the far-fields into the near-fields and vice versa. □

4.3 Stored Field Energies and Radiation Quality Factor

The stored EM field energies are important quantities in the study of wave propagation and energy storage in electrical and optical engineering, and many efforts have been placed on their evaluation in various media [40–63]. In electric circuits, power is defined as the amount of energy transferred (supplied or absorbed) per unit time. Consider a one-port network shown in Figure 4.7. Assume that the power p flowing into the network, with the passive sign convention, can be expressed as

$$p(t) = v(t)i(t) = \frac{dw(t)}{dt}, \tag{4.113}$$

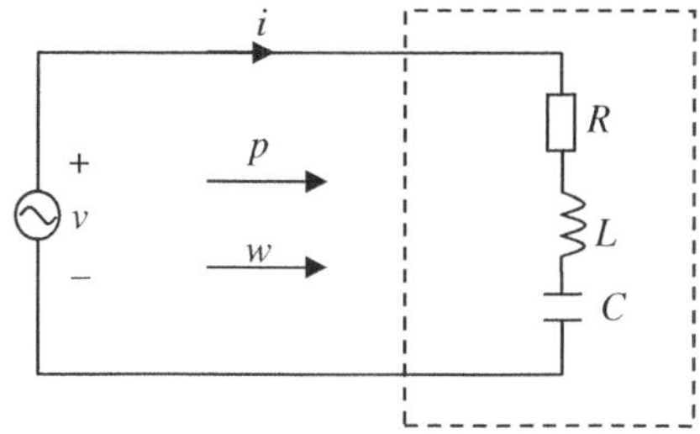

Figure 4.7 One-port network: an RLC circuit.

where w denotes the energy absorbed by the network. It should be mentioned that the instantaneous power p is an algebraic quantity. When p is positive, the network absorbs power; when p is negative, the network supplies power. The energy absorbed or supplied by the network in the time interval $t_0 < t < t_1$ is then given by

$$w(t_0, t_1) = \int_{t_0}^{t_1} p(t)dt. \tag{4.114}$$

Since the power p is an algebraic quantity, the energy $w(t_0, t_1)$ is also an algebraic quantity and can be negative. Similar to the power p, a positive (or negative) energy implies that the network absorbs (or supplies) energy. Let us examine a typical RLC circuit shown in Figure 4.7. According to Kirchhoff's voltage law, one may write

$$v = iR + L\frac{di}{dt} + v_c, \tag{4.115}$$

where $v_c = (1/C)\int_{-\infty}^{t} idt$ denotes the voltage across the capacitor. The time-domain power balance relation for the RLC circuit can be easily obtained from (4.115) as follows:

$$p = vi = p_R + \frac{dw_L}{dt} + \frac{dw_C}{dt}, \tag{4.116}$$

where $p_R = i^2R$, $w_L = Li^2/2$, and $w_C = Cv_c^2/2$ are the dissipated power in the resistor R, the stored magnetic energy in the inductor L, and the stored electric energy in the capacitor C, respectively. The left-hand side of (4.116) is the input power of the RLC circuit from the voltage source v, and the right-hand side of (4.116) denotes the rate of increase of energy in the RLC circuit, which can be decomposed into the sum of the rate of energy absorbed by the resistor, the rate of magnetic energy stored in the inductor, and the rate of electric energy stored in the capacitor. In general, it is impossible to find an energy function w_R such that the dissipated power can be written as the time derivative of w_R, i.e. $p_R = dw_R/dt$ since the integral $w_R = \int_{-\infty}^{t} i^2 R dt$ may not exist for an arbitrary time dependence. Equation (4.116) thus indicates that the input power of the RLC circuit is not always expressible as a complete differential of an energy function due to the heat loss.

4.3.1 Stored Field Energies in General Materials

A decomposition similar to (4.116) exists for EM fields. The field energy densities for time-harmonic fields may be derived from the time-domain Poynting theorem in a source-free region

$$-\nabla \cdot \mathbf{S}(\mathbf{r}, t) = \mathbf{E}(\mathbf{r}, t) \cdot \frac{\partial \mathbf{D}(\mathbf{r}, t)}{\partial t} + \mathbf{H}(\mathbf{r}, t) \cdot \frac{\partial \mathbf{B}(\mathbf{r}, t)}{\partial t}, \tag{4.117}$$

where $\mathbf{S}(\mathbf{r}, t) = \mathbf{E}(\mathbf{r}, t) \times \mathbf{H}(\mathbf{r}, t)$ is the Poynting vector. The left-hand side of (4.117) denotes the instantaneous density of inflow power from external sources. The right-hand side of (4.117) stands for the work done per second by the inflow power, and can thus be interpreted as the density of the total field energy per second required to establish the EM fields. Analogous to (4.116), the right-hand side of (4.117) may be decomposed into two parts. One part is converted into heat loss, and the other part is stored in the medium and is recoverable in the form of EM field energy, as described below

$$\mathbf{E}(\mathbf{r}, t) \cdot \frac{\partial \mathbf{D}(\mathbf{r}, t)}{\partial t} = p_e(\mathbf{r}, t) + \frac{dw_e(\mathbf{r}, t)}{dt}, \tag{4.118}$$

$$\mathbf{H}(\mathbf{r}, t) \cdot \frac{\partial \mathbf{B}(\mathbf{r}, t)}{\partial t} = p_m(\mathbf{r}, t) + \frac{dw_m(\mathbf{r}, t)}{dt}, \tag{4.119}$$

where $p_e(\mathbf{r}, t)$ and $p_m(\mathbf{r}, t)$ denote the rates of dissipated electric and magnetic field energy densities, respectively, while $w_e(\mathbf{r}, t)$ and $w_m(\mathbf{r}, t)$ stand for the stored electric and magnetic field energy density, respectively. Accordingly, (4.117) can be rewritten as

$$-\nabla \cdot \mathbf{S}(\mathbf{r}, t) = p_e(\mathbf{r}, t) + p_m(\mathbf{r}, t) + \frac{dw_e(\mathbf{r}, t)}{dt} + \frac{dw_m(\mathbf{r}, t)}{dt}. \tag{4.120}$$

The above equation is similar to the power balance relation (4.116) for the RLC circuit. Based on the above decomposition, the (time averaged) stored electric field energy density $w_e(\mathbf{r})$, the rate of dissipated electric field energy density $p_e(\mathbf{r})$, the stored magnetic field energy density $w_m(\mathbf{r})$, and the rate of dissipated magnetic field energy density $p_m(\mathbf{r})$ for a time-harmonic EM field with operating frequency ω in a general material have been obtained by the author in [40]

$$w_e(\mathbf{r}) = \frac{1}{4}\operatorname{Re}\mathbf{E}(\mathbf{r})\cdot\overline{\mathbf{D}}(\mathbf{r}) + \frac{1}{4}\omega\operatorname{Re}\left[\mathbf{E}(\mathbf{r})\cdot\frac{\partial\overline{\mathbf{D}}(\mathbf{r})}{\partial\omega} - \overline{\mathbf{D}}(\mathbf{r})\cdot\frac{\partial\mathbf{E}(\mathbf{r})}{\partial\omega}\right], \quad (4.121)$$

$$p_e(\mathbf{r}) = \frac{\omega}{2}\operatorname{Im}\left[\mathbf{E}(\mathbf{r})\cdot\overline{\mathbf{D}}(\mathbf{r})\right], \quad (4.122)$$

$$w_m(\mathbf{r}) = \frac{1}{4}\operatorname{Re}\mathbf{H}(\mathbf{r})\cdot\overline{\mathbf{B}}(\mathbf{r}) + \frac{1}{4}\omega\operatorname{Re}\left[\mathbf{H}(\mathbf{r})\cdot\frac{\partial\overline{\mathbf{B}}(\mathbf{r})}{\partial\omega} - \overline{\mathbf{B}}(\mathbf{r})\cdot\frac{\partial\mathbf{H}(\mathbf{r})}{\partial\omega}\right], \quad (4.123)$$

$$p_m(\mathbf{r}) = \frac{\omega}{2}\operatorname{Im}\left[\mathbf{H}(\mathbf{r})\cdot\overline{\mathbf{B}}(\mathbf{r})\right]. \quad (4.124)$$

The above expressions do not explicitly contain the constitutive relations and microscopic models, and therefore are applicable to any medium. In the derivation of the above expressions, the sinusoidal field is extended to the complex frequency domain by replacing ω with the complex frequency $s = \alpha + j\omega$, where α is a real parameter. By using the Cauchy–Riemann conditions and the field expansions for sufficiently small α, the energy densities (4.121)–(4.124) can be obtained through the decompositions (4.118) and (4.119) in the time domain.

In the conventional study of the stored field energy expressions for a sinusoidal (time-harmonic) field, a NB analysis is usually utilized by assuming that the field has frequency components in a narrow range about some central value ω_0 [e.g. 41–46]

$$\mathbf{E}(\mathbf{r},t) = \operatorname{Re}\mathbf{E}_a(\mathbf{r},t)e^{j\omega_0 t}, \quad (4.125)$$

where $\mathbf{E}_a(\mathbf{r}, t)$ is referred to as the complex envelope and is supposed to be a slowly varying function of time compared with the carrier wave $e^{j\omega_0 t}$, and ω_0 is the mid-band frequency, also called the **carrier frequency**. The complex envelope $\mathbf{E}_a(\mathbf{r}, t)$ reduces to the conventional phasor for a pure sinusoidal field if it is independent of time.

Remark 4.4 (Traditional NB Approach)
In the traditional NB approach, an assumption is often made that the medium is dispersive and has negligible losses [e.g. 44–46]. By means of the NB representation (4.125), the left-hand side of (4.118) can be written as

$$\mathbf{E}\cdot\frac{\partial\mathbf{D}}{\partial t} = \frac{1}{4}\left(\mathbf{E}_a e^{j\omega_0 t} + \overline{\mathbf{E}}_a e^{-j\omega_0 t}\right)\cdot\frac{\partial}{\partial t}\left(\mathbf{D}_a e^{j\omega_0 t} + \overline{\mathbf{D}}_a e^{-j\omega_0 t}\right). \quad (4.126)$$

The complex envelope $\mathbf{E}_a$ is then expanded as an integral of the Fourier components $\tilde{\mathbf{E}}_a e^{j\omega t}$, where $\tilde{\mathbf{E}}_a$ is independent of time. According to the NB assumption, the Fourier frequency components must satisfy $\omega \ll \omega_0$ to ensure that the complex envelope varies slowly. A specific constitutive relation such as $\mathbf{D} = \varepsilon\mathbf{E}$, where ε is the dielectric constant and is real for a lossless medium, can be introduced, and one thus has the following component-wise expression:

$$\begin{aligned}\frac{\partial \mathbf{D}}{\partial t} &= \frac{\partial}{\partial t}\left(\tilde{\mathbf{D}}_a e^{j(\omega+\omega_0)t}\right) = j(\omega+\omega_0)\tilde{\mathbf{D}}_a e^{j(\omega+\omega_0)t} \\ &= j(\omega+\omega_0)\varepsilon(\omega+\omega_0)\tilde{\mathbf{E}}_a e^{j(\omega+\omega_0)t}.\end{aligned}$$

Upon expanding the factor $(\omega+\omega_0)\varepsilon(\omega+\omega_0)$ about the carrier frequency ω_0 up to the linear term (first-order approximation), the above expression can be written as

$$\frac{\partial \mathbf{D}}{\partial t} = j\omega_0\varepsilon(\omega_0)\tilde{\mathbf{E}}_a e^{j(\omega+\omega_0)t} + j\omega\frac{d[\omega_0\varepsilon(\omega_0)]}{d\omega_0}\tilde{\mathbf{E}}_a e^{j(\omega+\omega_0)t}.$$

Summing the Fourier components yields

$$\frac{\partial \mathbf{D}}{\partial t} = j\omega_0\varepsilon(\omega_0)\mathbf{E}_a e^{j\omega_0 t} + \frac{d[\omega_0\varepsilon(\omega_0)]}{d\omega_0}\frac{\partial \mathbf{E}_a}{\partial t}e^{j\omega_0 t}. \tag{4.127}$$

Substituting (4.127) into (4.126), one may find

$$\begin{aligned}\mathbf{E}\cdot\frac{\partial \mathbf{D}}{\partial t} &= \frac{1}{4}\frac{d[\omega_0\varepsilon(\omega_0)]}{d\omega_0}\mathbf{E}_a\frac{\partial \overline{\mathbf{E}}_a}{\partial t} + \frac{1}{4}\frac{d[\omega_0\varepsilon(\omega_0)]}{d\omega_0}\overline{\mathbf{E}}_a\frac{\partial \mathbf{E}_a}{\partial t} + \text{terms containing}\, e^{\pm j2\omega_0 t} \\ &= \frac{\partial}{\partial t}\left[\frac{1}{4}\frac{d[\omega_0\varepsilon(\omega_0)]}{d\omega_0}|\mathbf{E}_a|^2\right] + \text{terms containing}\, e^{\pm j2\omega_0 t}.\end{aligned} \tag{4.128}$$

The terms containing $e^{\pm j2\omega_0 t}$ can be ignored since they are approximately zero after the time average over one period of the carrier wave. Comparing (4.128) with (4.118), one may find the stored electric field energy in time domain

$$w_e(t) = \frac{1}{4}\frac{d[\omega_0\varepsilon(\omega_0)]}{d\omega_0}|\mathbf{E}_a|^2.$$

Once this is averaged over one period of the carrier wave, the stored electric field energy in frequency domain can be written as

$$w_e(\omega_0) = \frac{1}{T_0}\int_0^{T_0} w_e(t)dt = \frac{1}{4}\frac{d[\omega_0\varepsilon(\omega_0)]}{d\omega_0}|\mathbf{E}_a|^2, \tag{4.129}$$

where the complex envelope $\mathbf{E}_a$ is assumed to be constant over $[0, T_0]$ with $T_0 = 2\pi/\omega_0$ representing the period of the carrier wave. As pointed out in [44], the complex envelope $\mathbf{E}_a$ in (4.129) is still considered to be time-dependent. An expression for the magnetic field in a medium with $\mathbf{B} = \mu\mathbf{H}$ can be found in a similar way

$$w_m(\omega_0) = \frac{1}{4}\frac{d[\omega_0\mu(\omega_0)]}{d\omega_0}|\mathbf{H}_a|^2, \tag{4.130}$$

where μ is the permeability of the medium. The above discussion occurs frequently in textbooks [e.g. 44–46] and has been a typical procedure to find the stored field energy expressions in a dispersive medium. □

The traditional NB analysis for stored field energies suffers from two major drawbacks. First, a specific constitutive relation has to be assumed, which imposes a restriction on the generality of the derived energy expressions. Second, the derived energy expressions include time-dependent complex envelopes and are thus incompatible with the conventional time-harmonic field theory where the time-independent phasors are commonly used. In what follows, the stored EM field energies in an arbitrary medium will be re-derived by introducing a new NB approach, the basic idea behind which is to treat the sinusoidal field as the limit of a NB field when its bandwidth tends to zero. In order to solve the above-mentioned problems with the traditional NB analysis, the NB field is first analyzed in Fourier frequency domain. By means of the expansion of the Fourier frequency spectrum (FFS) around the mid-band frequency, explicit expressions for the complex envelopes of the NB field can be found. When the complex envelopes are introduced into the time-domain Poynting theorem, the field energy expressions, which are valid for an arbitrary medium and expressed in terms of phasors, can be identified. Let us consider the following Fourier transform pair for the electric field $\mathbf{E}(\mathbf{r}, t)$:

$$\mathbf{E}(\mathbf{r},t) = \frac{1}{2\pi}\int_{-\infty}^{\infty}\tilde{\mathbf{E}}(\mathbf{r},\omega)e^{j\omega t}d\omega, \tilde{\mathbf{E}}(\mathbf{r},\omega) = \int_{-\infty}^{\infty}\mathbf{E}(\mathbf{r},t)e^{-j\omega t}dt, \tag{4.131}$$

where $\tilde{\mathbf{E}}(\mathbf{r},\omega)$ is the FFS of the electric field $\mathbf{E}(\mathbf{r}, t)$. Since $\mathbf{E}(\mathbf{r}, t)$ is a real function of time, the following symmetry property of the FFS about the frequency holds

$$\tilde{\mathbf{E}}(\mathbf{r},-\omega) = \overline{\tilde{\mathbf{E}}}(\mathbf{r},\omega). \tag{4.132}$$

The time-domain electric field $\mathbf{E}(\mathbf{r}, t)$ can thus be expressed in terms of FFS as an integral along the positive frequency axis

$$\mathbf{E}(\mathbf{r},t) = \mathrm{Re}\,\frac{1}{\pi}\int_{0}^{\infty}\tilde{\mathbf{E}}(\mathbf{r},\omega)e^{j\omega t}d\omega. \tag{4.133}$$

Now assume that the FFS of the electric field is narrow banded, centered at the mid-band frequency ω_0, and confined in the narrow range $(\pm\omega_0 - \Omega, \pm\,\omega_0 + \Omega)$ as illustrated in Figure 4.8. If the half-bandwidth Ω is sufficiently small, the integrand

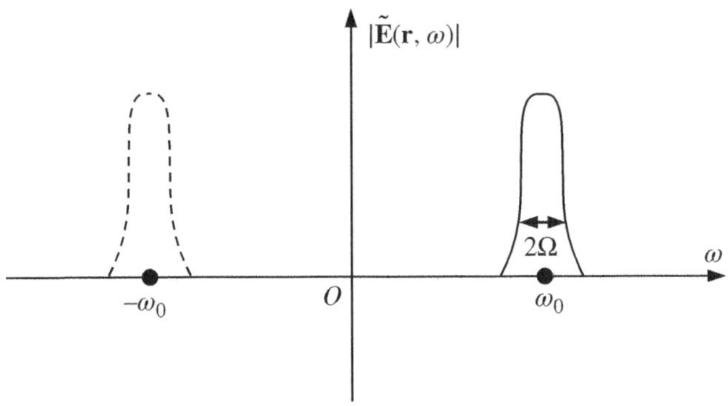

Figure 4.8 FSS of NB vector field.

in (4.133) may be considered as a constant and therefore can be approximated by its value at the central frequencyω_0. Thus,

$$\mathbf{E}(\mathbf{r}, t) = \mathrm{Re}\, \frac{1}{\pi} \tilde{\mathbf{E}}(\mathbf{r}, \omega_0) e^{j\omega_0 t} \int_{\omega_0 - \Omega}^{\omega_0 + \Omega} d\omega = \mathrm{Re}\, \mathbf{E}(\mathbf{r}) e^{j\omega_0 t}, \tag{4.134}$$

where

$$\mathbf{E}(\mathbf{r}) = \frac{2\Omega}{\pi} \tilde{\mathbf{E}}(\mathbf{r}, \omega_0) \tag{4.135}$$

is the phasor for the NB electric field $\mathbf{E}(\mathbf{r}, t)$. Similar expressions exist for other field quantities $\mathbf{D}$, $\mathbf{H}$, and $\mathbf{B}$. As $\Omega \to 0$, $\tilde{\mathbf{E}}(\mathbf{r}, \omega_0)$ approaches to infinity but the phasor (4.135) remains finite for (4.134) must be finite.

The zeroth-order approximation (4.134) is commonly used in the study of NB systems. It is often tacitly assumed that the time-harmonic (sinusoidal) field theory is also applicable to a NB field. For the study of the stored field energies in a general medium, one needs to consider the time derivative of the field involved in the quantity $\mathbf{E}(\mathbf{r}, t) \cdot \dfrac{\partial \mathbf{D}(\mathbf{r}, t)}{\partial t}$, which has to be decomposed according to (4.118). In this case, the zeroth-order approximation (4.134) for the NB electric field is no longer accurate enough and the time variation of the complex envelope must be taken into account. For this reason, one introduces the first-order approximation for the FFS of the NB field around the mid-band frequency ω_0:

$$\tilde{\mathbf{E}}(\mathbf{r}, \omega) \approx \tilde{\mathbf{E}}(\mathbf{r}, \omega_0) + (\omega - \omega_0) \frac{\partial \tilde{\mathbf{E}}(\mathbf{r}, \omega_0)}{\partial \omega}. \tag{4.136}$$

Inserting (4.136) into (4.133), one finds

$$\mathbf{E}(\mathbf{r},t)=\frac{1}{\pi}\mathrm{Re}\left[\tilde{\mathbf{E}}(\mathbf{r},\omega_0)\int\limits_{\omega_0-\Omega}^{\omega_0+\Omega}e^{j\omega t}d\omega\right]+\frac{1}{\pi}\mathrm{Re}\left[\frac{\partial\tilde{\mathbf{E}}(\mathbf{r},\omega_0)}{\partial\omega}\int\limits_{\omega_0-\Omega}^{\omega_0+\Omega}(\omega-\omega_0)e^{j\omega t}d\omega\right].\tag{4.137}$$

By means of the following calculations

$$\int\limits_{\omega_0-\Omega}^{\omega_0+\Omega}e^{j\omega t}d\omega=\frac{2\sin\Omega t}{t}e^{j\omega_0 t},\quad\int\limits_{\omega_0-\Omega}^{\omega_0+\Omega}(\omega-\omega_0)e^{j\omega t}d\omega=-je^{j\omega_0 t}\frac{d}{dt}\left(\frac{2\sin\Omega t}{t}\right),$$

(4.137) can be written in the form of (4.125):

$$\mathbf{E}(\mathbf{r},t)=\mathrm{Re}\,\mathbf{E}_a(\mathbf{r},t)e^{j\omega_0 t},\tag{4.138}$$

where

$$\mathbf{E}_a(\mathbf{r},t)=\tilde{\mathbf{E}}(\mathbf{r},\omega_0)g(t)-j\frac{\partial\tilde{\mathbf{E}}(\mathbf{r},\omega_0)}{\partial\omega}g'(t),\tag{4.139}$$

$$g(t)=\frac{2\Omega}{\pi}\frac{\sin\Omega t}{\Omega t}.\tag{4.140}$$

The complex envelope (4.139) is now composed of two terms and is more informative than just being slowly varying function of time as assumed in previous NB approaches. The first term $\tilde{\mathbf{E}}(\mathbf{r},\omega_0)g(t)$ tends to the complex envelope (4.135) as $\Omega\to 0$, while the second term contains the frequency derivative of the field in the frequency domain and tends to zero as $\Omega\to 0$. The FFS of the NB field (4.138) can be written as

$$\tilde{\mathbf{E}}(\mathbf{r},\omega)=\frac{1}{2}\left[\tilde{\mathbf{E}}_a(\mathbf{r},\omega-\omega_0)+\bar{\tilde{\mathbf{E}}}_a(\mathbf{r},-\omega-\omega_0)\right],\tag{4.141}$$

where $\tilde{\mathbf{E}}_a(\mathbf{r},\omega)$ is the FFS of the complex envelope $\mathbf{E}_a(\mathbf{r},t)$. Clearly the FFS $\tilde{\mathbf{E}}(\mathbf{r},\omega)$ is the sum of two shifted version of $\tilde{\mathbf{E}}_a(\mathbf{r},\omega)$, which are centered at $\pm\omega_0$, as sketched in Figure 4.8.

Example 4.3 Let us consider a **pulse-modulated sinusoidal signal** (PMSS, a truncated sinusoidal wave) shown in Figure 4.9a

$$x(t)=p_T(t)\cos\omega_0 t,\tag{4.142}$$

where envelope $p_T(t)$ is a rectangular pulse of width $2T$ defined by

$$p_T(t)=\begin{cases}0,|t|>0\\1,|t|<T\end{cases}.$$

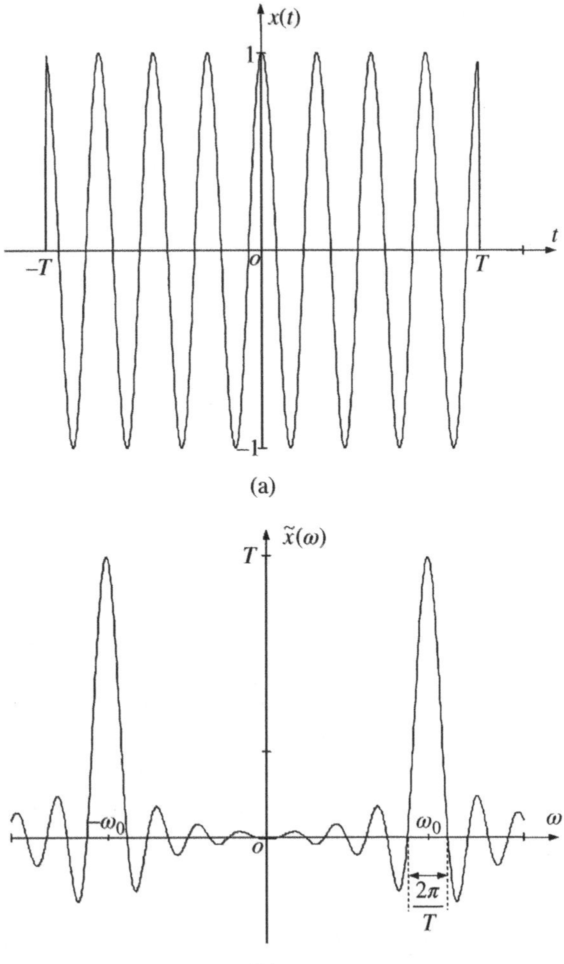

Figure 4.9 PMSS and its FFS. (a) PMSS. (b) FFS.

The FFS of the envelope is a sinc pulse and the FFS of the PMSS is thus composed of two sinc pulses centered at $\omega = \pm\omega_0$ with peak value T and width $2\Omega = 2\pi/T$ as depicted in Figure 4.9b, and is given by

$$\tilde{x}(\omega) = T\frac{\sin(\omega - \omega_0)T}{(\omega - \omega_0)T} + T\frac{\sin(\omega + \omega_0)T}{(\omega + \omega_0)T}. \tag{4.143}$$

As the duration T of the rectangular pulse grows, the FFS $\tilde{x}(\omega)$ of the PMSS becomes more concentrated about $\pm\omega_0$ but the product of the peak value T of the two sinc pulses and their width $2\Omega = 2\pi/T$ remains constant. In the limit of $T \to \infty$, the PMSS tends to a pure sinusoidal wave and its FFS (4.143) becomes the sum of two delta functions centered at $\pm\omega_0$:

$$\tilde{x}(\omega) = \pi[\delta(\omega - \omega_0) + \delta(\omega + \omega_0)]. \tag{4.144}$$

□

In order to find the rate of the dissipated electric field energy and the stored electric field energy in a medium, one needs to split $\mathbf{E}(\mathbf{r},t)\cdot\dfrac{\partial \mathbf{D}(\mathbf{r},t)}{\partial t}$ according to (4.118). By use of (4.138) and the similar representation for the electric induction intensity $\mathbf{D}$, the left-hand side of (4.118) can be expressed by

$$\begin{aligned}
\mathbf{E}(\mathbf{r},t)\cdot\frac{\partial \mathbf{D}(\mathbf{r},t)}{\partial t} = {} & \frac{1}{4}\left\{\tilde{\mathbf{E}}(\mathbf{r},\omega_0)g(t) - j\frac{\partial \tilde{\mathbf{E}}(\mathbf{r},\omega_0)}{\partial \omega}g'(t)\right\}\cdot\left\{\bar{\tilde{\mathbf{D}}}(\mathbf{r},\omega_0)g'(t) + j\frac{\partial \bar{\tilde{\mathbf{D}}}(\mathbf{r},\omega_0)}{\partial \omega}g''(t)\right\} \\
& - \frac{j\omega_0}{4}\left\{\tilde{\mathbf{E}}(\mathbf{r},\omega_0)g(t) - j\frac{\partial \tilde{\mathbf{E}}(\mathbf{r},\omega_0)}{\partial \omega}g'(t)\right\}\cdot\left\{\bar{\tilde{\mathbf{D}}}(\mathbf{r},\omega_0)g(t) + j\frac{\partial \bar{\tilde{\mathbf{D}}}(\mathbf{r},\omega_0)}{\partial \omega}g'(t)\right\} \\
& + \frac{1}{4}\left\{\bar{\tilde{\mathbf{E}}}(\mathbf{r},\omega_0)g(t) + j\frac{\partial \bar{\tilde{\mathbf{E}}}(\mathbf{r},\omega_0)}{\partial \omega}g'(t)\right\}\cdot\left\{\tilde{\mathbf{D}}(\mathbf{r},\omega_0)g'(t) - j\frac{\partial \tilde{\mathbf{D}}(\mathbf{r},\omega_0)}{\partial \omega}g''(t)\right\} \\
& + \frac{j\omega_0}{4}\left\{\bar{\tilde{\mathbf{E}}}(\mathbf{r},\omega_0)g(t) + j\frac{\partial \bar{\tilde{\mathbf{E}}}(\mathbf{r},\omega_0)}{\partial \omega}g'(t)\right\}\cdot\left\{\tilde{\mathbf{D}}(\mathbf{r},\omega_0)g(t) - j\frac{\partial \tilde{\mathbf{D}}(\mathbf{r},\omega_0)}{\partial \omega}g'(t)\right\} \\
& + \text{terms containing}\, e^{\pm j2\omega_0 t}.
\end{aligned}$$

The terms containing $e^{\pm j2\omega_0 t}$ can be ignored as they vanish after averaged over one period of the carrier wave $e^{j\omega_0 t}$. Rearranging the terms leads to

$$\begin{aligned}
\mathbf{E}(\mathbf{r},t)\cdot\frac{\partial \mathbf{D}(\mathbf{r},t)}{\partial t} = {} & \frac{1}{4}\frac{\partial}{\partial t}\mathrm{Re}\left[\tilde{\mathbf{E}}(\mathbf{r},\omega_0)\cdot\bar{\tilde{\mathbf{D}}}(\mathbf{r},\omega_0)g^2(t)\right] + \frac{1}{4}\frac{\partial}{\partial t}\mathrm{Re}\,\omega_0\left[\tilde{\mathbf{E}}(\mathbf{r},\omega_0)\cdot\frac{\partial \bar{\tilde{\mathbf{D}}}(\mathbf{r},\omega_0)}{\partial \omega}g^2(t)\right] \\
& - \frac{1}{4}\frac{\partial}{\partial t}\mathrm{Re}\left[\omega_0\bar{\tilde{\mathbf{D}}}(\mathbf{r},\omega_0)\cdot\frac{\partial \tilde{\mathbf{E}}(\mathbf{r},\omega_0)}{\partial \omega}g^2(t)\right] + \frac{1}{2}\omega_0\mathrm{Im}\left[\tilde{\mathbf{E}}(\mathbf{r},\omega_0)\cdot\bar{\tilde{\mathbf{D}}}(\mathbf{r},\omega_0)g^2(t)\right] \\
& - \frac{1}{2}\mathrm{Im}\left[\tilde{\mathbf{D}}(\mathbf{r},\omega_0)\cdot\frac{\partial \bar{\tilde{\mathbf{E}}}(\mathbf{r},\omega_0)}{\partial \omega}g'(t)g'(t)\right] - \frac{1}{2}\omega_0\mathrm{Im}\left[\frac{\partial \bar{\tilde{\mathbf{E}}}(\mathbf{r},\omega_0)}{\partial \omega}\frac{\partial \tilde{\mathbf{D}}(\mathbf{r},\omega_0)}{\partial \omega}g'(t)g'(t)\right] \\
& + \frac{1}{2}\mathrm{Re}\left[\frac{\partial \tilde{\mathbf{E}}(\mathbf{r},\omega_0)}{\partial \omega}\cdot\frac{\partial \bar{\tilde{\mathbf{D}}}(\mathbf{r},\omega_0)}{\partial \omega}g'(t)g''(t)\right] - \frac{1}{2}\mathrm{Im}\left[\tilde{\mathbf{E}}(\mathbf{r},\omega_0)\cdot\frac{\partial \bar{\tilde{\mathbf{D}}}(\mathbf{r},\omega_0)}{\partial \omega}g(t)g''(t)\right].
\end{aligned} \tag{4.145}$$

The first and second derivative of $g(t)$ in the above expressions are given by

$$\begin{aligned}
g'(t) &= \frac{2\Omega}{\pi}\Omega\frac{\Omega t\cos\Omega t - \sin\Omega t}{(\Omega t)^2}, \\
g''(t) &= \frac{2\Omega}{\pi}\Omega^3 t\frac{-(\Omega t)^2\sin\Omega t - 2(\Omega t\cos\Omega t - \sin\Omega t)}{(\Omega t)^4}.
\end{aligned} \tag{4.146}$$

As $\Omega \to 0$, the following asymptotic relations can be easily obtained from (4.140) and (4.146):

$$g(t) \approx \frac{2\Omega}{\pi} \propto \Omega,\ g'(t) \approx -\frac{2\Omega^3 t}{3\pi} \propto \Omega^2,\ g''(t) \approx -\frac{2\Omega^3}{3\pi} \propto \Omega^3. \tag{4.147}$$

As a result, the terms containing $g^2(t)$ in (4.145) remain finite as $\Omega \to 0$ and become

$$\begin{aligned}
&\tilde{\mathbf{E}}(\mathbf{r},\omega_0)\cdot\bar{\tilde{\mathbf{D}}}(\mathbf{r},\omega_0)g^2(t) \to \mathbf{E}(\mathbf{r})\cdot\bar{\mathbf{D}}(\mathbf{r}),\\
&\tilde{\mathbf{E}}(\mathbf{r},\omega_0)\cdot\frac{\partial\bar{\tilde{\mathbf{D}}}(\mathbf{r},\omega_0)}{\partial\omega}g^2(t) \to \mathbf{E}(\mathbf{r})\cdot\frac{\partial\bar{\mathbf{D}}(\mathbf{r})}{\partial\omega},\\
&\bar{\tilde{\mathbf{D}}}(\mathbf{r},\omega_0)\cdot\frac{\partial\tilde{\mathbf{E}}(\mathbf{r},\omega_0)}{\partial\omega}g^2(t) \to \bar{\mathbf{D}}(\mathbf{r})\cdot\frac{\partial\mathbf{E}(\mathbf{r})}{\partial\omega},
\end{aligned}$$

where $\mathbf{E}(\mathbf{r}) = (2\Omega/\pi)\tilde{\mathbf{E}}(\mathbf{r},\omega_0)$ and $\mathbf{D}(\mathbf{r}) = (2\Omega/\pi)\tilde{\mathbf{D}}(\mathbf{r},\omega_0)$ are phasors. The terms containing $g'(t)g'(t)$, $g(t)g''(t)$, and $g'(t)g''(t)$ in (4.145) approach to zero as $\Omega \to 0$, and can thus be ignored. So, (4.145) can be written as

$$\begin{aligned}
\mathbf{E}(\mathbf{r},t)\cdot\frac{\partial\mathbf{D}(\mathbf{r},t)}{\partial t} &= \frac{\omega_0}{2}\operatorname{Im}\left[\tilde{\mathbf{E}}(\mathbf{r},\omega_0)\cdot\bar{\tilde{\mathbf{D}}}(\mathbf{r},\omega_0)g^2(t)\right]\\
&+\frac{d}{dt}\operatorname{Re}\left[\frac{1}{4}\tilde{\mathbf{E}}(\mathbf{r},\omega_0)\cdot\bar{\tilde{\mathbf{D}}}(\mathbf{r},\omega_0)g^2(t)\right]\\
&+\frac{d}{dt}\operatorname{Re}\left[\frac{1}{4}\omega_0\tilde{\mathbf{E}}(\mathbf{r},\omega_0)\cdot\frac{\partial\bar{\tilde{\mathbf{D}}}(\mathbf{r},\omega_0)}{\partial\omega}g^2(t)-\frac{1}{4}\omega_0\bar{\tilde{\mathbf{D}}}(\mathbf{r},\omega_0)\cdot\frac{\partial\tilde{\mathbf{E}}(\mathbf{r},\omega_0)}{\partial\omega}g^2(t)\right].
\end{aligned} \tag{4.148}$$

Comparing (4.148) with (4.118), the stored electric field energy density and the rate of dissipated electric field energy density are easily identified to be

$$\begin{aligned}
w_e(\mathbf{r},t) &= \frac{1}{4}\operatorname{Re}\left[\tilde{\mathbf{E}}(\mathbf{r},\omega_0)\cdot\bar{\tilde{\mathbf{D}}}(\mathbf{r},\omega_0)g^2(t)\right]\\
&+\frac{1}{4}\omega_0\operatorname{Re}\left[\tilde{\mathbf{E}}(\mathbf{r},\omega_0)\cdot\frac{\partial\bar{\tilde{\mathbf{D}}}(\mathbf{r},\omega_0)}{\partial\omega}g^2(t)-\bar{\tilde{\mathbf{D}}}(\mathbf{r},\omega_0)\cdot\frac{\partial\tilde{\mathbf{E}}(\mathbf{r},\omega_0)}{\partial\omega}g^2(t)\right],
\end{aligned} \tag{4.149}$$

$$p_{ed}(\mathbf{r},t) = \frac{\omega_0}{2}\operatorname{Im}\left[\tilde{\mathbf{E}}(\mathbf{r},\omega_0)\cdot\bar{\tilde{\mathbf{D}}}(\mathbf{r},\omega_0)g^2(t)\right]. \tag{4.150}$$

By taking the time average of the above expressions over one period of the mid-band frequency ω_0, and using the calculation

$$\frac{1}{T_0}\int_0^{T_0} g^2(t)dt = \left(\frac{2\Omega}{\pi}\right)^2,$$

the time-averaged stored electric field energy density (4.149) and the rate of dissipated electric field energy density (4.150) can be expressed in terms of the complex envelopes as follows:

$$\begin{aligned} w_e(\mathbf{r}) &= \frac{1}{4}\,\mathrm{Re}\,\mathbf{E}(\mathbf{r})\cdot\overline{\mathbf{D}}(\mathbf{r}) \\ &+ \frac{1}{4}\omega_0\,\mathrm{Re}\left[\mathbf{E}(\mathbf{r},)\cdot\frac{\partial\overline{\mathbf{D}}(\mathbf{r})}{\partial\omega} - \overline{\mathbf{D}}(\mathbf{r})\cdot\frac{\partial\mathbf{E}(\mathbf{r})}{\partial\omega}\right], \end{aligned} \tag{4.151}$$

$$p_{ed}(\mathbf{r}) = \frac{\omega_0}{2}\,\mathrm{Im}\left[\mathbf{E}(\mathbf{r})\cdot\overline{\mathbf{D}}(\mathbf{r})\right]. \tag{4.152}$$

A similar discussion applies to $\mathbf{H}(\mathbf{r},t)\cdot\dfrac{\partial\mathbf{B}(\mathbf{r},t)}{\partial t}$. The time-averaged stored magnetic field energy density and the rate of dissipated magnetic field energy density are found as follows:

$$\begin{aligned} w_m(\mathbf{r}) &= \frac{1}{4}\,\mathrm{Re}\,\mathbf{H}(\mathbf{r})\cdot\overline{\mathbf{B}}(\mathbf{r}) \\ &+ \frac{1}{4}\omega_0\,\mathrm{Re}\left[\mathbf{H}(\mathbf{r})\cdot\frac{\partial\overline{\mathbf{B}}(\mathbf{r})}{\partial\omega} - \overline{\mathbf{B}}(\mathbf{r})\cdot\frac{\partial\mathbf{H}(\mathbf{r})}{\partial\omega}\right], \end{aligned} \tag{4.153}$$

$$p_{ed}(\mathbf{r}) = \frac{\omega_0}{2}\,\mathrm{Im}\left[\mathbf{H}(\mathbf{r})\cdot\overline{\mathbf{B}}(\mathbf{r})\right]. \tag{4.154}$$

The expressions (4.151)–(4.154) are identical to (4.121)–(4.124) and they do not explicitly contain medium parameters, and are therefore universally applicable. The NB approach presented above is physically more intuitive compared with the complex domain approach in [40] and may be thought of as a general technique for studying the time-harmonic field problems where an NB approximation in frequency domain must be assumed.

It is noted that all the field quantities in (4.151)–(4.154) are phasors for the sinusoidal field rather than the time-dependent complex envelopes in traditional NB analysis shown in [44–46]. A perplexing question that may arise is why the frequency derivatives occur in the stored field energy expressions as a pure sinusoidal field has only a single frequency component. One answer to the question is that all EM field quantities depend on frequency in microwave regime, and the frequency can thus be taken as a parameter and is allowed to vary with the purpose of examining how a field quantity changes with frequency. As a matter of fact, the frequency derivatives of field quantities (or circuit parameters) for a sinusoidal field appear frequently in the study of the stored field energy and group delay [64–66]. On the other hand, the new NB approach clearly reveals how the frequency derivatives of the fields appear from a different perspective. It indicates that the frequency derivatives of the fields are required essentially in order for the NB field to be accurately represented in the Fourier frequency domain.

4.3.2 Stored Field Energies of Antenna

Instinctively, a transmitting antenna can be represented by an equivalent RLC circuit as sketched in Figure 4.7, where R is related to the radiated power from the antenna, L and C are, respectively, related to the stored magnetic and electric field energies of the antenna. The equivalent circuit of receiving antenna is more complicated than its counterpart of transmitting antenna. For a detailed study of both, please refer to Section 5.4.3 in [67]. In order to determine the element values L and C in the RLC equivalent circuit, one needs to know the stored field energies of antenna. Since the antenna radiates the fields into the infinite space, the integration of the stored field energy (4.121) or (4.123) over all the space will be infinite. For this reason, the stored field energy of antenna needs to be properly defined.

Conventionally the stored field energy of an antenna is defined as the total field energy subtracted by the radiated field energy. This definition was first proposed by Counter [68], and has been widely used by antenna society. According to Counter, the total EM fields can be divided into the "local" and "radiation" fields, and the radiated fields stand for a flow energy and must be subtracted from the total field energy in order to obtain the stored field energy of antenna. Since both the total field energy and the radiated field energy are unbounded, their difference is unnecessarily a finite number. Consequently, the stored field energy founded on Counter's idea is based on an unproven hypothesis that the infinity in the total field energy is created by the energy flow associated with radiated power, as pointed out in [69].

An energy conservation law that contains the field energy densities (4.121)–(4.124) has been given in Chapter 1 and its real part is (1.23). By integrating (1.23) over the region V bounded by S containing the source region V_0, one may find

$$\tilde{W} = W_e(V) + W_m(V) - W^{rad}(S), \tag{4.155}$$

where

$$\tilde{W} = \frac{1}{4}\mathrm{Im}\int_{V_0}\left(\mathbf{E}\cdot\frac{\partial\bar{\mathbf{J}}}{\partial\omega} - \frac{\partial\mathbf{E}}{\partial\omega}\cdot\bar{\mathbf{J}}\right)dV, \tag{4.156}$$

$$W^{rad}(S) = \frac{1}{4}\mathrm{Im}\int_{S}\left(\mathbf{E}\times\frac{\partial\bar{\mathbf{H}}}{\partial\omega} - \frac{\partial\mathbf{E}}{\partial\omega}\times\bar{\mathbf{H}}\right)\cdot\mathbf{u}_n dS, \tag{4.157}$$

$$W_e(V) = \int_V w_e dV,\ W_m(V) = \int_V w_m dV. \tag{4.158}$$

The sum $W_e(V) + W_m(V)$ on the right-hand side of (4.155) represents the total stored field energy inside the region V; the third term $W^{rad}(S)$ is a surface integral and has the same dimension as the energy, and may therefore be interpreted as the **radiated field energy** flowing out of the surface S. Hence the right-hand side of (4.155) is the total stored field energy inside the region V subtracted by the radiated field energy flowing out of the surface S. By definition, this gives the **stored field energy of antenna**. Consequently, the left-hand side of (4.155) or (4.156) can be considered as the definition of the stored field energy of antenna. Physically, the field energy is emanated from the source and spreads into the region V and then flows out of its boundary S. Although both the total stored field energy inside V and the radiated field energy out of S change as S varies, the stored field energy $\tilde{W}$ defined by (4.156) remains constant as long as the source region V_0 is contained in S, which implies that the stored field energy of antenna is determined solely by the source distribution. It is noted that the definition of the stored field energy of antenna given by (4.156) applies to an arbitrary medium while all the previous studies have to assume that the medium is lossless. Also notice the difference between the stored field energy in a medium and the stored field energy of antenna. By combining (4.155) and the imaginary part of the integral form of Poynting theorem shown below

$$-\frac{1}{2}\mathrm{Im}\int_{V_0}\mathbf{E}\cdot\bar{\mathbf{J}}dV = \mathrm{Im}\int_S\frac{1}{2}(\mathbf{E}\times\bar{\mathbf{H}})\cdot\mathbf{u}_n dV + 2\omega[W_m(V)-W_e(V)], \tag{4.159}$$

one may find

$$\tilde{W}_e = W_e(V) - \frac{1}{2}W^{rad}(S) - \frac{1}{8\omega}\mathrm{Im}\int_S(\mathbf{E}\times\bar{\mathbf{H}})\cdot\mathbf{u}_n dS, \tag{4.160}$$

$$\tilde{W}_m = W_m(V) - \frac{1}{2}W^{rad}(S) + \frac{1}{8\omega}\mathrm{Im}\int_S(\mathbf{E}\times\bar{\mathbf{H}})\cdot\mathbf{u}_n dS, \tag{4.161}$$

where $\tilde{W}_e$ and $\tilde{W}_m$ are, respectively, defined by

$$\tilde{W}_e = \frac{1}{8}\mathrm{Im}\int_{V_0}\left(\mathbf{E}\cdot\frac{\partial\bar{\mathbf{J}}}{\partial\omega} - \frac{\partial\mathbf{E}}{\partial\omega}\cdot\bar{\mathbf{J}}\right)dV + \frac{1}{8\omega}\mathrm{Im}\int_{V_0}\mathbf{E}\cdot\bar{\mathbf{J}}dV, \tag{4.162}$$

$$\tilde{W}_m = \frac{1}{8}\mathrm{Im}\int_{V_0}\left(\mathbf{E}\cdot\frac{\partial\bar{\mathbf{J}}}{\partial\omega} - \frac{\partial\mathbf{E}}{\partial\omega}\cdot\bar{\mathbf{J}}\right)dV - \frac{1}{8\omega}\mathrm{Im}\int_{V_0}\mathbf{E}\cdot\bar{\mathbf{J}}dV. \tag{4.163}$$

From (4.160) and (4.161), one obtains

$$\begin{aligned}\tilde{W}_m + \tilde{W}_e &= \tilde{W},\\ \tilde{W}_m - \tilde{W}_e &= W_m(V) - W_e(V) - \frac{1}{4\omega}\mathrm{Im}\int_S(\mathbf{E}\times\bar{\mathbf{H}})\cdot\mathbf{u}_n dS.\end{aligned} \tag{4.164}$$

From the energy balance relations (4.155) and (4.164), one may interpret $\tilde{W}_e$ and $\tilde{W}_m$ as the electric and magnetic parts of the stored field energy of antenna, respectively. Both of them are independent of the choices of the closed surface S if the source is confined in S.

The stored electric and magnetic field energies of antenna can be expressed in terms of the source distributions only. Assume that the medium around the antenna is isotropic, homogeneous, and lossless. Using the integral representation for the electric field and the Green's identity $\nabla\cdot(\phi\mathbf{a}) = \mathbf{a}\cdot\nabla\phi+\phi\nabla\cdot\mathbf{a}$, one may find

$$
\begin{aligned}
\mathrm{Im}\int_{V_0}\mathbf{E}(\mathbf{r})\cdot\frac{\partial\bar{\mathbf{J}}(\mathbf{r})}{\partial\omega}dV(\mathbf{r}) &= \eta c\,\mathrm{Re}\int_{V_0}\int_{V_0}\bar{\rho}(\mathbf{r})\rho(\mathbf{r}')G(\mathbf{r},\mathbf{r}')dV(\mathbf{r})dV(\mathbf{r}')\\
&+c^2\eta k\mathrm{Im}\int_{V_0}\int_{V_0}j\frac{\partial\bar{\rho}(\mathbf{r})}{\partial\omega}\rho(\mathbf{r}')G(\mathbf{r},\mathbf{r}')dV(\mathbf{r})dV(\mathbf{r}')\\
&-k\eta\mathrm{Im}\int_{V_0}\int_{V_0}jG(\mathbf{r},\mathbf{r}')\mathbf{J}(\mathbf{r}')\cdot\frac{\partial\bar{\mathbf{J}}(\mathbf{r})}{\partial\omega}dV(\mathbf{r})dV(\mathbf{r}'),
\end{aligned}
\tag{4.165}
$$

$$
\begin{aligned}
\mathrm{Im}\int_{V_0}\bar{\mathbf{J}}(\mathbf{r})\cdot\frac{\partial\mathbf{E}(\mathbf{r})}{\partial\omega}dV(\mathbf{r}) &= c^2k\eta\mathrm{Im}\int_{V_0}\int_{V_0}j\bar{\rho}(\mathbf{r})\frac{\partial\rho(\mathbf{r}')}{\partial\omega}G(\mathbf{r},\mathbf{r}')dV(\mathbf{r})dV(\mathbf{r}')\\
&+ck\eta\mathrm{Im}\int_{V}\int_{V}\bar{\rho}(\mathbf{r})\rho(\mathbf{r}')RG(\mathbf{r},\mathbf{r}')dV(\mathbf{r})dV(\mathbf{r}')\\
&-\frac{\eta}{c}\mathrm{Im}\int_{V_0}\int_{V_0}j\bar{\mathbf{J}}(\mathbf{r})\cdot\mathbf{J}(\mathbf{r}')G(\mathbf{r},\mathbf{r}')dV(\mathbf{r})dV(\mathbf{r}')\\
&-k\eta\mathrm{Im}\int_{V_0}\int_{V_0}j\bar{\mathbf{J}}(\mathbf{r})\cdot\frac{\partial\mathbf{J}(\mathbf{r}')}{\partial\omega}G(\mathbf{r},\mathbf{r}')dV(\mathbf{r})dV(\mathbf{r}')\\
&-\frac{k\eta}{c}\mathrm{Im}\int_{V_0}\int_{V_0}\bar{\mathbf{J}}(\mathbf{r})\cdot\mathbf{J}(\mathbf{r}')RG(\mathbf{r},\mathbf{r}')dV(\mathbf{r})dV(\mathbf{r}'),
\end{aligned}
\tag{4.166}
$$

where $G(\mathbf{r},\mathbf{r}') = \left(e^{-jk|\mathbf{r}-\mathbf{r}'|}/4\pi|\mathbf{r}-\mathbf{r}'|\right)$, $c = \left(1/\sqrt{\mu\varepsilon}\right)$, and $\eta = \sqrt{(\mu/\varepsilon)}$. Substituting (4.165) and (4.166) into (4.162) and (4.163), and omitting the tedious derivation process, the stored field energies $\tilde{W}_e$ and $\tilde{W}_m$ are found to be

$$
\begin{aligned}
\tilde{W}_e &= \frac{\eta c}{16\pi}\int\limits_{V_0}\int\limits_{V_0}\bar{\rho}(\mathbf{r})\rho(\mathbf{r}')\frac{\cos kR}{R}dV(\mathbf{r})dV(\mathbf{r}')\\
&+\frac{ck\eta}{32\pi}\int\limits_{V_0}\int\limits_{V_0}\left[\bar{\rho}(\mathbf{r})\rho(\mathbf{r}')-\frac{1}{c^2}\bar{\mathbf{J}}(\mathbf{r})\cdot\mathbf{J}(\mathbf{r}')\right]\sin kR dV(\mathbf{r})dV(\mathbf{r}')\\
&+\frac{c^2k\eta}{16\pi}\int\limits_{V_0}\int\limits_{V_0}\mathrm{Im}\left[\rho(\mathbf{r})\frac{\partial\bar{\rho}(\mathbf{r}')}{\partial\omega}\right]\frac{\sin kR}{R}dV(\mathbf{r})dV(\mathbf{r}')\\
&-\frac{c^2k\eta}{16\pi}\int\limits_{V_0}\int\limits_{V_0}\frac{1}{c^2}\mathrm{Im}\left[\mathbf{J}(\mathbf{r})\cdot\frac{\partial\bar{\mathbf{J}}(\mathbf{r}')}{\partial\omega}\right]\frac{\sin kR}{R}dV(\mathbf{r})dV(\mathbf{r}'),
\end{aligned}
\tag{4.167}
$$

$$
\begin{aligned}
\tilde{W}_m &= \frac{\eta c}{16\pi}\int\limits_{V_0}\int\limits_{V_0}\frac{1}{c^2}\bar{\mathbf{J}}(\mathbf{r})\cdot\mathbf{J}(\mathbf{r}')\frac{\cos kR}{R}dV(\mathbf{r})dV(\mathbf{r}')\\
&+\frac{ck\eta}{32\pi}\int\limits_{V_0}\int\limits_{V_0}\left[\bar{\rho}(\mathbf{r})\rho(\mathbf{r}')-\frac{1}{c^2}\bar{\mathbf{J}}(\mathbf{r})\cdot\mathbf{J}(\mathbf{r}')\right]\sin kR dV(\mathbf{r})dV(\mathbf{r}')\\
&+\frac{c^2k\eta}{16\pi}\int\limits_{V_0}\int\limits_{V_0}\mathrm{Im}\left[\rho(\mathbf{r})\frac{\partial\bar{\rho}(\mathbf{r}')}{\partial\omega}\right]\frac{\sin kR}{R}dV(\mathbf{r})dV(\mathbf{r}')\\
&-\frac{c^2k\eta}{16\pi}\int\limits_{V_0}\int\limits_{V_0}\frac{1}{c^2}\mathrm{Im}\left[\mathbf{J}(\mathbf{r})\cdot\frac{\partial\bar{\mathbf{J}}(\mathbf{r}')}{\partial\omega}\right]\frac{\sin kR}{R}dV(\mathbf{r})dV(\mathbf{r}').
\end{aligned}
\tag{4.168}
$$

The expressions (4.167) and (4.168) agree with those obtained in [70], where a different approach is adopted. For small antennas, the above expressions reduce to the results obtained in [71].

Remark 4.5 A common mistake made in antenna theory is to assume that the source or the feeding current is independent of frequency, which may lead to erroneous results. For example, if one assumes that the current **J** is independent of frequency [72], the terms containing the frequency derivatives of the sources in (4.167) and (4.168) will disappear. For a detailed discussion about the common mistake, please refer to [73]. □

Example 4.4 (Minimization of Antenna QF)
According to (4.167), (4.168), and (4.29), both the stored field energy of antenna (source) and the radiated power can be expressed as an inner product

$$
\begin{aligned}
\omega\left(\tilde{W}_m+\tilde{W}_e\right) &= \left(\hat{A}\mathbf{J},\mathbf{J}\right),\\
p^{rad} &= \left(\hat{B}\mathbf{J},\mathbf{J}\right),
\end{aligned}
\tag{4.169}
$$

where

$$\begin{aligned}\hat{A}\mathbf{J}(\mathbf{r}) &= \frac{k\eta}{16\pi}\int_V \left\{\mathbf{J}(\mathbf{r}')\frac{\cos kR}{R} - \frac{1}{k^2}[\mathbf{J}(\mathbf{r}')\cdot\nabla']\nabla'\frac{\cos kR}{R}\right\}dV(\mathbf{r}')\\ &+ \frac{k\eta}{16\pi}\int_V \left\{\mathbf{J}(\mathbf{r}')k\sin kR + \frac{1}{k}[\mathbf{J}(\mathbf{r}')\cdot\nabla']\nabla'\sin kR\right\}dV(\mathbf{r}')\\ &+ \frac{k\eta}{8\pi}\mathrm{Im}\int_V \left\{\frac{k\partial\mathbf{J}(\mathbf{r}')}{\partial k}\frac{\sin kR}{R} + \frac{1}{k}\left[\frac{\partial\mathbf{J}(\mathbf{r}')}{\partial k}\cdot\nabla'\right]\nabla'\frac{\sin kR}{R}\right\}dV(\mathbf{r}'),\end{aligned} \tag{4.170}$$

$$\hat{B}\mathbf{J} = \frac{k\eta}{8\pi}\int_V \left\{\mathbf{J}(\mathbf{r}') + \frac{1}{k^2}[\mathbf{J}(\mathbf{r}')\cdot\nabla']\nabla'\right\}\frac{\sin kR}{R}dV(\mathbf{r}'). \tag{4.171}$$

Thus, the radiation QF defined by (4.30) can be written as a Rayleigh quotient

$$Q_I = \frac{\omega\left(\tilde{W}_e + \tilde{W}_m\right)}{P^{rad}} = \frac{\left(\hat{A}\mathbf{J},\mathbf{J}\right)}{\left(\hat{B}\mathbf{J},\mathbf{J}\right)}. \tag{4.172}$$

An optimal solution **J** that minimizes the QF in a permission region consisting of real vector functions for a small antenna can be obtained from the Rayleigh quotient. For a small antenna, the frequency derivatives of the sources in (4.170) can be neglected and the operator $\hat{A}$ becomes a symmetric operator. By variational analysis, if (4.172) is stationary, the current **J** must satisfy the generalized eigenvalue equation

$$\hat{B}\mathbf{J} = \alpha\hat{A}\mathbf{J}, \tag{4.173}$$

where $\alpha = 1/Q_I$. The largest eigenvalue α gives the smallest value of QF. Numerical examples can be found in [70]. □

Example 4.5 (Maximization of the Ratio of Gain Over QF)
From (4.12), the radiation intensity for the source distribution **J** in a region V_0 can be expressed as

$$U(\mathbf{u}_r) = \left(\hat{B}'\mathbf{J},\mathbf{J}\right), \tag{4.174}$$

where $(\cdot,\ \cdot)$ denotes the inner product and $\hat{B}'$ is an integral operator defined by

$$\begin{aligned}\hat{B}'\mathbf{J}(\mathbf{r}'') &= \frac{k^2\eta}{32\pi^2}\int_{V_0} e^{-jk(R'-R'')}\mathbf{J}(\mathbf{r}')dV(\mathbf{r}')\\ &- \frac{k^2\eta}{32\pi^2}\int_{V_0} e^{-jk(R'-R'')}[\mathbf{J}(\mathbf{r}')\cdot\mathbf{u}_{R''}]\mathbf{u}_{R''}dV(\mathbf{r}')\end{aligned}$$

with $\eta = \sqrt{\mu/\varepsilon}, R' = |\mathbf{r} - \mathbf{r}'|, R'' = |\mathbf{r} - \mathbf{r}''|$. Apparently, the operator $\hat{B}'$ is symmetric, i.e.

$$\left(\hat{B}'\mathbf{J}, \mathbf{J}\right) = \left(\mathbf{J}, \hat{B}'\mathbf{J}\right).$$

In terms of (4.169), the ratio of gain over QF can be written as the form of Rayleigh quotient

$$\frac{G}{Q_I} = \frac{(\hat{B}\mathbf{J}, \mathbf{J})}{(\hat{A}\mathbf{J}, \mathbf{J})}, \tag{4.175}$$

where the operator $\hat{B}$ is defined by $\hat{B} = 4\pi\hat{B}'$. By variational analysis, the ratio (4.175) reaches maximum if the current $\mathbf{J}$ satisfies the generalized eigenvalue equation

$$\hat{B}\mathbf{J} = \alpha\hat{A}\mathbf{J}, \tag{4.176}$$

where

$$\alpha = \max \frac{G}{Q_I}. \tag{4.177}$$

Numerical examples can be found in [74]. It is noted that the generalized eigenvalue equation for the ratio of gain to QF has a unique solution, while the generalized eigenvalue equation (4.173) has more than one solution. □

4.3.3 Radiated Field Energy

The surface integral (4.157) frequently occurs in the study of antenna input reactance [75–77], and is now interpreted as the radiated field energy from the radiator. When the approximate expressions for the far-field quantities are substituted into (4.157), some interesting properties can be revealed. Two different approximations to the surface integral (4.157) will be examined here. In what follows, the medium around antenna will be assumed to be lossless and isotropic. The magnetic field generated by the current source can be represented by

$$\mathbf{H}(\mathbf{r}) = \int_{V_0} \mathbf{J}(\mathbf{r}') \times \nabla' G(\mathbf{r}, \mathbf{r}') dV(\mathbf{r}'). \tag{4.178}$$

Its frequency derivative is then given by

$$\frac{\partial \mathbf{H}(\mathbf{r})}{\partial \omega} = \int_{V_0} \frac{\partial \mathbf{J}(\mathbf{r}')}{\partial \omega} \times \nabla' G(\mathbf{r}, \mathbf{r}') dV(\mathbf{r}') - \frac{j}{4\pi c} \int_{V_0} \mathbf{J}(\mathbf{r}') \times \nabla' e^{-jkR} dV(\mathbf{r}'). \tag{4.179}$$

In the far-field region, (4.178) and (4.179) can be approximated by

$$\mathbf{H}^{rad}(\mathbf{r}) \approx \frac{jk}{4\pi}\frac{e^{-jkr}}{r}\int\limits_{V_0}[\mathbf{J}(\mathbf{r}')\times\mathbf{u}_r]e^{jk\mathbf{u}_r\cdot\mathbf{r}'}dV(\mathbf{r}'), \tag{4.180}$$

$$\frac{\partial\mathbf{H}^{rad}(\mathbf{r})}{\partial\omega} \approx -j\frac{r}{c}\mathbf{H}^{rad}(\mathbf{r}) + \frac{jk}{4\pi}\frac{e^{-jkr}}{r}\int\limits_{V_0}\left[\frac{\partial\mathbf{J}(\mathbf{r}')}{\partial\omega}\times\mathbf{u}_r\right]e^{jk\mathbf{u}_r\cdot\mathbf{r}'}dV(\mathbf{r}'). \tag{4.181}$$

In the far-field region, the integrand in (4.157) can be written as

$$\mathrm{Im}\left(\mathbf{E}^{rad}\times\frac{\partial\overline{\mathbf{H}}^{rad}}{\partial\omega} - \frac{\partial\mathbf{E}^{rad}}{\partial\omega}\times\overline{\mathbf{H}}^{rad}\right) = -2\eta\mathbf{u}_r\mathrm{Im}\left(\overline{\mathbf{H}}^{rad}\cdot\frac{\partial\mathbf{H}^{rad}}{\partial\omega}\right). \tag{4.182}$$

In view of (4.180) and (4.181), one may find

$$\begin{aligned}\overline{\mathbf{H}}^{rad}(\mathbf{r})\cdot\frac{\partial\mathbf{H}^{rad}(\mathbf{r})}{\partial\omega} &= -j\frac{r}{c}\left|\mathbf{H}^{rad}(\mathbf{r})\right|^2 \\ &+ \frac{k^2}{16\pi^2r^2}\int\limits_{V_0}\int\limits_{V_0}\left[\overline{\mathbf{J}}(\mathbf{r}')\cdot\left(\overleftrightarrow{\mathbf{I}}-\mathbf{u}_r\mathbf{u}_r\right)\cdot\frac{\partial\mathbf{J}(\mathbf{r}'')}{\partial\omega}\right]e^{jk\mathbf{u}_r\cdot(\mathbf{r}''-\mathbf{r}')}dV(\mathbf{r}')dV(\mathbf{r}'').\end{aligned} \tag{4.183}$$

By making use of (4.182) and (4.183), the integral (4.157) on the sphere S_∞ of radius r can be expressed by

$$W^{rad}(S_\infty) = \frac{r}{c}\int\limits_{S_\infty}\frac{\eta}{2}\left|\mathbf{H}^{rad}(\mathbf{r})\right|^2 dS - \frac{k^2\eta}{8\pi}\int\limits_{V_0}\int\limits_{V_0}\mathrm{Im}\left[\overline{\mathbf{J}}(\mathbf{r}')\cdot\overleftrightarrow{\mathbf{U}}\cdot\frac{\partial\mathbf{J}(\mathbf{r}'')}{\partial\omega}\right]dV(\mathbf{r}')dV(\mathbf{r}''), \tag{4.184}$$

where $\overleftrightarrow{\mathbf{U}}$ is a dyadic defined by

$$\overleftrightarrow{\mathbf{U}} = \frac{1}{4\pi}\int\limits_{S_\infty}\frac{1}{r^2}\left(\overleftrightarrow{\mathbf{I}}-\mathbf{u}_r\mathbf{u}_r\right)e^{jk\mathbf{u}_r\cdot(\mathbf{r}''-\mathbf{r}')}dS(\mathbf{r}), \tag{4.185}$$

with $\overleftrightarrow{\mathbf{I}}$ being the unit dyadic. The integral (4.185) can be carried out by the Funk–Hecke formula in the theory of spherical harmonics [78, 79]

$$\overleftrightarrow{\mathbf{U}} = \left[j_0(kR) - \frac{j_1(kR)}{kR}\right]\overleftrightarrow{\mathbf{I}} + j_2(kR)\mathbf{u}_R\mathbf{u}_R, \tag{4.186}$$

where $R = |\mathbf{r}'' - \mathbf{r}'|$, $\mathbf{u}_R = (\mathbf{r}'' - \mathbf{r}')/R$, and j_n is the nth-order spherical Bessel function of the first kind. Equation (4.184) can be written as

$$W^{rad}(S_\infty) = \frac{r}{c}P^{rad} - W_d, \tag{4.187}$$

where W_d stands for the energy term caused by the frequency derivative of the source

$$W_d = \frac{k^2\eta}{8\pi}\int_{V_0}\int_{V_0}\mathrm{Im}\left[\bar{\mathbf{J}}(\mathbf{r}')\cdot\overleftrightarrow{\mathbf{U}}\cdot\frac{\partial\mathbf{J}(\mathbf{r}'')}{\partial\omega}\right]dV(\mathbf{r}')dV(\mathbf{r}''). \tag{4.188}$$

The first term on the right-hand side of (4.187) is equal to the integration of radiated field energy density over the spherical region V_∞ bounded by the sphere S_∞ of radius r, denoted by $W_{em}^{rad}(V_\infty)$. In fact,

$$\begin{aligned} W_{em}^{rad}(V_\infty) &= \int_{V_\infty}\left(w_e^{rad} + w_m^{rad}\right)dV = \int_0^r dr\int_\Omega\left(\frac{1}{4}\varepsilon|\mathbf{E}_\infty|^2 + \frac{1}{4}\mu|\mathbf{H}_\infty|^2\right)d\Omega \\ &= r\int_\Omega\left(\frac{1}{4}\varepsilon|\mathbf{E}_\infty|^2 + \frac{1}{4}\mu|\mathbf{H}_\infty|^2\right)d\Omega = \frac{r}{c}P^{rad}, \end{aligned} \tag{4.189}$$

where $d\Omega = \sin\theta d\theta d\varphi$ is the differential solid angle, Ω is the unit sphere, and

$$w_e^{rad} = \frac{1}{4}\varepsilon\left|\mathbf{E}^{rad}(\mathbf{r})\right|^2, w_m^{rad} = \frac{1}{4}\mu\left|\mathbf{H}^{rad}(\mathbf{r})\right|^2 \tag{4.190}$$

are, respectively, the radiated electric and magnetic field energy densities. Combining (4.155) and (4.187), one may find

$$\tilde{W} = W_e(V_\infty) + W_m(V_\infty) - W_{em}^{rad}(V_\infty) + W_d. \tag{4.191}$$

As indicated by (4.189), the term $W_{em}^{rad}(V_\infty)$ is equal to the radiated power multiplied by the time that the radiated wave takes to travel from the origin of the coordinate system to the sphere S_∞, and therefore it stands for the radiated field energy from a point source centered at $r = 0$. The term W_d in (4.191) may thus be considered as a correction to the point source to take account of the influence of the source distribution in the source region V_0. For small radiator for which $(\partial\mathbf{J}(\mathbf{r})/\partial\omega) \approx 0$, this term can be neglected. Substituting (4.186) into (4.188) and simplifying the resultant yield

$$\begin{aligned} W_d = &-\frac{c^2k\eta}{8\pi}\int_{V_0}\int_{V_0}\mathrm{Im}\left[\bar{\rho}(\mathbf{r}')\frac{\partial\rho(\mathbf{r}'')}{\partial\omega}\right]\frac{\sin kR}{R}dV(\mathbf{r}')dV(\mathbf{r}'') \\ &+\frac{k\eta}{8\pi}\int_{V_0}\int_{V_0}\mathrm{Im}\left[\bar{\mathbf{J}}(\mathbf{r}')\cdot\frac{\partial\mathbf{J}(\mathbf{r}'')}{\partial\omega}\right]\frac{\sin kR}{R}dV(\mathbf{r}')dV(\mathbf{r}''). \end{aligned} \tag{4.192}$$

This gives the sum of the last two terms on the right-hand side of (4.167) and (4.168).

Instead of using (4.179) to find the frequency derivative $(\partial \mathbf{H}^{rad}(\mathbf{r})/\partial \omega)$ in the far-field region, a common approach is to directly take the frequency derivative from the far-field expression $\mathbf{H}^{rad}(\mathbf{r}) = (e^{-jkr}/r)\mathbf{H}_{\infty}(\mathbf{u}_r)$ to get

$$\frac{\partial \mathbf{H}^{rad}(\mathbf{r})}{\partial \omega} = -j\frac{r}{c}\frac{e^{-jkr}}{r}\mathbf{H}_{\infty}(\mathbf{u}_r) + \frac{e^{-jkr}}{r}\frac{\partial \mathbf{H}_{\infty}(\mathbf{u}_r)}{\partial \omega}.$$

In this case, the integrand of (4.157) becomes

$$\begin{aligned}
&\mathrm{Im}\left(\mathbf{E}^{rad} \times \frac{\partial \overline{\mathbf{H}}^{rad}}{\partial \omega} - \frac{\partial \mathbf{E}^{rad}}{\partial \omega} \times \overline{\mathbf{H}}^{rad}\right) \\
&= \frac{r}{c}\frac{j2\eta \mathbf{u}_r}{r^2}|\mathbf{H}_{\infty}(\mathbf{u}_r)|^2 - \frac{j2\eta \mathbf{u}_r}{r^2}\mathrm{Im}\left[\overline{\mathbf{H}}_{\infty}(\mathbf{u}_r) \cdot \frac{\partial \mathbf{H}_{\infty}(\mathbf{u}_r)}{\partial \omega}\right].
\end{aligned} \tag{4.193}$$

In the far-field region, the surface integral (4.157) on the sphere S_{∞} can thus be expressed by

$$W^{rad}(S_{\infty}) = \frac{r}{c}P^{rad} - \frac{\eta}{2}\int_{\Omega}\mathrm{Im}\left[\overline{\mathbf{H}}_{\infty}(\mathbf{u}_r) \cdot \frac{\partial \mathbf{H}_{\infty}(\mathbf{u}_r)}{\partial \omega}\right]d\Omega. \tag{4.194}$$

The second term on the right-hand side of (4.194) often occurred in previous studies [76, 77], whose physical meaning has bewildered researchers for many years. Comparing (4.187) with (4.194), the following identification can be made

$$\begin{aligned}
W_d &= \frac{k^2\eta}{8\pi}\int_{V_0}\int_{V_0}\mathrm{Im}\left[\mathbf{J}(\mathbf{r}') \cdot \overleftrightarrow{\mathbf{U}} \cdot \frac{\partial \mathbf{J}(\mathbf{r}'')}{\partial \omega}\right]dV(\mathbf{r}')dV(\mathbf{r}'') \\
&\approx \frac{\eta}{2}\int_{\Omega}\mathrm{Im}\left[\overline{\mathbf{H}}_{\infty}(\mathbf{u}_r) \cdot \frac{\partial \mathbf{H}_{\infty}(\mathbf{u}_r)}{\partial \omega}\right]d\Omega.
\end{aligned} \tag{4.195}$$

Therefore, the second term on the right-hand side of (4.194) is also negligible for small radiators. For some typical antennas whose radiation patterns are real or imaginary (such as a dipole, loop, rectangular or circular aperture, rectangular or circular microstrip patch), (4.195) vanishes whether the antenna is big or small. On the other hand, if the radiation pattern is a slowly varying function of the frequency, (4.195) can also be neglected.

4.3.4 Evaluation of Radiation Quality Factor

For a high QF antenna, the QF can be shown to be reciprocal of antenna fractional bandwidth for the input impedance [80]. The antenna QF is a field quantity and is more convenient for theoretical research while the antenna bandwidth requires more information on the frequency behavior of the input impedance. There are two different methods for evaluating antenna QF. One is based on the current source distribution (field approach), in terms of which the stored field energies of antenna can be determined from (4.156) or (4.167) and (4.168). An alternative approach is to use the antenna input impedance (circuit approach). It has been demonstrated in [80] that the Foster reactance theorem is approximately valid for an ideal (lossless) metal antenna, and the stored field energies of a lossless metal antenna can be expressed in terms of the antenna input reactance as follows:

$$\tilde{W}_e = \frac{1}{8}|I|^2\left(\frac{\partial X}{\partial \omega} - \frac{X}{\omega}\right), \quad \tilde{W}_m = \frac{1}{8}|I|^2\left(\frac{\partial X}{\partial \omega} + \frac{X}{\omega}\right), \tag{4.196}$$

where I is the antenna terminal current and X is the input reactance defined by

$$X = \frac{4\omega}{|I|^2}\left(\tilde{W}_m - \tilde{W}_e\right). \tag{4.197}$$

It can be shown that both the field and circuit methods give identical results [70]. Based on the stored field energies (4.196) and the radiation resistance, an ideal transmitting antenna without ohmic loss can be equivalent to a series RLC circuit with the element values given by [80]

$$R^{rad} = \frac{2P^{rad}}{|I|^2}, L = \frac{4\tilde{W}_m}{|I|^2}, C = \frac{|I|^2}{4\omega^2\tilde{W}_e}. \tag{4.198}$$

The expressions (4.196) for the stored field energies can be easily generalized to an antenna array with N input ports. Let $[V] = [V_1, V_2, \ldots, V_N]^T$ and $[I] = [I_1, I_2, \ldots, I_N]^T$, respectively, denote the terminal voltage and current vectors of the antenna array, where T denotes the transpose operation. The voltage and current vectors are related by $[V] = [Z][I]$, where $[Z]$ is the impedance matrix of the array. For the antenna array, (4.196) should be replaced by

$$\tilde{W}_m = \frac{1}{8}[I]^H\left(\frac{d[X]}{d\omega} + \frac{[X]}{\omega}\right)[I], \quad \tilde{W}_e = \frac{1}{8}[I]^H\left(\frac{d[X]}{d\omega} - \frac{[X]}{\omega}\right)[I], \tag{4.199}$$

where $[X] = (1/2j)\left([Z] - \overline{[Z]}\right)$ is the reactance matrix of the array and the superscript H denotes the conjugate transpose. The radiated power of the antenna array is

$$P^{rad} = \frac{1}{2}\,\mathrm{Re}\,[I]^H[V] = \frac{1}{2}\,\mathrm{Re}\,[I]^H[Z][I] = \frac{1}{4}[I]^H\left[[Z] + [Z]^H\right][I]. \tag{4.200}$$

In terms of (4.30), (4.199), and (4.200), the QF for the antenna array can be expressed as

$$Q_I = \frac{\omega\tilde{W}}{P^{rad}} = \frac{1}{2}\,\frac{[I]^H\left[\omega\left(\frac{dX}{d\omega}\right)\right][I]}{[I]^H\left\{[Z] + [Z]^H\right\}[I]}. \tag{4.201}$$

This gives the Rayleigh quotient for the QF, and can be used to find the optimal distribution of excitations for the antenna array to minimize the QF.

Example 4.6 Consider a dipole antenna of radius a_0 and length $2a$. The current distribution on the wire surface is assumed to be

$$\mathbf{J}(\mathbf{r}) = \mathbf{u}_z \frac{I_0}{2\pi a_0} \sin k(a - |z|), \; -a < z < a, r = a_0$$

and the terminal current is given by $I = I_0 \sin ka$. The radiation resistance and reactance of the dipole antenna can be found as follows:

$$\begin{aligned} R^{rad} &= \frac{\eta}{2\pi \sin^2(ka)} \left\{ C + \ln 2ka - \text{ci}(2ka) + \frac{1}{2} \sin 2ka[\text{si}(4ka) - 2\text{si}(2ka)] \right. \\ &\quad \left. + \frac{1}{2} \cos 2ka[C + \ln(ka) + \text{ci}(4ka) - 2\text{ci}(2ka)] \right\}, \\ X &= \frac{\eta}{4\pi \sin^2(ka)} \left\{ 2\text{si}(2ka) + \cos 2ka[2\text{si}(2ka) - \text{si}(2ka)] \right. \\ &\quad \left. - \sin 2ka \left[2\text{ci}(2ka) - \text{ci}(4ka) - \text{ci}\left(\frac{ka_0^2}{a}\right) \right] \right\}, \end{aligned}$$

where si and ci are the sine and cosine integrals defined by

$$\text{si}(x) = \int_0^x \frac{\sin y}{y} dy, \quad \text{ci}(x) = \int_\infty^x \frac{\cos y}{y} dy.$$

The stored field energies, equivalent circuit element values, and QF can be evaluated from (4.196) and (4.198), and are shown in Figures 4.10, 4.11, and 4.12, respectively. □

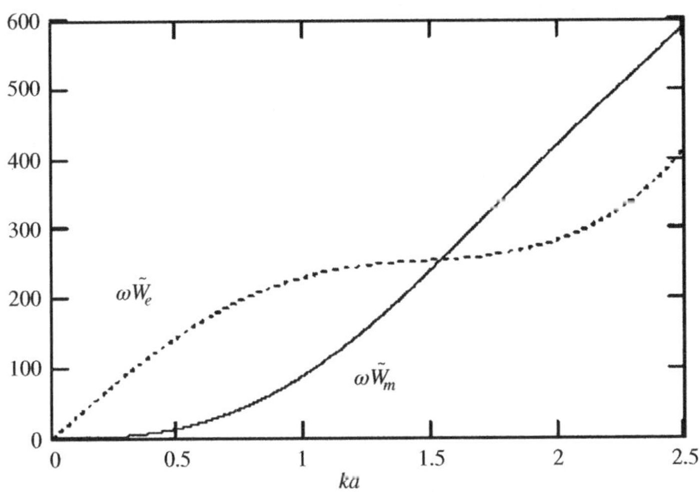

Figure 4.10 Stored field energies of dipole antenna.

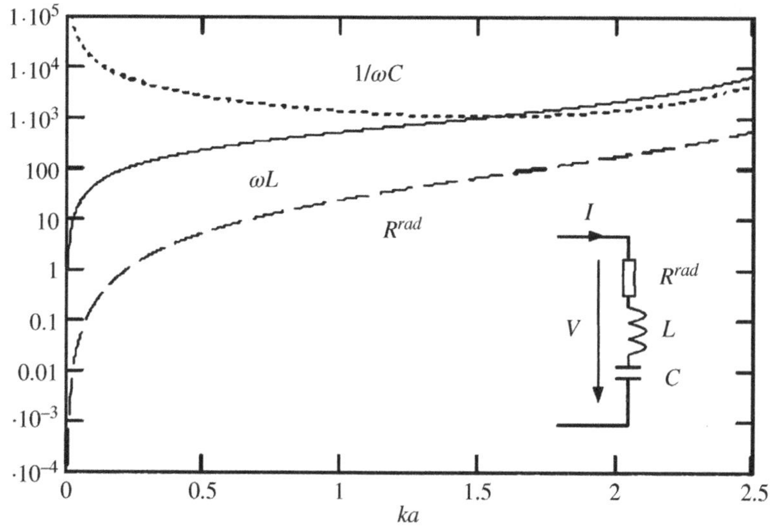

Figure 4.11 Element values of equivalent RLC circuit for the dipole.

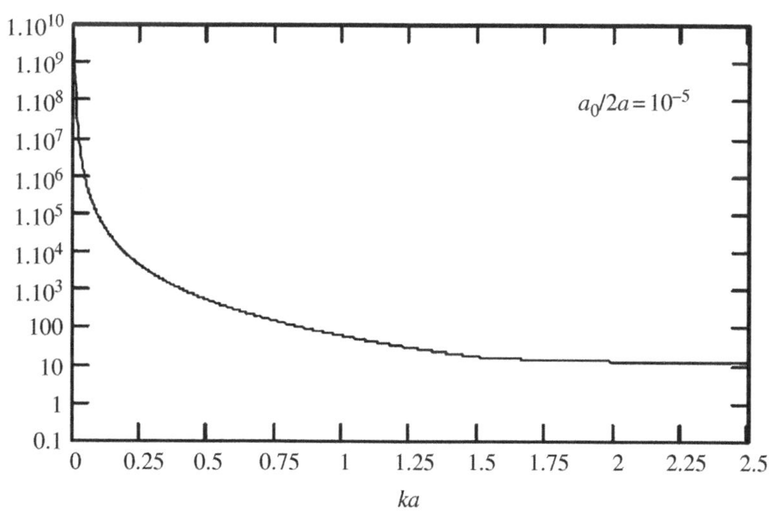

Figure 4.12 Radiation QF of the dipole antenna.

4.4 Modal Quality Factors

The concept of modal QF first appeared in Chu's work [81]. Chu's analysis was based on the theory of spherical harmonics. By examining a special case that the antenna is omnidirectional and radiates either TE or TM modes, Chu obtained the bounds of antenna QF and the ratio of gain over QF. In Chu's analysis, the total EM field energy inside the circumscribing sphere of the antenna was ignored in the study of QF. One technical difficulty encountered in Chu's analysis is that the total EM field energy outside the circumscribing sphere of antenna is infinite. To overcome this difficulty, Chu had to truncate the series expansions for the gain, QF, and the ratio of gain over QF. In this way, the infinity problem was eliminated. Therefore, Chu's analysis and results are only approximately valid and some concerns were raised about the reasonability of his derivations. Collin and Rothschild adopted a different method for calculating the modal QF [82]. Their method is established on the proposition made by Counter that the stored EM field energy is defined as the total field energy outside the circumscribing sphere of the antenna subtracted from the radiated field energy [68]. By adopting the same definition of stored field energy and ignoring the stored field energy inside the circumscribing sphere of the antenna, Fante was able to find a general expression of QF for an arbitrary ideal antenna [76].

4.4.1 Stored Field Energies Outside the Circumscribing Sphere of Antenna

Suppose that the current source **J** is confined in the region V_0. The stored EM field energy of the source is given by (4.191):

$$\tilde{W} = W_e(V_\infty) + W_m(V_\infty) - W_{em}^{rad}(V_\infty), \tag{4.202}$$

where the term W_d defined by (4.195) has been ignored for it is negligible for a small antenna. The study of antenna QF was usually based on the spherical wave function expansion outside the circumscribing sphere of the antenna. Such an approach results in an antenna QF that is much lower than its real value, denoted by Q_{real}, as the stored field energy inside the circumscribing sphere has been neglected. Let the antenna be enclosed by the circumscribing sphere V_a of radius a, bounded by S_a, as illustrated in Figure 4.13. The total stored field energy outside the circumscribing sphere can be evaluated in a spherical coordinate system through (4.89):

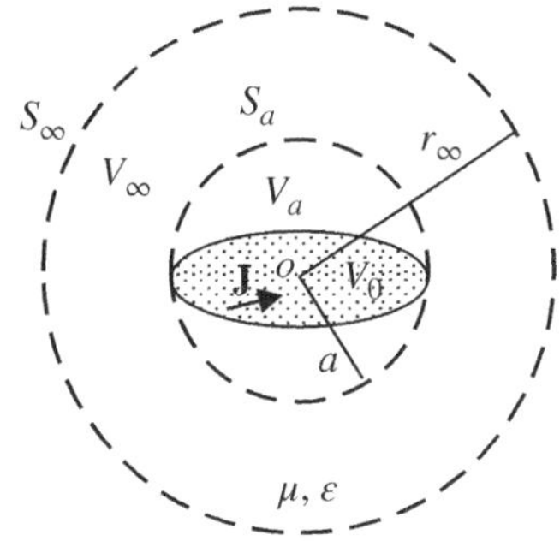

Figure 4.13 An arbitrary antenna.

$$\tilde{W}_e + \tilde{W}_m = \int_{V_\infty - V_a} (\tilde{w}_e + \tilde{w}_m) dV = \frac{\varepsilon a}{4k^2} \sum_{n,m,l} N_{nm}^2 \left(\left| \alpha_{nml}^{(2)} \right|^2 + \left| \beta_{nml}^{(2)} \right|^2 \right)$$
$$\cdot \left\{ 2 - (ka)^2 \left[\left| h_n^{(2)}(ka) \right|^2 - j_{n-1}(ka) j_{n+1}(ka) - n_{n-1}(ka) n_{n+1}(ka) \right] \right.$$
$$\left. - \left| h_n^{(2)}(ka) \right|^2 - (ka) \left[j_n(ka) \dot{j}_n(ka) + n_n(kr) \dot{n}_n(ka) \right] \right\}. \tag{4.203}$$

To determine the stored field energies $\tilde{W}_e$ and $\tilde{W}_m$, another equation is needed. From Poynting theorem, one may find the difference of stored energies

$$\tilde{W}_m - \tilde{W}_e = \frac{1}{4\omega} \operatorname{Im} \int_{S_a} (\mathbf{E} \times \overline{\mathbf{H}}) \cdot \mathbf{u}_n dS. \tag{4.204}$$

Introducing the field expansions (4.89) into the above equation, one may find

$$\tilde{W}_m - \tilde{W}_e = \frac{\varepsilon a}{4k^2} \sum_{n,m,l} N_{nm}^2 \left(\left| \beta_{nml}^{(2)} \right|^2 - \left| \alpha_{nml}^{(2)} \right|^2 \right)$$
$$\cdot \left\{ \left| h_n^{(2)}(ka) \right|^2 + (ka) \left[j_n(ka) \dot{j}_n(ka) + n_n(ka) \dot{n}_n(ka) \right] \right\}. \tag{4.205}$$

Combining (4.203) and (4.205) yields

$$\begin{aligned} \tilde{W}_m &= \frac{\varepsilon}{4k^3} \sum_{n,m,l} N_{nm}^2 \left(\left| \alpha_{nml}^{(2)} \right|^2 Q_n + \left| \beta_{nml}^{(2)} \right|^2 Q_n' \right), \\ \tilde{W}_e &= \frac{\varepsilon}{4k^3} \sum_{n,m,l} N_{nm}^2 \left(\left| \alpha_{nml}^{(2)} \right|^2 Q_n' + \left| \beta_{nml}^{(2)} \right|^2 Q_n \right), \end{aligned} \tag{4.206}$$

where

$$\begin{aligned} Q_n &= ka - \left|h_n^{(2)}(ka)\right|^2 \left[\frac{1}{2}(ka)^3 + ka(n+1)\right] - \frac{1}{2}(ka)^3 \left|h_{n+1}^{(2)}(ka)\right|^2 \\ &\quad + \frac{1}{2}(ka)^2(2n+3)[j_n(ka)j_{n+1}(ka) + n_n(ka)n_{n+1}(ka)], \\ Q_n' &= ka - \frac{1}{2}(ka)^3 \left[\left|h_n^{(2)}(ka)\right|^2 - j_{n-1}(ka)j_{n+1}(ka) - n_{n-1}(ka)n_{n+1}(ka)\right] \end{aligned} \tag{4.207}$$

are called modal QFs. Substituting (4.112) and (4.206) into (4.30) and (4.31), one finds

$$Q_I = \frac{1}{2} \frac{\sum\limits_{n,m,l} N_{nm}^2 \left(\left|\alpha_{nml}^{(2)}\right|^2 + \left|\beta_{nml}^{(2)}\right|^2\right)(Q_n + Q_n')}{\sum\limits_{n,m,l} N_{nm}^2 \left(\left|\alpha_{nml}^{(2)}\right|^2 + \left|\beta_{nml}^{(2)}\right|^2\right)}, \tag{4.208}$$

and

$$Q_{II} = \begin{cases} \dfrac{\sum\limits_{n,m,l} N_{nm}^2 \left(\left|\alpha_{nml}^{(2)}\right|^2 Q_n + \left|\beta_{nml}^{(2)}\right|^2 Q_n'\right)}{\sum\limits_{n,m,l} N_{nm}^2 \left(\left|\alpha_{nml}^{(2)}\right|^2 + \left|\beta_{nml}^{(2)}\right|^2\right)}, \tilde{W}_m > \tilde{W}_e \\ \dfrac{\sum\limits_{n,m,l} N_{nm}^2 \left(\left|\alpha_{nml}^{(2)}\right|^2 Q_n' + \left|\beta_{nml}^{(2)}\right|^2 Q_n\right)}{\sum\limits_{n,m,l} N_{nm}^2 \left(\left|\alpha_{nml}^{(2)}\right|^2 + \left|\beta_{nml}^{(2)}\right|^2\right)}, \tilde{W}_e > \tilde{W}_m \end{cases}. \tag{4.209}$$

Equations (4.208) and (4.209) can be written as

$$Q_I = \frac{\sum\limits_n (a_n^2 + b_n^2) \frac{1}{2}(Q_n + Q_n')}{\sum\limits_n (a_n^2 + b_n^2)}, \tag{4.210}$$

$$Q_{II} = \max\left\{ \frac{\sum\limits_{n=1}^{\infty} (a_n^2 Q_n + b_n^2 Q_n')}{\sum\limits_{n=1}^{\infty} (a_n^2 + b_n^2)}, \frac{\sum\limits_{n=1}^{\infty} (a_n^2 Q_n' + b_n^2 Q_n)}{\sum\limits_{n=1}^{\infty} (a_n^2 + b_n^2)} \right\}, \tag{4.211}$$

where $a_n^2 = \sum_{m,l} N_{nm}^2 \left|\alpha_{nml}^{(2)}\right|^2$ and $b_n^2 = \sum_{m,l} N_{nm}^2 \left|\beta_{nml}^{(2)}\right|^2$.

Remark 4.6 The total field energies for the spherical modal fields defined by (4.107), stored in the region bounded by two spheres of radius a and b with $b > a$, are

$$
\begin{aligned}
W^{TE^{(2)}_{nml}} &= \frac{1}{4}\varepsilon \int_a^b dr \int_S \left|\mathbf{E}^{TE^{(2)}_{nml}}\right|^2 dS + \frac{1}{4}\mu \int_a^b dr \int_S \left|\mathbf{H}^{TE^{(2)}_{nml}}\right|^2 dS, \\
W^{TM^{(2)}_{nml}} &= \frac{1}{4}\varepsilon \int_a^b dr \int_S \left|\mathbf{E}^{TM^{(2)}_{nml}}\right|^2 dS + \frac{1}{4}\mu \int_a^b dr \int_S \left|\mathbf{H}^{TM^{(2)}_{nml}}\right|^2 dS.
\end{aligned}
\tag{4.212}
$$

Applying the orthogonality properties of the SVWFs, one may find

$$
\begin{aligned}
W^{TE^{(2)}_{nml}} &= \frac{N^2_{nm}\varepsilon}{4k^2} \int_a^b \left\{ \left[1 + \frac{n(n+1)}{(kr)^2}\right] \left|krh^{(2)}_n(kr)\right|^2 + \left|\dot{\tilde{h}}^{(2)}_n(kr)\right|^2 \right\} dr, \\
W^{TM^{(2)}_{nml}} &= \frac{N^2_{nm}\varepsilon}{4k^2} \int_a^b \left\{ \left[1 + \frac{n(n+1)}{(kr)^2}\right] \left|krh^{(2)}_n(kr)\right|^2 + \left|\dot{\tilde{h}}^{(2)}_n(kr)\right|^2 \right\} dr.
\end{aligned}
\tag{4.213}
$$

The total stored field energies in (4.213) become infinite as $b \to \infty$ since both of them contain the radiated field energy. The radiated field energy in the spherical shell $a < r < b$ may be evaluated using (4.212) with the SVWFs replaced by their far-field expressions (4.106). The results are

$$
W^{rad}_{TE^{(2)}_{nml}} = W^{rad}_{TM^{(2)}_{nml}} = \frac{N^2_{nm}\varepsilon}{4k^2} \int_a^b 2dr. \tag{4.214}
$$

The radiated field energies also become infinite as $b \to \infty$. The stored field energy of antenna on the outside sphere of radius a is defined as the total field energy subtracted by the radiated field energy. So, the stored field energies of antenna that radiates the spherical modal fields are

$$
\begin{aligned}
\tilde{W}^{TE^{(2)}_{nml}} &= W^{TE^{(2)}_{nml}} - W^{rad}_{TE^{(2)}_{nml}} \\
&= \frac{N^2_{nm}\varepsilon}{4k^3} \int_a^\infty \left\{ \left[1 + \frac{n(n+1)}{(kr)^2}\right] \left|krh^{(2)}_n(kr)\right|^2 + \left|\dot{\tilde{h}}^{(2)}_n(kr)\right|^2 - 2 \right\} dkr, \\
\tilde{W}^{TM^{(2)}_{nml}} &= W^{TM^{(2)}_{nml}} - W^{rad}_{TM^{(2)}_{nml}} \\
&= \frac{N^2_{nm}\varepsilon}{4k^3} \int_a^\infty \left\{ \left[1 + \frac{n(n+1)}{(kr)^2}\right] \left|krh^{(2)}_n(kr)\right|^2 + \left|\dot{\tilde{h}}^{(2)}_n(kr)\right|^2 - 2 \right\} dkr,
\end{aligned}
\tag{4.215}
$$

where the radius b has been set as ∞. The integrals in (4.215) turn out to be finite and can be readily found

$$\tilde{W}^{TE^{(2)}_{nml}} = \tilde{W}^{TM^{(2)}_{nml}} = \frac{N^2_{nm}\varepsilon}{2k^3} Q_n(ka).$$

The QF for the spherical modal fields is then given by

$$Q = \frac{\omega \tilde{W}^{TE^{(2)}_{nml}}}{P^{rad}_{TE^{(2)}_{nml}}} = \frac{\omega \tilde{W}^{TM^{(2)}_{nml}}}{P^{rad}_{TM^{(2)}_{nml}}} = Q_n(ka), \tag{4.216}$$

where (4.108) has been used. □

Remark 4.7 A direct calculation shows that the squared fields averaged in all directions are

$$\langle |\mathbf{E}|^2 \rangle = \int_0^{2\pi} d\varphi \int_0^{\pi} |\mathbf{E}|^2 \sin\theta d\theta = \sum_{n,m,l} N^2_{nm} \left\{ \left|\alpha^{(2)}_{nml}\right|^2 \left|h^{(2)}_n(kr)\right|^2 + \left|\beta^{(2)}_{nml}\right|^2 \gamma^2_n \right\}, \tag{4.217}$$

$$\langle |\mathbf{H}|^2 \rangle = \int_0^{2\pi} d\varphi \int_0^{\pi} |\mathbf{H}|^2 \sin\theta d\theta = \sum_{n,m,l} N^2_{nm} \left\{ \left|\beta^{(2)}_{nml}\right|^2 \left|h^{(2)}_n(kr)\right|^2 + \left|\alpha^{(2)}_{nml}\right|^2 \gamma^2_n \right\}, \tag{4.218}$$

where $\gamma^2_n = \left(\left|\dot{\tilde{h}}^{(2)}_n(kr)/kr\right|^2\right) + n(n+1)\left(\left|h^{(2)}_n(kr)/kr\right|^2\right)$. □

4.4.2 Two Inequalities for Spherical Hankel Functions

In order to establish the properties of modal QFs, one first needs to prove two inequalities for spherical Hankel functions. If ν is not an integer, a useful equation for Lommel's polynomial may be obtained as follows [83]:

$$J_{\nu+n}(x)J_{n+1-\nu}(x) + J_{-\nu-n}(x)J_{-n-1+\nu}(x) = \frac{2}{\pi x}(-1)^n \sin\nu\pi R_{2n,\nu-n}(x), \tag{4.219}$$

where $J_\nu(x)$ is the Bessel functions and $R_{2n,\,\nu-n}(x)$ is the Lommel's polynomial defined by

$$R_{2n,\nu-n}(x) = \sum_{m=0}^{n} \frac{(-1)^m(2n-m)!\Gamma(\nu-n+2n-m)}{m!(2n-2m)!\Gamma(\nu-n+m)} \left(\frac{x}{2}\right)^{-2n+2m},$$

with Γ being the Gamma function. For $\nu = 1/2$, (4.219) gives the square modulus of the spherical Hankel function of the second kind

$$\left|h_n^{(2)}(x)\right|^2 = \frac{1}{x^2}\sum_{m=0}^{n}\frac{(2n-m)!(2n-2m)!}{[(n-m)!]^2 m!}(2x)^{2(m-n)}. \tag{4.220}$$

This can be written as

$$\left|h_n^{(2)}(x)\right|^2 = \frac{1}{x^2}[1+a(x)],$$

where $a(x)$ is positive for $x > 0$. Thus, one obtains the first inequality for spherical Hankel functions of the second kind

$$\left|h_n^{(2)}(x)\right|^2 > \frac{1}{x^2}. \tag{4.221}$$

According to (4.220), one may write

$$\left|h_{n-1}^{(2)}(x)\right|^2 = \frac{1}{x^2}\sum_{m=0}^{n-1}\frac{[2(n-1)-2m]![2(n-1)-m]!}{[(n-1-m)!]^2 m!}(2x)^{2(m-n+1)},$$

which can be rearranged as

$$\left|h_{n-1}^{(2)}(x)\right|^2 = \frac{1}{x^2}\sum_{m=1}^{n}\frac{(2n-2m)!(2n-m)!}{(2n/m-1)[(n-m)!]^2 m!}(2x)^{2(m-n)}.$$

Since $(2n/m-1) \geq 1$ for $1 \leq m \leq n$, it follows that

$$\begin{aligned}\left|h_{n-1}^{(2)}(x)\right|^2 &= \frac{1}{x^2}\sum_{m=1}^{n}\frac{(2n-2m)!(2n-m)!}{(2n/m-1)[(n-m)!]^2 m!}(2x)^{2(m-n)}\\ &\leq \frac{1}{x^2}\sum_{m=1}^{n}\frac{(2n-2m)!(2n-m)!}{[(n-m)!]^2 m!}(2x)^{2(m-n)}\\ &= \left|h_n^{(2)}(x)\right|^2 - \left[\frac{(2n)!}{n!}\right]^2(2x)^{-2n},\end{aligned} \tag{4.222}$$

which implies

$$\left|h_n^{(2)}(x)\right|^2 > \left|h_{n-1}^{(2)}(x)\right|^2. \tag{4.223}$$

This is the second inequality for spherical Hankel functions. It indicates that the absolute value of spherical Hankel functions increases with the index n. The inequality (4.223) was first conjectured by Fante and Mayhan through extensive numerical tabulation, and proven to be useful in the study of the bounds of the radiated field outside the radiating system [84].

Example 4.7 Consider the minimization problem of the ratio $\langle|\mathbf{E}|^2\rangle/P^{rad}$. From (4.112) and (4.217), the ratio can be expressed by

$$\frac{\langle|\mathbf{E}|^2\rangle}{P^{rad}} = \frac{\sum\limits_{n,m,l} N_{nm}^2\left\{\left|\alpha_{nml}^{(2)}\right|^2\left|h_n^{(2)}(kr)\right|^2 + \left|\beta_{nml}^{(2)}\right|^2\gamma_n^2\right\}}{\frac{1}{2k^2\eta}\sum\limits_{n,m,l} N_{nm}^2\left(\left|\alpha_{nml}^{(2)}\right|^2 + \left|\beta_{nml}^{(2)}\right|^2\right)}. \tag{4.224}$$

On use of (4.223) and the similar properties for γ_n^2, i.e. $\gamma_n^2 > \gamma_{n-1}^2$ and $\gamma_1^2 > \left|h_1^{(2)}\right|^2$, one may obtain

$$\frac{\langle|\mathbf{E}|^2\rangle}{P^{rad}} > \frac{\left|h_1^{(2)}(kr)\right|^2\sum\limits_{n,m,l} N_{nm}^2\left\{\left|\alpha_{nml}^{(2)}\right|^2 + \left|\beta_{nml}^{(2)}\right|^2\right\}}{\frac{1}{2k^2\eta}\sum\limits_{n,m,l} N_{nm}^2\left(\left|\alpha_{nml}^{(2)}\right|^2 + \left|\beta_{nml}^{(2)}\right|^2\right)} = 2k^2\eta\left|h_1^{(2)}(kr)\right|^2. \tag{4.225}$$

This gives the minimum squared electric field averaged in all directions

$$\min\langle|\mathbf{E}|^2\rangle = 2k^2\eta P^{rad}\left|h_1^{(2)}(kr)\right|^2. \tag{4.226}$$

The above equation can be used to derive the lower bound for the maximum electric field averaged in all directions [84]. □

4.4.3 Properties of Modal Quality Factors

Both the modal QFs Q_n and Q'_n are important in theoretical studies. They have been extensively used to investigate the properties of antennas such as the upper bounds of the ratio of gain to QF and the lower bounds of QF. Some interesting properties of the Q_n and Q'_n have been discussed by Fante through numerical tabulations [76] and are listed in Table 4.3. Properties 1 and 2 state that both Q_n and Q'_n

Table 4.3 Properties of modal QFs I.

Property 1	$Q_n(x) > 0$.
Property 2	$Q'_n(x) > 0$.
Property 3	$Q_{n+1}(x) > Q_n(x)$.
Property 4	$Q'_{n+1}(x) > Q'_n(x)$.
Property 5	$Q_n(x) > Q'_n(x)$.

Table 4.4 Properties of modal QFs II.

Property 6	$Q_{n+1}(x) > \frac{2n+3}{2n+1} Q_n(x).$
Property 7	$Q'_{n+1}(x) > \frac{2n+3}{2n+1} Q'_n(x).$
Property 8	$Q_{n+1}(x) = Q'_n(x) + x(n+1)\left\lvert h_{n+1}^{(2)}(x)\right\rvert^2.$
Property 9	$Q'_{n+1}(x) = Q_n(x) + x(n+1)\left\lvert h_n^{(2)}(x)\right\rvert^2.$
Property 10	$Q_n(x) = \sum_{m=0}^{n} \frac{q(n,m)}{x^{2(n-m)+1}}.$
Property 11	$Q'_n(x) = \sum_{m=0}^{n} \frac{q'(n,m)}{x^{2(n-m)+1}}.$

are positive, properties 3 and 4 indicate that both Q_n and Q'_n are monotonically increasing with respect to the index n, and property 5 simply states that Q_n is always greater than Q'_n.

Some new properties for the modal QFs will be derived below and they are listed in Table 4.4, where properties 6 and 7 are stronger than properties 3 and 4; properties 8 and 9 give the alternating recurrence relations for the modal QFs; properties 10 and 11 are the finite power series expansions for the modal QFs, and they are more informative and concise than their original definitions (4.207). The expansion coefficients in properties 10 and 11 are given by

$$q(n,m) = \frac{\left[n(n+2) + (n-m)^2 - m\right](2n-m)!(2n-2m)!}{2^{2(n-m)+1}[(n-m)!]^2(n+1-m)m!}, \tag{4.227}$$

$$q'(n,m) = \frac{m(2n+1-m)(2n-m)!(2n-2m)!}{2^{2(n-m)+1}[(n-m)!]^2(n+1-m)m!}. \tag{4.228}$$

Most of the properties of the modal QFs can be proved by the method of mathematical induction. Since the proofs are purely mathematical, all the properties of the modal QFs essentially belong to the spherical Bessel functions.

4.4.3.1 Proof of Properties 2, 4, and 7

For the technique reason, the property 2 will be established first by the method of mathematical induction. As the first step of the proof, one needs to show that $Q'_n(x)$ is positive for $n = 1$. This can be easily done from the definition of spherical Bessel functions. Indeed, it is easy to show that $Q'_1(x) = 1/x > 0$ for $x = ka > 0$. As the inductive step, $Q'_n(x)$ is assumed to be positive for an arbitrary$n$:

$$Q'_n(x) > 0, \tag{4.229}$$

and the rest is to show that $Q'_n(x)$ is positive for $n+1$. According to the definition (4.207), one may write

$$Q'_{n+1}(x) = x - \frac{x^3}{2}\left[\left|h^{(2)}_{n+1}(x)\right|^2 - j_n(x)j_{n+2}(x) - n_n(x)n_{n+2}(x)\right]. \tag{4.230}$$

By use of the recurrence relations of spherical Bessel functions, one may find

$$\begin{aligned}&j_n(x)j_{n+2}(x) + n_n(x)n_{n+2}(x)\\ &= -\left|h^{(2)}_n(x)\right|^2 + \frac{2n+3}{2n+1}\left[\left|h^{(2)}_{n+1}(x)\right|^2 + j_{n-1}(x)j_{n+1}(x) + n_{n-1}(x)n_{n+1}(x)\right].\end{aligned}$$

Thus,

$$\begin{aligned}&\left|h^{(2)}_{n+1}(x)\right|^2 - j_n(x)j_{n+2}(x) - n_n(x)n_{n+2}(x)\\ &= -\frac{2}{2n+1}\left[\left|h^{(2)}_{n+1}(x)\right|^2 + \left|h^{(2)}_n(x)\right|^2\right]\\ &\quad + \frac{2n+3}{2n+1}\left[\left|h^{(2)}_n(x)\right|^2 - j_{n-1}(x)j_{n+1}(x) - n_{n-1}(x)n_{n+1}(x)\right].\end{aligned}$$

Substituting this into (4.230) and rearranging terms yield

$$Q'_{n+1}(x) = \frac{2n+3}{2n+1}Q'_n(x) + \frac{x^3}{2n+1}\left[\left|h^{(2)}_{n+1}(x)\right|^2 + \left|h^{(2)}_n(x)\right|^2\right] - \frac{2}{2n+1}x, \tag{4.231}$$

which is the recurrence relation for the modal Q'_n. It follows from (4.221) that

$$Q'_{n+1}(x) > \frac{2n+3}{2n+1}Q'_n(x) + \frac{x^3}{x^2}\frac{2}{2n+1} - \frac{2}{2n+1}x = \frac{2n+3}{2n+1}Q'_n(x), \tag{4.232}$$

which gives property 7, and thus also proves property 4. From (4.232) and the assumption (4.229), one immediately gets

$$Q'_{n+1}(x) > 0.$$

The proof of property 2 is completed.

A question may be raised whether there exists a constant c greater than $(2n+3)/(2n+1)$ in (4.232), such that

$$Q'_{n+1}(x) > cQ'_n(x). \tag{4.233}$$

To answer this question, one may use (4.231) and the asymptotic expressions of spherical Hankel functions to find

$$Q'_{n+1}(x) \to \frac{2n+3}{2n+1}Q'_n(x), \text{as } x \to \infty.$$

Therefore, the coefficient $(2n+3)/(2n+1)$ in (4.232) is the maximum possible value that satisfies (4.233).

4.4.3.2 Proof of Properties 1, 3, 6, 8, and 9

Similarly, property 1 can be established by the method of mathematical induction. As the first step of the proof, it is easy to find that $Q_1(x) = x^{-1} + x^{-3} > 0$ for $x > 0$ from the definition of spherical Bessel functions. As the inductive step, the property 1 will be assumed to hold for an arbitrary n:

$$Q_n(x) > 0. \tag{4.234}$$

To show that the above inequality is also valid for $n+1$, one may write

$$\begin{aligned} Q_{n+1}(x) &= x - \left|h_{n+1}^{(2)}(x)\right|^2 \left[x(n+2) + \frac{x^3}{2}\right] - \frac{x^3}{2}\left|h_{n+2}^{(2)}(x)\right|^2 \\ &\quad + x^2\frac{2n+5}{2}[j_{n+1}(x)j_{n+2}(x) + n_{n+1}(x)n_{n+2}(x)]. \end{aligned} \tag{4.235}$$

By the recurrence relations for spherical Bessel functions, one may find

$$\begin{aligned} \left|h_{n+2}^{(2)}(x)\right|^2 &= \frac{(2n+3)^2}{x^2}\left|h_{n+1}^{(2)}(x)\right|^2 + \left|h_n^{(2)}(x)\right|^2 \\ &\quad - \frac{2(2n+3)}{x}[j_{n+1}(x)j_n(x) + n_{n+1}(x)n_n(x)] \end{aligned}$$

and

$$\begin{aligned} &j_{n+1}(x)j_{n+2}(x) + n_{n+1}(x)n_{n+2}(x) \\ &= \frac{2n+3}{x}\left|h_{n+1}^{(2)}(x)\right|^2 - j_n(x)j_{n+1}(x) - n_n(x)n_{n+1}(x). \end{aligned}$$

Substituting these into (4.235) and ignoring the tedious process yield

$$\begin{aligned} Q_{n+1}(x) &= Q_n(x) + x(n+1)\left[\left|h_{n+1}^{(2)}(x)\right|^2 + \left|h_n^{(2)}(x)\right|^2\right] \\ &\quad - x^2[j_{n+1}(x)j_n(x) + n_{n+1}(x)n_n(x)]. \end{aligned} \tag{4.236}$$

By the recurrence relations for spherical Bessel functions, it is readily found from (4.207) that

$$\begin{aligned} Q_n(x) - Q_n'(x) &= \left|h_n^{(2)}(x)\right|^2 nx - x^2[j_n(x)j_{n-1}(x) + n_n(x)n_{n-1}(x)] \\ &= -x(n+1)\left|h_n^{(2)}(x)\right|^2 + x^2[j_n(x)j_{n+1}(x) + n_n(x)n_{n+1}(x)]. \end{aligned} \tag{4.237}$$

By means of (4.236) and (4.237) as well as the property 2, it is easy to find

$$Q_{n+1}(x) = Q_n'(x) + x(n+1)\left|h_{n+1}^{(2)}(x)\right|^2 > 0.$$

This completes the proof of property 1 as well as property 8. According to (4.237), one may write

$$Q_{n+1} - Q'_{n+1} = \left|h_{n+1}^{(2)}(x)\right|^2 (n+1)x - x^2[j_{n+1}(x)j_n(x) + n_{n+1}(x)n_n(x)].$$

Insertion of this into (4.236) gives property 9. It follows from properties 7 and 8 that

$$Q'_n(x) > \frac{2n+1}{2n-1}Q'_{n-1}(x) > \frac{2n+3}{2n+1}Q'_{n-1}(x), \tag{4.238}$$

$$Q_n(x) = Q'_{n-1}(x) + xn\left|h_n^{(2)}(x)\right|^2. \tag{4.239}$$

Multiplying (4.239) by $(2n+3)/(2n+1)$ and subtracting the resultant equation from property 8 yield

$$\begin{aligned} Q_{n+1}(x) - \frac{2n+3}{2n+1}Q_n(x) &= Q'_n(x) - \frac{2n+3}{2n+1}Q'_{n-1}(x) \\ &+ x(n+1)\left|h_{n+1}^{(2)}(x)\right|^2 - x\left(n + \frac{2n}{2n+1}\right)\left|h_n^{(2)}(x)\right|^2. \end{aligned} \tag{4.240}$$

By use of property 7, (4.223), and (4.238), one may find

$$Q_{n+1}(x) > \frac{2n+3}{2n+1}Q_n(x). \tag{4.241}$$

Thus, the property 6 and thus property 3 have been established. Similarly, it can be shown that the number $(2n+3)/(2n+1)$ is the maximum possible value for the inequality (4.241) to hold. By means of (4.231), (4.240) can be expressed by

$$\begin{aligned} Q_{n+1}(x) - \frac{2n+3}{2n+1}Q_n(x) &= x(n+1)\left|h_{n+1}^{(2)}(x)\right|^2 - x\left(n + \frac{2n}{2n+1}\right)\left|h_n^{(2)}(x)\right|^2 \\ &+ \frac{x^3}{2n+1}\left[\left|h_{n+1}^{(2)}(x)\right|^2 + \left|h_n^{(2)}(x)\right|^2\right] - \frac{2}{2n+1}x. \end{aligned} \tag{4.242}$$

This is the recurrence relation for the modal $Q_n(x)$.

4.4.3.3 Proof of Property 5

To demonstrate property 5, one has to resort to the following equation for the Lommel's polynomial [83]:

$$J_{\nu+n+1}(x)J_{-\nu+n+1}(x) - J_{-\nu-n-1}(x)J_{\nu-n-1}(x) = \frac{2}{\pi z}(-1)^n \sin\nu\pi R_{2n+1,\nu-n}(x), \tag{4.243}$$

where $R_{2n+1,\,\nu-n}(x)$ is the Lommel's polynomial defined by

$$R_{2n+1,\nu-n}(x) = \sum_{m=0}^{n} \frac{(-1)^m(2n+1-m)!\Gamma(\nu-n+2n+1-m)}{m!(2n+1-2m)!\Gamma(\nu-n+m)}\left(\frac{x}{2}\right)^{2(m-n)-1}.$$

Here, ν is a non-integer. For $\nu = 1/2$, (4.243) can be written as

$$\begin{aligned} & x^2[j_{n+1}(x)j_n(x) + n_{n+1}(x)n_n(x)] \\ & = 2\sum_{m=0}^{n} \frac{(2n+1-m)(2n-m)!(2n-2m)!}{[(n-m)!]^2 m!}(2x)^{2(m-n)-1}. \end{aligned} \tag{4.244}$$

It follows from (4.220) that

$$x(n+1)\left|h_n^{(2)}(x)\right|^2 = 2\sum_{m=0}^{n} \frac{(n+1)(2n-2m)!(2n-m)!}{[(n-m)!]^2 m!}(2x)^{2(m-n)-1}. \tag{4.245}$$

From (4.237), (4.244), and (4.245), one may find

$$\begin{aligned} & Q_n(x) - Q_n'(x) \\ & = 2\sum_{m=0}^{n} \frac{(n-m)(2n-m)!(2n-2m)!}{[(n-m)!]^2 m!}(2x)^{2(m-n)-1} > 0 \end{aligned} \tag{4.246}$$

since $m \le n$. Property 5 is thus validated.

4.4.3.4 Proof of Properties 10 and 11

Properties 10 and 11 can be derived from (4.220), (4.244), and (4.246). In fact, it follows from (4.220) and (4.244) that

$$\begin{aligned} \left|h_n^{(2)}(x)\right|^2\left[\frac{x^3}{2} + x(n+1)\right] = & \frac{x}{2} + 2\frac{(n+1)[(2n)!]^2}{[(n)!]^2}(2x)^{-2n-1} \\ & + \frac{1}{4}\sum_{m=1}^{n} \frac{(2n+1-m)!(2n+2-2m)!}{[(n+1-m)!]^2(m-1)!}(2x)^{2(m-n)-1} \\ & + 2\sum_{m=1}^{n} \frac{(n+1)(2n-2m)!(2n-m)!}{[(n-m)!]^2 m!}(2x)^{2(m-n)-1}, \end{aligned}$$

$$\begin{aligned} \frac{x^3}{2}\left|h_{n+1}^{(2)}(x)\right|^2 = & \frac{x}{2} + \frac{1}{4}\sum_{m=1}^{n} \frac{(2n+2-m)!(2n+2-2m)!}{[(n+1-m)!]^2 m!}(2x)^{2(m-n)-1} \\ & + \frac{1}{4}\frac{[(2n+2)!]^2}{[(n+1)!]^2}(2x)^{-2n-1}, \end{aligned}$$

$$\frac{2n+3}{2}x^2[j_{n+1}(x)j_n(x)+n_{n+1}(x)n_n(x)]$$
$$=\sum_{m=1}^{n}\frac{(2n+3)(2n+1-m)(2n-m)!(2n-2m)!}{[(n-m)!]^2 m!}(2x)^{2(m-n)-1}$$
$$+\frac{(2n+1)(2n+3)(2n!)^2}{(n!)^2}(2x)^{-2n-1}.$$

Insertion of these into the first expression of (4.207) yields

$$Q_n(x)=C_1(2x)^{-2n-1}+\sum_{m=1}^{n}C_2(2x)^{2(m-n)-1}. \tag{4.247}$$

Ignoring the tedious simplifying process, the coefficients C_1 and C_2 are found to be

$$C_1=\frac{(2n!)^2 2n}{(n!)^2(2x)^{2n+1}},$$
$$C_2=\frac{(2n-m)!(2n-2m)!}{[(n-m)!]^2(n+1-m)m!}\left[n(n+2)+(n-m)^2-m\right].$$

On substitution of these into (4.247), one immediately obtains property 10. Combining property 10 and (4.246) gives property 11.

It can be seen from the properties 10 and 11 that both the modal Q_n and Q_n' are monotonically decreasing functions of the variable $x=ka$. For demonstrating their properties, the modal Q_n and Q_n' are plotted in Figure 4.14.

4.4.4 Lower Bound for Antenna Quality Factor

Since $Q_n>Q_n'$, $Q_{n+1}>Q_n$, and $Q_{n+1}'>Q_n'$, it follows from first definition of antenna QF (4.210) that

$$Q_I>\frac{1}{2}\frac{\sum_n\left(a_n^2+b_n^2\right)\left(Q_1+Q_1'\right)}{\sum_n\left(a_n^2+b_n^2\right)}=\frac{1}{2}\left(Q_1+Q_1'\right).$$

Therefore, the minimum possible value for Q_I is given by

$$\min Q_I=\frac{Q_1+Q_1'}{2}=\frac{1}{ka}+\frac{1}{2(ka)^3}. \tag{4.248}$$

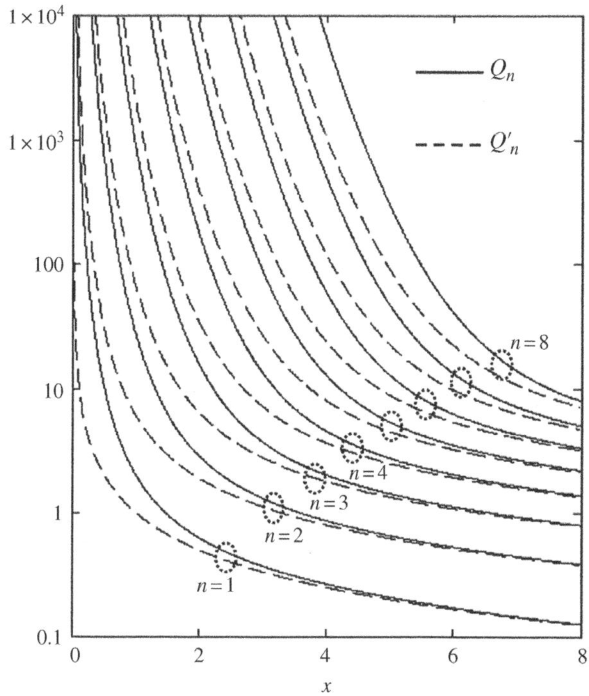

Figure 4.14 Modal QFs.

The above lower limit for the QF is also valid if the second definition (4.31) is used. In fact, if the first term on the right-hand side of (4.211) is assumed to be the largest, the antenna Q_{II} is

$$Q_{II} = \frac{\sum_{n=1}^{\infty} \left(a_n^2 Q_n + b_n^2 Q'_n\right)}{\sum_{n=1}^{\infty} \left(a_n^2 + b_n^2\right)}.$$

To ensure that the first term on the right-hand side of (4.211) is always larger than the second during the optimizing process, a constraint on the coefficients a_n and b_n is needed. This can be achieved by assuming $a_n \geq b_n$. Under this condition, one may find

$$Q_{II} = \frac{\sum_{n=1}^{\infty} \left(a_n^2 Q_n + b_n^2 Q'_n\right)}{\sum_{n=1}^{\infty} \left(a_n^2 + b_n^2\right)} \geq \frac{\sum_{n=1}^{\infty} \left(a_n^2 Q_1 + b_n^2 Q'_1\right)}{\sum_{n=1}^{\infty} \left(a_n^2 + b_n^2\right)} = C\left(Q_1 - Q'_1\right) + Q'_1,$$

where $C = \sum_{n=1}^{\infty} a_n^2 / \sum_{n=1}^{\infty} (a_n^2 + b_n^2) \geq \frac{1}{2}$ since $a_n \geq b_n$. The right-hand side of the above expression can be minimized by setting $C = 1/2$ or $a_n = b_n (n \geq 1)$. Therefore,

$$\min Q_{II} = \frac{Q_1 + Q_1'}{2} = \frac{1}{ka} + \frac{1}{2(ka)^3}, \tag{4.249}$$

which can be achieved by setting $a_1 = b_1$ and $a_n = b_n = 0$ for $n \geq 2$. If the second expression in (4.211) is the largest, the exactly same result can be obtained by interchanging a_n and b_n. Thus, the minimum QF problem has a unique lowest limit although the QF is defined conditionally.

The existence of a lower bound for antenna QF implies that the stored field energy of antenna can never be made zero. Once the maximum antenna size is given, this lower bound is then determined. For a small antenna ($ka < 1$), equation (4.248) can be approximated by

$$\min Q \approx \frac{1}{2(ka)^3}. \tag{4.250}$$

Since the real Q_{real} is always greater than the QF defined by (4.208) or (4.209), (4.248) may be considered as the minimum possible value for Q_{real}.

4.5 Upper Bounds for the Products of Gain and Bandwidth

From time to time, there arises a question of how to achieve better antenna performance than previously obtained. In most applications, one needs to maximize antenna gain and bandwidth simultaneously. For this reason, a reasonable performance index for characterizing antenna would be the product of antenna gain and bandwidth, or the ratio of antenna gain to antenna Q_{real}. In order to seek the maximum possible ratio of gain over antenna QF, one may use the QF defined by (4.208) or (4.209) to replace Q_{real} in the optimization process.

The upper limit for the ratio of gain over antenna QF shares similarity with Shannon's capacity theorem in communications theory and gives maximum possible PGB once the antenna size and its operating frequency are specified. The most general upper bounds for the ratio of gain over QF for both omnidirectional and directional antennas were found by the author in 2003 [85] and re-derived later via two different methods. One was based on the conventional definition of antenna QF and a novel algebraic inequality [86], which eliminated an unnecessary assumption made in the earlier study [85], and the other was based on the general definition of QF from IEEE standards [74]. Three independent derivations gave exactly the same results. These upper bounds are of fundamental importance in antenna theory and design. They can be served as the guidelines for antenna designers to determine the maximum possible PGB once the antenna size and operating frequency are specified, or equivalently, determine the minimum possible antenna size when the PGB is specified. They are especially useful in small antenna designs.

4.5.1 Directive Antenna

Without loss of generality, it will be assumed that the antenna is placed in a spherical coordinate system (r, θ, φ) and enclosed by the smallest circumscribing sphere of radius a, and the spherical coordinate system is oriented in such a way that the maximum radiation is in $(\theta, \varphi) = (0, 0)$ direction. From (4.92), the transverse radiated fields outside the circumscribing sphere of the antenna can be expanded in terms of spherical harmonics as follows:

$$\begin{aligned} kr\mathbf{E}_t &= \sum_{n,m,l} N_{nm}\left(\alpha_{nml}^{(2)}\tilde{h}_n^{(2)}\mathbf{h}_{nml} - \beta_{nml}^{(2)}\dot{\tilde{h}}_n^{(q)}\mathbf{e}_{nml}\right), \\ kr\mathbf{H}_t &= \frac{1}{j\eta}\sum_{n,m,l} N_{nm}\left(\alpha_{nml}^{(2)}\dot{\tilde{h}}_n^{(2)}\mathbf{e}_{nml} - \beta_{nml}^{(2)}\tilde{h}_n^{(2)}\mathbf{h}_{nml}\right), \end{aligned} \tag{4.251}$$

where the vector modal functions $\mathbf{e}_{nml}$ are explicitly given by

$$\begin{aligned} \mathbf{e}_{nml} &= -\frac{\mathbf{u}_\theta}{N_{nm}\sin\theta} f_{ml}(\varphi)\left[(n+1)\cos\theta P_n^m(\cos\theta) - (n-m+1)P_{n+1}^m(\cos\theta)\right] \\ &\quad + \frac{\mathbf{u}_\varphi}{N_{nm}\sin\theta}\left[P_n^m(\cos\theta)f'_{ml}(\varphi)\right], \\ \mathbf{h}_{nml} &= \mathbf{u}_r \times \mathbf{e}_{nml}. \end{aligned}$$

By definition of the associated Legendre function, it is easy to see that only $m = 1$ contributes to the field in the direction of $\theta = 0$. By the expression

$$\lim_{\theta\to 0}\frac{P_n^1(\cos\theta)}{\sin\theta} = \frac{1}{2}n(n+1),$$

one may find the expressions for the vector modal function in the direction of $\theta = 0$:

$$\begin{aligned} \lim_{\theta\to 0}\mathbf{e}_{n1l} &= -\frac{1}{2N_{n1}}(n+1)n\left[\mathbf{u}_\theta f_{1l}(\varphi) + \mathbf{u}_\varphi f'_{1l}(\varphi)\right], \\ \lim_{\theta\to 0}\mathbf{h}_{n1l} &= -\frac{1}{2N_{n1}}(n+1)n\left[\mathbf{u}_\varphi f_{1l}(\varphi) - \mathbf{u}_\theta f'_{1l}(\varphi)\right]. \end{aligned}$$

It follows from (4.251) that the far-field components in the direction of $(\theta, \varphi) = (0, 0)$ are

$$
\begin{aligned}
E_\theta &= \frac{1}{2kr}\sum_n (n+1)n\left[\dot{\tilde{h}}_n^{(2)}(kr)\beta_{n1e}^{(2)} + \tilde{h}_n^{(2)}(kr)\alpha_{n1o}^{(2)}\right],\\
H_\varphi &= -\frac{1}{j2\eta kr}\sum_n (n+1)n\left[-\tilde{h}_n^{(2)}(kr)\beta_{n1e}^{(2)} + \dot{\tilde{h}}_n^{(2)}(kr)\alpha_{n1o}^{(2)}\right],\\
E_\varphi &= \frac{1}{2kr}\sum_n (n+1)n\left[\dot{\tilde{h}}_n^{(2)}(kr)\beta_{n1o}^{(2)} - \tilde{h}_n^{(2)}(kr)\alpha_{n1e}^{(2)}\right],\\
H_\theta &= -\frac{1}{j2\eta kr}\sum_n (n+1)n\left[\tilde{h}_n^{(2)}(kr)\beta_{n1o}^{(2)} + \dot{\tilde{h}}_n^{(2)}(kr)\alpha_{n1e}^{(2)}\right].
\end{aligned}
$$

In the far-field region, the radiation intensity can be represented by

$$
\begin{aligned}
\frac{1}{2}r^2\,\mathrm{Re}\left(\mathbf{E}\times\overline{\mathbf{H}}\right)\cdot\mathbf{u}_r &= \frac{1}{2}r^2\,\mathrm{Re}\left(E_\theta\overline{H}_\varphi - E_\varphi\overline{H}_\theta\right)\\
&= \frac{1}{8k^2\eta}\left|\sum_n (n+1)nj^n\left(\beta_{n1e}^{(2)} + j\alpha_{n1o}^{(2)}\right)\right|^2\\
&\quad + \frac{1}{8k^2\eta}\left|\sum_n (n+1)nj^n\left(\beta_{n1o}^{(2)} - j\alpha_{n1e}^{(2)}\right)\right|^2.
\end{aligned}
$$

The directivity in the direction of $(\theta, \varphi) = (0, 0)$ is then given by

$$
\begin{aligned}
G &= 4\pi r^2\frac{\frac{1}{2}\mathrm{Re}\left(\mathbf{E}\times\overline{\mathbf{H}}\right)\cdot\mathbf{u}_r}{p^{rad}}\\
&= \pi\frac{\left|\sum_n (n+1)nj^n\left(\beta_{n1e}^{(2)} + j\alpha_{n1o}^{(2)}\right)\right|^2 + \left|\sum_n (n+1)nj^n\left(\beta_{n1o}^{(2)} - j\alpha_{n1e}^{(2)}\right)\right|^2}{\sum_{n,m,l} N_{nm}^2\left(\left|\alpha_{nml}^{(2)}\right|^2 + \left|\beta_{nml}^{(2)}\right|^2\right)},
\end{aligned}
\tag{4.252}
$$

where use has been made of (4.112). For simplicity, (4.208) will be used to maximize the ratio of gain over QF. In terms of (4.252), the ratio of gain over QF can be expressed as

$$
\begin{aligned}
\left.\frac{G}{Q}\right|_{dir} &= \frac{2\pi}{\sum_{n,m,l} N_{nm}^2\left(\left|\alpha_{nml}^{(2)}\right|^2 + \left|\beta_{nml}^{(2)}\right|^2\right)(Q_n + Q_n')}\cdot\left[\left|\sum_{n=1}^{\infty}(n+1)nj^n\left(\beta_{n1e}^{(2)} + j\alpha_{n1o}^{(2)}\right)\right|^2\right.\\
&\quad\left. + \left|\sum_{n=1}^{\infty}(n+1)nj^n\left(\beta_{n1o}^{(2)} - j\alpha_{n1e}^{(2)}\right)\right|^2\right].
\end{aligned}
\tag{4.253}
$$

Since only $\alpha_{n1l}^{(2)}$ and $\beta_{n1l}^{(2)}$ contribute to the numerator, the ratio (4.253) can be increased by setting $\alpha_{nml}^{(2)} = \beta_{nml}^{(2)} = 0 (m \neq 1)$. Thus,

$$\left.\frac{G}{Q}\right|_{dir} = \frac{\left|\sum_{n=1}^{\infty}(A_{on} + B_{en})\right|^2 + \left|\sum_{n=1}^{\infty}(A_{en} + B_{on})\right|^2}{\sum_{n=1}^{\infty}\frac{(Q_n + Q_n')}{2n+1}\left[\left(|A_{en}|^2 + |B_{en}|^2\right) + \left(|A_{on}|^2 + |B_{on}|^2\right)\right]}, \tag{4.254}$$

where

$$\begin{cases} A_{on} = j^{n+1}n(n+1)\alpha_{n1o}^{(2)} \\ A_{en} = -j^{n+1}n(n+1)\alpha_{n1e}^{(2)} \end{cases}, \begin{cases} B_{en} = j^{n}n(n+1)\beta_{n1e}^{(2)} \\ B_{on} = j^{n}n(n+1)\beta_{n1o}^{(2)} \end{cases}.$$

The denominator of (4.254) depends only on the magnitudes of A_n and B_n. If the phase of A_n and B_n are adjusted such that they are in phase to maximize the numerator, the denominator will not change. Therefore, (4.254) can be maximized as follows:

$$\left.\frac{G}{Q}\right|_{dir} = \frac{\left[\sum_{n=1}^{\infty}(|A_{on}| + |B_{en}|)\right]^2 + \left[\sum_{n=1}^{\infty}(|A_{en}| + |B_{on}|)\right]^2}{\sum_{n=1}^{\infty}\frac{(Q_n + Q_n')}{2n+1}\left[\left(|A_{en}|^2 + |B_{en}|^2\right) + \left(|A_{on}|^2 + |B_{on}|^2\right)\right]}. \tag{4.255}$$

If the inequality $(a+b)^2 \leq 2(a^2+b^2)$ is applied, one may find

$$\left.\frac{G}{Q}\right|_{dir} \leq 2\frac{\left(\sum_{n=1}^{\infty}a_n\right)^2 + \left(\sum_{n=1}^{\infty}b_n\right)^2}{\sum_{n=1}^{\infty}\frac{(Q_n + Q_n')}{2n+1}(a_n^2 + b_n^2)} = 2\frac{(\boldsymbol{\zeta}, \mathbf{C}_a)_E^2 + (\boldsymbol{\zeta}, \mathbf{C}_b)_E^2}{(\mathbf{C}_a, \mathbf{C}_a)_E + (\mathbf{C}_b, \mathbf{C}_b)_E}, \tag{4.256}$$

where

$$a_n = |A_{on}| + |B_{en}|,\ b_n = |A_{en}| + |B_{on}|,$$
$$\boldsymbol{\zeta} = (\zeta_1, \zeta_2, \ldots), \mathbf{C}_{a(b)} = \left(C_{a(b)1}, C_{a(b)2}, \ldots\right),$$
$$\zeta_n = \sqrt{\frac{2n+1}{Q_n + Q_n'}}, C_{an} = \frac{|a_n|}{\zeta_n}, C_{bn} = \frac{|b_n|}{\zeta_n},$$

and both $\boldsymbol{\zeta}$ and $\mathbf{C}_{a(b)}$ are vectors in the Euclidean space consisting of all vectors of infinite dimension with the inner product and norm defined by $(\boldsymbol{\zeta}, \mathbf{C})_E = \sum_{n=1}^{\infty}\zeta_n C_n$ and $\|\boldsymbol{\zeta}\| = (\boldsymbol{\zeta}, \boldsymbol{\zeta})_E^{1/2}$, respectively. It follows from (4.256) and Schwartz inequality that

$$\left.\frac{G}{Q}\right|_{dir} \leq 2\|\boldsymbol{\zeta}\|_E^2. \tag{4.257}$$

The equality holds if $\mathbf{C}_a = \mathbf{C}_b = c_1\zeta$. Hence the upper bound of ratio of gain to QF for a directional antenna is

$$\max \left.\frac{G}{Q}\right|_{dir} = 2\|\zeta\|_E^2 = \sum_{n=1}^{\infty} \frac{2(2n+1)}{Q_n + Q_n'}. \tag{4.258}$$

This is the maximum possible ratio of gain to QF for directive antenna.

4.5.2 OmniDirectional Antenna

For an omnidirectional antenna, the field is independent of φ. Without loss of generality, the maximum possible ratio of gain to QF in the direction of $\theta = \pi/2$ will be considered. Since the field is independent of φ, the vector modal functions in (4.251) can be chosen as

$$\mathbf{e}_{n0e} = \mathbf{u}_\theta \frac{1}{N_{n0}} P_n^1(\cos\theta), \mathbf{h}_{n0e} = \mathbf{u}_\varphi \frac{1}{N_{n0}} P_n^1(\cos\theta).$$

The field components produced by the omnidirectional antenna are given by

$$\begin{aligned}
E_\theta &= \frac{1}{kr}\sum_n \beta_{n0e}^{(2)} \dot{\tilde{h}}_n^{(2)}(kr) P_n^1(\cos\theta), \\
E_\varphi &= -\frac{1}{kr}\sum_n \alpha_{n0e}^{(2)} \tilde{h}_n^{(2)}(kr) P_n^1(\cos\theta), \\
H_\theta &= -\frac{1}{j\eta kr}\sum_n \alpha_{n0e}^{(2)} \dot{\tilde{h}}_n^{(2)}(kr) P_n^1(\cos\theta), \\
H_\varphi &= \frac{1}{j\eta kr}\sum_n \beta_{n0e}^{(2)} \tilde{h}_n^{(2)}(kr) P_n^1(\cos\theta).
\end{aligned}$$

For sufficiently large r, the radiation intensity may be represented by

$$\begin{aligned}
\frac{1}{2} r^2 \operatorname{Re} \mathbf{E} \times \overline{\mathbf{H}} \cdot \mathbf{u}_r &= \frac{1}{2} r^2 \operatorname{Re}\left(E_\theta \overline{H}_\varphi - E_\varphi \overline{H}_\theta\right) \\
&= \frac{1}{2\eta k^2}\left(\left|\sum_n j^n \beta_{n0e}^{(2)} P_n^1(\cos\theta)\right|^2 + \left|\sum_n j^{n+1} \alpha_{n0e}^{(2)} P_n^1(\cos\theta)\right|^2\right).
\end{aligned}$$

By use of (4.112), the directivity for an omnidirectional antenna can thus be expressed by

$$G = 4\pi \frac{\left|\sum\limits_{n=1}^{\infty} j^n \beta_{n0e}^{(2)} P_n^1(0)\right|^2 + \left|\sum\limits_{n=1}^{\infty} j^{n+1} \alpha_{n0e}^{(2)} P_n^1(0)\right|^2}{\sum\limits_{n,m,l} N_{nm}^2 \left(\left|\alpha_{nml}^{(2)}\right|^2 + \left|\beta_{nml}^{(2)}\right|^2\right)}. \tag{4.259}$$

It follows from (4.208) and (4.259) that

$$\left.\frac{G}{Q}\right|_{omn} = 8\pi \frac{\left|\sum_n j^n \beta_{n0e}^{(2)} P_n^1(0)\right|^2 + \left|\sum_n j^{n+1} \alpha_{n0e}^{(2)} P_n^1(0)\right|^2}{\sum_{n,m,l} N_{nm}^2 \left(\left|\alpha_{nml}^{(2)}\right|^2 + \left|\beta_{nml}^{(2)}\right|^2\right)(Q_n + Q_n')}. \tag{4.260}$$

Only $\alpha_{n0e}^{(2)}$ and $\beta_{n0e}^{(2)}$ contribute to the numerator. As a result, the ratio (4.260) can be increased by setting $\alpha_{nml}^{(2)} = \beta_{nml}^{(2)} = 0 \quad (m \neq 0)$, $\alpha_{n0o}^{(2)} = \beta_{n0o}^{(2)} = 0$. By letting $A_n = j^{n+1}\alpha_{n0e}^{(2)}$ and $B_n = j^n \beta_{n0e}^{(2)}$, (4.260) becomes

$$\left.\frac{G}{Q}\right|_{omn} = 8\pi \frac{\left|\sum_{n=1}^{\infty} A_n P_n^1(0)\right|^2 + \left|\sum_{n=1}^{\infty} B_n P_n^1(0)\right|^2}{\sum_{n=1}^{\infty} N_{n0}^2 \left(|A_n|^2 + |B_n|^2\right)(Q_n + Q_n')}. \tag{4.261}$$

Since the denominator of (4.261) depends only on the magnitude of A_n and B_n, the denominator is not changed if the phases of A_n and B_n are adjusted to maximize the ratio of gain to Q. If the phases of A_n and B_n are chosen to be the negative of $P_n^1(0)$, the terms in the numerator will be added in phase. Thus, (4.261) can be maximized as

$$\left.\frac{G}{Q}\right|_{omn} = 8\pi \frac{\left(\sum_{n=1}^{\infty} |A_n|\left|P_n^1(0)\right|\right)^2 + \left(\sum_{n=1}^{\infty} |B_n|\left|P_n^1(0)\right|\right)^2}{\sum_{n=1}^{\infty} N_{n0}^2 \left(|A_n|^2 + |B_n|^2\right)(Q_n + Q_n')}.$$

Setting $a_n = |A_n|\left|P_n^1(0)\right|$, $b_n = |B_n|\left|P_n^1(0)\right|$, one obtains

$$\begin{aligned}
\left.\frac{G}{Q}\right|_{omn} &= 8\pi \frac{\left(\sum_{n=1}^{\infty} a_n\right)^2 + \left(\sum_{n=1}^{\infty} b_n\right)}{\sum_{n=1}^{\infty}\left(\frac{N_{n0}^2 (a_n^2 + b_n^2)(Q_n + Q_n')}{\left|P_n^1(0)\right|^2}\right)} \\
&= 8\pi \frac{(\boldsymbol{\xi}, \mathbf{D}_a)_E^2 + (\boldsymbol{\xi}, \mathbf{D}_b)_E^2}{(\mathbf{D}_a, \mathbf{D}_a)_E + (\mathbf{D}_b, \mathbf{D}_b)_E} \leq 8\pi \|\boldsymbol{\xi}\|_E^2,
\end{aligned} \tag{4.262}$$

where $\boldsymbol{\xi} = (\xi_1, \xi_2, \ldots)$ and $\mathbf{D}_{a(b)} = (\mathbf{D}_{a(b)1}, \mathbf{D}_{a(b)2}, \ldots)$with

$$\xi_n = \frac{\left|P_n^1(0)\right|}{N_{n0}\sqrt{Q_n + Q_n'}}, D_{an} = \frac{|a_n|}{\xi_n}, D_{bn} = \frac{|b_n|}{\xi_n}.$$

The ratio (4.262) reaches maximum if $\mathbf{D}_a = \mathbf{D}_b = c_1\boldsymbol{\xi}$. As a result, the upper bound of the ratio of gain to QF for an omnidirectional antenna is

$$\max \left.\frac{G}{Q}\right|_{omn} = 8\pi\|\xi\|_E^2 = \sum_{n=1}^{\infty} \frac{2(2n+1)\left|P_n^1(0)\right|^2}{n(n+1)\left(Q_n + Q_n'\right)}. \tag{4.263}$$

Remark 4.8 The same upper bounds (4.258) and (4.263) may be obtained if the traditional definition (4.31) for QF is used [85]. A rigorous approach based on (4.31) is to use an inequality invented by the author in [86]

$$\frac{\sum_{n=1}^{N}(x_n + y_n)}{\sum_{n=1}^{N}(a_n x_n + b_n y_n)} \le 2,$$

where $x_n \ge y_n > 0$, $a_n > b_n > 0$, and $a_n + b_n = 1$. □

Remark 4.9 Chu has shown that the maximum ratio of gain to QF for an omnidirectional antenna is [81]

$$\max \left.\frac{G}{Q}\right|_{omn}^{Chu} = \sum_{n=1}^{N} \frac{(2n+1)\left|P_n^1(0)\right|^2}{n(n+1)Q_n^{Chu}}. \tag{4.264}$$

In the above, Q_n^{Chu} stands for the QF of nth TM modes and is a function of ka, which is different from the Q_n in (4.263). Also note that Q_n' does not appear in Chu's limit (4.264). Chu only studied an omnidirectional antenna that radiates either TE or TM modes, and his derivation relied on the truncation of the infinite series expansions for various quantities related to the stored field energy to avoid the infinity problem with the total field energy outside the circumscribing sphere of antenna. □

4.5.3 Best Possible Antenna Performance-Guidelines for Small Antenna Design

It is known that the larger the antenna size, the better its performance. In many cases, antenna must be made small due to the space limitations placed by the application environment. For this reason, understanding various limits of antenna performances are of fundamental importance. It will now be shown how the upper bounds on the PGB can be used as a guideline for antenna design. As proved in [80], the antenna fractional bandwidth B_f is reciprocal to antenna Q_{real} if Q_{real} is not very small. For this reason, the PGB can be expressed as $GB_f \approx G/Q_{real}$. Since the antenna QF in (4.258) and (4.263) has excluded the stored field energies inside the circumscribing sphere of antenna, it is smaller than the antenna Q_{real}.

It follows from (4.258) and (4.263) that the PGBs for an arbitrary antenna of dimension $2a$ are bounded by

$$GB_f\big|_{dir} \le \max GB_f\big|_{dir} = \sum_{n=1}^{\infty} \frac{2(2n+1)}{Q_n(ka) + Q'_n(ka)},$$
$$GB_f\big|_{omn} \le \max GB_f\big|_{omn} = \sum_{n=1}^{\infty} \frac{2(2n+1)\left|P_n^1(0)\right|^2}{n(n+1)\left[Q_n(ka) + Q'_n(ka)\right]}. \quad (4.265)$$

The first expression applies for directional antenna, and the second for the omnidirectional antenna. It should be notified that the right-hand sides of (4.265) are finite numbers. Equation (4.248) shows that the fractional bandwidth of an arbitrary antenna of dimension $2a$ has an upper bound

$$B_f \le \max B_f = \frac{2(ka)^3}{2(ka)^2 + 1}. \quad (4.266)$$

The expressions in (4.265) indicate that there exists a trade-off between the gain and bandwidth in order to achieve a specified value of GB_f. One can sacrifice gain to enhance bandwidth, but the latter is limited by the upper bound (4.266). One can also sacrifice bandwidth to get a high gain antenna. Theoretically, there is no upper bound on the antenna gain, and a super gain antenna can be achieved if the bandwidth becomes very narrow.

The upper bounds $\max GB_f|_{dir}$, $\max GB_f|_{omn}$, and $\max B_f$ are all monotonically increasing function of ka. Therefore, the antenna performance intends to improve as the antenna size increases. The upper bounds (4.265) and (4.266) are plotted in Figure 4.15, and they are the guidelines for various antenna designs. When an antenna project is initiated, the first question that needs to be answered is how much space is required to accommodate the antenna so that the prescribed antenna performance can be achieved. The solution can be easily found from Figure 4.15. For example, if one needs to design a directional antenna with $GB_f|_{dir} = 10$, the minimum antenna size must satisfy $ka > 2.25$ as illustrated in the figure.

Analytical solutions to the minimum antenna size that is required to achieve a prescribed performance can be derived from (4.265) for small antennas. Various definitions exist for small antennas, such as [87–92]

$$ka < 0.5, \quad (4.267)$$

or

$$ka < 1. \quad (4.268)$$

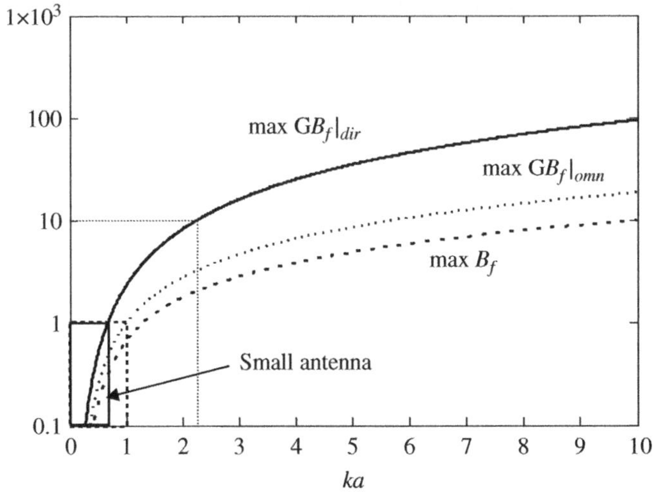

Figure 4.15 Upper bounds of antenna performances.

In other words, an antenna is said to be small if its maximum size is less than $\lambda/2\pi$ or λ/π. It is noted that higher order spherical modes are evanescent for antennas of the size λ/π [90].

Since the PGB is the most important performance index and is monotonically increasing with the antenna size, the definition of small antenna can also be given in terms of the PGB. One may propose that an antenna is said to be small if its PGB is less than one

$$GB_f < 1. \tag{4.269}$$

For a directive antenna, it can be found from (4.265) that the condition (4.269) requires

$$ka|_{dir} < 0.667, \tag{4.270}$$

while for an omnidirectional antenna, it implies

$$ka|_{omn} < 1. \tag{4.271}$$

Note that the definition (4.271) for small omnidirectional antenna is identical with the usual definition (4.268).

For small antennas, only the first term in either expression of (4.265) is dominant. According to (4.265) and (4.266), the PGBs for the small directional and omnidirectional antennas, and the fractional bandwidth for an arbitrary antenna are bounded by

$$GB_f\Big|_{dir}^{small} \le \frac{6(ka)^3}{2(ka)^2 + 1},$$
$$GB_f\Big|_{omn}^{small} \le \frac{3(ka)^3}{2(ka)^2 + 1}, \tag{4.272}$$
$$B_f \le \frac{2(ka)^3}{2(ka)^2 + 1}.$$

It can be seen that the maximum possible PGB for a small directional antenna is twice that for a small omnidirectional antenna of the same size. The above inequalities show that the maximum possible gain for a small directional (or omnidirectional) antenna is 3 (or 1.5) once its bandwidth has reached the maximum possible value. By solving the inequalities in (4.272), one finds

$$ka \ge \frac{2}{3}P + \sqrt[3]{\frac{8}{27}P^3 + \frac{1}{2}P + P\sqrt{\frac{8}{27}P^2 + \frac{1}{4}}} + \sqrt[3]{\frac{8}{27}P^3 + \frac{1}{2}P - P\sqrt{\frac{8}{27}P^2 + \frac{1}{4}}}, \tag{4.273}$$

where

$$P = \begin{cases} \frac{1}{6}GB_f\Big|_{dir}^{small}, & \text{for small directional antenna} \\ \frac{1}{3}GB_f\Big|_{omn}^{small}, & \text{for small omnidirectional antenna} \\ \frac{1}{2}B_f, & \text{for small antenna} \end{cases}.$$

The right-hand side of (4.273) gives the minimum antenna size required to achieve the specified PGB or bandwidth alone.

Using (4.269) as the definition of small antenna is very convenient in small antenna design. In practice, the gain and bandwidth are usually specified, from which one can immediately tell if the antenna under design is small or not.

Example 4.8 Consider the design of a directional antenna with gain required to be greater than -1 dBi, covering the sub-6 GHz frequency band N78: 3.3 – 3.8 GHz. From the requirements of gain and bandwidth, one may find $G = 0.79$ and $B_f = (3.8 - 3.3)/3.55 = 0.14$. The wavelength at the center frequency 3.55 GHz is $\lambda_0 = 8.45$ cm. The PGB to be realized is

$$GB_f\big|_{dir} = 0.794 \times 0.141 = 0.112.$$

Therefore, the antenna under design falls into the category of small antennas. Plugging the value of $GB_f|_{dir}$ into (4.273) yields $ka > 0.278$. The minimum antenna diameter required to achieve the specified gain and bandwidth must then be greater than $2a = 0.278\lambda_0/\pi = 7.5$ mm. □

Example 4.9 Consider wristwatch with built-in Global Positioning System (GPS) capability. The receiving frequency for GPS watch is 1575.42 MHz. The embedded GPS antenna is placed inside the watch and is assumed to be directive. The maximum possible size of the antenna is approximately equal to the diameter of the watch denoted by $2a$. The typical size of the watch is $a = 0.1\lambda$, which gives $ka = 0.628$. The best overall antenna performances one can achieve for the watch are $G = 3$ and $B_f = 0.28$. The maximum possible antenna bandwidth is 436 MHz. Since the bandwidth requirement for a GPS antenna is 2 MHz, there is potential to trade more gain from the bandwidth. □

It is noted that the performance of small antenna is very sensitive to its size. As indicated by Figure 4.15, the curves for the PGB have a big positive slope when antenna is small ($ka < 1$), which implies that a significant improvement of antenna performance can be achieved for a small increment of antenna size or a severe performance degradation may be resulted from a slight reduction of antenna size. This phenomenon has been widely observed in the process of small antenna designs. As the antenna size increases, the slope of the curves drops quickly outside the region of small antenna ($ka > 1$) and the rate of increase of the PGB becomes very slow. For this reason, a reasonable decision on the antenna size must be made so that the antenna performance can be achieved while the design is kept cost effective.

4.6 Expansions of the Radiated Fields in Time Domain

Since the vector modal functions in the spherical waveguide are independent of frequency, they can be equally applied to study the radiation problems in time domain. The treatment in time domain is similar to the time-harmonic theory for the spherical waveguide. In the spherical coordinate system (r, θ, φ), the time-domain EM fields can be decomposed into the transverse components and the radial components

$$\mathbf{E} = \mathbf{E}_t + \mathbf{u}_r E_r, \mathbf{H} = \mathbf{H}_t + \mathbf{u}_r H_r.$$

Taking the vector and scalar product of the time-domain Maxwell equations

$$\nabla \times \mathbf{H}(\mathbf{r}, t) = \varepsilon \frac{\partial \mathbf{E}(\mathbf{r}, t)}{\partial t} + \mathbf{J}(\mathbf{r}, t),$$
$$\nabla \times \mathbf{E}(\mathbf{r}, t) = -\mu \frac{\partial \mathbf{H}(\mathbf{r}, t)}{\partial t} - \mathbf{J}_m(\mathbf{r}, t),$$

with the vector $\mathbf{r}$, one obtains two sets of equations which connect the transverse and longitudinal field components

$$\begin{aligned}\varepsilon\frac{\partial}{\partial t}(\mathbf{r}\times\mathbf{E})+\mathbf{r}\times\mathbf{J}&=\nabla(\mathbf{r}\cdot\mathbf{H})-(\mathbf{r}\cdot\nabla)\mathbf{H}-\mathbf{H},\\-\mu\frac{\partial}{\partial t}(\mathbf{r}\times\mathbf{H})-\mathbf{r}\times\mathbf{J}_m&=\nabla(\mathbf{r}\cdot\mathbf{E})-(\mathbf{r}\cdot\nabla)\mathbf{E}-\mathbf{E},\end{aligned}\tag{4.274}$$

$$\begin{aligned}-\nabla\cdot(\mathbf{r}\times\mathbf{H}_t)&=\varepsilon\frac{\partial(\mathbf{r}\cdot\mathbf{E})}{\partial t}+\mathbf{r}\cdot\mathbf{J},\\\nabla\cdot(\mathbf{r}\times\mathbf{E}_t)&=\mu\frac{\partial(\mathbf{r}\cdot\mathbf{H})}{\partial t}+\mathbf{r}\cdot\mathbf{J}_m.\end{aligned}\tag{4.275}$$

Comparing the transverse components of (4.274) yields

$$\begin{aligned}\frac{1}{r}\nabla_{\theta\varphi}(\mathbf{r}\cdot\mathbf{H})-r\frac{\partial\mathbf{H}_t}{\partial r}-\mathbf{H}_t&=\varepsilon\frac{\partial}{\partial t}(\mathbf{r}\times\mathbf{E}_t)+\mathbf{r}\times\mathbf{J}_t,\\\frac{1}{r}\nabla_{\theta\varphi}(\mathbf{r}\cdot\mathbf{E})-r\frac{\partial\mathbf{E}_t}{\partial r}-\mathbf{E}_t&=-\mu\frac{\partial}{\partial t}(\mathbf{r}\times\mathbf{H}_t)-\mathbf{r}\times\mathbf{J}_{mt},\end{aligned}\tag{4.276}$$

while (4.275) can be written as

$$\begin{aligned}-\frac{1}{r}\nabla_{\theta\varphi}\cdot(\mathbf{r}\times\mathbf{H}_t)&=\varepsilon\frac{\partial(\mathbf{r}\cdot\mathbf{E})}{\partial t}+\mathbf{r}\cdot\mathbf{J},\\\frac{1}{r}\nabla_{\theta\varphi}\cdot(\mathbf{r}\times\mathbf{E}_t)&=\mu\frac{\partial(\mathbf{r}\cdot\mathbf{H})}{\partial t}+\mathbf{r}\cdot\mathbf{J}_m.\end{aligned}\tag{4.277}$$

Similar to the time-harmonic theory, the transverse and longitudinal field components in time domain may be expanded by the vector modal functions as follows:

$$\begin{aligned}r\mathbf{E}_t(\mathbf{r},t)&=\sum_{n,m,l}\left[V_{nml}^{TM}(r,t)\mathbf{e}_{nml}(\theta,\varphi)+V_{nml}^{TE}(r,t)\mathbf{h}_{nml}(\theta,\varphi)\right],\\r\mathbf{H}_t(\mathbf{r},t)&=\sum_{n,m,l}\left[I_{nml}^{TM}(r,t)\mathbf{h}_{nml}(\theta,\varphi)-I_{nml}^{TE}(r,t)\mathbf{e}_{nml}(\theta,\varphi)\right],\end{aligned}\tag{4.278}$$

$$\begin{aligned}\mathbf{r}\cdot\mathbf{E}(\mathbf{r},t)&=\sum_{n,m,l}\frac{1}{N_{nm}}E_{nml}(r,t)Y_{nml}(\theta,\varphi),\\\mathbf{r}\cdot\mathbf{H}(\mathbf{r},t)&=\sum_{n,m,l}\frac{1}{N_{nm}}H_{nml}(r,t)Y_{nml}(\theta,\varphi).\end{aligned}\tag{4.279}$$

Taking the vector product of (4.278) with $\mathbf{u}_r$ yields

$$\begin{aligned}\mathbf{r}\times\mathbf{E}_t(\mathbf{r},t)&=\sum_{n,m,l}\left[V_{nml}^{TM}(r,t)\mathbf{h}_{nml}(\theta,\varphi)-V_{nml}^{TE}(r,t)\mathbf{e}_{nml}(\theta,\varphi)\right],\\\mathbf{r}\times\mathbf{H}_t(\mathbf{r},t)&=\sum_{n,m,l}\left[-I_{nml}^{TM}(r,t)\mathbf{e}_{nml}(\theta,\varphi)-I_{nml}^{TE}(r,t)\mathbf{h}_{nml}(\theta,\varphi)\right].\end{aligned}\tag{4.280}$$

Substituting (4.278)–(4.280) into (4.276) and (4.277), one may find

$$
\begin{aligned}
&\sum_{n,m,l}\frac{1}{r}H_{nml}\mathbf{e}_{nml}-\sum_{n,m,l}\left(\frac{\partial I_{nml}^{TM}}{\partial r}\mathbf{h}_{nml}-\frac{\partial I_{nml}^{TE}}{\partial r}\mathbf{e}_{nml}\right)\\
&=\sum_{n,m,l}\varepsilon\left(\frac{\partial V_{nml}^{TM}}{\partial t}\mathbf{h}_{nml}-\frac{\partial V_{nml}^{TE}}{\partial t}\mathbf{e}_{nml}\right)+\mathbf{r}\times\mathbf{J}_t,\\
&\sum_{n,m,l}\frac{1}{r}E_{nml}\mathbf{e}_{nml}-\sum_{n,m,l}\left(\frac{\partial V_{nml}^{TM}}{\partial r}\mathbf{e}_{nml}+\frac{\partial V_{nml}^{TE}}{\partial r}\mathbf{h}_{nml}\right)\\
&=\sum_{n,m,l}\mu\left(\frac{\partial I_{nml}^{TM}}{\partial t}\mathbf{e}_{nml}+\frac{\partial I_{nml}^{TE}}{\partial t}\mathbf{h}_{nml}\right)-\mathbf{r}\times\mathbf{J}_{mt},
\end{aligned}
\tag{4.281}
$$

$$
\begin{aligned}
&-\frac{1}{r}\sum_{n,m,l}\frac{1}{N_{nm}}I_{nml}^{TM}n(n+1)Y_{nml}=\varepsilon\frac{\partial}{\partial t}\sum_{n,m,l}\frac{1}{N_{nm}}E_{nml}Y_{nml}+\mathbf{r}\cdot\mathbf{J},\\
&\frac{1}{r}\sum_{n,m,l}\frac{1}{N_{nm}}V_{nml}^{TE}n(n+1)Y_{nml}=\mu\frac{\partial}{\partial t}\sum_{n,m,l}\frac{1}{N_{nm}}H_{nml}Y_{nml}+\mathbf{r}\cdot\mathbf{J}_m.
\end{aligned}
\tag{4.282}
$$

By equating the coefficients before the vector modal functions and making use of (4.47), the equations for the expansion coefficients in (4.278) and (4.279) can be found as follows:

$$
\begin{aligned}
&\frac{1}{r}H_{nml}+\frac{\partial I_{nml}^{TE}}{\partial r}+\varepsilon\frac{\partial V_{nml}^{TE}}{\partial t}=\int_{S'}\mathbf{r}\times\mathbf{J}\cdot\mathbf{e}_{nml}d\Omega,\\
&-\frac{\partial I_{nml}^{TM}}{\partial r}-\varepsilon\frac{\partial V_{nml}^{TM}}{\partial t}=\int_{S'}\mathbf{r}\times\mathbf{J}\cdot\mathbf{h}_{nml}d\Omega,\\
&\frac{1}{r}E_{nml}-\frac{\partial V_{nml}^{TM}}{\partial r}-\mu\frac{\partial I_{nml}^{TM}}{\partial t}=-\int_{S'}\mathbf{r}\times\mathbf{J}_m\cdot\mathbf{e}_{nml}d\Omega,\\
&-\frac{\partial V_{nml}^{TE}}{\partial r}-\mu\frac{\partial I_{nml}^{TE}}{\partial t}=-\int_{S'}\mathbf{r}\times\mathbf{J}_m\cdot\mathbf{h}_{nml}d\Omega,\\
&\frac{1}{r}n(n+1)I_{nml}^{TM}+\varepsilon\frac{\partial}{\partial t}E_{nml}=\int_{S'}(\mathbf{r}\cdot\mathbf{J})\nabla_{\theta\varphi}\cdot\mathbf{e}_{nml}d\Omega,\\
&\frac{1}{r}n(n+1)V_{nml}^{TE}-\mu\frac{\partial}{\partial t}H_{nml}=-\int_{S'}(\mathbf{r}\cdot\mathbf{J}_m)\nabla_{\theta\varphi}\cdot\mathbf{e}_{nml}d\Omega,
\end{aligned}
\tag{4.283}
$$

where Ω is a sphere of unit radius and $d\Omega$ is the differential element of the solid angle. The equations for V_{nml}^{TE} and I_{nml}^{TM} can be derived from (4.283) by eliminating other expansion coefficients

$$\left[\frac{\partial^2}{\partial r^2} - \frac{1}{c^2}\frac{\partial^2}{\partial t^2} - \frac{n(n+1)}{r^2}\right]V_{nml}^{TE} = -\mu\frac{\partial}{\partial t}\int_{S'} \mathbf{r}\times\mathbf{J}\cdot\mathbf{e}_{nml}d\Omega$$
$$+\frac{1}{r}\int_{S'}(\mathbf{r}\cdot\mathbf{J}_m)\nabla_{\theta\varphi}\cdot\mathbf{e}_{nml}d\Omega + \frac{\partial}{\partial r}\int_{S'}\mathbf{r}\times\mathbf{J}_m\cdot\mathbf{h}_{nml}d\Omega,$$
$$\left[\frac{\partial^2}{\partial r^2} - \frac{1}{c^2}\frac{\partial^2}{\partial t^2} - \frac{n(n+1)}{r^2}\right]I_{nml}^{TM} = -\varepsilon\frac{\partial}{\partial t}\int_{S'}\mathbf{r}\times\mathbf{J}_m\cdot\mathbf{e}_{nml}(\theta,\varphi)d\Omega$$
$$-\frac{1}{r}\int_{S'}(\mathbf{r}\cdot\mathbf{J})\nabla_{\theta\varphi}\cdot\mathbf{e}_{nml}d\Omega - \frac{\partial}{\partial r}\int_{S'}\mathbf{r}\times\mathbf{J}\cdot\mathbf{h}_{nml}d\Omega. \tag{4.284}$$

where $c = 1/\sqrt{\mu\varepsilon}$. The equations in (4.284) are the **time-domain spherical transmission line equations**. To solve these equations, the method of Green's function can be used. Consider the retarded Green's function defined by

$$\left[\frac{\partial^2}{\partial r^2} - \frac{1}{c^2}\frac{\partial^2}{\partial t^2} - \frac{n(n+1)}{r^2}\right]G_n(r,t;r',t') = -\delta(r-r')\delta(t-t'). \tag{4.285}$$

Taking the Fourier transform with respect to time t gives

$$\left[\frac{\partial^2}{\partial r^2} + k^2 - \frac{n(n+1)}{r^2}\right]\tilde{G}_n(r,\omega;r',t') = -\delta(r-r')e^{-j\omega t'}, \tag{4.286}$$

where $k = \omega/c$. The corresponding homogeneous equation has two independent solutions $x_1 = rj_n(kr)$ and $x_2 = rn_n(kr)$, where $j_n(kr)$ and $n_n(kr)$ are spherical Bessel functions of first and second kind, respectively. Therefore, the solution of (4.286) can be written as

$$\tilde{G}_n(r,\omega;r',t') = \begin{cases} a_1x_1 + a_2x_2, r < r' \\ b_1x_1 + b_2x_2, r > r' \end{cases}$$

The constants $a_i(i = 1, 2)$ and $b_i(i = 1, 2)$ may be determined by the behavior of $\tilde{G}_n(r,\omega;r',t')$ at $r = r'$. Ignoring the details, the solution is

$$\tilde{G}_n(r,\omega;r',t') = -jke^{-j\omega t'}rr'j_n(kr_<)h_n^{(2)}(kr_>), \tag{4.287}$$

where $h_n^{(2)}(kr)$ is the spherical Hankel function of the second kind, and $r_<$ ($r_>$) designates the minimum (maximum) of r and r'. By use of the expansion for the free-space Green's function

$$\frac{e^{-jk|\mathbf{r}-\mathbf{r}'|}}{4\pi|\mathbf{r}-\mathbf{r}'|} = -\frac{jk}{4\pi}\sum_{n=0}^{\infty}(2n+1)j_n(kr_<)h_n^{(2)}(kr_>)P_n(\cos\gamma), \tag{4.288}$$

where γ is the angle between $\mathbf{r}$ and $\mathbf{r}'$ and $P_n(\gamma)$ is Legendre polynomial, one may find

$$-jkj_n(kr_<)h_n^{(2)}(kr_>) = \frac{1}{2}\int_{-1}^{1}\frac{e^{-jk|\mathbf{r}-\mathbf{r}'|}}{|\mathbf{r}-\mathbf{r}'|}P_n(x)dx, \tag{4.289}$$

with $x = \cos\gamma$. Thus, (4.287) can be written as

$$\tilde{G}_n(r,\omega;r',t') = \frac{rr'e^{-j\omega t'}}{2}\int_{-1}^{1}\frac{e^{-jk|\mathbf{r}-\mathbf{r}'|}}{|\mathbf{r}-\mathbf{r}'|}P_n(x)dx. \tag{4.290}$$

Taking the inverse Fourier transform gives

$$\begin{aligned} G_n(r,t;r',t') &= \frac{1}{2\pi}\int_{-\infty}^{\infty}\tilde{G}_n(r,\omega;r',t')e^{j\omega t}d\omega \\ &= \frac{rr'}{4\pi}\int_{-1}^{1}\frac{P_n(x)}{|\mathbf{r}-\mathbf{r}'|}dx\int_{-\infty}^{\infty}e^{j\omega(t-t')-j\omega|\mathbf{r}-\mathbf{r}'|/c}d\omega \\ &= \frac{rr'}{2}\int_{-1}^{1}\frac{P_n(x)}{|\mathbf{r}-\mathbf{r}'|}\delta\left(t-t'-\frac{1}{c}|\mathbf{r}-\mathbf{r}'|\right)dx. \end{aligned} \tag{4.291}$$

Let $u = |\mathbf{r}-\mathbf{r}'|/c = (r^2+r'^2-2rr'x)^{1/2}/c$. Then,

$$x = \frac{r^2+r'^2-c^2u^2}{2rr'}, dx = -\frac{cudu}{rr'}.$$

As a result, (4.291) can be expressed as

$$\begin{aligned} G_n(r,t;r',t') &= \frac{1}{2}\int_{|r-r'|/c}^{|r+r'|/c}P_n\left(\frac{r^2+r'^2-c^2u^2}{2rr'}\right)\delta(t-t'-u)du \\ &= \frac{1}{2}P_n\left[\frac{r^2+r'^2-c^2(t-t')^2}{2rr'}\right]\left[H\left(t-t'-\frac{1}{c}|r-r'|\right)-H\left(t-t'-\frac{1}{c}|r+r'|\right)\right]. \end{aligned} \tag{4.292}$$

where $H(t)$ is the unit step function. Expression (4.292) is the Green's function for the spherical transmission line equation. Note that the Green's function satisfies the causality condition. The solution of the inhomogeneous spherical transmission line equation

$$\left[\frac{\partial^2}{\partial r^2}-\frac{1}{c^2}\frac{\partial^2}{\partial t^2}-\frac{n(n+1)}{r^2}\right]u_n(r,t) = f_n(r,t). \tag{4.293}$$

is readily found by using the Green's function (4.292)

$$u_n(r,t) = -\int_0^{\infty} dr' \int_{-\infty}^{\infty} G_n(r,t;r',t') f_n(r',t') dt'. \tag{4.294}$$

According to (4.294), the solutions of the equations in (4.284) are given by

$$\begin{aligned} V_{nml}^{TE}(r,t) = \mu \int_0^{\infty} dr' \int_{-\infty}^{\infty} dt' G_n(r,t;r',t') \\ \times \left\{ \int_{\Omega} \left[\frac{\partial}{\partial t'} [\mathbf{r}' \times \mathbf{J}(\mathbf{r}',t')] + \frac{1}{\mu r'} \nabla_{\theta'\varphi'} [\mathbf{r}' \cdot \mathbf{J}_m(\mathbf{r}',t')] \right] \cdot \mathbf{e}_{nml}(\theta',\varphi') d\Omega(\mathbf{r}') \right. \\ \left. - \frac{1}{\mu} \frac{\partial}{\partial r'} \int_{\Omega} [\mathbf{r}' \times \mathbf{J}_m(\mathbf{r}',t')] \cdot \mathbf{h}_{nml}(\theta',\varphi') d\Omega(\mathbf{r}') \right\}, \end{aligned} \tag{4.295}$$

$$\begin{aligned} I_{nml}^{TM}(r,t) = -\varepsilon \int_0^{\infty} dr' \int_{-\infty}^{\infty} dt' G_n(r,t;r',t') \\ \times \left\{ \int_{\Omega} \left[-\frac{\partial}{\partial t'} [\mathbf{r}' \times \mathbf{J}_m(\mathbf{r}',t')] + \frac{1}{\varepsilon r'} \nabla_{\theta'\varphi'} [\mathbf{r}' \cdot \mathbf{J}(\mathbf{r}',t')] \right] \cdot \mathbf{e}_{nml}(\theta',\varphi') d\Omega(\mathbf{r}') \right. \\ \left. - \frac{1}{\varepsilon} \frac{\partial}{\partial r'} \int_{\Omega} [\mathbf{r}' \times \mathbf{J}(\mathbf{r}',t')] \cdot \mathbf{h}_{nml}(\theta',\varphi') d\Omega(\mathbf{r}') \right\}. \end{aligned} \tag{4.296}$$

Once the modal voltage V_{nml}^{TE} and the modal current I_{nml}^{TM} are determined, other modal voltages and currents can be readily determined from (4.283).

Example 4.10 (Steady-State Response)
Assume that the excitation source is of electric type and is a separable function of space and time

$$\mathbf{J}(\mathbf{r},t) = \mathbf{J}(\mathbf{r}) f(t), \quad \mathbf{J}_m(\mathbf{r},t) = 0.$$

Let $f(t) = H(t)\sin\omega t$ be a sinusoidal wave turned on at $t = 0$. To determine the steady-state responses of the modal voltage $V_{nml}^{TE}(r,t)$ and current $I_{nml}^{TM}(r,t)$, one needs to evaluate the following integrals:

$$\begin{aligned} I_n(r,t,r') &= \int_{-\infty}^{\infty} G_n(r,t;r',t')f(t')dt', \\ I'_n(r,t,r') &= \int_{-\infty}^{\infty} G_n(r,t;r',t')f'(t')dt'. \end{aligned} \tag{4.297}$$

On account of (4.292), the first expression of (4.297) can be written as

$$\begin{aligned} I_n(r,t,r') &= \int_{t-\frac{1}{c}|r+r'|}^{t-\frac{1}{c}|r-r'|} \frac{1}{2}P_n\left[\frac{r^2+r'^2-c^2(t-t')^2}{2rr'}\right]f(t')dt' \\ &= \int_{c^{-1}|r-r'|}^{c^{-1}|r+r'|} \frac{1}{2}P_n\left(\frac{r^2+r'^2-c^2u^2}{2rr'}\right)f(t-u)du. \end{aligned}$$

By use of the transformations

$$x = \frac{r^2+r'^2-c^2u^2}{2rr'}, u = \frac{1}{c}\sqrt{r^2+r'^2-2rr'x}, du = -\frac{rr'}{cu}dx,$$

one may find

$$I_n(r,t,r') = \int_{-1}^{1} \frac{1}{2}f\left(t-\frac{1}{c}\sqrt{r^2+r'^2-2rr'x}\right)\frac{rr'P_n(x)}{\sqrt{r^2+r'^2-2rr'x}}dx. \tag{4.298}$$

Similarly, one may obtain

$$I'_n(r,t,r') = \int_{-1}^{1} \frac{1}{2}f'\left(t-\frac{1}{c}\sqrt{r^2+r'^2-2rr'x}\right)\frac{rr'P_n(x)}{\sqrt{r^2+r'^2-2rr'x}}dx. \tag{4.299}$$

For the steady-state response, the time t may be assumed to be sufficiently large so that $t > c^{-1}|r+r'|$. Thus, $H\left(t-c^{-1}\sqrt{r^2+r'^2-2rr'x}\right) = 1$ and (4.298) becomes

$$
\begin{aligned}
I_n(r,t,r')|_{t>c^{-1}|r+r'|} &= \frac{rr'}{2}\int_{-1}^{1}\sin\left(\omega t - k\sqrt{r^2+r'^2-2rr'x}\right)\frac{P_n(x)}{\sqrt{r^2+r'^2-2rr'x}}dx \\
&= \frac{rr'}{2}\sin\omega t\int_{-1}^{1}\cos\left(k\sqrt{r^2+r'^2-2rr'x}\right)\frac{P_n(x)}{\sqrt{r^2+r'^2-2rr'x}}dx \\
&\quad -\frac{rr'}{2}\cos\omega t\int_{-1}^{1}\sin\left(k\sqrt{r^2+r'^2-2rr'x}\right)\frac{P_n(x)}{\sqrt{r^2+r'^2-2rr'x}}dx.
\end{aligned}
$$

The integrals can be carried out by using the following relations [93]:

$$
\begin{aligned}
&\int_{-1}^{1}\frac{\sin\left[\lambda\sqrt{a^2+b^2-2abx}\right]}{\sqrt{a^2+b^2-2abx}}P_n(x)dx = \frac{\pi}{\sqrt{ab}}J_{n+\frac{1}{2}}(\lambda a)J_{n+\frac{1}{2}}(\lambda b), a>0, b>0, \\
&\int_{-1}^{1}\frac{\cos\left[\lambda\sqrt{a^2+b^2-2abx}\right]}{\sqrt{a^2+b^2-2abx}}P_n(x)dx = -\frac{\pi}{\sqrt{ab}}J_{n+\frac{1}{2}}(\lambda a)Y_{n+\frac{1}{2}}(\lambda b), 0\le a<b,
\end{aligned}
$$

to yield

$$
I_n(r,t,r')|_{t>c^{-1}|r+r'|} = -\begin{cases} krr'j_n(kr)\,\mathrm{Re}\left[e^{j\omega t}h_n^{(2)}(kr')\right], r<r' \\ krr'j_n(kr')\,\mathrm{Re}\left[e^{j\omega t}h_n^{(2)}(kr)\right], r>r' \end{cases}. \tag{4.300}
$$

Similarly, one may find

$$
I'_n(r,t,r')\big|_{t>c^{-1}|r+r'|} = -\omega\begin{cases} krr'j_n(kr)\,\mathrm{Re}\left[e^{j\omega t}jh_n^{(2)}(kr')\right], r<r' \\ krr'j_n(kr')\,\mathrm{Re}\left[e^{j\omega t}jh_n^{(2)}(kr)\right], r>r' \end{cases}. \tag{4.301}
$$

If the observation point is outside the source region so that $r > r'$, by combining (4.295), (4.296), (4.300), and (4.301), the steady-state responses of the modal voltage and current may be found as follows:

$$
\begin{aligned}
V_{nml}^{TE}(r,t)\big|_{Steady} &= \mu\int_{V_0}\frac{1}{r'^2}I'_n(r,t,r')[\mathbf{r}'\times\mathbf{J}_t(\mathbf{r}')]\cdot\mathbf{e}_{nml}(\theta',\varphi')dV(\mathbf{r}') \\
&= \mathrm{Re}\left[e^{j\omega t}V_{nml}^{TE}(r)\right],
\end{aligned} \tag{4.302}
$$

$$
\begin{aligned}
I_{nml}^{TM}(r,t)\big|_{Steady} &= -\int_{V_0} \frac{1}{r'}\frac{dI_n(r,t,r')}{dr'}\mathbf{J}_t(\mathbf{r}')\cdot\mathbf{e}_{nml}(\theta',\varphi')dV(\mathbf{r}') \\
&\quad -\frac{n(n+1)}{N_{nm}}\int_{V_0}\frac{1}{r'^2}I_n(r,t,r')[\mathbf{u}_{r'}\cdot\mathbf{J}(\mathbf{r}')]Y_{nm}^{l}(\theta',\varphi')dV(\mathbf{r}') \\
&= \mathrm{Re}\left[e^{j\omega t}I_{nml}^{TM}(r)\right],
\end{aligned}
\tag{4.303}
$$

where $V_{nml}^{TE}(r)$ and $I_{nml}^{TM}(r)$ are the phasors

$$
V_{nml}^{TE}(r) = jk\eta\tilde{h}_n^{(2)}(kr)\int_{V_0} j_n(kr')\mathbf{J}_t(\mathbf{r}')\cdot\mathbf{h}_{nml}(\theta',\varphi')dV(\mathbf{r}'), \tag{4.304}
$$

$$
\begin{aligned}
I_{nml}^{TM}(r) &= \tilde{h}_n^{(2)}(kr)\int_{V_0}\frac{1}{r'}\dot{\tilde{j}}_n(kr')\mathbf{J}_t(\mathbf{r}')\cdot\mathbf{e}_{nml}(\mathbf{r}')dV(\mathbf{r}') \\
&\quad + \frac{n(n+1)}{N_{nm}}\tilde{h}_n^{(2)}(kr)\int_{V_0}\frac{1}{r'}j_n(kr')[\mathbf{u}_{r'}\cdot\mathbf{J}(\mathbf{r}')]Y_{nm}^{l}(\theta',\varphi')dV(\mathbf{r}').
\end{aligned}
\tag{4.305}
$$

These agree with (4.81) and (4.82), validating the time-domain theory. □

Physics is essentially an intuitive and concrete science. Mathematics is only a means for expressing the laws that govern phenomena.

–Albert Einstein

References

1 Aharoni, J., *Antennas*, Oxford, Clarendon Press, 1946.

2 Silver, S., *Microwave Antenna Theory and Design*, New York, Dover publications, 1949.

3 Schelkunoff, S. A., *Antennas: Theory and Practice*, John Wiley & Sons, 1952.

4 Müller, C., *Foundations of the Mathematical Theory of Electromagnetic Waves*, Springer, 1969.

5 Jones, D. S., *Methods in Electromagnetic Wave Propagation*, Oxford, Clarendon Press, 1979.

6 Stutzman, W. L. and G. A. Thiele, *Antenna Theory and Design*, New York, John Wiley & Sons, 1981.

7 Elliott, R. S., *Antenna Theory and Design*, New York, Prentice-Hall, 1981.

8 Popvić, B. D., M. B. Dragovic, and A. R. Djordjevic, *Analysis and Synthesis of Wire Antennas*, New York, Research Studies Press, John Wiley & Sons, 1982.

9 Lee, K. F., *Principles of Antenna Theory*, New York, John Wiley & Sons, 1984.

10 Collin, R. E., *Antennas and Radio Wave Propagation*, New York, McGraw-Hill, 1985.

11 Lo, Y. T. and S. W. Lee, *Antenna Handbook-Theory, Applications, and Design*, Van Nostrand Reinhold, 1988.

12 Harrington, R. F., *Field Computation by Moment Methods*, IEEE Press, 1993.

13 Hoop, T. A., *Handbook of Radiation and Scattering of Waves: Acoustic Waves in Fluids, Elastic Waves in Solids, Electromagnetic Waves*, Academic Press, 1995.

14 Balanis, C. A., *Antenna Theory: Analysis and Design*, 2nd Ed., John Wiley & Sons, 1997.

15 Kraus, J. D. and R. J. Marhefka, *Antennas for All Applications*, McGraw-Hill, 2001.

16 Fujimoto, K. and J. R. James, *Mobile Antenna Systems Handbook*, Artech House, 2001.

17 Luk, K. M., K. W. Leung, and J. R. James, *Dielectric Resonator Antennas*, Research Studies Press, 2002.

18 Milligan, T. A., *Modern Antenna Design*, 2nd Ed., Wiley, 2005.

19 Volakis, J. L., *Antenna Engineering Handbook*, 4th Ed., McGraw-Hill, 2007.

20 Rabinovich, V., N. Alexandrov, and B. Alkhateeb, *Automotive Antenna Design and Applications*, CRC Press, 2010.

21 Fang, D. G., *Antenna Theory and Microstrip Antennas*, CRC Press, 2010.

22 Zhang, Z., *Antenna Design for Mobile Devices*, Wiley, 2011.

23 Poisel, R., *Antenna Systems and Electronic Warfare Applications*, Artech House, 2012.

24 Chen, Z. N., D. Liu, H. Nakano, X. Qing, and T. Zwick, *Handbook of Antenna Technologies*, Springer, 2016.

25 Rahmat-Samii, Y. and E. Topsakal, *Antenna and Sensor Technologies in Modern Medical Applications*, Wiley, 2021.

26 Geyi, W., *Foundations of Applied Electrodynamics*, New York, Wiley, 2010.

27 Hansen, W. W., "A new type of expansion in radiation problems", *Phys. Rev.*, Vol. 47, pp. 139–143, 1935.

28 Stratton, J. A., *Electromagnetic Theory*, New York, McGraw-Hill, 1941.

29 Morse, P. M. and H. Feshbach, *Methods of Theoretical Physics*, McGraw-Hill, 1953.

30 Harrington, R. F., *Time-Harmonic Electromagnetic Fields*, McGraw-Hill Book Company, Inc., 1961.

31 Papas, C. H., *Theory of Electromagnetic Wave Propagation*, McGraw-Hill, 1965.

32 Jones, M. N., *Spherical Harmonics and Tensors for Classical Field Theory*, Research Studies Press, 1985.

33 Calderon, A. P., "The multiple expansion of radiation fields", *J. Rational Mech. Anal.*, Vol. 3, pp. 523–537, 1954.

34 Wilcox, C. H., "Debye potentials", *J. Math. Mech.*, Vol. 6, pp. 167–202, 1957.
35 Aydin, K. and A. Hizal, "On the completeness of the spherical vector wave functions", *J. Math. Anal. Appl.*, Vol. 117, pp. 428–440, 1986.
36 Bladel, J. V., *Electromagnetic Fields*, IEEE Press, 2007.
37 Wood, P. J., "Spherical harmonic expansions in near-field aerial problems", *Electron. Lett.*, Vol. 6, pp. 535–536, 1970.
38 Wood, P. J., "Spherical waves in antenna problems", *Marconi Rev.*, Vol. 34, pp. 149–172, 1971.
39 Rudge, A. W., K. Milne, A. D. Olver, and P. Knight, *Handbook of Antenna Design*, Vol. 1-2, Peter Peregrinus Ltd., 1982.
40 Geyi, W., "Stored electromagnetic field energies in general materials", *J. Opt. Soc. Am. B*, Vol. 36, No. 4, pp. 917–925, 2019.
41 Brillouin, L., *Wave Propagation and Group Velocity*, Academic Press, 1960.
42 Ginzburg, V. L., *Applications of Electrodynamics in Theoretical Physics and Astrophysics*, Chapter 13, CRC Press, 1989.
43 Ginzburg, V. L., *The Propagation of Electromagnetic Waves in Plasmas*, Pergman Press, 1964.
44 Landau, L. D., E. M. Lifshitz, and L. P. Pitaevskii, *Electrodynamics of Continuous Media*, 2nd Ed., Oxford, Pergamon Press, pp. 274–276, 1960.
45 Felsen, L. B. and N. Marcuvitz, *Radiation and Scattering of Waves*, IEEE Press, 1994.
46 Jackson, J. D., *Classical Electrodynamics*, 3rd Ed., New York, John Wiley & Sons, 1998.
47 Tonning, A., "Energy density in continuous electromagnetic media", *IEEE Trans. Antennas Propag.*, Vol. 8, pp. 428–434, 1960.
48 Zheng, X. and P. Palffy-Muhoray, "Electrical energy storage and dissipation in materials", *Phys. Lett. A*, Vol. 379, pp. 1853–1856, 2015.
49 Askne, J. and B. Lind, "Energy of electromagnetic waves in the presence of absorption and dispersion", *Phys. Rev. A*, Vol. 2, pp. 2335–2340, 1970.
50 Ziolkowski, R. W., "Superluminal transmission of information through an electromagnetic medium", *Phys. Rev. E*, Vol. 63, 046604, 2001.
51 Cui, T. J. and J. A. Kong, "Time-domain electromagnetic energy in a frequency-dispersive left-handed medium", *Phys. Rev. B*, Vol. 70, 205106, 2004.
52 Boardman, A. D. and K. Marinov, "Electromagnetic energy in a dispersive metamaterial", *Phys. Rev. B*, Vol. 73, 165110, 2006.
53 Luan, P. G., "Power loss and electromagnetic energy density in a dispersive metamaterial medium", *Phys. Rev. E*, Vol. 80, 046601, 2009.
54 Ruppin, R., "Electromagnetic energy density in a dispersive and absorptive material", *Phys. Lett. A*, Vol. 299, pp. 309–312, 2002.
55 Vorobyev, O. B., "Energy density of macroscopic electric and magnetic fields in dispersive medium with losses", *Prog. Electromagn. Res. B*, Vol. 40, pp. 343–360, 2012.

56 Loudon, R., “The propagation of electromagnetic energy through an absorbing dielectric”, *J. Phys. A: Gen. Phys.*, Vol. 3, No. 4, pp. 233–245, 1970.

57 Carcione, J. M., “On energy definition in electromagnetism: An analogy with viscoelasticity”, *J. Acoust. Soc. Am.*, Vol. 105, No. 2, pp. 626–632, 1999.Pt. 1

58 Webb, K. J. and S. Shivanand, “Electromagnetic field energy in dispersive materials”, *J. Opt. Soc. Am. B*, Vol. 27, pp. 1215–1220, 2010.

59 Nunes, F. D., T. C. Vasconcelos, M. Bezerra, and J. Weiner, “Electromagnetic energy density in dispersive and dissipative media”, *J. Opt. Soc. Am. B*, Vol. 28, pp. 1544–1552, 2011.

60 Tretyakov, S. A., “Electromagnetic field energy density in artificial microwave materials with strong dispersion and loss”, *Phys. Lett. A*, Vol. 343, pp. 231–237, 2005.

61 Ikonen, P. and S. Tretyakov, “Determination of generalized permeability function and field energy density in artificial magnetics using the equivalent circuit model”, *IEEE Trans. Microw. Theory Tech.*, Vol. 55, pp. 92–99, 2007.

62 Fung, P. C. W. and K. Young, “Electric energy density in a dissipative medium by circuit analog”, *Am. J. Phys.*, Vol. 46, pp. 57–59, 1978.

63 Geyi, W., “A derivation of stored electromagnetic field energies in an arbitrary medium”, *Tech Rxiv*, 2022, DOI:10.36227/techrxiv.19248254.v1.

64 Montgomery, C. G., R. H. Dicke, and E. M. Purcell, *Principles of Microwave Circuits*, McGraw-Hill, pp. 151–156, 1948.

65 Carlin, H. J., “Network theory without circuit elements”, *Proc. IEEE*, Vol. 55, pp. 482–497, 1967.

66 Penfield, P. Jr., R. Spencer, and S. Duinker, *Tellegen’s Theorem and Electrical Networks*, The MIT Press, pp. 63–67, 1970.

67 Geyi, W., *Foundations for Radio Frequency Engineering*, World Scientific, 2015.

68 Counter, V. A., “Miniature cavity antennas,” Rep. No. 2, Contract No. W28-099-ac-382, Microwave Lab., Stanford University, 1948.

69 Rhodes, D. R., “Observable stored energies of electromagnetic systems”, *J. Frankl. Inst.*, Vol. 302, No. 3, pp. 225–237, 1976.

70 Geyi, W., “Stored energies and radiation Q”, *IEEE Trans. Antennas Propag.*, Vol. 63, No. 2, pp. 636–645, 2015.

71 Geyi, W., “A method for the evaluation of small antenna Q”, *IEEE Trans. Antennas Propag.*, Vol. 51, pp. 2124–2129, 2003.

72 Vandenbosch, G. A. E., “Reactive energies, impedance, Q factor of radiating structures”, *IEEE Trans. Antennas Propag.*, Vol. 58, No. 4, pp. 1112–1127, 2010.

73 Geyi, W., “Reply to “Comments on “Stored Energies and Radiation Q”””, *IEEE Trans. Antennas Propag.*, Vol. 64, No. 10, pp. 4577–4580, 2016.

74 Geyi, W., “Optimization of the ratio of gain to Q”, *IEEE Trans. Antennas Propag.*, Vol. 61, No. 4, pp. 3488–3490, 2013.

75 Levis, C. A., “A reactance theorem for antennas”, *Proc. IRE*, Vol. 45, pp. 1128–1134, 1957.

76 Fante, R. L., "Quality factor of general idea antennas", *IEEE Trans. Antennas Propag.*, Vol. 17, pp. 151–155, 1969.

77 Yaghjian, A. D. and S. R. Best, "Impedance, bandwidth and Q of antennas", *IEEE Trans. Antennas Propag.*, Vol. 53, No. 4, pp. 1298–1324, 2005.

78 Colton, D. and R. Kress, *Inverse Acoustic and Electromagnetic Scattering Theory*, 2nd Ed., Springer, 32, 1998.

79 Jonsson, B. L. G. and M. Gustafsson, "Stored energies for electric and magnetic current densities", *arXiv: 1604.08572v2* [physics.class-ph], 2016.

80 Geyi, W., P. Jarmuszewski, and Y. Qi, "Foster reactance theorems for antennas and radiation Q", *IEEE Trans. Antennas Propag.*, Vol. 48, pp. 401–408, 2000.

81 Chu, L. J., "Physical limitations of omni-directional antennas", *J. Appl. Phys.*, Vol. 19, pp. 1163–1175, 1948.

82 Collin, R. E. and S. Rothschild, "Evaluation of antenna Q", *IEEE Trans. Antennas Propag.*, Vol. 12, pp. 23–27, 1964.

83 Watson, G. N., *A Treatise on the Theory of Bessel Functions*, Cambridge, UK, Cambridge University Press, pp. 294–301, 1922.

84 Fante, R. L. and J. T. Mayhan, "Bounds on the electric field outside a radiating system", *IEEE Trans. Antennas Propag.*, Vol. 16, pp. 712–717, 1968.

85 Geyi, W., "Physical limitations of antenna", *IEEE Trans. Antennas Propag.*, Vol. 51, pp. 2116–2123, 2003.

86 Geyi, W., "A New derivation of the upper bounds for the ratio of gain to Q", *IEEE Trans. Antennas Propag.*, Vol. 60, No. 7, pp. 1916–1922, 2012.

87 Volakis, J., C.-C. Chen, and K. Fujimoto, *Small Antennas: Miniaturization Techniques and Applications*, McGraw Hill, 2009.

88 Wheeler, H. A., "Small antennas", *IEEE Trans. Antennas Propag.*, Vol. 23, No. 4, pp. 462–469, 1975.

89 Fujimoto, K. and H. Morishita, *Modern Small Antennas*, Cambridge University Press, 2014.

90 Hansen, R. C., "Fundamental limitations in antennas", *Proc. IEEE*, Vol. 69, No. 2, pp. 170–182, 1981.

91 Hansen, R. C., *Electrically Small, Superdirective and Superconducting Antennas*, John Wiley & Sons, 2006.

92 Miron, D., *Small Antenna Design*, Elsevier, 2006.

93 Gradsheyn, L. S. and I. M. Ryzhik, *Tables of Integrals, Series, and Products*, Academic Press, 1994.

5

Radiation in Free Space (II)

Modal Analysis

The art of doing mathematics consists in finding that special case which contains all the germs of generality.

–David Hilbert (German mathematician, 1862–1943)

An antenna is equivalent to electric and/or magnetic current sources. Once the equivalent sources are known, the conventional antenna analysis is to express the radiated fields in terms of the known sources through integral representations. The sources are either specified or to be determined from an external incident field. In both cases, the sources are assumed to be independent of the radiated fields produced by themselves. For this reason, the antenna radiation theory is only valid approximately. In most cases, one must resort to various numerical techniques to first determine the current distributions. An alternative approach for antenna analysis is to treat the space as a spherical waveguide and express the radiated fields as a series expansion of spherical vector wave functions (SVWFs) or vector modal functions. Such an approach is justified by the fact that the antenna can be considered as a point source in the far-field region, and therefore the radiated fields by the antenna is the superposition of spherical waves.

It has been shown in Chapter 4 that the electromagnetic (EM) fields generated by current sources can be expanded in terms SVWFs. Using dyadic notations, the series expansions for the EM fields can be written in a compact form as [see (4.85)]

$$\begin{aligned}\mathbf{E}(\mathbf{r}) &= -jk\eta\int\limits_{V_0}\overleftrightarrow{\mathbf{G}}_e(\mathbf{r},\mathbf{r}')\cdot\mathbf{J}(\mathbf{r}')dV(\mathbf{r}') - \int\limits_{V_0}\overleftrightarrow{\mathbf{G}}_m(\mathbf{r},\mathbf{r}')\cdot\mathbf{J}_m(\mathbf{r}')dV(\mathbf{r}'),\\ \mathbf{H}(\mathbf{r}) &= -j\frac{k}{\eta}\int\limits_{V_0}\overleftrightarrow{\mathbf{G}}_e(\mathbf{r},\mathbf{r}')\cdot\mathbf{J}_m(\mathbf{r}')dV(\mathbf{r}') + \int\limits_{V_0}\overleftrightarrow{\mathbf{G}}_m(\mathbf{r},\mathbf{r}')\cdot\mathbf{J}(\mathbf{r}')dV(\mathbf{r}'),\end{aligned} \tag{5.1}$$

Foundations of Antenna Radiation Theory: Eigenmode Analysis, First Edition. Wen Geyi.

Published 2023 by John Wiley & Sons, Inc.

where

$$\overleftrightarrow{\mathbf{G}}_e(\mathbf{r},\mathbf{r}') = \sum_{n,m,l} \frac{-jk}{N_{nm}^2} \begin{cases} \mathbf{M}_{nml}^{(1)}(\mathbf{r})\mathbf{M}_{nml}^{(2)}(\mathbf{r}') + \mathbf{N}_{nml}^{(1)}(\mathbf{r})\mathbf{N}_{nml}^{(2)}(\mathbf{r}'), \ r < r' \\ \mathbf{M}_{nml}^{(2)}(\mathbf{r})\mathbf{M}_{nml}^{(1)}(\mathbf{r}') + \mathbf{N}_{nml}^{(2)}(\mathbf{r})\mathbf{N}_{nml}^{(1)}(\mathbf{r}'), \ r > r' \end{cases} \\ - \frac{1}{k^2}\delta(\mathbf{r}-\mathbf{r}')\mathbf{u}_r\mathbf{u}_{r'}, \tag{5.2}$$

$$\overleftrightarrow{\mathbf{G}}_m(\mathbf{r},\mathbf{r}') = \sum_{n,m,l} \frac{-jk^2}{N_{nm}^2} \begin{cases} \mathbf{N}_{nml}^{(1)}(\mathbf{r})\mathbf{M}_{nml}^{(2)}(\mathbf{r}') + \mathbf{M}_{nml}^{(1)}(\mathbf{r})\mathbf{N}_{nml}^{(2)}(\mathbf{r}'), \ r < r' \\ \mathbf{N}_{nml}^{(2)}(\mathbf{r})\mathbf{M}_{nml}^{(1)}(\mathbf{r}') + \mathbf{M}_{nml}^{(2)}(\mathbf{r})\mathbf{N}_{nml}^{(1)}(\mathbf{r}'), \ r > r' \end{cases}, \tag{5.3}$$

are the electric and magnetic dyadic Green's functions, respectively. Outside the circumscribing sphere of the sources, the EM fields in (5.1) are outgoing and can be written as

$$\begin{aligned} \mathbf{E} &= -\sum_{n,m,l} \left(\alpha_{nml}^{(2)}\mathbf{M}_{nml}^{(2)} + \beta_{nml}^{(2)}\mathbf{N}_{nml}^{(2)}\right), \\ \mathbf{H} &= \frac{1}{j\eta}\sum_{n,m,l} \left(\alpha_{nml}^{(2)}\mathbf{N}_{nml}^{(2)} + \beta_{nml}^{(2)}\mathbf{M}_{nml}^{(2)}\right), \end{aligned} \tag{5.4}$$

where $\alpha_{nml}^{(2)}$ and $\beta_{nml}^{(2)}$ are constants determined by the sources

$$\begin{aligned} \alpha_{nml}^{(2)} &= \frac{k^2\eta}{N_{nm}^2}\int_{V_0} \mathbf{J}(\mathbf{r}')\cdot\mathbf{M}_{nml}^{(1)}(\mathbf{r}')dV(\mathbf{r}') - \frac{jk^2}{N_{nm}^2}\int_{V_0} \mathbf{J}_m(\mathbf{r}')\cdot\mathbf{N}_{nml}^{(1)}(\mathbf{r}')dV(\mathbf{r}'), \\ \beta_{nml}^{(2)} &= \frac{k^2\eta}{N_{nm}^2}\int_{V_0} \mathbf{J}(\mathbf{r}')\cdot\mathbf{N}_{nml}^{(1)}(\mathbf{r}')dV(\mathbf{r}') - \frac{jk^2}{N_{nm}^2}\int_{V_0} \mathbf{J}_m(\mathbf{r}')\cdot\mathbf{M}_{nml}^{(1)}(\mathbf{r}')dV(\mathbf{r}'). \end{aligned} \tag{5.5}$$

In the far-field region, the SVWFs are transverse

$$\mathbf{M}_{nml}^{(2)} \approx -\frac{N_{nm}}{kr}j^{n+1}e^{-jkr}\mathbf{h}_{nml}, \ \mathbf{N}_{nml}^{(2)} \approx \frac{N_{nm}}{kr}j^{n}e^{-jkr}\mathbf{e}_{nml}, \tag{5.6}$$

and the fields in (5.4) reduce to

$$\mathbf{E}(\mathbf{r}) = \frac{e^{-jkr}}{r}\mathbf{E}_\infty(\mathbf{u}_r), \ \mathbf{H}(\mathbf{r}) = \frac{e^{-jkr}}{r}\mathbf{H}_\infty(\mathbf{u}_r), \tag{5.7}$$

where

$$\begin{aligned}
\mathbf{E}_{\infty}(\mathbf{u}_r) &= \frac{1}{k}\sum_{n,m,l} j^n N_{nm}\left(j\alpha_{nml}^{(2)}\mathbf{h}_{nml} - \beta_{nml}^{(2)}\mathbf{e}_{nml}\right),\\
\mathbf{H}_{\infty}(\mathbf{u}_r) &= \frac{1}{k\eta}\sum_{n,m,l} j^n N_{nm}\left(-j\alpha_{nml}^{(2)}\mathbf{e}_{nml} - \beta_{nml}^{(2)}\mathbf{h}_{nml}\right),
\end{aligned} \tag{5.8}$$

are the electric and magnetic far-field patterns, respectively.

In this chapter, the modal expansions in (5.1) are applied to the formulations of integral equations for the antenna of composite structures as well as to the analysis of typical antennas. A brief introduction to the microstrip patch antenna is presented for it provides an excellent example of how the theory of eigenfunctions is applied to the antenna design and how a resonant antenna actually works. The resonant modal theory (RMT), which is based on the expressions (4.167) and (4.168) for the stored field energies of antenna, is also introduced, and its applications to the design of resonant antennas are expounded.

5.1 Basic Antenna Types

Antenna typically consists of a scatterer connected to the transmitter or receiver through a feeding waveguide, and can be classified according to their radiation patterns (isotropic, omnidirectional, directional, beam scanning, multiple beams, etc.), feeding mechanisms (balanced, unbalanced), sources (electric current, magnetic current, mixed type of both electric and magnetic currents), physical structures (wires, patches, slots, apertures, reflectors, lens, arrays, fluids, etc.), and applications (mobile phones, RFID, medical systems, vehicles, radars, missiles, satellites, air-crafts, space-crafts, ships, submarines, ground-based broadcast, etc.). The working principles of many antennas encountered in practice may be understood through several basic antennas of simple geometry.

Table 5.1 shows some basic antenna types and their inventors. The dipole and loop are the first antennas introduced by Hertz to show the existence of EM waves predicted by Maxwell equations. The typical size of a dipole antenna is half wavelength. The monopole antenna is usually one half of the dipole, and is one-quarter wavelength long with the other side connected to ground plane. To increase the directivity of the antenna, additional antenna elements can be introduced to form an antenna array. The antenna array can be used to generate different radiation patterns by properly selecting the excitation distribution of amplitudes and phases at the element terminals. The Yagi–Uda antenna and Log-periodic antenna are typical array antennas suitable for applications where high directivity is required. Greater directivity can be achieved by using horn or parabolic reflector.

Table 5.1 Basic antenna types and their inventors.

Types	Inventers
Dipole and loop antenna	Invented by Hertz in 1886.
Monopole antenna	Invented by G. Marconi in 1895.
Yagi–Uda antenna	Invented by S. Uda and H. Yagi in 1926.
Log-periodic antenna	Invented by R. H. DuHamel and D. E. Isbell in 1957.
Horn antenna	Invented by J. C. Bose (1858–1937) in 1897; first experimental research by G. C. Southworth and (1890–1972) and W. L. Barrow (1903–1975) in 1936; theoretical analysis by Barrow and L. J. Chu (1913–1973) in 1939; corrugated horn invented by A. F. Kay in 1962.
Parabolic reflector antenna	Invented by Hertz in 1888; Cassegrain antenna was developed in Japan in 1963 by NTT, KDDI, and Mitsubishi Electric.
Microstrip patch antenna	Invented by R. E. Munson in 1972.

Many antennas are designed on the basis of resonance, at which the stored electric field energy of antenna is approximately equal to the stored magnetic field energy across a frequency band.

5.2 Equivalent Current Distributions of Antenna

Figure 5.1 shows the typical configurations of antennas and their equivalent circuit. Figure 5.1a–c demonstrate an antenna excited by an external incident field, fed by a waveguide and a point source, respectively. The point source approximates a source applied to a pair of terminals separated by a distance much smaller than a wavelength. Figure 5.1d is the equivalent circuit for a general transmitting antenna system. For the derivation of the equivalent circuits for the transmitting antenna and the source, please refer to sections 4.4 and 8.7 in [1]. For the derivation of the equivalent circuit for a receiving antenna, please refer to section 5.4.3.2 in [2].

Let us consider the canonical antenna configuration shown in Figure 5.2a. The impressed electric current $\mathbf{J}_{imp}$ and magnetic current $\mathbf{J}_{m,imp}$, confined in the source region V_0, generate an incident field $(\mathbf{E}^{in}, \mathbf{H}^{in})$. The incident field propagates through a feeding aperture (antenna terminal or reference plane) and induces currents on the scatterer (designated by the shaded region). The induced currents

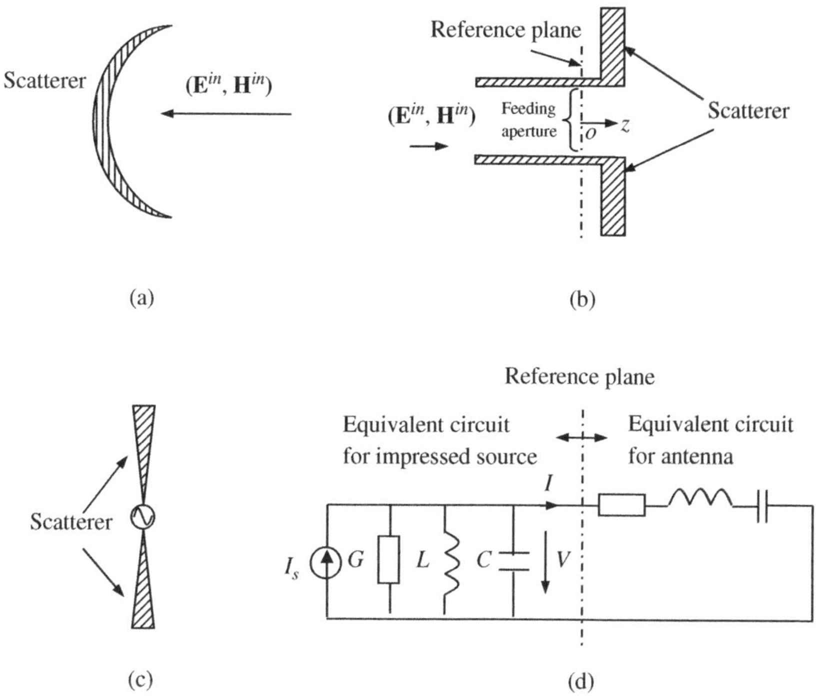

Figure 5.1 Antennas and their equivalent circuits. (a) Antenna excited by incident field. (b) Antenna fed by waveguide. (c) Antenna fed by point source. (d) Equivalent circuit for antenna system.

together with the impressed sources, all confined in the source region V_0, produce the fields radiated into space. In many cases, the contribution to the radiated fields from the impressed sources is negligible. According to Schelkunoff–Love equivalence principle, the radiated fields outside the source region V_0 are equal to those generated by the equivalent surface sources distributed on the boundary of the source region V_0 as sketched in Figure 5.2b

$$\mathbf{J} = \mathbf{u}_n \times \mathbf{H}, \ \ \mathbf{J}_m = -\mathbf{u}_n \times \mathbf{E}, \ \ \text{on}\, \partial V_0, \tag{5.9}$$

where $\mathbf{u}_n$ is the unit outward normal to the boundary ∂V_0 of V_0.

Figure 5.2c shows another possible antenna configuration, where the impressed sources are outside the scatterer. The incident field $(\mathbf{E}^{in}, \mathbf{H}^{in})$, generated by the impressed sources, induces the currents on the scatterer. Again, the induced currents together with the impressed sources produce the radiation fields. In this case, the field generated by the induced currents is called **scattered field**, denoted by $(\mathbf{E}^s, \mathbf{H}^s)$. The total field $(\mathbf{E}, \mathbf{H})$ outside the scatterer is the sum of incident field and the scattered field

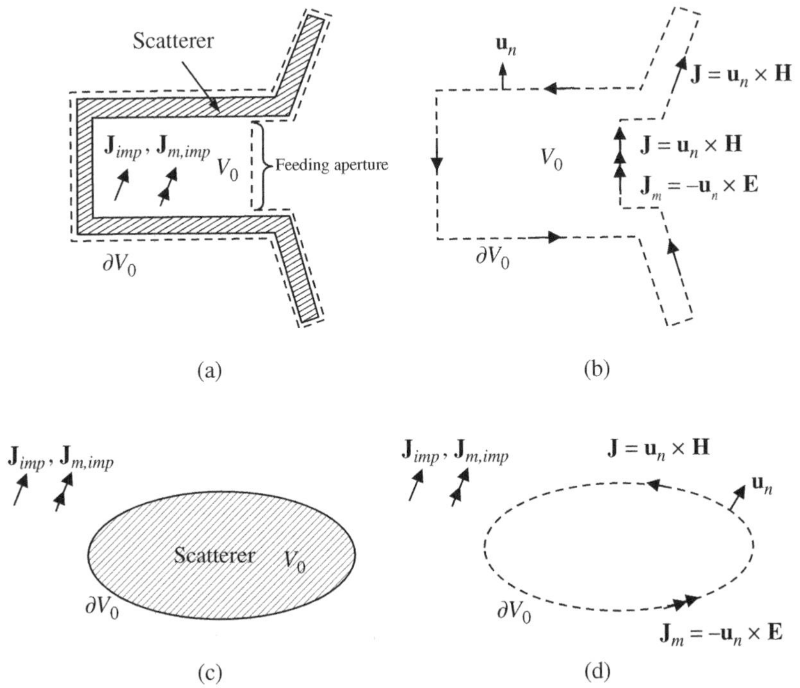

Figure 5.2 Antenna equivalent to current sources. (a) Antenna configuration. (b) Equivalent surface sources. (c) Antenna configuration. (d) Equivalent surface sources.

$$\mathbf{E} = \mathbf{E}^{in} + \mathbf{E}^{s}, \mathbf{H} = \mathbf{H}^{in} + \mathbf{H}^{s}. \tag{5.10}$$

By means of Schelkunoff–Love equivalence principle, the total radiated field outside the scatterer is produced by the impressed sources outside the scatterer together with the equivalent sources distributed on the boundary as illustrated in Figure 5.2d.

Example 5.1 Consider a patch antenna consisting of a metallic patch bonded to an insulating dielectric substrate with a metal layer (ground) bonded to the opposite side of the substrate, as depicted in Figure 5.3. The impressed source of the patch antenna can be represented by a current distribution $\mathbf{J}_{imp} = \mathbf{u}_z J$, which is assumed to be independent of z if $h << \lambda$. This implies that the charge distribution $\rho = 0$. The source region V_0 is bounded by a closed surface ∂V_0 consisting of the top patch, the bottom ground plane, and the side wall. Since the electric field inside the source region has only z-component, the side wall has no tangential

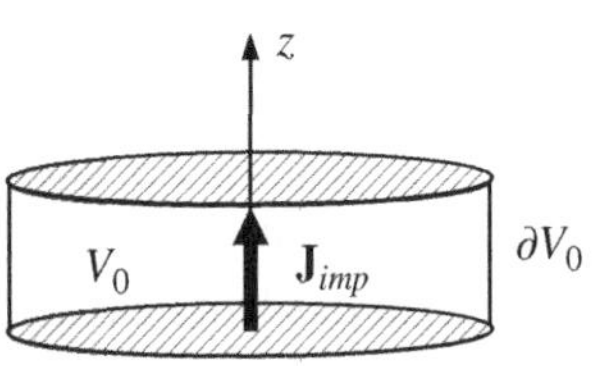

Figure 5.3 Patch antenna.

Figure 5.4 A cylindrical dipole excited by delta gap.

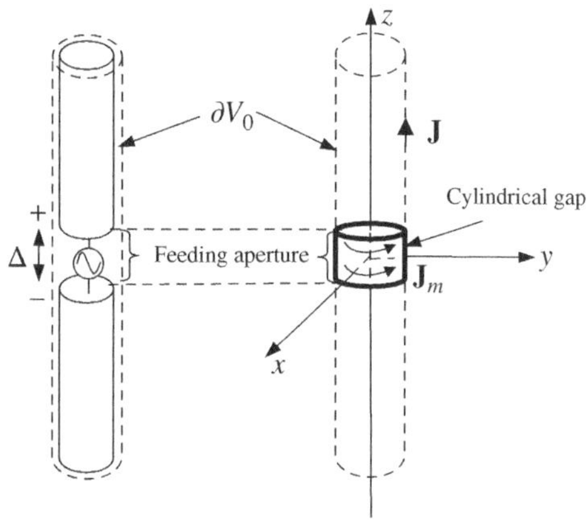

magnetic field (neglecting the fringing effect). As a result, the equivalent magnetic current exists on the side wall only. □

Example 5.2 Consider a cylindrical dipole with a small gap Δ, as sketched in Figure 5.4. A constant voltage source V^{in} is applied to the two feeding terminals of the dipole. The incident field generated by the voltage source is also a constant given by $\mathbf{E}^{in} = -\mathbf{u}_z V^{in}/\Delta$. For the cylindrical dipole, the equivalent magnetic current in (5.9) only appears on the surface of the cylindrical gap. Neglecting the scattered fields in the gap region, the equivalent magnetic current on the cylindrical gap can be written as

$$\mathbf{J}_m \approx -\mathbf{u}_\rho \times \mathbf{E}^{in}\delta(\rho-\rho_0) = \mathbf{u}_\varphi \frac{V^{in}}{\Delta}\delta(\rho-\rho_0),$$

where $\mathbf{u}_\rho$ and $\mathbf{u}_\varphi$ are unit vectors along the coordinate curves in the cylindrical coordinate system (ρ, φ, z). The equivalent electric current on the conducting surface of the dipole can be determined by solving an integral equation. □

5.3 Antenna as a Waveguide Junction

Theoretically, an antenna can be viewed as a device that transforms the guided waves in the feeding line into the spherical waves in the spherical waveguide. Based on this understanding, the antenna may be treated as a discontinuity or waveguide junction as illustrated in Figure 5.5a. Suppose that the antenna is excited by the dominant mode in the feeding waveguide. One may introduce an antenna–air interface (e.g. the circumscribing sphere of the antenna) to serve as the reference plane in the spherical waveguide and the fields outside the reference

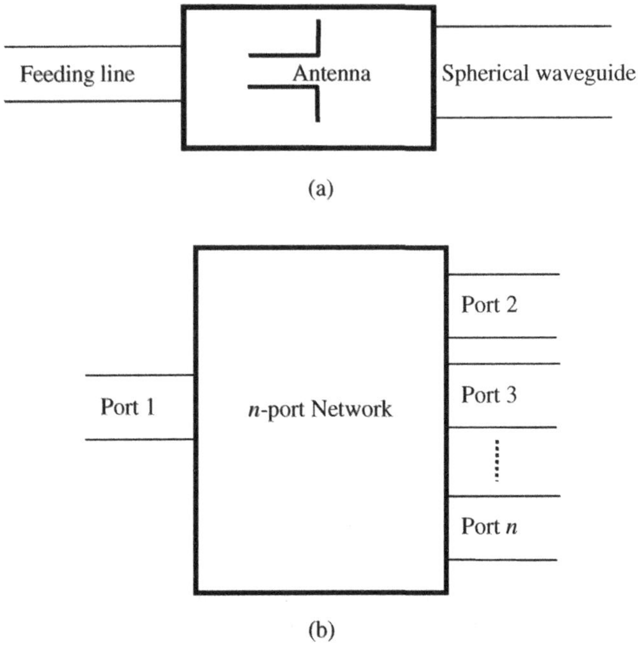

Figure 5.5 Antenna is treated as a waveguide junction and multi-port network. (a) Antenna as a waveguide junction. (b) Equivalent network.

plane can be analyzed by the theory of SVWFs. As demonstrated in Chapters 2 and 4, the modal voltages and currents in both the feeding line and the spherical waveguide at the reference planes are proportional to the transverse electric and magnetic fields, respectively. According to the uniqueness theorem of EM fields [1], the modal voltages can be determined by the modal currents and they are linearly related if the medium is linear. As a result, the antenna (the waveguide junction) can be characterized by a multi-port network as illustrated in Figure 5.5b. The total number of the ports of the network equals the number of the propagating modes in the feeding line plus that of the propagating modes in the spherical waveguide.

5.4 Integral Equation Formulations

A physical problem may be locally characterized by a differential equation and globally by an integral equation. The former can be numerically solved by a domain method (such as the finite element method and the finite difference method), yielding a sparse matrix; while the latter can be solved by a boundary method (such as the moment method and the boundary element method), yielding a dense matrix. The order of the sparse matrix resulted from the differential equation formulation is usually much higher than the dense matrix from the integral equation formulation. When the differential equation is defined in a domain of infinite extent, an artificial boundary (truncated boundary) must be introduced to truncate the domain to keep the number of unknowns finite. To simulate the original field behavior on the truncated boundary without introducing too much distortion, an appropriate boundary condition called absorbing boundary condition must be imposed to diminish the artificial reflections from the truncated boundary.

The integral equations offer some unique features that the differential equations do not have. The integral equation formulation is most appropriate for solving a field problem whose domain extends to infinity, such as the radiation and scattering problems. The boundary condition at infinity is automatically incorporated in the integral equation formulation, and the unbounded-domain problem is transformed into a bounded-domain problem. Since the unknowns are restricted on the boundary of the physical problem, the dimension of the problem is decreased by one. As a result, the number of unknowns is reduced and the numerical accuracy is improved when the integral equation is solved numerically. For a detailed description of integral formulations for EM field problems, please refer to [1, 3].

5.4.1 Compensation Theorem for Time-Harmonic Fields

In circuit theory, an element can be replaced by an ideal current source of the same current intensity as in the element. This property is called the **compensation theorem**. The general form of the compensation theorem in EM fields states that the influence of substance on the fields can partly or completely be compensated by appropriate distribution of impressed currents. Let us consider a perfect conductor in free space, which occupies a region V_e bounded by S_e and is illuminated by the incident field $(\mathbf{E}^{in}, \mathbf{H}^{in})$ generated by the impressed source $\mathbf{J}_{imp}$, as illustrated in Figure 5.6. The incident field induces a surface current $\mathbf{J}_s$ on the conductor, and the latter generates the scattered field $(\mathbf{E}^s, \mathbf{H}^s)$. The incident field and the scattered field satisfy the equations

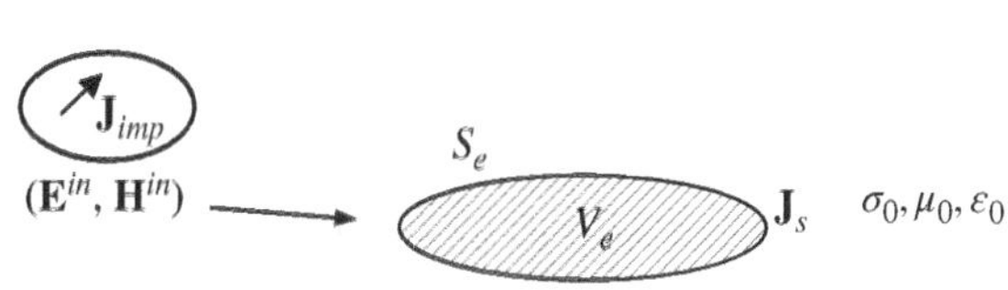

Figure 5.6 Scattering by a perfect conductor.

$$\begin{aligned}\nabla \times \mathbf{E}^{in} &= -j\omega\mu_0\mathbf{H}^{in},\\ \nabla \times \mathbf{H}^{in} &= (\sigma_0 + j\omega\varepsilon_0)\mathbf{E}^{in} + \mathbf{J}_{imp},\end{aligned} \tag{5.11}$$

$$\begin{aligned}\nabla \times \mathbf{E}^{s} &= -j\omega\mu_0\mathbf{H}^{s},\\ \nabla \times \mathbf{H}^{s} &= (\sigma_0 + j\omega\varepsilon_0)\mathbf{E}^{s} + \mathbf{J}_{s},\end{aligned} \tag{5.12}$$

respectively. The equations for the total field $(\mathbf{E}, \mathbf{H})$ in the surrounding medium can be obtained by adding (5.11) for the incident field and (5.12) for the scattered field

$$\begin{aligned}\nabla \times \mathbf{E} &= -j\omega\mu_0\mathbf{H},\\ \nabla \times \mathbf{H} &= (\sigma_0 + j\omega\varepsilon_0)\mathbf{E} + \mathbf{J}_{imp} + \mathbf{J}_{s}.\end{aligned} \tag{5.13}$$

If the parameters of the surrounding medium are changed to $\mu(\mathbf{r})$, $\varepsilon(\mathbf{r})$, and $\sigma(\mathbf{r})$, the total field will be changed to $(\mathbf{E}', \mathbf{H}')$ and governed by

$$\begin{aligned}\nabla \times \mathbf{E}' &= -j\omega\mu\mathbf{H}',\\ \nabla \times \mathbf{H}' &= (\sigma + j\omega\varepsilon)\mathbf{E}'(\mathbf{r}) + \mathbf{J}_{imp} + \mathbf{J}'_{s},\end{aligned} \tag{5.14}$$

where $\mathbf{J}'_s$ is the induced current on the conductor with the new surrounding material. Equations (5.14) can be rewritten as

$$\begin{aligned}\nabla \times \mathbf{E}' &= -j\omega\mu_0\mathbf{H}' - \mathbf{J}'_{m},\\ \nabla \times \mathbf{H}' &= (\sigma_0 + j\omega\varepsilon_0)\mathbf{E}' + \mathbf{J}' + \mathbf{J}_{imp} + \mathbf{J}'_{s},\end{aligned} \tag{5.15}$$

where $\mathbf{J}'_m = R_c\mathbf{H}'$ and $\mathbf{J}' = Y_c\mathbf{E}'$ are the equivalent current sources introduced in the region in which medium parameters change, and

$$\begin{aligned}R_c &= j\omega(\mu - \mu_0),\\ Y_c &= \sigma - \sigma_0 + j\omega(\varepsilon - \varepsilon_0).\end{aligned} \tag{5.16}$$

Consequently, the perturbed field can be determined by introducing the equivalent current sources $\mathbf{J}'$ and $\mathbf{J}'_m$ as if the medium parameters had not changed. This is one of the forms of the compensation theorem in EM field theory. For a more general form of the compensation theorem, please refer to [1].

5.4.2 Integral Equations for Composite Structure

Consider a scatterer composed of a perfectly conductor occupying the region V_e bounded by S_e and a dielectric body occupying the region V_d bounded by S_d. The scatterer is illuminated by an incident field $(\mathbf{E}^{in}, \mathbf{H}^{in})$ as illustrated in Figure 5.7. According to the compensation theorem, the impressed current $\mathbf{J}_{imp}$, the induced current $\mathbf{J}_s$ on the conducting surface, and the following volume sources

$$\mathbf{J}_m = R_c\mathbf{H}, \quad \mathbf{J} = Y_c\mathbf{E}, \tag{5.17}$$

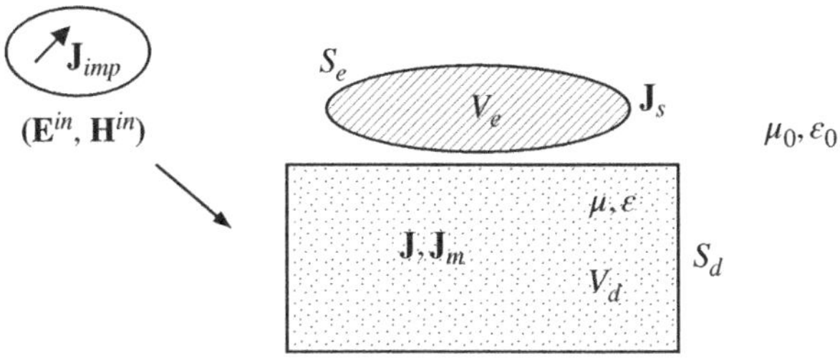

Figure 5.7 Scattering by a composite structure composed of a perfect conductor and a lossy dielectric.

in the dielectric region generate the total field $(\mathbf{E}, \mathbf{H})$, which can be determined by (5.1).

$$\begin{aligned}\mathbf{E}(\mathbf{r}) &= \mathbf{E}^{in}(\mathbf{r}) - jk_0\eta_0 \int_{S_e} \overleftrightarrow{\mathbf{G}}_e(\mathbf{r}, \mathbf{r}') \cdot \mathbf{J}_s(\mathbf{r}')dV(\mathbf{r}') \\ &\quad - jk_0\eta_0 \int_{V_d} \overleftrightarrow{\mathbf{G}}_e(\mathbf{r}, \mathbf{r}') \cdot \mathbf{J}(\mathbf{r}')dV(\mathbf{r}') - \int_{V_d} \overleftrightarrow{\mathbf{G}}_m(\mathbf{r}, \mathbf{r}') \cdot \mathbf{J}_m(\mathbf{r}')dV(\mathbf{r}'),\end{aligned} \tag{5.18}$$

$$\begin{aligned}\mathbf{H}(\mathbf{r}) &= \mathbf{H}^{in}(\mathbf{r}) + \int_{S_e} \overleftrightarrow{\mathbf{G}}_m(\mathbf{r}, \mathbf{r}') \cdot \mathbf{J}_s(\mathbf{r}')dV(\mathbf{r}') \\ &\quad - j\frac{k_0}{\eta_0} \int_{V_d} \overleftrightarrow{\mathbf{G}}_e(\mathbf{r}, \mathbf{r}') \cdot \mathbf{J}_m(\mathbf{r}')dV(\mathbf{r}') + \int_{V_d} \overleftrightarrow{\mathbf{G}}_m(\mathbf{r}, \mathbf{r}') \cdot \mathbf{J}(\mathbf{r}')dV(\mathbf{r}').\end{aligned} \tag{5.19}$$

If the observation point is selected on the surface of the conductor or in the dielectric body, one may obtain three integral equations

$$\begin{aligned}&jk_0\eta_0\mathbf{u}_n(\mathbf{r}) \times \int_{S_e} \overleftrightarrow{\mathbf{G}}_e(\mathbf{r}, \mathbf{r}') \cdot \mathbf{J}_s(\mathbf{r}')dV(\mathbf{r}') + jk_0\eta_0\mathbf{u}_n(\mathbf{r}) \times \int_{V_d} \overleftrightarrow{\mathbf{G}}_e(\mathbf{r}, \mathbf{r}') \cdot \mathbf{J}(\mathbf{r}')dV(\mathbf{r}') + \mathbf{u}_n(\mathbf{r}) \\ &\quad \times \int_{V_d} \overleftrightarrow{\mathbf{G}}_m(\mathbf{r}, \mathbf{r}') \cdot \mathbf{J}_m(\mathbf{r}')dV(\mathbf{r}') = \mathbf{u}_n(\mathbf{r}) \times \mathbf{E}^{in}(\mathbf{r}),\ \mathbf{r} \in S_e, \\ &\mathbf{J}(\mathbf{r}) + jk_0\eta_0 Y_c \int_{S_e} \overleftrightarrow{\mathbf{G}}_e(\mathbf{r}, \mathbf{r}') \cdot \mathbf{J}_s(\mathbf{r}')dV(\mathbf{r}') + jk_0\eta_0 Y_c \int_{V_d} \overleftrightarrow{\mathbf{G}}_e(\mathbf{r}, \mathbf{r}') \cdot \mathbf{J}(\mathbf{r}')dV(\mathbf{r}') \\ &\quad + Y_c \int_{V_d} \overleftrightarrow{\mathbf{G}}_m(\mathbf{r}, \mathbf{r}') \cdot \mathbf{J}_m(\mathbf{r}')dV(\mathbf{r}') = Y_c\mathbf{E}^{in}(\mathbf{r}),\ \mathbf{r} \in V_d, \\ &\mathbf{J}_m(\mathbf{r}) - R_c \int_{S_e} \overleftrightarrow{\mathbf{G}}_m(\mathbf{r}, \mathbf{r}') \cdot \mathbf{J}_s(\mathbf{r}')dV(\mathbf{r}') + jk_0\frac{R_c}{\eta_0} \int_{V_d} \overleftrightarrow{\mathbf{G}}_e(\mathbf{r}, \mathbf{r}') \cdot \mathbf{J}_m(\mathbf{r}')dV(\mathbf{r}') \\ &\quad - R_c \int_{V_d} \overleftrightarrow{\mathbf{G}}_m(\mathbf{r}, \mathbf{r}') \cdot \mathbf{J}(\mathbf{r}')dV(\mathbf{r}') = R_c\mathbf{H}^{in}(\mathbf{r}),\ \mathbf{r} \in V_d.\end{aligned} \tag{5.20}$$

These integral equations contain the current densities on the conductor and the equivalent current densities inside the dielectric body as unknowns and can be solved numerically [4]. If the scatterer is a perfect conductor, (5.20) reduces to an integral equation for the surface current on the conductor

$$j\omega\mu\mathbf{u}_n(\mathbf{r}) \times \int_{S_e} \overleftrightarrow{\mathbf{G}}_e(\mathbf{r},\mathbf{r}') \cdot \mathbf{J}_s(\mathbf{r}')dS(\mathbf{r}') = \mathbf{u}_n(\mathbf{r}) \times \mathbf{E}^{in}(\mathbf{r}). \tag{5.21}$$

If the dielectric body is nonmagnetic, (5.20) reduces to

$$\begin{aligned} &jk_0\eta_0\mathbf{u}_n(\mathbf{r}) \times \int_{S_e} \overleftrightarrow{\mathbf{G}}_e(\mathbf{r},\mathbf{r}') \cdot \mathbf{J}_s(\mathbf{r}')dV(\mathbf{r}') + jk_0\eta_0\mathbf{u}_n(\mathbf{r}) \\ &\quad \times \int_{V_d} \overleftrightarrow{\mathbf{G}}_e(\mathbf{r},\mathbf{r}') \cdot \mathbf{J}(\mathbf{r}')dV(\mathbf{r}') = \mathbf{u}_n(\mathbf{r}) \times \mathbf{E}^{in}(\mathbf{r}),\ \mathbf{r} \in S_e, \\ &\mathbf{J}(\mathbf{r}) + jk_0\eta_0 Y_c \int_{S_e} \overleftrightarrow{\mathbf{G}}_e(\mathbf{r},\mathbf{r}') \cdot \mathbf{J}_s(\mathbf{r}')dV(\mathbf{r}') \\ &\quad + jk_0\eta_0 Y_c \int_{V_d} \overleftrightarrow{\mathbf{G}}_e(\mathbf{r},\mathbf{r}') \cdot \mathbf{J}(\mathbf{r}')dV(\mathbf{r}') = Y_c\mathbf{E}^{in}(\mathbf{r}),\ \mathbf{r} \in V_d, \end{aligned} \tag{5.22}$$

where only electric current densities are involved.

5.4.3 Integral Equation for Wire Antenna

As a special case, let us consider the integral equation for the wire antenna shown in Figure 5.8. It will be assumed that the wire is a circular conducting cylinder of radius a and length l. In this case, only the integral equation (5.21) needs to be solved and it reduces to

$$j\omega\mu\mathbf{u}_n(\mathbf{r}) \times \int_0^l dz' \int_C \overleftrightarrow{\mathbf{G}}_e(\mathbf{r},\mathbf{r}') \cdot \mathbf{J}_s(\mathbf{r}')dC(\mathbf{r}') = \mathbf{u}_n \times \mathbf{E}^{in}(\mathbf{r}), \tag{5.23}$$

where C is the circumference of the wire. The surface current density $\mathbf{J}$ generally has a component locally parallel to the wire axis and a circumferential component. If the wire is very thin, the latter component is usually very small, except at the antenna discontinuities, which will not be considered for the moment. So, it can be assumed that the current density is concentrated on the wire axis

$$\mathbf{J}_s(\mathbf{r}) = \frac{I(z)}{2\pi a}\mathbf{u}_z,$$

Figure 5.8 A wire antenna.

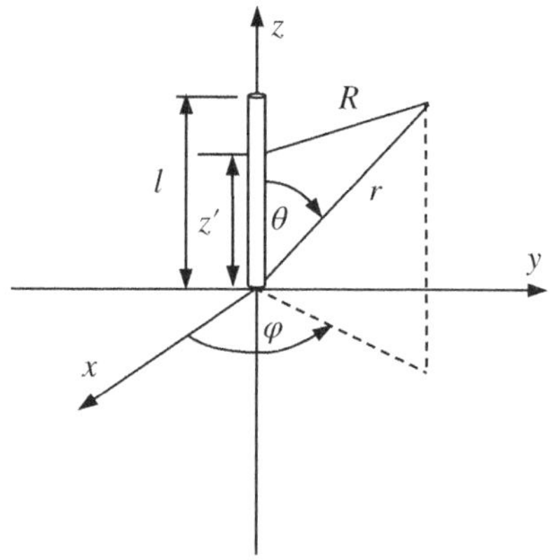

where $I(z)$ is the axial current. Thus, (5.23) becomes

$$j\omega\mu\mathbf{u}_n(\mathbf{r}) \times \int_0^l I(z')dz' \frac{1}{2\pi a} \int_C \overleftrightarrow{\mathbf{G}}_e(\mathbf{r}, \mathbf{r}') \cdot \mathbf{u}_{z'} dC(\mathbf{r}') = \mathbf{u}_n \times \mathbf{E}^{in}(\mathbf{r}). \tag{5.24}$$

If the observation point $\mathbf{r}$ is set on the surface of the wire so that $\mathbf{r} \neq \mathbf{r}'$, the singular term of the electric dyadic Green's function in (5.2) vanishes. Since the scattered field is independent of the azimuthal angle, the electric dyadic Green's function in (5.24) reduces to

$$\overleftrightarrow{\mathbf{G}}_e(\mathbf{r}, \mathbf{r}') \cdot \mathbf{u}_{z'} = \sum_n \frac{-jk}{N_{n0}^2} \begin{cases} \mathbf{M}_{n0e}^{(1)}(\mathbf{r})\mathbf{M}_{n0e}^{(2)}(\mathbf{r}') \cdot \mathbf{u}_{z'} + \mathbf{N}_{n0e}^{(1)}(\mathbf{r})\mathbf{N}_{n0e}^{(2)}(\mathbf{r}') \cdot \mathbf{u}_{z'}, \ r < r' \\ \mathbf{M}_{n0e}^{(2)}(\mathbf{r})\mathbf{M}_{n0e}^{(1)}(\mathbf{r}') \cdot \mathbf{u}_{z'} + \mathbf{N}_{n0e}^{(2)}(\mathbf{r})\mathbf{N}_{n0e}^{(1)}(\mathbf{r}') \cdot \mathbf{u}_{z'}, \ r > r' \end{cases}. \tag{5.25}$$

For the thin wire, one may let $\mathbf{r}' = (0, 0, z')$. The SVWFs can then be calculated as follows:

$$\begin{aligned} \mathbf{M}_{n0e}^{(1)}(\mathbf{r}') \cdot \mathbf{u}_{z'} &= -j_n(kr')P_n^1(1)\mathbf{u}_{z'} \cdot \mathbf{u}_{\varphi'} = 0, \\ \mathbf{N}_{n0e}^{(1)}(\mathbf{r}') \cdot \mathbf{u}_{z'} &= \frac{n(n+1)}{kr'} j_n(kr')P_n(1)\mathbf{u}_{z'} \cdot \mathbf{u}_{r'} - \frac{1}{kb}\hat{j}_n(kr')P_n^1(1)\mathbf{u}_{z'} \cdot \mathbf{u}_{\theta'} \\ &= \frac{n(n+1)}{kr'} j_n(kr'), \\ \mathbf{M}_{n0e}^{(2)}(\mathbf{r}') \cdot \mathbf{u}_{z'} &= 0, \\ \mathbf{N}_{n0e}^{(2)}(\mathbf{r}') \cdot \mathbf{u}_{z'} &= \frac{n(n+1)}{kr'} h_n^{(2)}(kr'). \end{aligned} \tag{5.26}$$

As a result, (5.25) reduces to

$$\overleftrightarrow{\mathbf{G}}_e(\mathbf{r},\mathbf{r}')\cdot\mathbf{u}_{z'}=\sum_n\frac{-j}{N_{n0}^2r'}\begin{cases}\mathbf{N}_{n0e}^{(1)}(\mathbf{r})n(n+1)h_n^{(2)}(kr'), & r<r'\\ \mathbf{N}_{n0e}^{(2)}(\mathbf{r})n(n+1)j_n(kr'), & r>r'\end{cases},\tag{5.27}$$

where

$$\mathbf{N}_{n0e}^{(1)}(\mathbf{r})=\frac{1}{k}\nabla\times\mathbf{M}_{n0e}^{(1)}(\mathbf{r}),\ \mathbf{N}_{n0e}^{(2)}(\mathbf{r})=\frac{1}{k}\nabla\times\mathbf{M}_{n0e}^{(2)}(\mathbf{r})\tag{5.28}$$

with

$$\mathbf{M}_{n0e}^{(1)}(\mathbf{r})=j_n(kr)P_n^1(\cos\theta)\mathbf{u}_\varphi,$$
$$\mathbf{M}_{n0e}^{(2)}(\mathbf{r})=h_n^{(2)}(kr)P_n^1(\cos\theta)\mathbf{u}_\varphi.$$

Substitution of (5.28) into (5.27) gives

$$\overleftrightarrow{\mathbf{G}}_e(\mathbf{r},\mathbf{r}')\cdot\mathbf{u}_{z'}=\frac{1}{k^2r'}\nabla\times\sum_n\mathbf{u}_\varphi\frac{-jk}{4\pi}\begin{cases}(2n+1)j_n(kr)h_n^{(2)}(kr')P_n^1(\cos\theta), & r<r'\\ (2n+1)j_n(kr')h_n^{(2)}(kr)P_n^1(\cos\theta), & r>r'\end{cases}.\tag{5.29}$$

By use of the series expansion for the free-space Green's function

$$G(R)=\frac{e^{-jkR}}{4\pi R}=\sum_n\frac{-jk}{4\pi}\begin{cases}(2n+1)j_n(kr)h_n^{(2)}(kr')P_n(\cos\theta), & r<r'\\ (2n+1)j_n(kr')h_n^{(2)}(kr)P_n(\cos\theta), & r>r'\end{cases},\tag{5.30}$$

with $R=|\mathbf{r}-\mathbf{r}'|=\sqrt{a^2+(z-z')^2}$, (5.29) can be rewritten as

$$\overleftrightarrow{\mathbf{G}}_e(\mathbf{r},\mathbf{r}')\cdot\mathbf{u}_{z'}=-\frac{1}{k^2r'}\nabla\times\left[\frac{\partial}{\partial\theta}G(R)\mathbf{u}_\varphi\right].\tag{5.31}$$

Consequently, the integral equation (5.24) is simplified to

$$-j\omega\mu\mathbf{u}_n\times\int_0^l dz'\frac{1}{k^2r'}I(z')\frac{1}{2\pi a}\int_C\nabla\times\left[\mathbf{u}_\varphi\frac{\partial}{\partial\theta}G(R)\right]dC(\mathbf{r}')=\mathbf{u}_n\times\mathbf{E}^{in}(\mathbf{r}).\tag{5.32}$$

Since the current is concentrated on the z-axis, one may let $x'=y'=0$ to get

$$\nabla\times\left[\mathbf{u}_\varphi\frac{\partial}{\partial\theta}G(R)\right]=-z'\frac{\partial}{\partial z}\left[\frac{\partial}{\partial\rho}G(R)\right]\mathbf{u}_\rho-z'\frac{\partial^2}{\partial z^2}G(R)\mathbf{u}_z-k^2z'G(R)\mathbf{u}_z.$$

In the above, the relation $(\nabla^2 + k^2)G(R) = 0$ for $\mathbf{r} \neq \mathbf{r}'$ has been used. Insertion of the above expression into (5.32) yields

$$\mathbf{u}_\rho \times \mathbf{u}_z \int_0^l I(z') \left(\frac{\partial^2}{\partial z^2} + k^2 \right) G(z, z') dz' = -j\frac{k}{\eta} \mathbf{u}_\rho \times \mathbf{E}^{in}(\mathbf{r}), \tag{5.33}$$

where

$$G(z, z') = \frac{1}{2\pi a} \int_C G(R) dC(\mathbf{r}') = G(R)$$

for the wire is very thin. If the incident field on the wire surface is z-directed and given by $\mathbf{E}^{in}(\mathbf{r}) = \mathbf{u}_z E^{in}|_{\rho = a}$, (5.33) becomes

$$\int_0^l I(z') \left(\frac{\partial^2}{\partial z^2} + k^2 \right) G(z, z') dz' = -j\frac{k}{\eta} E^{in}\Big|_{\rho = a}. \tag{5.34}$$

This is the well-known **Pocklington's integral-differential equation** [5].

5.5 Vertical Dipole

Consider a current element (a dipole) oriented along the z-axis and centered at $\mathbf{r}_0 = (0, 0, b)$

$$\mathbf{J}(\mathbf{r}) = I_0 l \delta(\mathbf{r} - \mathbf{r}_0) \mathbf{u}_z, \tag{5.35}$$

where I_0 is the current in the element and l its length, as illustrated in Figure 5.9.

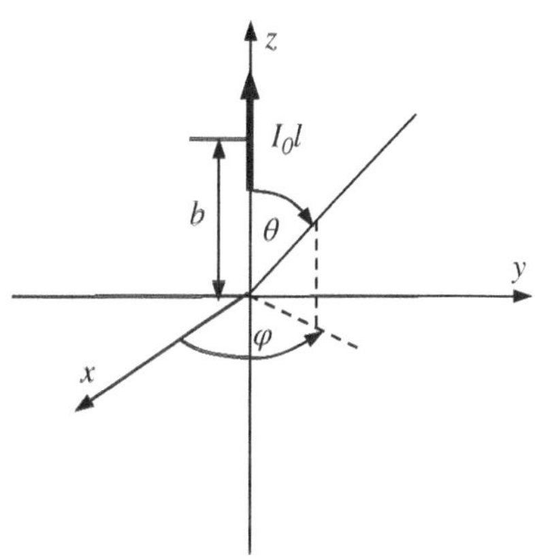

Figure 5.9 A vertical dipole.

5.5.1 Fields in the Region $r > b$

In the region $r > b$, the radiated electric field is given by (5.1):

$$\mathbf{E}(\mathbf{r}) = -jk\eta \int_{V_0} \overleftrightarrow{\mathbf{G}}_e(\mathbf{r}, \mathbf{r}') \cdot \mathbf{J}(\mathbf{r}') dV(\mathbf{r}') = -jk\eta I_0 l \overleftrightarrow{\mathbf{G}}_e(\mathbf{r}, \mathbf{r}_0) \cdot \mathbf{u}_z. \tag{5.36}$$

Since the fields must be outgoing in this region, one must select the lower expression of (5.2) for the electric dyadic Green's function

$$\overleftrightarrow{\mathbf{G}}_e(\mathbf{r}, \mathbf{r}_0) \cdot \mathbf{u}_z = -\sum_{n,m,l} \frac{jk}{N_{nm}^2} \mathbf{M}_{nml}^{(2)}(\mathbf{r}) \mathbf{M}_{nml}^{(1)}(\mathbf{r}_0) \cdot \mathbf{u}_z - \sum_{n,m,l} \frac{jk}{N_{nm}^2} \mathbf{N}_{nml}^{(2)}(\mathbf{r}) \mathbf{N}_{nml}^{(1)}(\mathbf{r}_0) \cdot \mathbf{u}_z.$$

In view of the symmetry of the vertical dipole source, the radiated fields are independent of the azimuth angle. So, one must have $m = 0$ and $l = e$. The above expression becomes

$$\overleftrightarrow{\mathbf{G}}_e(\mathbf{r}, \mathbf{r}_0) \cdot \mathbf{u}_{z'} = -\sum_{n} \frac{jk}{N_{n0}^2} \mathbf{M}_{n0e}^{(2)}(\mathbf{r}) \mathbf{M}_{n0e}^{(1)}(\mathbf{r}_0) \cdot \mathbf{u}_z - \sum_{n} \frac{jk}{N_{n0}^2} \mathbf{N}_{n0e}^{(2)}(\mathbf{r}) \mathbf{N}_{n0e}^{(1)}(\mathbf{r}_0) \cdot \mathbf{u}_z. \tag{5.37}$$

The SVWFs at the source point $\mathbf{r}_0$ are given by

$$\begin{aligned}
\mathbf{u}_z \cdot \mathbf{M}_{n0e}^{(1)}(\mathbf{r}_0) &= -j_n(kb) P_n^1(1) \mathbf{u}_z \cdot \mathbf{u}_\varphi(\mathbf{r}_0) = 0, \\
\mathbf{u}_z \cdot \mathbf{N}_{n0e}^{(1)}(\mathbf{r}_0) &= \frac{n(n+1)}{kb} j_n(kb) P_n(1) \mathbf{u}_z \cdot \mathbf{u}_r(\mathbf{r}_0) - \frac{1}{kb} \hat{j}_n(kb) P_n^1(1) \mathbf{u}_z \cdot \mathbf{u}_\theta(\mathbf{r}_0) \\
&= \frac{n(n+1)}{kb} j_n(kb).
\end{aligned} \tag{5.38}$$

The first sum on the right-hand side of (5.37) thus vanishes, yielding

$$\overleftrightarrow{\mathbf{G}}_e(\mathbf{r}, \mathbf{r}_0) \cdot \mathbf{u}_z = -\frac{j}{4\pi b} \sum_n (2n+1) j_n(kb) \mathbf{N}_{n0e}^{(2)}(\mathbf{r}), \tag{5.39}$$

where

$$\mathbf{N}_{n0e}^{(2)}(\mathbf{r}) = \frac{n(n+1)}{kr} h_n^{(2)}(kr) P_n(\cos\theta) \mathbf{u}_r - \frac{1}{kr} \hat{h}_n^{(2)}(kr) P_n^1(\cos\theta) \mathbf{u}_\theta. \tag{5.40}$$

Hence, the radiated field (5.36) can be written as

$$\mathbf{E} = -\frac{k\eta I_0 l}{4\pi b} \sum_n (2n+1) j_n(kb) \mathbf{N}_{n0e}^{(2)}. \tag{5.41}$$

On account of the far-field expressions (5.6) for the SVWFs, (5.41) can be approximated by

$$\mathbf{E} \approx -\frac{e^{-jkr}}{4\pi r}\frac{\eta I_0 l}{b}\sum_n j^n(2n+1)N_{n0}j_n(kb)\mathbf{e}_{n0e}, \tag{5.42}$$

in the far-field region. The vector effective length of the vertical dipole is then given by

$$\mathbf{L}(\mathbf{u}_r) = \frac{l}{jkb}\sum_n j^n(2n+1)N_{n0}j_n(kb)\mathbf{e}_{n0e}. \tag{5.43}$$

Figure 5.10 shows the first four vector modal functions for $n = 1, 2, 3, 4$ and $m = 0$, and they are all omnidirectional. As n increases, the pattern starts to split into n sub-patterns in the θ-direction.

As $b \to 0$, only the term $n = 1$ contributes to the field expansion (5.41). In this case, the radiated electric field (5.41) reduces to

$$\mathbf{E} = -\frac{k^2\eta I_0 l}{4\pi}\mathbf{N}_{10e}^{(2)}, \tag{5.44}$$

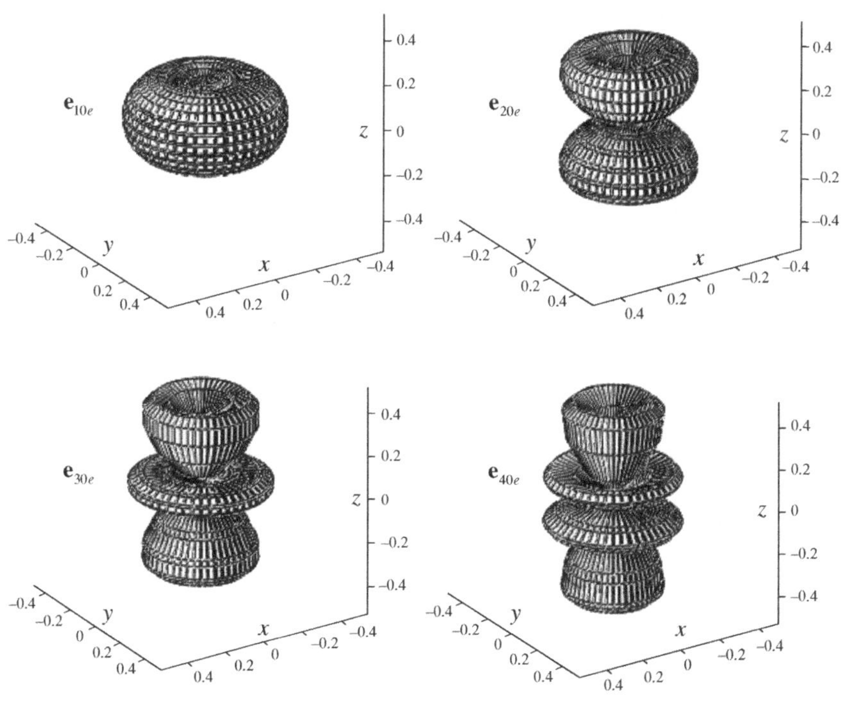

Figure 5.10 Vector modal functions.

where

$$\mathbf{N}_{10e}^{(2)} = \frac{2}{kr} h_1^{(2)}(kr) \cos\theta \mathbf{u}_r - \frac{1}{kr} \dot{\hat{h}}_1^{(2)}(kr) \sin\theta \mathbf{u}_\theta,$$

with

$$\begin{aligned} h_1^{(2)}(kr) &= -\left[\frac{1}{kr} - j\frac{1}{(kr)^2}\right] e^{-jkr}, \\ \dot{\hat{h}}_1^{(2)}(kr) &= j\left[1 - j\frac{1}{kr} - \frac{1}{(kr)^2}\right] e^{-jkr}. \end{aligned} \tag{5.45}$$

Explicitly, the electric field (5.44) can be represented by

$$\begin{aligned} \mathbf{E} =& \frac{2k\eta I_0 l \cos\theta e^{-jkr}}{4\pi r}\left[\frac{1}{kr} - j\frac{1}{(kr)^2}\right]\mathbf{u}_r \\ &+ \frac{jk\eta I_0 l \sin\theta e^{-jkr}}{4\pi r}\left[1 - j\frac{1}{kr} - \frac{1}{(kr)^2}\right]\mathbf{u}_\theta. \end{aligned} \tag{5.46}$$

As $b \to 0$, the vector effective length (5.43) reduces to

$$\mathbf{L}(\mathbf{u}_r) = -l \sin\theta \mathbf{u}_\theta. \tag{5.47}$$

The magnetic field in the region $r > b$ can be obtained from the second expression of (5.1):

$$\mathbf{H}(\mathbf{r}) = \int_{V_0} \overleftrightarrow{\mathbf{G}}_m(\mathbf{r}, \mathbf{r}') \cdot \mathbf{J}(\mathbf{r}') dV(\mathbf{r}') = I_0 l \overleftrightarrow{\mathbf{G}}_m(\mathbf{r}, \mathbf{r}_0) \cdot \mathbf{u}_z, \tag{5.48}$$

where the lower expression of (5.3) must be selected as the magnetic dyadic Green's function

$$\begin{aligned} \overleftrightarrow{\mathbf{G}}_m(\mathbf{r}, \mathbf{r}_0) \cdot \mathbf{u}_z =& -\sum_{n,m,l} \frac{jk^2}{N_{nm}^2} \mathbf{N}_{nml}^{(2)}(\mathbf{r}) \mathbf{M}_{nml}^{(1)}(\mathbf{r}_0) \cdot \mathbf{u}_z \\ &- \sum_{n,m,l} \frac{jk^2}{N_{nm}^2} \mathbf{M}_{nml}^{(2)}(\mathbf{r}) \mathbf{N}_{nml}^{(1)}(\mathbf{r}_0) \cdot \mathbf{u}_z. \end{aligned}$$

For the same reasons explained above, only the second sum on the right-hand side contributes to the radiated fields, and only the terms with ($m = 0$, $l = e$) remain. Thus,

$$\overleftrightarrow{\mathbf{G}}_m(\mathbf{r}, \mathbf{r}_0) \cdot \mathbf{u}_z = -\frac{jk}{4\pi b} \sum_n (2n+1) j_n(kb) \mathbf{M}_{n0e}^{(2)}(\mathbf{r}), \tag{5.49}$$

where

$$\mathbf{M}_{n0e}^{(2)}(\mathbf{r}) = h_n^{(2)}(kr)P_n^1(\cos\theta)\mathbf{u}_\varphi. \tag{5.50}$$

The radiated magnetic field (5.48) is then simplified to

$$\mathbf{H} = -\frac{jkI_0 l}{4\pi b}\sum_n (2n+1)j_n(kb)\mathbf{M}_{n0e}^{(2)}. \tag{5.51}$$

Similarly, when the dipole is centered at the origin ($b \to 0$), all the terms in the sum vanish except for $n = 1$, and the radiated magnetic field can be written as

$$\mathbf{H} = -\frac{jk^2 I_0 l}{4\pi}\mathbf{M}_{10e}^{(2)} = \frac{jkI_0 l \sin\theta e^{-jkr}}{4\pi r}\left(1 - j\frac{1}{kr}\right)\mathbf{u}_\varphi. \tag{5.52}$$

Equations (5.41) and (5.51) indicate that a vertical dipole only generates TM_{n0e} modes. Note that (5.46) and (5.52) agree with the conventional analysis [6].

5.5.2 Fields in the Region $r < b$

In the region $r < b$, the fields are still given by (5.36) and (5.48), with the dyadic Green's functions replaced by the upper expressions of (5.2) and (5.3), respectively, and again only TM_{n0e} modes are excited. Thus,

$$\mathbf{E} = -\frac{k\eta I_0 l}{4\pi b}\sum_n (2n+1)h_n^{(2)}(kb)\mathbf{N}_{n0e}^{(1)}, \tag{5.53}$$

$$\mathbf{H} = -\frac{jkI_0 l}{4\pi b}\sum_n (2n+1)h_n^{(2)}(kb)\mathbf{M}_{n0e}^{(1)}, \tag{5.54}$$

where

$$\begin{aligned}
\mathbf{N}_{n0e}^{(1)} &= \frac{n(n+1)}{kr}j_n(kr)P_n(\cos\theta)\mathbf{u}_r - \frac{1}{kr}\dot{\hat{j}}_n(kr)P_n^1(\cos\theta)\mathbf{u}_\theta, \\
\mathbf{M}_{n0e}^{(1)} &= j_n(kr)P_n^1(\cos\theta)\mathbf{u}_\varphi.
\end{aligned} \tag{5.55}$$

5.6 Horizontal Dipole

A dipole oriented transverse to z, centered at $\mathbf{r}_0 = (0, 0, b)$, is shown in Figure 5.11. Its current density is defined by

$$\mathbf{J}(\mathbf{r}) = I_0 l\delta(\mathbf{r} - \mathbf{r}_0)\mathbf{u}_y. \tag{5.56}$$

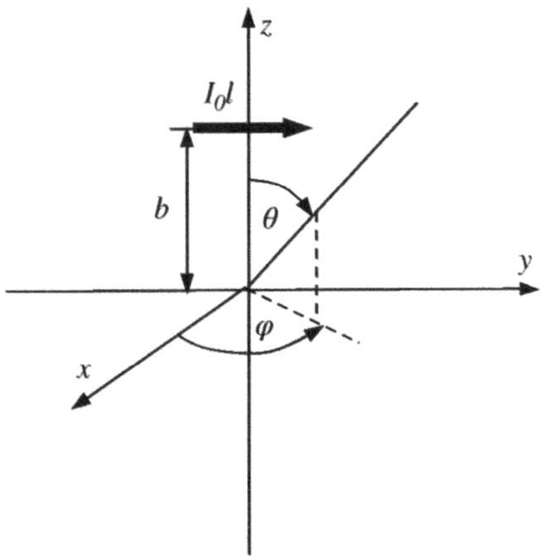

Figure 5.11 A horizontal dipole.

5.6.1 Fields in the Region *r* > *b*

In the region $r > b$, the radiated electric field is given by the first expression of (5.1):

$$\mathbf{E}(\mathbf{r}) = -jk\eta \int_{V_0} \overleftrightarrow{\mathbf{G}}_e(\mathbf{r}, \mathbf{r}') \cdot \mathbf{J}(\mathbf{r}')dV(\mathbf{r}') = -jk\eta I_0 l \overleftrightarrow{\mathbf{G}}_e(\mathbf{r}, \mathbf{r}_0) \cdot \mathbf{u}_y. \tag{5.57}$$

The field must be outgoing in this region, and one must select the lower expression of (5.2) for the electric dyadic Green's function $\overleftrightarrow{\mathbf{G}}_e$:

$$\begin{aligned}\overleftrightarrow{\mathbf{G}}_e(\mathbf{r}, \mathbf{r}_0) \cdot \mathbf{u}_y = &- \sum_{n,m,l} \frac{jk}{N_{nm}^2} \mathbf{M}_{nml}^{(2)}(\mathbf{r}) \mathbf{M}_{nml}^{(1)}(\mathbf{r}_0) \cdot \mathbf{u}_y \\ &- \sum_{n,m,l} \frac{jk}{N_{nm}^2} \mathbf{N}_{nml}^{(2)}(\mathbf{r}) \mathbf{N}_{nml}^{(1)}(\mathbf{r}_0) \cdot \mathbf{u}_y,\end{aligned}$$

where

$$\begin{aligned}\mathbf{u}_y \cdot \mathbf{M}_{nml}^{(1)}(\mathbf{r}_0) &= j_n(kb) \frac{\cos\theta}{\sin\theta} \frac{\partial Y_{nml}(\theta,\varphi)}{\partial\varphi}\bigg|_{\theta=0} \sin\varphi \\ &\quad - j_n(kb) \frac{\partial Y_{nml}(\theta,\varphi)}{\partial\theta}\bigg|_{\theta=0} \cos\varphi, \\ \mathbf{u}_y \cdot \mathbf{N}_{nml}^{(1)}(\mathbf{r}_0) &= \frac{n(n+1)}{kb} j_n(kb) Y_{nml}(\theta,\varphi) \sin\theta|_{\theta=0} \sin\varphi \\ &\quad + \frac{1}{kb} \frac{d[krj_n(kr)]}{dkr}\bigg|_{r=b} \cos\theta \frac{\partial Y_{nml}(\theta,\varphi)}{\partial\theta}\bigg|_{\theta=0} \sin\varphi \\ &\quad + \frac{1}{kb} \frac{d[krj_n(kr)]}{dkr}\bigg|_{r=b} \frac{1}{\sin\theta} \frac{\partial Y_{nml}(\theta,\varphi)}{\partial\varphi}\bigg|_{\theta=0} \cos\varphi.\end{aligned} \tag{5.58}$$

The azimuthal angle at $\mathbf{r}_0$ can be arbitrarily set. For convenience, one may choose $\varphi = 0$ at $\mathbf{r}_0$ and the above equations are simplified to

$$\begin{aligned}\mathbf{u}_y \cdot \mathbf{M}_{nml}^{(1)}(\mathbf{r}_0) &= -j_n(kb)\left.\frac{\partial Y_{nml}(\theta,\varphi)}{\partial \theta}\right|_{\theta=0}, \\ \mathbf{u}_y \cdot \mathbf{N}_{nml}^{(1)}(\mathbf{r}_0) &= \left.\frac{1}{kb}\frac{d[krj_n(kr)]}{dkr}\right|_{r=b}\left.\frac{1}{\sin\theta}\frac{\partial Y_{nml}(\theta,\varphi)}{\partial \varphi}\right|_{\theta=0}.\end{aligned} \tag{5.59}$$

Taking the following calculations into account

$$\begin{aligned}Y_{nml}(0,\varphi) &= \begin{cases} 0, m \neq 0 \\ f_{1l}(\varphi), m = 0 \end{cases}, \\ \lim_{\theta\to 0}\frac{1}{\sin\theta}\frac{\partial Y_{nml}(\theta,\varphi)}{\partial\varphi} &= \begin{cases} 0, & m \neq 1 \\ \dfrac{n(n+1)}{2}\dfrac{\partial f_{1l}(\varphi)}{\partial\varphi}, & m = 1 \end{cases}, \\ \lim_{\theta\to 0}\frac{\partial Y_{nml}(\theta,\varphi)}{\partial\theta} &= \begin{cases} 0, & m \neq 1 \\ \dfrac{n(n+1)}{2} f_{1l}(\varphi), m = 1 \end{cases},\end{aligned} \tag{5.60}$$

one only needs to retain the terms for $m = 1$ in the field expansions. As a result, only the following terms in (5.58) remain

$$\begin{aligned}\mathbf{u}_y \cdot \mathbf{M}_{n1e}^{(1)}(\mathbf{r}_0) &= -\frac{n(n+1)}{2}j_n(kb), \\ \mathbf{u}_y \cdot \mathbf{N}_{n1o}^{(1)}(\mathbf{r}_0) &= \frac{n(n+1)}{2kb}\dot{\hat{j}}_n(kb).\end{aligned} \tag{5.61}$$

The radiated electric field (5.57) in the region $r > b$ is then reduced to

$$\mathbf{E} = \frac{k^2\eta I_0 l}{2}\sum_{n=1}^{\infty}\frac{n(n+1)}{N_{n1}^2}j_n(kb)\mathbf{M}_{n1e}^{(2)} - \frac{k\eta I_0 l}{2b}\sum_{n=1}^{\infty}\frac{n(n+1)}{N_{n1}^2}\dot{\hat{j}}_n(kb)\mathbf{N}_{n1o}^{(2)}, \tag{5.62}$$

where

$$\begin{aligned}\mathbf{M}_{n1e}^{(2)} &= h_n^{(2)}(kr)\frac{1}{\sin\theta}\frac{\partial Y_{n1e}(\theta,\varphi)}{\partial\varphi}\mathbf{u}_\theta - h_n^{(2)}(kr)\frac{\partial Y_{n1e}(\theta,\varphi)}{\partial\theta}\mathbf{u}_\varphi, \\ \mathbf{N}_{n1o}^{(2)} &= \frac{n(n+1)}{kr}h_n^{(2)}(kr)Y_{n1o}(\theta,\varphi)\mathbf{u}_r + \frac{1}{kr}\frac{d\left[krh_n^{(2)}(kr)\right]}{dkr}\frac{\partial Y_{n1o}(\theta,\varphi)}{\partial\theta}\mathbf{u}_\theta \\ &\quad + \frac{1}{kr}\frac{d\left[krh_n^{(2)}(kr)\right]}{dkr}\frac{1}{\sin\theta}\frac{\partial Y_{n1o}(\theta,\varphi)}{\partial\varphi}\mathbf{u}_\varphi.\end{aligned} \tag{5.63}$$

Explicitly, the components of the electric field are

$$
\begin{aligned}
E_r &= -\frac{\eta I_0 l}{4\pi r b}\sum_{n=1}^{\infty}(2n+1)\dot{\hat{j}}_n(kb)h_n^{(2)}(kr)P_n^1(\cos\theta)\sin\varphi,\\
E_\theta &= -\frac{k^2\eta I_0 l}{4\pi}\sum_{n=1}^{\infty}\frac{2n+1}{n(n+1)}j_n(kb)h_n^{(2)}(kr)\frac{P_n^1(\cos\theta)}{\sin\theta}\sin\varphi\\
&\quad -\frac{\eta I_0 l}{4\pi r b}\sum_{n=1}^{\infty}\frac{2n+1}{n(n+1)}\dot{\hat{j}}_n(kb)\dot{\hat{h}}_n^{(2)}(kr)\frac{dP_n^1(\cos\theta)}{d\theta}\sin\varphi,\\
E_\varphi &= -\frac{k^2\eta I_0 l}{4\pi}\sum_{n=1}^{\infty}\frac{2n+1}{n(n+1)}j_n(kb)h_n^{(2)}(kr)\frac{dP_n^1(\cos\theta)}{d\theta}\cos\varphi\\
&\quad -\frac{\eta I_0 l}{4\pi r b}\sum_{n=1}^{\infty}\frac{2n+1}{n(n+1)}\dot{\hat{j}}_n(kb)\dot{\hat{h}}_n^{(2)}(kr)\frac{P_n^1(\cos\theta)}{\sin\theta}\cos\varphi.
\end{aligned}
\tag{5.64}
$$

In the far-field region, the radiated electric field (5.62) can be written as

$$
\begin{aligned}
\mathbf{E} \approx &-\frac{e^{-jkr}}{r}\frac{k\eta I_0 l}{2}\sum_{n=1}^{\infty}\frac{j^{n+1}n(n+1)}{N_{n1}}j_n(kb)\mathbf{h}_{n1e}\\
&-\frac{e^{-jkr}}{r}\frac{\eta I_0 l}{2b}\sum_{n=1}^{\infty}\frac{j^n n(n+1)}{N_{n1}}\dot{\hat{j}}_n(kb)\mathbf{e}_{n1o},
\end{aligned}
\tag{5.65}
$$

which is composed of two types of vector modal functions $\mathbf{e}_{n1o}$ and $\mathbf{h}_{n1e}$. Figure 5.12 shows the first three vector modal functions for $n = 1, 2, 3$ and $m = 1$. The vector effective length is given by

$$
\mathbf{L}(\mathbf{u}_r) = 2\pi l\sum_{n=1}^{\infty}\frac{j^n n(n+1)}{N_{n1}}j_n(kb)\mathbf{h}_{n1e} + 2\pi l\frac{1}{kb}\sum_{n=1}^{\infty}\frac{j^{n-1}n(n+1)}{N_{n1}}\dot{\hat{j}}_n(kb)\mathbf{e}_{n1o}.
\tag{5.66}
$$

The radiated magnetic field in the region $r > b$ is determined by the second expression of (5.1):

$$
\mathbf{H}(\mathbf{r}) = \int_{V_0}\overleftrightarrow{\mathbf{G}}_m(\mathbf{r},\mathbf{r}')\cdot\mathbf{J}(\mathbf{r}')dV(\mathbf{r}') = I_0 l\overleftrightarrow{\mathbf{G}}_m(\mathbf{r},\mathbf{r}_0)\cdot\mathbf{u}_y.
\tag{5.67}
$$

Similarly, one must select the lower expression of (5.3) for the magnetic dyadic Green's function $\overleftrightarrow{\mathbf{G}}_m$:

$$
\overleftrightarrow{\mathbf{G}}_m(\mathbf{r},\mathbf{r}_0)\cdot\mathbf{u}_y = \sum_{n,m,l}\frac{-jk^2}{N_{nm}^2}\left[\mathbf{N}_{nml}^{(2)}(\mathbf{r})\mathbf{M}_{nml}^{(1)}(\mathbf{r}_0)\cdot\mathbf{u}_y + \mathbf{M}_{nml}^{(2)}(\mathbf{r})\mathbf{N}_{nml}^{(1)}(\mathbf{r}_0)\cdot\mathbf{u}_y\right].
$$

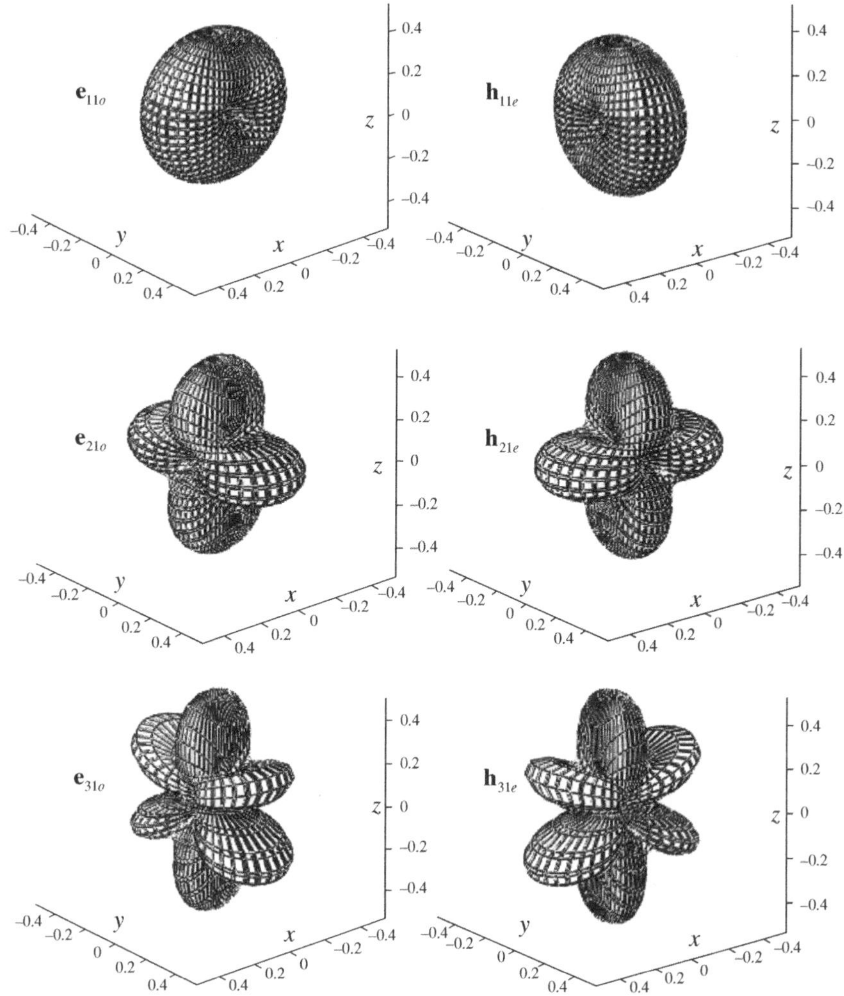

Figure 5.12 Vector modal functions.

Only the terms shown in (5.61) need to be considered in the above expansion, and (5.67) can be written as

$$\mathbf{H} = \frac{jk^2 I_0 l}{2} \sum_{n=1}^{\infty} \frac{n(n+1)}{N_{n1}^2} j_n(kb) \mathbf{N}_{n1e}^{(2)} - \frac{jkI_0 l}{2b} \sum_{n=1}^{\infty} \frac{n(n+1)}{N_{n1}^2} \dot{j}_n(kb) \mathbf{M}_{n1o}^{(2)}, \tag{5.68}$$

where

$$
\begin{aligned}
\mathbf{M}_{n1o}^{(2)} &= h_n^{(2)}(kr)\frac{P_n^1(\cos\theta)}{\sin\theta}\cos\varphi\mathbf{u}_\theta - h_n^{(2)}(kr)\frac{dP_n^1(\cos\theta)}{d\theta}\sin\varphi\mathbf{u}_\varphi,\\
\mathbf{N}_{n1e}^{(2)} &= \frac{n(n+1)}{kr}h_n^{(2)}(kr)P_n^1(\cos\theta)\cos\varphi\mathbf{u}_r + \frac{1}{kr}\overset{\therefore}{h}_n^{(2)}(kr)\frac{dP_n^1(\cos\theta)}{d\theta}\cos\varphi\mathbf{u}_\theta\\
&\quad - \frac{1}{kr}\overset{\therefore}{h}_n^{(2)}(kr)\frac{P_n^1(\cos\theta)}{\sin\theta}\sin\varphi\mathbf{u}_\varphi.
\end{aligned}
\tag{5.69}
$$

Substitution of (5.69) into (5.68) gives the components of the magnetic field

$$
\begin{aligned}
H_r &= \frac{jkI_0l}{4\pi r}\sum_{n=1}^{\infty}(2n+1)j_n(kb)h_n^{(2)}(kr)P_n^1(\cos\theta)\cos\varphi,\\
H_\theta &= \frac{jkI_0l}{4\pi r}\sum_{n=1}^{\infty}\frac{2n+1}{n(n+1)}j_n(kb)\overset{\therefore}{h}_n^{(2)}(kr)\frac{dP_n^1(\cos\theta)}{d\theta}\cos\varphi\\
&\quad - \frac{jkI_0l}{4\pi b}\sum_{n=1}^{\infty}\frac{2n+1}{n(n+1)}\overset{\therefore}{j}_n(kb)h_n^{(2)}(kr)\frac{P_n^1(\cos\theta)}{\sin\theta}\cos\varphi,\\
H_\varphi &= -\frac{jkI_0l}{4\pi r}\sum_{n=1}^{\infty}\frac{2n+1}{n(n+1)}j_n(kb)\overset{\therefore}{h}_n^{(2)}(kr)\frac{P_n^1(\cos\theta)}{\sin\theta}\sin\varphi\\
&\quad + \frac{jkI_0l}{4\pi b}\sum_{n=1}^{\infty}\frac{2n+1}{n(n+1)}\overset{\therefore}{j}_n(kb)h_n^{(2)}(kr)\frac{dP_n^1(\cos\theta)}{d\theta}\sin\varphi.
\end{aligned}
\tag{5.70}
$$

When the dipole is located at the origin ($b = 0$), only the term $n = 1$ is nonzero. In this case, the EM fields are simplified to

$$
\begin{aligned}
E_r &= -\frac{k\eta I_0 l}{2\pi r}h_1^{(2)}(kr)\sin\theta\sin\varphi,\\
E_\theta &= -\frac{k\eta I_0 l}{4\pi r}\overset{\therefore}{h}_1^{(2)}(kr)\cos\theta\sin\varphi,\\
E_\varphi &= -\frac{k\eta I_0 l}{4\pi r}\overset{\therefore}{h}_1^{(2)}(kr)\cos\varphi,
\end{aligned}
\tag{5.71}
$$

$$
\begin{aligned}
H_r &= 0,\\
H_\theta &= -\frac{jk^2I_0l}{4\pi}h_1^{(2)}(kr)\cos\varphi,\\
H_\varphi &= \frac{jk^2I_0l}{4\pi}h_1^{(2)}(kr)\cos\theta\sin\varphi.
\end{aligned}
\tag{5.72}
$$

In deriving (5.71) and (5.72), the following relation has been used:

$$
\frac{1}{kb}\overset{\therefore}{j}_1(kb)\xrightarrow[kb\to 0]{}\frac{2}{3}.
$$

Therefore, a horizontal dipole placed at the origin of the coordinate system only radiates TM fields. Especially, the components of the magnetic field in the rectangular coordinate system can be found from (5.72):

$$\begin{aligned} H_x &= \frac{jkI_0 l}{4\pi r}\left(1+\frac{1}{jkr}\right)e^{-jkr}\cos\theta, \\ H_y &= 0, \\ H_z &= -\frac{jkI_0 l}{4\pi r}\left(1+\frac{1}{jkr}\right)e^{-jkr}\sin\theta\cos\varphi. \end{aligned} \tag{5.73}$$

These agree with the conventional analysis [6].

5.6.2 Fields in the Region $r < b$

The fields in the region $r < b$ can be determined in a similar way, and they are given by

$$\begin{aligned} \mathbf{E} &= \frac{k^2\eta I_0 l}{2}\sum_{n=1}^{\infty}\frac{n(n+1)}{N_{n1}^2}h_n^{(2)}(kb)\mathbf{M}_{n1e}^{(1)} - \frac{k\eta I_0 l}{2b}\sum_{n=1}^{\infty}\frac{n(n+1)}{N_{n1}^2}\dot{\hat{h}}_n^{(2)}(kb)\mathbf{N}_{n1o}^{(1)}, \\ \mathbf{H} &= \frac{jk^2 I_0 l}{2}\sum_{n=1}^{\infty}\frac{n(n+1)}{N_{n1}^2}h_n^{(2)}(kb)\mathbf{N}_{n1e}^{(1)} - \frac{jkI_0 l}{2b}\sum_{n=1}^{\infty}\frac{n(n+1)}{N_{n1}^2}\dot{\hat{h}}_n^{(2)}(kb)\mathbf{M}_{n1o}^{(1)}. \end{aligned} \tag{5.74}$$

5.7 Loop

Consider a small circular current loop located at the (x, y)-plane and centered at the origin of the coordinate system

$$\mathbf{J}(\mathbf{r}) = I\delta(\rho - a)\delta(z)\mathbf{u}_\varphi, \tag{5.75}$$

where I is the current in the loop and a its radius, as illustrated in Figure 5.13. Let S denote the loop area. The loop current source is equivalent to a magnetic dipole defined by

$$\mathbf{J}_m(\mathbf{r}) = I_m l\delta(\mathbf{r})\mathbf{u}_z, \tag{5.76}$$

where $I_m l = j\omega\mu IS$. The fields generated by the magnetic dipole (5.76) can be obtained in a similar way to the electric dipole. The radiated electric field is determined by the first expression of (5.1):

$$\mathbf{E} = -\int_{V_0}\overleftrightarrow{\mathbf{G}}_m(\mathbf{r},\mathbf{r}')\cdot\mathbf{J}_m(\mathbf{r}')dV(\mathbf{r}') = -I_m l\overleftrightarrow{\mathbf{G}}_m(\mathbf{r},0)\cdot\mathbf{u}_z.$$

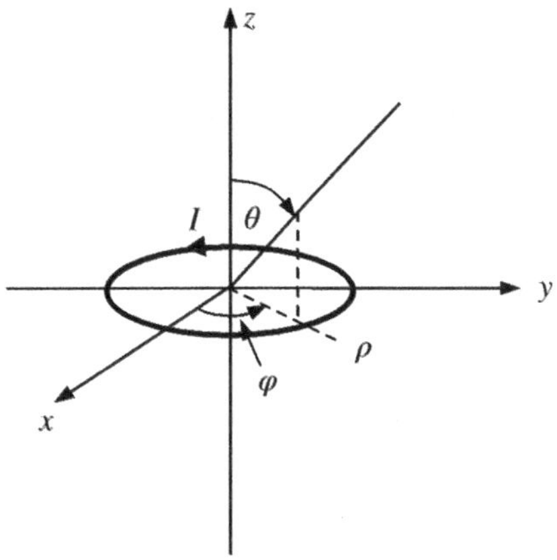

Figure 5.13 A current loop.

It can be found from (5.3) (lower expression) that

$$\overleftrightarrow{\mathbf{G}}_m(\mathbf{r},0)\cdot\mathbf{u}_z = -\frac{jk^2}{N_{10}^2}\mathbf{u}_z\cdot\mathbf{N}_{10e}^{(1)}(0)\mathbf{M}_{10e}^{(2)}(\mathbf{r}) = \frac{jk\sin\theta e^{-jkr}}{4\pi r}\left(1-j\frac{1}{kr}\right)\mathbf{u}_\varphi. \tag{5.77}$$

The radiated electric field from the loop current is then given by

$$\mathbf{E} = -\frac{jkI_m l\sin\theta e^{-jkr}}{4\pi r}\left(1-j\frac{1}{kr}\right)\mathbf{u}_\varphi. \tag{5.78}$$

The vector effective length of the loop is

$$\mathbf{L}(\mathbf{u}_r) = jkS\sin\theta\mathbf{u}_\varphi. \tag{5.79}$$

The radiated magnetic field can be determined from the second expression of (5.1):

$$\mathbf{H}(\mathbf{r}) = -j\frac{k}{\eta}\int_{V_0}\overleftrightarrow{\mathbf{G}}_e(\mathbf{r},\mathbf{r}')\cdot\mathbf{J}_m(\mathbf{r}')dV(\mathbf{r}') = -j\frac{k}{\eta}I_m l\overleftrightarrow{\mathbf{G}}_e(\mathbf{r},0)\cdot\mathbf{u}_z.$$

Using the lower expression of (5.2), one may find

$$\begin{aligned}\overleftrightarrow{\mathbf{G}}_e(\mathbf{r},0)\cdot\mathbf{u}_z &= \frac{-jk}{N_{10}^2}\mathbf{u}_{z'}\cdot\mathbf{N}_{10e}^{(1)}(0)\mathbf{N}_{10e}^{(2)}(\mathbf{r}) \\ &= \frac{j2\cos\theta e^{-jkr}}{4\pi r}\left[\frac{1}{kr}-j\frac{1}{(kr)^2}\right]\mathbf{u}_r - \frac{\sin\theta e^{-jkr}}{4\pi r}\left[1-j\frac{1}{kr}-\frac{1}{(kr)^2}\right]\mathbf{u}_\theta.\end{aligned} \tag{5.80}$$

The radiated magnetic field can be written as

$$\mathbf{H} = \frac{2kI_m l\cos\theta}{\eta}\frac{e^{-jkr}}{4\pi r}\left[\frac{1}{kr} - j\frac{1}{(kr)^2}\right]\mathbf{u}_r + j\frac{kI_m l\sin\theta}{\eta}\frac{e^{-jkr}}{4\pi r}\left[1 - j\frac{1}{kr} - \frac{1}{(kr)^2}\right]\mathbf{u}_\theta. \tag{5.81}$$

The above analysis indicates that a circular loop current only generates TE_{10e} modes. The field expressions (5.78) and (5.81) for the loop can also be obtained from the radiated dipole fields (5.46) and (5.52) by the principle of duality.

5.8 Spherical Dipole

Figure 5.14 shows a spherical dipole antenna, which is obtained by dividing a conducting sphere into halves and using them as the two arms of the dipole antenna. It will be assumed that the gap between the two hemispheres is sufficiently small so that the voltage source applied to the gap can be considered as impulse function

$$\mathbf{E}_t|_{r=a} = \frac{V}{a}\delta\left(\theta - \frac{\pi}{2}\right)\mathbf{u}_\theta, \tag{5.82}$$

where a is the radius of the conducting sphere. Since the antenna is rotationally symmetric, the radiated fields must be independent of the azimuth angle. In the field expansions (5.4), only the $\mathbf{M}_{n0e}^{(2)}$ and $\mathbf{N}_{n0e}^{(2)}$ modes are excited

$$\begin{aligned}\mathbf{M}_{n0e}^{(2)} &= h_n^{(2)}(kr)P_n^1(\cos\theta)\mathbf{u}_\varphi,\\ \mathbf{N}_{n0e}^{(2)} &= \frac{n(n+1)}{kr}h_n^{(2)}(kr)P_n(\cos\theta)\mathbf{u}_r - \frac{1}{kr}\hat{h}_n^{(2)}(kr)P_n^1(\cos\theta)\mathbf{u}_\theta.\end{aligned} \tag{5.83}$$

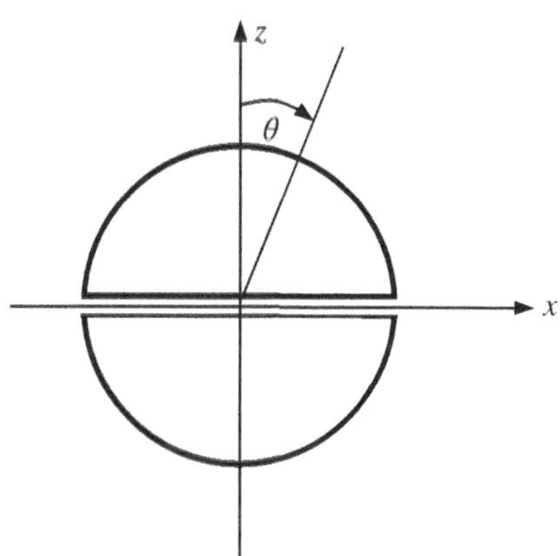

Figure 5.14 A spherical dipole.

Thus, the radiated fields by the spherical dipole may be expressed by

$$\begin{aligned}\mathbf{E} &= -\sum_n \left(\alpha_{n0e}^{(2)} \mathbf{M}_{n0e}^{(2)} + \beta_{n0e}^{(2)} \mathbf{N}_{n0e}^{(2)}\right),\\ \mathbf{H} &= \frac{1}{j\eta} \sum_n \left(\alpha_{n0e}^{(2)} \mathbf{N}_{nml}^{(2)} + \beta_{n0e}^{(2)} \mathbf{M}_{n0e}^{(2)}\right).\end{aligned} \tag{5.84}$$

On account of the excitation condition (5.82), the φ-component of the electric field must vanish, yielding $\alpha_{n0e}^{(2)} = 0$ and

$$\mathbf{E} = -\sum_n \beta_{n0e}^{(2)} \mathbf{N}_{n0e}^{(2)}, \quad \mathbf{H} = \frac{1}{j\eta} \sum_n \beta_{n0e}^{(2)} \mathbf{M}_{n0e}^{(2)}. \tag{5.85}$$

The field expressions in (5.85) indicate that a spherical dipole only radiates TM_{n0e} modes. In the far-field region, the above expressions reduce to

$$\begin{aligned}\mathbf{E} &= -\frac{e^{-jkr}}{kr} \sum_n j^n N_{nm} \beta_{n0e}^{(2)} \mathbf{e}_{n0e},\\ \mathbf{H} &= -\frac{1}{j\eta} \frac{e^{-jkr}}{kr} \sum_n j^{n+1} N_{nm} \beta_{n0e}^{(2)} \mathbf{h}_{n0e}.\end{aligned} \tag{5.86}$$

The transverse electric field at the spherical boundary $r = a$ has only an E_θ component

$$E_\theta\big|_{r=a} = \frac{1}{ka} \sum_n \beta_{n0e}^{(2)} \dot{\hat{h}}_n^{(2)}(ka) P_n^1(\cos\theta). \tag{5.87}$$

The expansion coefficients $\beta_{n0e}^{(2)}$ can be determined by multiplying both sides by $P_m^1(\cos\theta)\sin\theta$ and integrating from 0 to π on θ, and making use of the orthogonality relation of spherical harmonics, as follows:

$$\beta_{n0e}^{(2)} = \frac{2n+1}{2n(n+1)} \frac{ka}{\dot{\hat{h}}_n^{(2)}(ka)} \int_0^\pi E_\theta\big|_{r=a} P_n^1(\cos\theta)\sin\theta d\theta.$$

Inserting the excitation source (5.82) into the above expression gives

$$\beta_{n0e}^{(2)} = \frac{2n+1}{2n(n+1)} \frac{ka}{\dot{\hat{h}}_n^{(2)}(ka)} \frac{V}{a} P_n^1(0). \tag{5.88}$$

The vector effective length of the spherical dipole is

$$\mathbf{L}(\mathbf{u}_r) = \frac{4\pi Z}{k\eta} \sum_n j^{n-1} \frac{2n+1}{2n(n+1)} \frac{N_{nm} P_n^1(0)}{\dot{\hat{h}}_n^{(2)}(ka)} \mathbf{e}_{n0e}, \tag{5.89}$$

where $Z = V/I$ is the input impedance of the spherical dipole.

5.9 Dipole Near Conducting Sphere

Figure 5.15 shows a radially directed electric dipole placed above a conducting sphere of radius a, whose far-zone fields have been investigated in [7]. The radiated fields from the dipole without the conducting sphere are now considered as the incident fields upon the conducting sphere. In the region $r < b$, the incident fields are determined by (5.53) and (5.54):

$$\mathbf{E}^{in} = -\frac{k\eta I_0 l}{4\pi b}\sum_n (2n+1)h_n^{(2)}(kb)\mathbf{N}_{n0e}^{(1)}, \tag{5.90}$$

$$\mathbf{H}^{in} = -\frac{jkI_0 l}{4\pi b}\sum_n (2n+1)h_n^{(2)}(kb)\mathbf{M}_{n0e}^{(1)}. \tag{5.91}$$

The induced current on the sphere by the incident fields produces a scattered field which, in view of the symmetry, may be represented by (5.4):

$$\begin{aligned}\mathbf{E}^s &= -\sum_n \left(\alpha_{n0e}^{(2)}\mathbf{M}_{n0e}^{(2)} + \beta_{n0e}^{(2)}\mathbf{N}_{n0e}^{(2)}\right),\\ \mathbf{H}^s &= \frac{1}{j\eta}\sum_n \left(\alpha_{n0e}^{(2)}\mathbf{N}_{n0e}^{(2)} + \beta_{n0e}^{(2)}\mathbf{M}_{noe}^{(2)}\right).\end{aligned} \tag{5.92}$$

The total electric field must satisfy the boundary condition on the conducting sphere

$$\mathbf{u}_r \times \left(\mathbf{E}^{in} + \mathbf{E}^s\right)\Big|_{r=a} = 0.$$

Substituting (5.90) and the first expression of (5.92) into the above equation yields

$$\left\{\frac{k^2\eta I_0 l}{kb}\sum_n \frac{2n+1}{4\pi}h_n^{(2)}(kb)\mathbf{u}_r\times\mathbf{N}_{n0e}^{(1)} + \sum_n\left(\alpha_{n0e}^{(2)}\mathbf{u}_r\times\mathbf{M}_{n0e}^{(2)} + \beta_{n0e}^{(2)}\mathbf{u}_r\times\mathbf{N}_{n0e}^{(2)}\right)\right\}_{r=a} = 0,$$

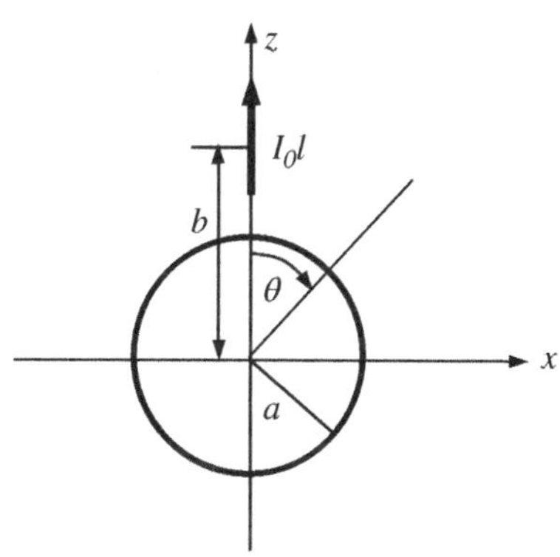

Figure 5.15 A dipole near conducting sphere.

which gives $\alpha_{n0e}^{(2)} = 0$ and

$$\frac{k\eta I_0 l}{4\pi b}\sum_n (2n+1)h_n^{(2)}(kb)\hat{\dot{j}}_n(ka)P_n^1(\cos\theta) + \sum_n \beta_{n0e}^{(2)}\hat{\dot{h}}_n^{(2)}(ka)P_n^1(\cos\theta) = 0.$$

The above equation can be used to determine the expansion coefficients $\beta_{n0e}^{(2)}$:

$$\beta_{n0e}^{(2)} = -\frac{k\eta I_0 l(2n+1)}{4\pi b}\frac{\hat{\dot{j}}_n(ka)}{\hat{\dot{h}}_n^{(2)}(ka)}h_n^{(2)}(kb).$$

The scattered fields (5.92) can thus be written as

$$\begin{aligned}
\mathbf{E}^s &= -\frac{k\eta I_0 l}{4\pi b}\sum_n \frac{\hat{\dot{j}}_n(ka)}{\hat{\dot{h}}_n^{(2)}(ka)}(2n+1)h_n^{(2)}(kb)\mathbf{N}_{n0e}^{(2)},\\
\mathbf{H}^s &= -\frac{jkI_0 l}{4\pi b}\sum_n \frac{\hat{\dot{j}}_n(ka)}{\hat{\dot{h}}_n^{(2)}(ka)}(2n+1)h_n^{(2)}(kb)\mathbf{M}_{noe}^{(2)}.
\end{aligned} \tag{5.93}$$

The total radiated fields in the region $r > b$ can be determined by adding the incident fields (5.41) and (5.51) to the scattered fields in (5.93):

$$\begin{aligned}
\mathbf{E} &= -\frac{k\eta I_0 l}{4\pi b}\sum_n \left[\frac{j_n(kb)}{h_n^{(2)}(kb)} - \frac{\hat{\dot{j}}_n(ka)}{\hat{\dot{h}}_n^{(2)}(ka)}\right](2n+1)h_n^{(2)}(kb)\mathbf{N}_{n0e}^{(2)},\\
\mathbf{H} &= -\frac{jkI_0 l}{4\pi b}\sum_n \left[\frac{j_n(kb)}{h_n^{(2)}(kb)} - \frac{\hat{\dot{j}}_n(ka)}{\hat{\dot{h}}_n^{(2)}(ka)}\right](2n+1)h_n^{(2)}(kb)\mathbf{M}_{noe}^{(2)}.
\end{aligned} \tag{5.94}$$

Equations (5.94) indicate that a dipole near a conducting sphere radiates TM_{n0e} only. Explicitly, the components of the electric field are

$$\begin{aligned}
E_r &= -\frac{\eta I_0 l}{4\pi br}\sum_n \left[\frac{j_n(kb)}{h_n^{(2)}(kb)} - \frac{\hat{\dot{j}}_n(ka)}{\hat{\dot{h}}_n^{(2)}(ka)}\right]n(n+1)(2n+1)h_n^{(2)}(kb)h_n^{(2)}(kr)P_n(\cos\theta),\\
E_\theta &= \frac{\eta I_0 l}{4\pi br}\sum_n \left[\frac{j_n(kb)}{h_n^{(2)}(kb)} - \frac{\hat{\dot{j}}_n(ka)}{\hat{\dot{h}}_n^{(2)}(ka)}\right](2n+1)h_n^{(2)}(kb)\hat{\dot{h}}_n^{(2)}(kr)P_n^1(\cos\theta),\\
H_\varphi &= -\frac{jkI_0 l}{4\pi b}\sum_n \left[\frac{j_n(kb)}{h_n^{(2)}(kb)} - \frac{\hat{\dot{j}}_n(ka)}{\hat{\dot{h}}_n^{(2)}(ka)}\right](2n+1)h_n^{(2)}(kb)h_n^{(2)}(kr)P_n^1(\cos\theta).
\end{aligned} \tag{5.95}$$

Using the asymptotic formulas of spherical Hankel functions, the far-field expressions are found to be

$$E_\theta = -\frac{j\eta I_0 l}{b}\frac{e^{-jkr}}{4\pi r}\sum_n\left[\frac{j_n(kb)}{h_n^{(2)}(kb)} - \frac{\hat{\dot{j}}_n(ka)}{\hat{\dot{h}}_n^{(2)}(ka)}\right]j^{n+1}(2n+1)h_n^{(2)}(kb)P_n^1(\cos\theta),$$

$$H_\varphi = -\frac{jI_0 l}{b}\frac{e^{-jkr}}{4\pi r}\sum_n\left[\frac{j_n(kb)}{h_n^{(2)}(kb)} - \frac{\hat{\dot{j}}_n(ka)}{\hat{\dot{h}}_n^{(2)}(ka)}\right]j^{n+1}(2n+1)h_n^{(2)}(kb)P_n^1(\cos\theta).$$

(5.96)

The vector effective length is given by

$$\mathbf{L}(\mathbf{u}_r) = \mathbf{u}_\theta\frac{l}{kb}\sum_n\left[\frac{j_n(kb)}{h_n^{(2)}(kb)} - \frac{\hat{\dot{j}}_n(ka)}{\hat{\dot{h}}_n^{(2)}(ka)}\right]j^{n+1}(2n+1)h_n^{(2)}(kb)P_n^1(\cos\theta). \tag{5.97}$$

5.10 Finite Length Wire Antenna

Figure 5.8 shows a finite length wire antenna with current distribution

$$\mathbf{J}(\mathbf{r}) = \mathbf{u}_z I(z)\delta(x)\delta(y), 0 < z < l. \tag{5.98}$$

The radiated fields are determined by (5.1):

$$\begin{aligned}\mathbf{E}(\mathbf{r}) &= -jk\eta\int_0^l I(z')\overleftrightarrow{\mathbf{G}}_e(\mathbf{r},\mathbf{r}')\cdot\mathbf{u}_{z'}dz',\\ \mathbf{H}(\mathbf{r}) &= \int_0^l I(z')\overleftrightarrow{\mathbf{G}}_m(\mathbf{r},\mathbf{r}')\cdot\mathbf{u}_{z'}dz'.\end{aligned} \tag{5.99}$$

5.10.1 Fields in the Region $r > l$

In the region $r > l/2$, the dyadic Green's functions must be selected as the lower expressions of (5.2) and (5.3). Since the radiated fields are independent of the azimuthal angle, the dyadic Green's functions in (5.99) can be written as

$$\overleftrightarrow{\mathbf{G}}_e(\mathbf{r},\mathbf{r}')\cdot\mathbf{u}_{z'} = \sum_n\frac{-jk}{N_{n0}^2}\left[\mathbf{M}_{n0e}^{(2)}(\mathbf{r})\mathbf{M}_{n0e}^{(1)}(\mathbf{r}')\cdot\mathbf{u}_{z'} + \mathbf{N}_{n0e}^{(2)}(\mathbf{r})\mathbf{N}_{n0e}^{(1)}(\mathbf{r}')\cdot\mathbf{u}_{z'}\right],$$

$$\overleftrightarrow{\mathbf{G}}_m(\mathbf{r},\mathbf{r}')\cdot\mathbf{u}_{z'} = \sum_n\frac{-jk^2}{N_{n0}^2}\left[\mathbf{N}_{n0e}^{(2)}(\mathbf{r})\mathbf{M}_{n0e}^{(1)}(\mathbf{r}')\cdot\mathbf{u}_{z'} + \mathbf{M}_{n0e}^{(2)}(\mathbf{r})\mathbf{N}_{n0e}^{(1)}(\mathbf{r}')\cdot\mathbf{u}_{z'}\right].$$

Similar to (5.26), one may find

$$\mathbf{u}_{z'} \cdot \mathbf{M}_{n0e}^{(1)}(0,0,z') = 0,$$
$$\mathbf{u}_{z'} \cdot \mathbf{N}_{n0e}^{(1)}(0,0,z') = \frac{n(n+1)}{kr'} j_n(kr').$$

The EM fields in (5.99) can be expressed as

$$\begin{aligned} \mathbf{E} &= -\frac{k^2\eta}{4\pi}\sum_{n=1}^{\infty}(2n+1)\mathbf{N}_{n0e}^{(2)}\int_0^l I(z')\frac{j_n(kz')}{kz'}dz', \\ \mathbf{H} &= -\frac{jk^2}{4\pi}\sum_n(2n+1)\mathbf{M}_{n0e}^{(2)}\int_0^l I(z')\frac{j_n(kz')}{kz'}dz', \end{aligned} \tag{5.100}$$

where

$$\begin{aligned} \mathbf{M}_{n0e}^{(2)} &= h_n^{(2)}(kr)P_n^1(\cos\theta)\mathbf{u}_\varphi, \\ \mathbf{N}_{n0e}^{(2)} &= \frac{n(n+1)}{kr}h_n^{(2)}(kr)P_n(\cos\theta)\mathbf{u}_r - \frac{1}{kr}\dot{h}_n^{(2)}(kr)P_n^1(\cos\theta)\mathbf{u}_\theta. \end{aligned} \tag{5.101}$$

Therefore, a dipole of finite length radiates TM_{n0e} modes. Upon introducing the first expression of (5.101) into the second expression of (5.100), one may rewrite the magnetic field as

$$\begin{aligned} \mathbf{H} &= -\frac{jk}{4\pi}\mathbf{u}_\varphi\int_0^l \frac{I(z')}{r'}\left[\sum_n(2n+1)j_n(kz')h_n^{(2)}(kr)P_n^1(\cos\theta)\right]dz' \\ &= \frac{jk}{4\pi}\mathbf{u}_\varphi\int_0^l \frac{1}{z'}I(z')\frac{\partial}{\partial\theta}\left[\sum_n(2n+1)j_n(kz')h_n^{(2)}(kr)P_n(\cos\theta)\right]dz' \\ &= -\mathbf{u}_\varphi\int_0^l \frac{I(z')}{z'}\frac{\partial}{\partial\theta}\frac{e^{-jkR}}{4\pi R}dz', \end{aligned} \tag{5.102}$$

where use has been made of the spherical wave expansion for the free-space Green's function

$$\frac{e^{-jkR}}{4\pi R} = -\frac{jk}{4\pi}\sum_n(2n+1)j_n(kz')h_n^{(2)}(kr)P_n(\cos\theta).$$

Let $\rho = \sqrt{x^2+y^2}$ be the radius in cylindrical coordinate system as illustrated in Figure 5.8. Then,

$$\frac{\partial}{\partial\theta}\frac{e^{-jkR}}{R} = \frac{xz}{\rho}\frac{\partial}{\partial x}\frac{e^{-jkR}}{R} + \frac{yz}{\rho}\frac{\partial}{\partial y}\frac{e^{-jkR}}{R} - \rho\frac{\partial}{\partial z}\frac{e^{-jkR}}{R} = z'\frac{\partial}{\partial\rho}\frac{e^{-jkR}}{R}.$$

The magnetic field (5.102) can thus be written as

$$\mathbf{H} = -\mathbf{u}_\varphi \frac{\partial}{\partial \rho} \int_0^l I(z') \frac{e^{-jkR}}{4\pi R} dz', \tag{5.103}$$

which agrees with the conventional analysis based on integral representation [8].

5.10.2 The Fields in the Region $r < l$

In this region, the dyadic Green's functions in (5.99) must be selected as the upper expressions of (5.2) and (5.3):

$$\overleftrightarrow{\mathbf{G}}_e(\mathbf{r}, \mathbf{r}') \cdot \mathbf{u}_{z'} = \sum_n \frac{-jk}{N_{n0}^2} \left[\mathbf{M}_{n0e}^{(1)}(\mathbf{r}) \mathbf{M}_{n0e}^{(2)}(\mathbf{r}') \cdot \mathbf{u}_{z'} + \mathbf{N}_{n0e}^{(1)}(\mathbf{r}) \mathbf{N}_{n0e}^{(2)}(\mathbf{r}') \cdot \mathbf{u}_{z'} \right],$$

$$\overleftrightarrow{\mathbf{G}}_m(\mathbf{r}, \mathbf{r}') \cdot \mathbf{u}_{z'} = \sum_n \frac{-jk^2}{N_{n0}^2} \left[\mathbf{N}_{n0e}^{(1)}(\mathbf{r}) \mathbf{M}_{n0e}^{(2)}(\mathbf{r}') \cdot \mathbf{u}_{z'} + \mathbf{M}_{n0e}^{(1)}(\mathbf{r}) \mathbf{N}_{n0e}^{(2)}(\mathbf{r}') \cdot \mathbf{u}_{z'} \right].$$

Similarly, one may find

$$\mathbf{u}_{z'} \cdot \mathbf{M}_{n0e}^{(2)}(0, 0, z') = 0,$$

$$\mathbf{u}_{z'} \cdot \mathbf{N}_{n0e}^{(2)}(0, 0, z') = \frac{n(n+1)}{kr'} h_n^{(2)}(kr').$$

The radiated fields are given by (5.99) and can be written as

$$\begin{aligned} \mathbf{E} &= -\frac{k^2 \eta}{4\pi} \sum_n (2n+1) \mathbf{N}_{n0e}^{(1)} \int_0^l I(z') \frac{h_n^{(2)}(kz')}{kz'} dz', \\ \mathbf{H} &= \frac{-jk^2}{4\pi} \sum_n (2n+1) \mathbf{M}_{n0e}^{(1)} \int_0^l I(z') \frac{h_n^{(2)}(kz')}{kz'} dz', \end{aligned} \tag{5.104}$$

where

$$\begin{aligned} \mathbf{M}_{n0e}^{(1)} &= j_n(kr) P_n^1(\cos\theta) \mathbf{u}_\varphi, \\ \mathbf{N}_{n0e}^{(1)} &= \frac{n(n+1)}{kr} j_n(kr) P_n(\cos\theta) \mathbf{u}_r - \frac{1}{kr} \dot{j}_n(kr) P_n^1(\cos\theta) \mathbf{u}_\theta. \end{aligned} \tag{5.105}$$

Following the same procedure, the magnetic field in (5.104) is found as follows:

$$\begin{aligned} \mathbf{H} &= \frac{-jk}{4\pi} \mathbf{u}_\varphi \int_0^l \frac{I(z')}{z'} \left[\sum_n (2n+1) j_n(kr) h_n^{(2)}(k|z'|) P_n^1(\cos\theta) \right] dz' \\ &= -\mathbf{u}_\varphi \frac{\partial}{\partial \rho} \int_0^l I(z') \frac{e^{-jkR}}{4\pi R} dz', \end{aligned}$$

which is identical with (5.103).

The electric field in both regions can be determined from

$$\mathbf{E} = \frac{1}{j\omega\varepsilon}\nabla \times \mathbf{H}. \tag{5.106}$$

Example 5.3 For an infinitesimal dipole, the electric current may be considered as a constant, i.e. $I(z') = I_0$ as $l \to 0$. The integral in (5.100) becomes

$$C_n = \int_0^l I(z')\frac{j_n(kz')}{kz'}dz' = \frac{I_0}{k}\int_0^{kl}\frac{j_n(z')}{z'}dz'.$$

As $l \to 0$, only $n = 1$ has significant contribution, and one may find $C_1 = (1/3)I_0 l$. Thus, (5.100) is simplified to

$$\mathbf{E} = -\frac{k^2\eta I_0 l}{4\pi}\mathbf{N}_{10e}^{(2)}, \quad \mathbf{H}(\mathbf{r}) = -\frac{jk^2 I_0 l}{4\pi}\mathbf{M}_{10e}^{(2)}. \tag{5.107}$$

These agree with (5.44) and (5.52). □

5.11 Aperture Antenna

Aperture antennas are often used at microwave and the millimeter wave frequency regime for their high directivity. The most commonly used aperture antennas are slot antenna, microstrip patch antenna, horn antenna, lenses, and reflector antenna. The geometry of the aperture may be rectangular, circular, or may have any other shape. Aperture antennas are widely used in wireless communication systems and can be flush mounted onto the surface of vehicle, aircraft, and spacecraft. The analysis of aperture antenna can be carried out by using the field equivalence principle, by means of which the actual aperture can be replaced by equivalent sources as illustrated in Figure 5.2. Consider an arbitrary aperture antenna fed by a waveguide shown in Figure 5.1b. According to (5.1) and (5.9), the radiated fields by the equivalent sources are

$$\begin{aligned}
\mathbf{E}(\mathbf{r}) &= -jk\eta\int_{\partial V_0}\overleftrightarrow{\mathbf{G}}_e(\mathbf{r},\mathbf{r}')\cdot\mathbf{J}(\mathbf{r}')dS(\mathbf{r}') - \int_{\partial V_0}\overleftrightarrow{\mathbf{G}}_m(\mathbf{r},\mathbf{r}')\cdot\mathbf{J}_m(\mathbf{r}')dS(\mathbf{r}'),\\
\mathbf{H}(\mathbf{r}) &= -j\frac{k}{\eta}\int_{\partial V_0}\overleftrightarrow{\mathbf{G}}_m(\mathbf{r},\mathbf{r}')\cdot\mathbf{J}_m(\mathbf{r}')dS(\mathbf{r}') + \int_{\partial V_0}\overleftrightarrow{\mathbf{G}}_m(\mathbf{r},\mathbf{r}')\cdot\mathbf{J}(\mathbf{r}')dS(\mathbf{r}'),
\end{aligned} \tag{5.108}$$

where the magnetic current only exists in the feeding aperture (i.e. the input terminal of antenna or the reference plane). To determine the equivalent magnetic current on the reference plane, one has to use the excitation conditions. The fields in the feeding waveguide may be expressed as [see (2.26)]

$$\begin{aligned} \mathbf{u}_z \times \mathbf{E}(\mathbf{r}) &= \sum_{n=1}^{\infty} \mathbf{u}_z \times \mathbf{e}_n(\mathbf{r}) V_n(z), \\ \mathbf{u}_z \times \mathbf{H}(\mathbf{r}) &= -\sum_{n=1}^{\infty} \mathbf{e}_n(\mathbf{r}) I_n(z), \end{aligned} \tag{5.109}$$

where $\mathbf{e}_n$ ($n = 1, 2, \ldots$) are the normalized vector modal functions of the feeding waveguide and

$$\begin{aligned} V_n(z) &= A_n e^{-j\beta_n z} + B_n e^{j\beta_n z}, \\ I_n(z) &= Z_{wn}^{-1}\left(A_n e^{-j\beta_n z} - B_n e^{j\beta_n z}\right), \end{aligned}$$

are the modal voltages and currents with

$$Z_{wn} = \begin{cases} \eta, \text{TEM mode} \\ \eta k/\beta_n, \text{TE mode} \\ \eta \beta_n/k, \text{TM mode} \end{cases},$$

$$\beta_n = \begin{cases} k, \text{TEM mode} \\ \sqrt{k^2 - k_{cn}^2}, \text{ TE or TM mode} \end{cases},$$

being the wave impedances and propagation constants of the modes. If the feeding waveguide is in a single-mode operation, the modal voltages and currents can be expressed by

$$\begin{aligned} V_n(z) &= A_{n1} e^{-j\beta_n z} + B_n e^{j\beta_n z}, \\ I_n(z) &= \left(A_{n1} e^{-j\beta_n z} - B_n e^{j\beta_n z}\right) Z_{wn}^{-1}, \end{aligned}$$

where $A_{n1} = \begin{cases} \delta, n = 1 \\ 0, n \neq 1 \end{cases}$, and δ is a constant (amplitude of the incident field) for a transmitting antenna excited by the dominant mode, and is set to zero for a receiving antenna. On the reference plane ($z = 0$), (5.109) may be expressed in equivalent currents as

$$\begin{aligned} \mathbf{J}_{ms}(\mathbf{r}) &= -\mathbf{u}_z \times \mathbf{e}_1(\mathbf{r})(\delta + B_1) - \sum_{n=2}^{\infty} \mathbf{u}_z \times \mathbf{e}_n(\mathbf{r}) B_n, \\ \mathbf{J}_s(\mathbf{r}) &= -\mathbf{e}_1(\mathbf{r})(\delta - B_1) Z_{w1}^{-1} + \sum_{n=2}^{\infty} \mathbf{e}_n(\mathbf{r}) B_n Z_{wn}^{-1}. \end{aligned}$$

By use of the orthogonality condition of the vector modal functions in the feeding waveguide, the expansion coefficients can be determined by the second equation

$$B_1 = \delta + Z_{w1} \int_{\Omega} \mathbf{J} \cdot \mathbf{e}_1 d\Omega, B_n = Z_{wn} \int_{\Omega} \mathbf{J} \cdot \mathbf{e}_n d\Omega.$$

Hence, the equivalent magnetic current on the feeding aperture may be expressed in terms of the equivalent electric current on the same feeding aperture

$$\begin{aligned} \mathbf{J}_m(\mathbf{r}) &= -2\delta \mathbf{u}_z \times \mathbf{e}_1(\mathbf{r}) - \sum_{n=1}^{\infty} \mathbf{u}_z \times \mathbf{e}_n(\mathbf{r}) Z_{wn} \int_{\Omega} \mathbf{J}(\mathbf{r}') \cdot \mathbf{e}_n(\mathbf{r}') d\Omega(\mathbf{r}') \\ &= -2\delta \mathbf{u}_z \times \mathbf{e}_1(\mathbf{r}) - \int_{\Omega} \mathbf{J}(\mathbf{r}') \cdot \sum_{n=1}^{\infty} [Z_{wn} \mathbf{e}_n(\mathbf{r}') \mathbf{u}_z \times \mathbf{e}_n(\mathbf{r})] d\Omega(\mathbf{r}'). \end{aligned} \tag{5.110}$$

The electric current $\mathbf{J}$ on the whole boundary of the source region can be determined by solving an integral equation that contains the equivalent electric current on the aperture and the induced electric current on the conducting surface.

Example 5.4 A magnetic field integral equation (MFIE) of the second kind for an aperture antenna can be obtained as follows [9]:

$$\begin{aligned} &-\frac{1}{2}\mathbf{J}(\mathbf{r}) + \mathbf{u}_n(\mathbf{r}) \times \int_{\partial V_0} \mathbf{J}(\mathbf{r}') \times \nabla' G(\mathbf{r}, \mathbf{r}') dS(\mathbf{r}') \\ &+ \mathbf{u}_n(\mathbf{r}) \times \sum_{n=1}^{\infty} Z_{wn} \mathbf{G}_{hn}(\mathbf{r}) \int_{A} \mathbf{J}(\mathbf{r}') \cdot \mathbf{e}_n(\mathbf{r}') d\Omega(\mathbf{r}') = \mathbf{F}_h(\mathbf{r}), \end{aligned} \tag{5.111}$$

where $G(\mathbf{r}, \mathbf{r}') = e^{-jk|\mathbf{r}-\mathbf{r}'|}/4\pi|\mathbf{r}-\mathbf{r}'|$, $\mathbf{F}_h(\mathbf{r}) = -2\delta \mathbf{u}_n(\mathbf{r}) \times \mathbf{G}_{h1}(\mathbf{r})$ and

$$\begin{aligned} \mathbf{G}_{hn}(\mathbf{r}) &= \frac{jk}{\eta} \left\{ \int_{\Omega} G(\mathbf{r}, \mathbf{r}') \mathbf{u}_z \times \mathbf{e}_n(\mathbf{r}') d\Omega(\mathbf{r}') \right. \\ &\left. - \frac{1}{k^2} \int_{\Omega} \nabla_s' \cdot [\mathbf{u}_z \times \mathbf{e}_n(\mathbf{r}')] \nabla' G(\mathbf{r}, \mathbf{r}') d\Omega(\mathbf{r}') \right\}. \end{aligned}$$

Let us consider an aperture antenna fed by a coaxial line consisting of an inner conductor of radius a and an outer conductor of radius b with $b = 2a$, as shown in

Figure 5.16. The infinite flange has been truncated to a finite annular region. The number of the quadrilateral elements used in the numerical calculation is $N = 445$. It is assumed that only dominant mode (TEM mode) is propagating in the coaxial line and the reference plane is right at the aperture ($L = 0$). The radiation resistance and reactance have been calculated by the MFIE and compared to the analytical results from [10]. A perfect agreement has been obtained as shown in Figures 5.17 and 5.18. Note that the operating frequency is limited between the cutoff frequency $k_c a = 0$ of dominant TEM mode and the cutoff frequency $k_c a \approx 0.68$ of the first higher order TE_{11} mode.

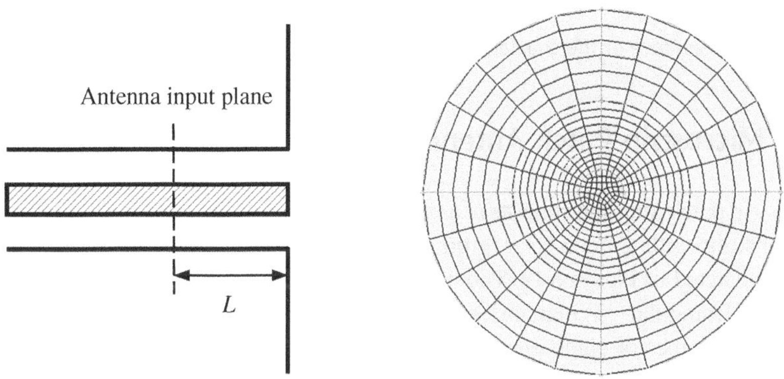

Figure 5.16 A coaxial aperture antenna with infinite conducting flange and mesh.

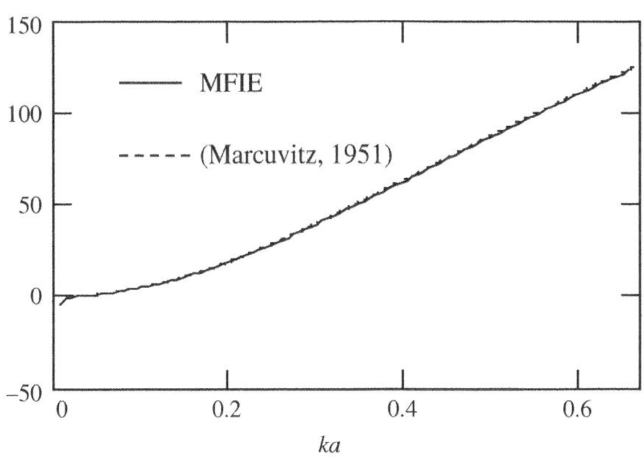

Figure 5.17 Radiation resistance of the coaxial aperture antenna with infinite flange (L = 0).

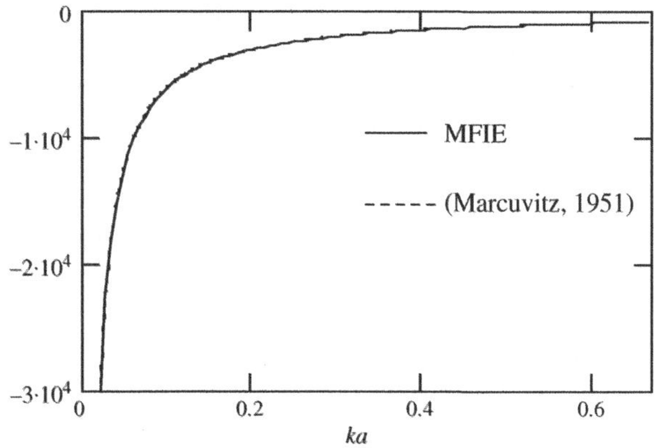

Figure 5.18 Reactance of the coaxial aperture antenna with infinite flange ($L = 0$).

5.12 Microstrip Patch Antenna

A microstrip patch antenna consists of a metallic patch bonded to an insulating dielectric substrate with a metal layer (ground) bonded to the opposite side of the substrate, as depicted in Figure 5.19. The metallic patch can take any shapes, such as rectangular, triangular, circular, disk sector, elliptical, annular ring, and square ring. Some typical shapes of patch elements are shown in Figure 5.20. The microstrip patch antennas have found wide applications for their advantages of low profile, low cost, and light weight; they can be shaped to conform to curved surfaces, and are easy to integrate with other circuits and form large array, and they allow both linear and circular polarizations [11–23]. The microstrip patch antennas also have some disadvantages such as low gain, low efficiency, low power-handling capability, and narrow bandwidth. Typical dimensions for a microstrip patch antenna are

$$\frac{\lambda}{3} < \text{patch size} < \frac{\lambda}{2}, \ 0.003\lambda < h < 0.05\lambda.$$

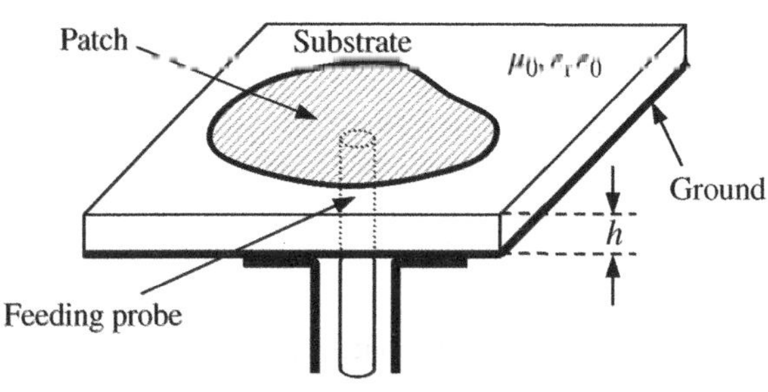

Figure 5.19 Patch antenna fed by a coaxial probe.

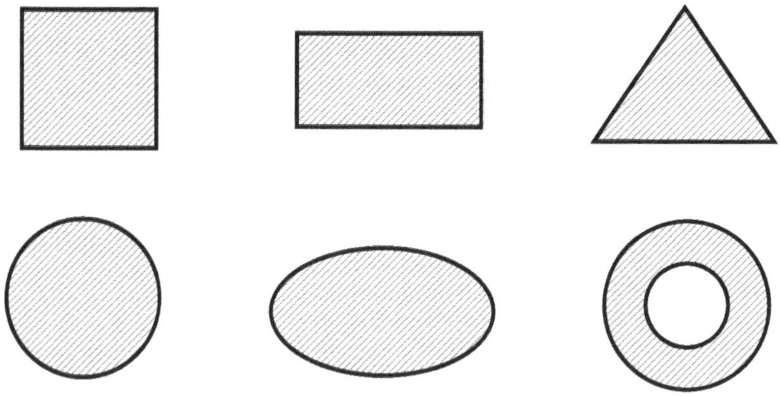

Figure 5.20 Typical shapes of patch antennas.

The dielectric constant ε_r of the substrate is usually selected in the range of $2.2 < \varepsilon_r < 12$. A patch antenna can be fed by a microstrip line or by a coaxial line with the inner conductor terminated on the patch. In both cases, the exciting source of the patch antennas can be represented by a current distribution $\mathbf{J} = \mathbf{u}_z J_z$, which is independent of z due to $h << \lambda$. This implies that the charge distribution $\rho = 0$. According to the equivalent principle, the radiated fields by the patch antenna are generated by the magnetic current on the side wall and the electric current on the upper surface of the patch along with the electric current on the side wall. If the patch is very thin (so that most of the current flows on the lower surface of the patch), only the radiated fields generated by the magnetic current on the side wall dominate, and a magnetic side wall may be introduced along the perimeter of the patch to simulate the open circuit, as illustrated in Figure 5.21. Let V_0 denote the source region, whose boundary ∂V_0 consists of the lower surface of the top patch, the upper surface of the bottom ground plane, and the side wall. In the interior of the source region V_0, the z component of the electric field satisfies

$$\left(\nabla^2 + k_e^2\right)E_z(x,y) = j\omega\mu_0 J_z(x,y), \tag{5.112}$$

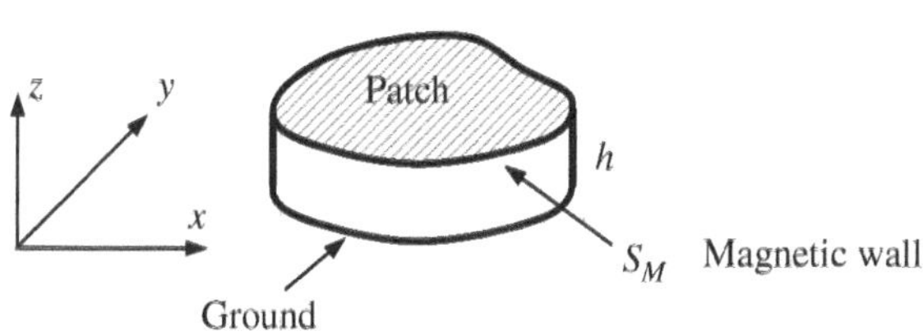

Figure 5.21 Cavity model.

where $k_e = \omega\sqrt{\mu_0\varepsilon_e}, \varepsilon_e = \varepsilon(1-j(\sigma/\omega\varepsilon)), \varepsilon = \varepsilon_r\varepsilon_0$. Equation (5.112) can be solved by using the orthonormal set of the eigenfunctions (modal functions) e_{pq} of the corresponding homogeneous equation for the electric field, which satisfy the equation

$$\begin{cases} \left(\nabla^2 + k_{pq}^2\right)e_{pq}(x,y) = 0, (x,y) \in P \\ \dfrac{\partial e_{pq}}{\partial n} = 0, (x,y) \in \partial P \end{cases}, \tag{5.113}$$

and the orthonormal condition

$$\int_P e_{pq}(x,y)e_{p'q'}(x,y)dxdy = \delta_{pp'}\delta_{qq'},$$

where P denotes the patch area and ∂P its boundary. The eigenvalue problem (5.113) is similar to that for TE modes in the waveguide theory. Once the eigenvalue problem (5.113) is solved, the magnetic modal field may be determined from

$$\mathbf{h}_{pq} = \frac{1}{jk\eta}\mathbf{u}_z \times \nabla e_{pq}. \tag{5.114}$$

Apparently, the modal fields are transverse magnetic with respect to z and they are usually termed as TM_{pq} modes. The resonant frequencies of these modes are given by

$$f_{pq} = \frac{k_{pq}}{2\pi\sqrt{\mu_0\varepsilon}}. \tag{5.115}$$

The induced modal currents on the lower surface of the patch are

$$\mathbf{J}_{spq} = -\mathbf{u}_z \times \mathbf{h}_{pq}. \tag{5.116}$$

The electric field inside the cavity may be expanded in terms of the modal functions as follows:

$$E_z(x,y) = \sum_{p,q} a_{pq}e_{pq}(x,y).$$

Substituting the expansion into (5.112) and using the orthogonal property of the eigenfunctions, one finds

$$E_z(x,y) = j\omega\mu_0\sum_{p,q}\frac{e_{pq}(x,y)}{k_e^2 - k_{pq}^2}\int_P J_z(x',y')e_{pq}(x',y')dx'dy'. \tag{5.117}$$

The magnetic field inside the cavity may be determined by

$$\mathbf{H}(x,y) = \frac{1}{jk\eta}\mathbf{u}_z \times \nabla E_z(x,y). \tag{5.118}$$

Once the EM fields in the cavity are known, the equivalent magnetic current on ∂V_0 can be determined by $\mathbf{J}_{ms} = -\mathbf{u}_n \times \mathbf{E}$, where $\mathbf{u}_n$ is the outward normal of ∂V_0. Apparently, $\mathbf{J}_{ms}$ vanishes on the top patch and ground. Ignoring the currents on the upper surface of the patch and the ground, and the bound sources in the dielectric substrate, the radiated fields by the microstrip patch antenna can be determined from the equivalent magnetic current on the side wall with the ground plane in place. The effect of the infinite ground plane can be taken into account by the method of image, and the total equivalent magnetic current is thus given by $\mathbf{J}_{ms} = -2\mathbf{u}_n \times \mathbf{E}$, which determines the radiated electric field in the far-field region $(z > 0)$.

$$\begin{aligned}\mathbf{E}(\mathbf{r}) &\approx \frac{jk_0}{4\pi}\frac{e^{-jk_0 r}}{r}\mathbf{u}_r \times \int_{S_M} e^{jk_0 \mathbf{u}_r \cdot \mathbf{r}'}\mathbf{J}_{ms}(\mathbf{r}')dS(\mathbf{r}') \\ &= -\frac{jk_0}{2\pi}\frac{e^{-jk_0 r}}{r}\mathbf{u}_r \times \int_{\partial P} e^{jk_0 \mathbf{u}_r \cdot \mathbf{r}'}\mathbf{u}_n \times \mathbf{u}_z V(\mathbf{r}')dl(\mathbf{r}') \\ &= -\frac{jk_0\eta_0 I}{4\pi}\frac{e^{-jk_0 r}}{r}\mathbf{L}(\mathbf{u}_r),\end{aligned} \tag{5.119}$$

where $\eta_0 = \sqrt{\mu_0/\varepsilon_0}$, $k_0 = \omega\sqrt{\mu_0\varepsilon_0}$, $V = hE_z$ is the voltage at the patch edge, I is the feeding current, S_M denotes the magnetic wall (the side wall), and

$$\mathbf{L}(\mathbf{u}_r) = \frac{2}{\eta_0 I}\mathbf{u}_r \times \int_{\partial P} e^{jk_0 \mathbf{u}_r \cdot \mathbf{r}'}\mathbf{u}_n \times \mathbf{u}_z V(\mathbf{r}')dl(\mathbf{r}') \tag{5.120}$$

is the vector effective length of the patch antenna. The total radiated power of the patch is given by the integration of Poynting vector over a spherical surface of radius r in the far-field region

$$P^{rad} = \int_0^{\pi/2} \sin\theta d\theta \int_0^{2\pi} \frac{r^2}{2\eta_0}|\mathbf{E}(\mathbf{r})|^2 d\varphi = \frac{|I|^2\eta_0}{8}\int_0^{\pi/2} \sin\theta d\theta \int_0^{2\pi} \left|\frac{\mathbf{L}(\mathbf{u}_r)}{\lambda_0}\right|^2 d\varphi. \tag{5.121}$$

The quality factors Q_α related to various losses are defined by

$$Q_\alpha = \frac{\omega_r \tilde{W}}{P^\alpha}, (\alpha = rad, c, d), \tag{5.122}$$

where $\tilde{W}$ is the total stored field energy of antenna at the resonance frequency ω_r, and P^α stands for the power loss related to radiation loss ($\alpha = rad$), or conductor loss ($\alpha = c$), or dielectric loss ($\alpha = d$). The total stored field energy of antenna is given by (4.155)

$$\tilde{W} = \frac{1}{2}\int_{V_P} \varepsilon|E_z|^2 dV = \frac{h}{2}\int_P \varepsilon|E_z|^2 dxdy = \frac{h}{2}\int_P \mu_0|\mathbf{H}|^2 dxdy. \tag{5.123}$$

It is noted that the surface integral representing the radiated energy in (4.155) vanishes due to the fact that the surface ∂V_0 enclosing V_0 consists of electric and magnetic walls. The conduction loss is determined by

$$P^c = 2\frac{1}{2}\int_P |\mathbf{J}_s|R^s dxdy = R^s\int_P |\mathbf{H}|dxdy = \frac{2R^s\tilde{W}}{\mu_0 h},$$

where $\mathbf{J_s} = \mathbf{u}_n \times \mathbf{H}$ is the surface current on the upper and lower surface of the patch, and R^s is the surface resistance of the patch

$$R^s = \sqrt{\frac{\omega_r\mu_0}{2\sigma_c}}, \tag{5.124}$$

with σ_c being the conductivity of the patch. The dielectric loss is

$$P^d = \frac{1}{2}\int_{V_P} \sigma_d|E_z|^2 dV = \frac{h}{2}\int_P \sigma_d|E_z|^2 dxdy = \frac{\sigma_d}{\varepsilon_r\varepsilon_0}\tilde{W}. \tag{5.125}$$

The quality factors related to the conduction loss and dielectric loss are then given by

$$Q_c = \frac{\omega_r\tilde{W}}{P^c} = \frac{\omega_r\mu_0 h}{2R^s}, Q_d = \frac{\omega_r\tilde{W}}{P^d} = \frac{\omega_r\varepsilon_r\varepsilon_0}{\sigma_d} = \frac{1}{\tan\delta}, \tag{5.126}$$

where $\tan\delta$ denotes the loss tangent of the dielectric substrate.

For a thin circular probe located at (x_0, y_0), one may assume that

$$J_z(x,y) = \frac{I}{2\pi a_0}\delta(x-x_0)\delta(y-y_0),$$

where a_0 denotes the radius of the probe. The voltage drop along the probe is

$$V = \int_0^h \mathbf{E}(x_0,y_0)\cdot\mathbf{u}_z dz = hE_z(x_0,y_0) = \frac{j\omega\mu_0 Ih}{2\pi a_0}\sum_{p,q}\frac{e_{pq}^2(x_0,y_0)}{k_e^2 - k_{pq}^2}. \tag{5.127}$$

The input impedance is defined by

$$Z = \frac{V}{-I} = -\frac{j\omega\mu_0 h}{2\pi a_0}\sum_{p,q}\frac{e_{pq}^2(x_0,y_0)}{k_e^2 - k_{pq}^2} = \sum_{p,q}\frac{he_{pq}^2(x_0,y_0)}{2\pi a_0}\frac{1}{\sigma + j\omega\varepsilon + \dfrac{k_{pq}^2}{j\omega\mu_0}}. \tag{5.128}$$

Figure 5.22 shows the equivalent network for the patch antenna. The TM_{00} mode is represented by a capacitance in series with a resistance while all other modes are represented by a parallel RLC resonant circuit.

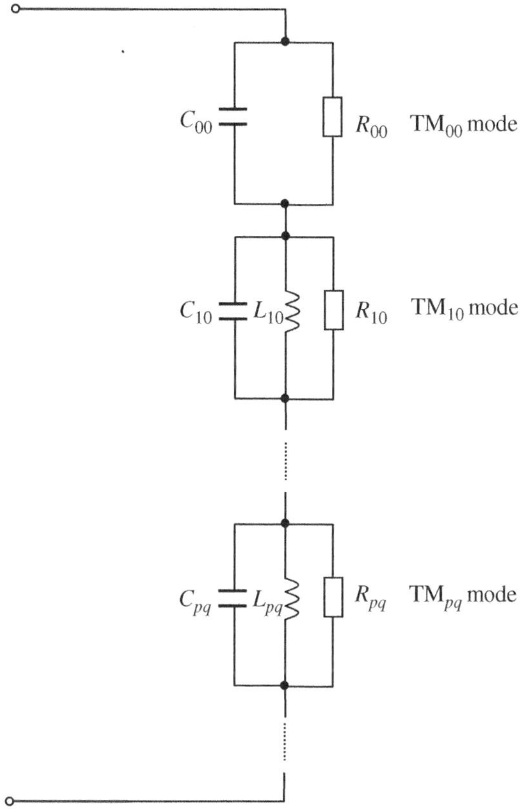

Figure 5.22 Equivalent circuit of patch antenna.

The cavity model is the foundation for the analysis of patch antenna. Additional modifications are usually needed for some particular shapes so that a reasonable accuracy can be achieved. There have been various methods to feed the patch antennas. Figure 5.23 shows some typical feed methods.

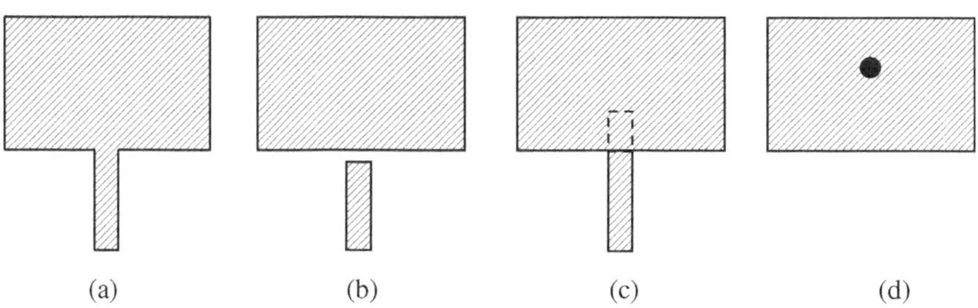

Figure 5.23 Typical feed methods. (a) Edge feed. (b) Edge feed with a capacitance gap. (c) Two-layer feed. (d) Coaxial line feed.

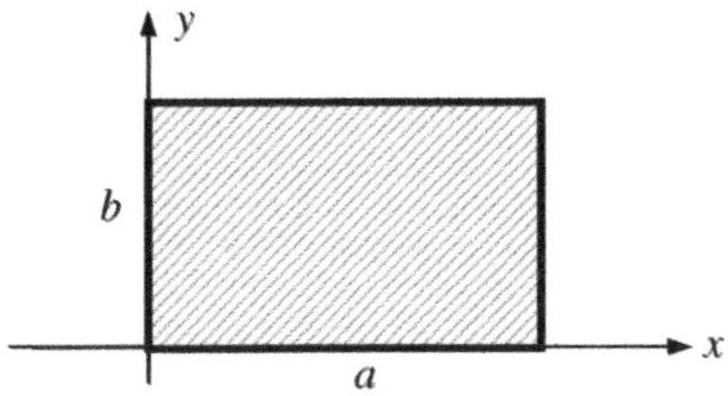

Figure 5.24 Rectangular patch.

The rectangular patch is the most commonly used in practice. For a rectangular patch and the rectangular coordinate system shown in Figure 5.24, the eigenfunctions of (5.113) are easily found to be

$$\begin{aligned} e_{pq}(x,y) &= N_{pq} \cos \frac{p\pi}{a} x \cos \frac{q\pi}{b} y, \\ k_{pq} &= \sqrt{\left(\frac{p\pi}{a}\right)^2 + \left(\frac{q\pi}{b}\right)^2}, \end{aligned} \tag{5.129}$$

where

$$N_{pq} = \frac{C_{pq}}{\sqrt{hab}}, C_{pq} = \begin{cases} 1, & p = q = 0 \\ \sqrt{2}, & p = 0 \text{ or } q = 0 \\ 2, & p \neq 0, q \neq 0 \end{cases}.$$

The indices p and q, respectively, denote the number of half-cycles of the field along the sides a and b. If only one TM_{pq} mode exists in the cavity, (5.117) reduces to

$$E_z(x,y) = \frac{V_c}{h} \cos \frac{p\pi}{a} x \cos \frac{q\pi}{b} y, \tag{5.130}$$

where V_c is the voltage at the corner (0, 0). The magnetic field is given by (5.114):

$$\begin{aligned} \mathbf{H}(x,y) &= \frac{1}{jk\eta}\left(\mathbf{u}_y \frac{\partial E_z}{\partial x} - \mathbf{u}_x \frac{\partial E_z}{\partial y}\right) \\ &= \frac{1}{jk\eta}\frac{V_c}{h}\frac{q\pi}{b} \cos \frac{p\pi}{a} x \sin \frac{q\pi}{b} y \mathbf{u}_x \\ &\quad - \frac{1}{jk\eta}\frac{V_c}{h}\frac{p\pi}{a} \sin \frac{p\pi}{a} x \cos \frac{q\pi}{b} y \mathbf{u}_y. \end{aligned} \tag{5.131}$$

The induced current on the lower surface of the patch is

$$\begin{aligned} \mathbf{J}_s &= -\frac{1}{jk\eta}\frac{V_c}{h}\frac{q\pi}{b} \cos \frac{p\pi}{a} x \sin \frac{q\pi}{b} y \mathbf{u}_y \\ &\quad - \mathbf{u}_x \frac{1}{jk\eta}\frac{V_c}{h}\frac{p\pi}{a} \sin \frac{p\pi}{a} x \cos \frac{q\pi}{b} y \mathbf{u}_x. \end{aligned} \tag{5.132}$$

The equivalent magnetic modal currents on the magnetic walls are given by

$$\mathbf{J}_{ms} = -\mathbf{u}_n \times \mathbf{u}_z E_z = \begin{cases} \frac{V_c}{h} \cos p\pi \cos \frac{q\pi}{b} y \mathbf{u}_y, & x = a \\ -\frac{V_c}{h} \cos \frac{q\pi}{b} y \mathbf{u}_y, & x = 0 \\ -\frac{V_c}{h} \cos q\pi \cos \frac{p\pi}{a} x \mathbf{u}_x, & y = b \\ \frac{V_c}{h} \cos \frac{p\pi}{a} x \mathbf{u}_x, & y = 0 \end{cases}. \tag{5.133}$$

If $a > b$, the lowest mode for the rectangular patch is TM_{10} mode, which has the lowest resonant frequency. The next higher order mode for the rectangular patch is TM_{01} mode. The directions of the equivalent magnetic currents along the four edges of the rectangular patch are depicted in Figure 5.25. It can been seen that, to generate the TM_{10} (or TM_{01}) mode, the feeding point must be selected along the centerline of dimension b (or a) according to the symmetry of the field distributions. For the TM_{10} mode, only the equivalent magnetic currents along the two b edges are in phase and therefore account for most of the radiation; while those along the two a edges are out of phase and their radiated fields largely cancel. The radiation pattern of this mode is linearly polarized and has a maximum in the direction normal to the patch, and the E-plane is perpendicular to the radiating edges. Similarly, for the TM_{01} mode, the magnetic currents are in phase along a and are out of phase along b. Consequently, only the two a edges are radiating for the TM_{01} mode. The voltages along the boundary of the patch can be determined by (5.130):

$$V(x,y) = \begin{cases} V_c \cos \frac{p\pi}{a} x, & y = 0 \\ V_c \cos \frac{p\pi}{a} x \cos q\pi, & y = b \\ V_c \cos \frac{q\pi}{b} y, & x = 0 \\ V_c \cos p\pi \cos \frac{q\pi}{b} y, & x = a \end{cases}. \tag{5.134}$$

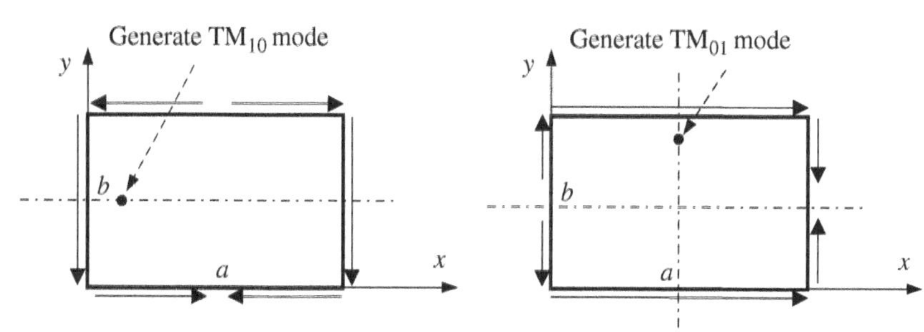

(a) (b)

Figure 5.25 Equivalent magnetic currents. (a) TE_{10} mode. (b) TE_{01} mode.

Combining (5.120) with (5.134) gives the vector effective length of the patch

$$\mathbf{L}(\mathbf{u}_r) = -j\frac{2V_c}{I\eta_0}\left(\cos p\pi e^{jk_0 a\sin\theta\cos\varphi} - 1\right)\left(\cos q\pi e^{jk_0 b\sin\theta\sin\varphi)} - 1\right)$$
$$\times \mathbf{u}_r \times \left[\frac{\mathbf{u}_x k_0 \sin\theta\cos\varphi}{k_0^2\sin^2\theta\cos^2\varphi - \left(\frac{p\pi}{a}\right)^2} - \mathbf{u}_y\frac{k_0\sin\theta\sin\varphi}{k_0^2\sin^2\theta\sin^2\varphi - \left(\frac{q\pi}{a}\right)^2}\right]. \tag{5.135}$$

In spherical coordinate system, this becomes

$$\mathbf{L}(\mathbf{u}_r) = -j\frac{2V_c}{I\eta_0}k_0\sin\theta\left(\cos p\pi e^{jk_0 a\sin\theta\cos\varphi} - 1\right)\left(\cos q\pi e^{jk_0 b\sin\theta\sin\varphi)} - 1\right)$$
$$\times\left\{\mathbf{u}_\theta\left[\frac{\sin\varphi\cos\varphi}{k_0^2\sin^2\theta\cos^2\varphi - \left(\frac{p\pi}{a}\right)^2} + \frac{\sin\varphi\cos\varphi}{k_0^2\sin^2\theta\sin^2\varphi - \left(\frac{q\pi}{b}\right)^2}\right]\right.$$
$$\left. + \mathbf{u}_\varphi\left[\frac{\cos\theta\cos^2\varphi}{k_0^2\sin^2\theta\cos^2\varphi - \left(\frac{p\pi}{a}\right)^2} - \frac{\cos\theta\sin^2\varphi}{k_0^2\sin^2\theta\sin^2\varphi - \left(\frac{q\pi}{b}\right)^2}\right]\right\}. \tag{5.136}$$

In the far-field region, the electric field is given by (5.119):

$$E_\theta(\mathbf{r}) \approx -\frac{k_0^2 V_c}{4\pi r}e^{-jk_0 r}\sin\theta\sin 2\varphi$$
$$\times\left(\cos p\pi e^{jk_0 a\sin\theta\cos\varphi} - 1\right)\left(\cos q\pi e^{jk_0 b\sin\theta\sin\varphi} - 1\right)$$
$$\times\left[\frac{1}{k_0^2\sin^2\theta\cos^2\varphi - \left(\frac{p\pi}{a}\right)^2} + \frac{1}{k_0^2\sin^2\theta\sin^2\varphi - \left(\frac{q\pi}{b}\right)^2}\right], \tag{5.137}$$

$$E_\varphi(\mathbf{r}) \approx -\frac{k_0^2 V_c}{4\pi r}e^{-jk_0 r}\sin 2\theta$$
$$\times\left(\cos p\pi e^{jk_0 a\sin\theta\cos\varphi} - 1\right)\left(\cos q\pi e^{jk_0 b\sin\theta\sin\varphi} - 1\right)$$
$$\times\left[\frac{\cos^2\varphi}{k_0^2\sin^2\theta\cos^2\varphi - \left(\frac{p\pi}{a}\right)^2} - \frac{\sin^2\varphi}{k_0^2\sin^2\theta\sin^2\varphi - \left(\frac{q\pi}{b}\right)^2}\right]. \tag{5.138}$$

The stored field energy of the patch can be obtained from (5.123):

$$\tilde{W} = \frac{\varepsilon_r\varepsilon_0 V_c^2}{2h}\int_0^a \cos^2\left(\frac{p\pi}{a}x\right)\cos^2\left(\frac{q\pi}{b}y\right)dxdy = \frac{\varepsilon_r\varepsilon_0 V_c^2 ab}{2h\varepsilon_p\varepsilon_q}, \tag{5.139}$$

where $\varepsilon_{p,q} = \begin{cases} 1, p,q=0 \\ 2, p,q \geq 1 \end{cases}$. Apparently, the stored field energy of the patch is proportional to the relative dielectric constant of the substrate and inversely proportional to the height of the substrate. One may thus (roughly) conclude that the bandwidth of the patch antenna increases as the height (or relative dielectric constant) increases (or decreases).

For the TM_{01} mode, the far-zone electric field components become

$$\begin{aligned} E_\theta(\mathbf{r}) &\approx \frac{k_0^2 V_c}{4\pi r} e^{-jk_0 r} \sin\theta \sin 2\varphi \times \left(e^{jk_0 a \sin\theta\cos\varphi} - 1\right)\left(e^{jk_0 b \sin\theta \sin\varphi} + 1\right) \\ &\quad \times \left[\frac{1}{k_0^2 \sin^2\theta\cos^2\varphi} + \frac{1}{k_0^2 \sin^2\theta \sin^2\varphi - (\pi/b)^2}\right], \\ E_\varphi(\mathbf{r}) &\approx \frac{k_0^2 V_c}{4\pi r} e^{-jk_0 r} \sin 2\theta\left(e^{jk_0 a\sin\theta\cos\varphi} - 1\right)\left(e^{jk_0 b\sin\theta\sin\varphi} + 1\right) \\ &\quad \times \left[\frac{\cos^2\varphi}{k_0^2 \sin^2\theta\cos^2\varphi} - \frac{\sin^2\varphi}{k_0^2 \sin^2\theta\sin^2\varphi - (\pi/b)^2}\right]. \end{aligned} \tag{5.140}$$

In the direction of $\theta = 0°$, these reduce to

$$\begin{aligned} E_\theta(\mathbf{r}) &= \frac{jk_0 a V_c}{\pi r} e^{-jk_0 r} \sin\varphi, \\ E_\varphi(\mathbf{r}) &= \frac{jk_0 a V_c}{\pi r} e^{-jk_0 r} \cos\varphi. \end{aligned} \tag{5.141}$$

For the TM_{10} mode, the far-zone electric field can be written as

$$\begin{aligned} E_\theta(\mathbf{r}) &\approx \frac{k_0^2 V_c}{4\pi r} e^{-jk_0 r} \sin\theta \sin 2\varphi\left(e^{jk_0 a \sin\theta\cos\varphi} + 1\right)\left(e^{jk_0 b \sin\theta \sin\varphi} - 1\right) \\ &\quad \times \left[\frac{1}{k_0^2 \sin^2\theta\cos^2\varphi - \left(\frac{\pi}{a}\right)^2} + \frac{1}{k_0^2 \sin^2\theta \sin^2\varphi}\right], \\ E_\varphi(\mathbf{r}) &\approx \frac{k_0^2 V_c}{4\pi r} e^{-jk_0 r} \sin 2\theta\left(e^{jk_0 a\sin\theta\cos\varphi} + 1\right)\left(e^{jk_0 b\sin\theta\sin\varphi} - 1\right) \\ &\quad \times \left[\frac{\cos^2\varphi}{k_0^2 \sin^2\theta\cos^2\varphi - \left(\frac{\pi}{a}\right)^2} - \frac{\sin^2\varphi}{k_0^2 \sin^2\theta\sin^2\varphi}\right]. \end{aligned} \tag{5.142}$$

In the direction of $\theta = 0°$, these reduce to

$$\begin{aligned} E_\theta(\mathbf{r}) &= j\frac{k_0 b V_0}{\pi r} \cos\varphi e^{-jk_0 r}, \\ E_\varphi(\mathbf{r}) &= -j\frac{k_0 b V_0}{\pi r} \sin\varphi e^{-jk_0 r}. \end{aligned} \tag{5.143}$$

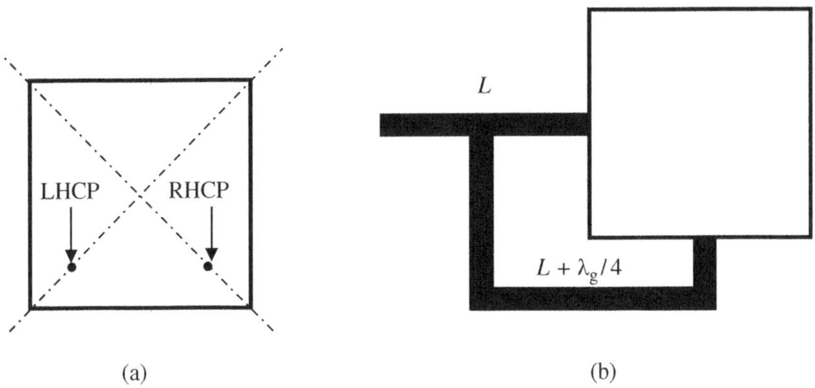

Figure 5.26 Circular polarization. (a) Single feed. (b) Dual feed.

Comparing (5.141) with (5.143), one may find that the far-zone electric field for the TM_{10} mode can be obtained through rotating the radiated field for the TM_{01} mode in the φ-plane by 90° if the patch is square with $a = b$. In other words, the radiated fields in $\theta = 0°$ direction for these two modes are orthogonal for a square patch. Therefore, a circular polarization can be generated in the $\theta = 0°$ direction if the two modes are excited with identical amplitudes and a 90° phase difference. There are two methods for generating a circularly polarized wave from a rectangular patch. One is to feed the patch at a single point and the other is to feed the patch using a feeding network that excites the patch with equal amplitudes and 90° phase difference, as illustrated in Figure 5.26. For a single feed technique, a perturbation segment such as a slot or truncated segments must be introduced so that the generated mode can be separated into two orthogonal modes [20].

5.13 Resonant Modal Theory for Antenna Design

An antenna is said to be **resonant** if its input reactance is zero, and the corresponding frequency at which the resonance occurs is called the **resonant frequency**. Physically, this implies that the stored electric field energy of antenna is equal to the stored magnetic field energy. A unified procedure for evaluating the resonant modes for an arbitrary current source region is proposed in [24], which is based on the expressions of stored field energies of the source (antenna) [see (4.167) and (4.168)]. When the antenna is assumed to be resonant, the condition that the difference between the stored electric and magnetic field energies vanishes leads to a homogeneous integral equation for the modal currents in the source region. After discretization, the integral equation is reduced to a real homogenous algebraic equation. By enforcing the condition that a nontrivial current distribution exits in the source region so that the determinant of the coefficient matrix of the algebraic equation must be zero, the resonant frequencies and the corresponding resonant modes can then be determined. Once the resonant modes are determined, the rest is to introduce a scatterer in the source region and a feeding mechanism to excite the resonant modes. The above procedure will be called RMT.

5.13.1 Formulations

In general, an arbitrary scatterer (without an antenna input terminal) is said to be resonant if the stored electric field energy of the scatterer (antenna) is equal to the stored magnetic field energy around the scatterer. Let V_0 denote an arbitrary source region bounded by ∂V_0. It follows from (4.167) and (4.168) that the difference between the stored electric and magnetic field energies of antenna can be expressed by

$$\tilde{W}_m - \tilde{W}_e = \frac{\eta}{16\pi c} \int\limits_{V_0}\int\limits_{V_0} \left[\bar{\mathbf{J}}(\mathbf{r}) \cdot \mathbf{J}(\mathbf{r}') + \frac{1}{k^2} \nabla \cdot \bar{\mathbf{J}}(\mathbf{r}) \nabla' \cdot \mathbf{J}(\mathbf{r}')\right] \frac{\cos(kR)}{R} dV(\mathbf{r}) dV(\mathbf{r}'). \tag{5.144}$$

This can be written as an inner product

$$\tilde{W}_m - \tilde{W}_e = \left(\hat{L}\mathbf{J}, \mathbf{J}\right), \tag{5.145}$$

where the integral operator $\hat{L}$ is defined by

$$\hat{L}\mathbf{J} = \frac{\eta}{16\pi v} \int\limits_{V_0} \left[\mathbf{J}(\mathbf{r}') + \frac{1}{k^2} (\nabla' \cdot \mathbf{J}(\mathbf{r}')) \nabla\right] \frac{\cos(kR)}{R} dV(\mathbf{r}'). \tag{5.146}$$

The source region V_0 is said to be resonant if (5.145) vanishes. In order to find the resonant frequency and the corresponding current distribution $\mathbf{J}$ that makes (5.145) vanish, a sufficient condition is

$$\hat{L}\mathbf{J} = 0, \tag{5.147}$$

which is valid throughout the source region. One can now follow the standard procedure of the method of moments to solve (5.147). For simplicity, the source region will be assumed to be a surface S_0, so that (5.147) becomes a surface integral equation. The current may be expanded in terms of the well-known Rao–Wilton–Glisson basis functions [25]

$$\mathbf{J} = \sum_{n=1}^{N} I_n \mathbf{f}_n. \tag{5.148}$$

Introducing (5.148) into (5.147), the following homogenous matrix equation by Galerkin method can be obtained

$$[L(\omega)][I] = 0, \tag{5.149}$$

where $[I] = [I_1, I_2, ..., I_N]^T$, and the matrix elements of $[L(\omega)]$ are given by

$$L_{mn} = \frac{\eta}{16\pi v} \int_{S_0}\int_{S_0} \left[\mathbf{f}_m(\mathbf{r}) \cdot \mathbf{f}_n(\mathbf{r}') - \frac{1}{k^2}\nabla' \cdot \mathbf{f}_n(\mathbf{r}')\nabla \cdot \mathbf{f}_m(\mathbf{r})\right] \frac{\cos(kR)}{R} dS(\mathbf{r})dS(\mathbf{r}'). \tag{5.150}$$

A sufficient condition for the existence of a nonzero solution of (5.149) is that the determinant of its coefficient matrix is zero

$$\det([L(\omega)]) = 0. \tag{5.151}$$

The above equation determines the resonant frequencies, and the corresponding resonant current modes can be found from (5.149).

It is noted that (5.151) is a sufficient condition that makes (5.145) vanish. It can be shown that (5.151) is also a necessary condition. In fact, after discretization, (5.145) can be written as a quadratic form

$$(\hat{L}\mathbf{J}, \mathbf{J}) = [I]^T[L][I]. \tag{5.152}$$

Since $[L]$ is symmetric (this property is not necessary), there exists an orthogonal matrix $[O]$ such that

$$[O]^T[L][O] = [D],$$

where $[D] = [\lambda_1, \lambda_2, ..., \lambda_N]^T$ is a diagonal matrix. Introducing a new vector $[Y] = [y_1, y_2, ..., y_n]^T$ defined by

$$[I] = [O][Y]$$

and substituting this into (5.152), one obtains

$$(\hat{L}\mathbf{J}, \mathbf{J}) = [Y]^T[D][Y] = \sum_{n=1}^{N} \lambda_n y_n^2. \tag{5.153}$$

If the above quadratic form is required to be zero for an arbitrary current distribution $\mathbf{J}$ (thus an arbitrary $[Y]$ by definition), one must have $\lambda_n = 0 (n = 1, 2, ..., N)$. This implies

$$\det[L] = \det[D] = \prod_{n=1}^{N} \lambda_n = 0 \tag{5.154}$$

since $[O]^T[O] = [I]$, where $[I]$ is the unit matrix of order N. Thus, (5.151) is also a necessary condition that makes (5.145) vanish.

5.13.2 Applications

The RMT directly determines the resonant frequencies from the difference of the stored field energies of antenna. Once the corresponding resonant current modes are determined, the next is to use various excitations to realize the resonant current modes in the selected source region.

5.13.2.1 Crossed-Dipole

Let the source region V_0 be a cross, consisting of two planar strips shown in Figure 5.27a, where the lengths of the horizontal and vertical strips are $l_1 = 118$ mm and $l_2 = 97$ mm, respectively, and the width of the strips is set as $w = 2$ mm. From (5.151), the first three resonant frequencies of the cross are found to be 1.190, 1.428, and 1.443 GHz. The corresponding modal currents are depicted in Figure 5.27b–d. As shown in Figure 5.27b, the fundamental (the first) modal

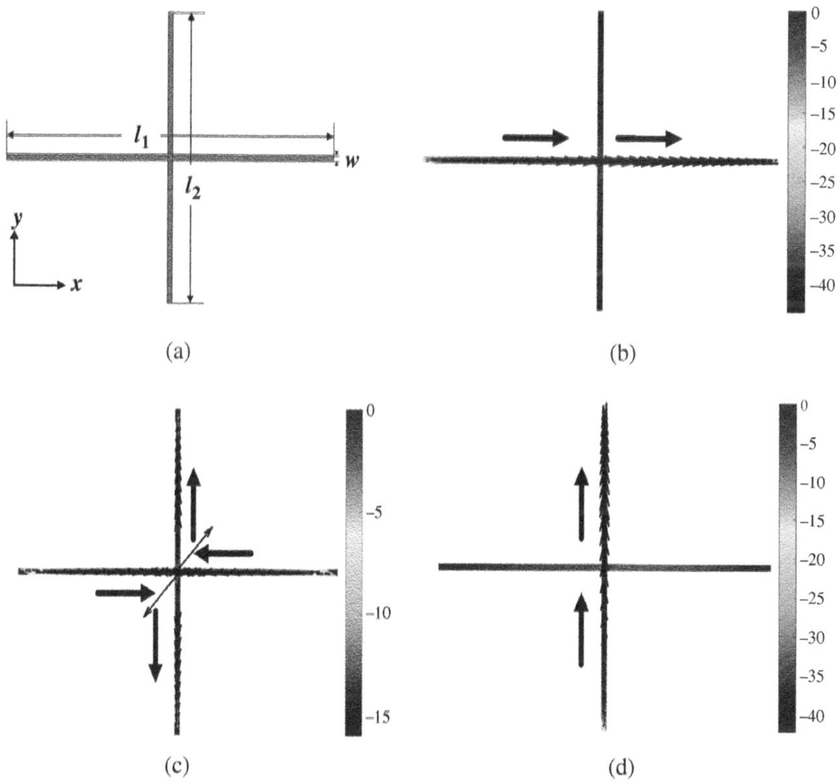

Figure 5.27 A cross and its modal currents. (a) Dimensions. (b) Modal current at 1.190 GHz. (c) Modal current at 1.427 GHz. (d) Modal current at 1.440 GHz.

current concentrates on the horizontal strip. In contrast, the third modal current is along the vertical dipole, as indicated by Figure 5.27d. The second modal current, shown in Figure 5.27c, has a distribution that is quite different from the first and the third. It has two different current paths. The first path is composed of the top vertical arm and the right horizontal arm of the cross, and the second path is composed of the bottom vertical arm and left horizontal arm.

The first and the third mode of the cross will now be used to build a circular polarized (CP) antenna. As indicated in Figure 5.27, the maximum values of the first and third modal currents are right at the center of the cross. To excite the two modes simultaneously, a voltage source must be introduced at the center of the cross. For this purpose, the cross is broken into two separate parts to form a crossed-dipole antenna. The first part consists of the top vertical arm and the right horizontal arm, and the second part consists of the bottom vertical arm and the left horizontal arm, with a small gap at the center of the cross (see the inset of Figure 5.30a). A discrete face port with 50 Ω is applied across the gap in the simulation with CST Studio to excite the first and third modes simultaneously. The fabricated crossed-dipole antenna is shown in Figure 5.28. Figure 5.29 is the reflection coefficient and axial ratio of the crossed-dipole antenna. Two distinct reflection zeros occur approximately around the first and third resonant frequencies in the simulated and measured results. The fractional bandwidth of the crossed-dipole antenna is 26.6% at −10 dB of the reflection coefficient. The bandwidth of the 3 dB axial ratio in the direction of ($\theta = 0°, \varphi = 0°$) is 6.1%. The antenna exhibits a bidirectional radiation pattern. In the direction of $\theta = 0°$, it generates a left-handed circularly polarized wave, while in the direction of $\theta = 180°$, the radiated wave exhibits a right-handed circularly polarized behavior.

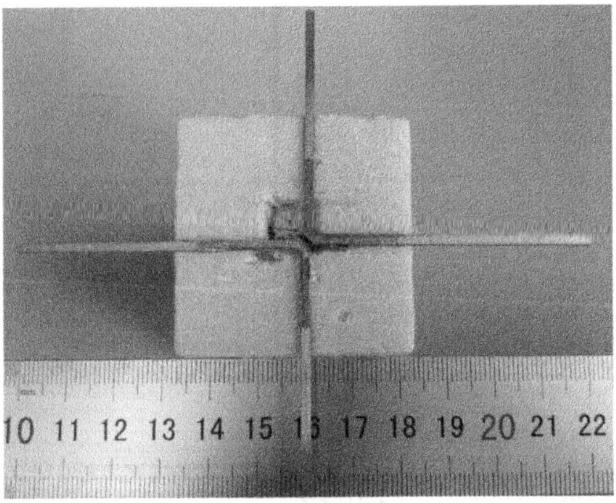

Figure 5.28 Photograph of the fabricated CP crossed-dipole antenna.

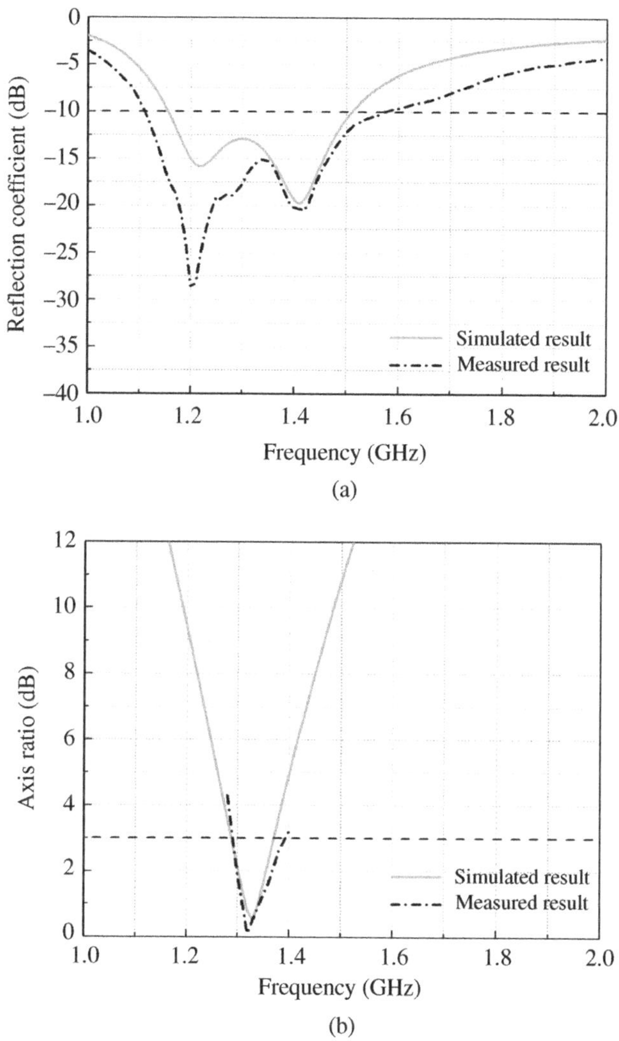

Figure 5.29 (a) Reflection coefficient. (b) Axial ratio in the direction ($\theta = 0°$, $\varphi = 0°$).

Figure 5.30 shows the simulated resonant current distributions of the crossed dipole at the two reflection zeros shown in Figure 5.29a. As expected, the resonant current at the first reflection zero mainly concentrates on the two horizontal arms (see Figure 5.30a) while the resonant current at the second reflection zero is mainly distributed on the vertical dipole (Figure 5.30b), agreeing with the simulated modal current distributions shown in Figure 5.27.

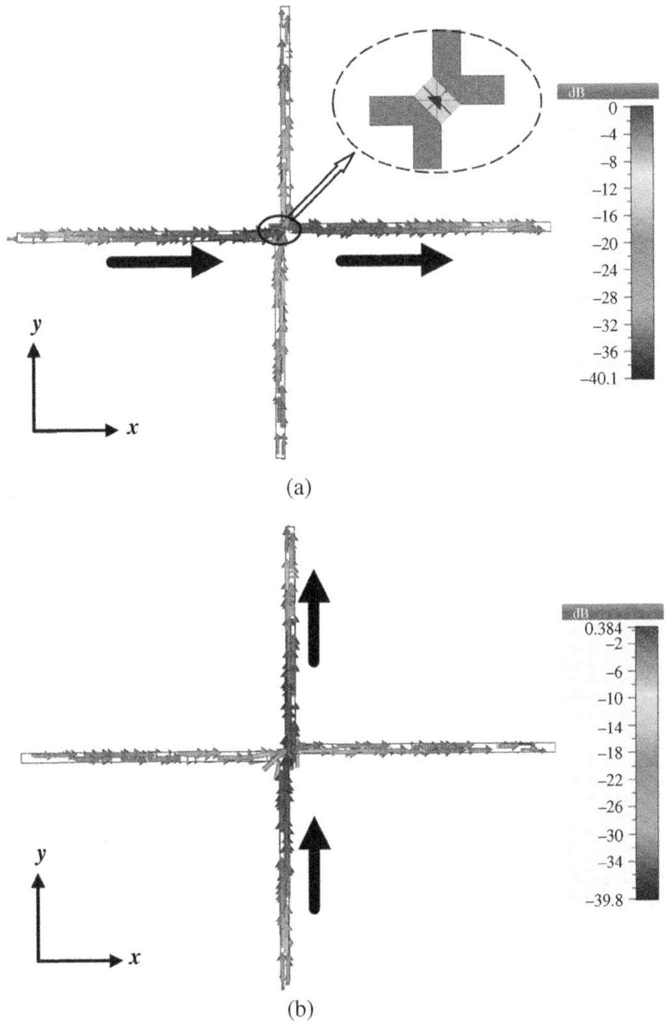

Figure 5.30 Simulated currents at (a) 1.22 GHz and (b) 1.42 GHz.

The radiation patterns at *xoz*-plane and *yoz*-plane for the first and third modal currents from the RMT are compared with the CST simulations and the measured ones in Figure 5.31. Figure 5.31a,b indicates that the radiation patterns from the RMT, the CST simulation, and measured ones agree well at the first resonant frequency. The maximum gains from the RMT, the CST simulation, and the measurement at the two cut planes are, respectively, 2.10, 2.14, and 2.17 dB. The radiation patterns in Figure 5.31a,b also indicate that the radiation at the first

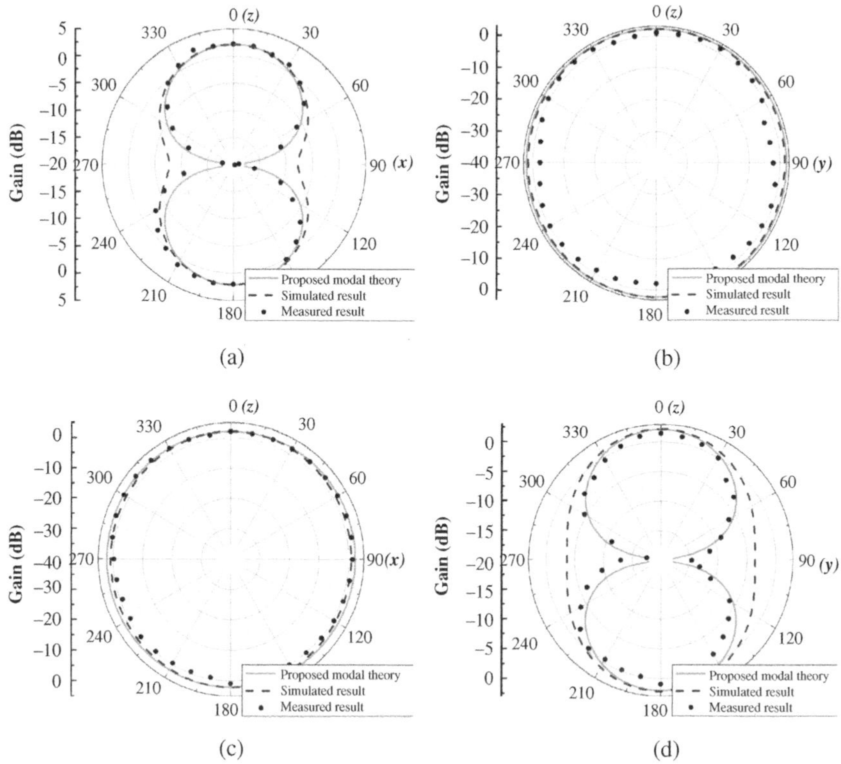

Figure 5.31 Radiation patterns. (a) Patterns at *xoz*-plane at first resonant frequency. (b) Patterns at *yoz*-plane at first resonant frequency. (c) Patterns at *xoz*-plane at third resonant frequency. (d) Patterns at *yoz*-plane at third resonant frequency.

resonant frequency mainly comes from the two horizontal arms. Similar results can be found in Figure 5.31c,d. In this case, radiation mainly comes from the two vertical arms.

5.13.2.2 Dual-Band Bowtie Antenna

A bowtie antenna is often used for wideband applications, and its wideband characteristic is determined by the flare angle and the size of the structure. The RMT will now be applied to the design of a dual-band bowtie antenna with a fixed flare angle and size. Consider a planar slotted bowtie (without a feeding mechanism) shown in Figure 5.32a, whose geometrical parameters are listed in Table 5.2.

The first three resonant frequencies are found to be 1.123, 2.521, and 3.568 GHz, and the corresponding modal currents are plotted in Figure 5.32b–d. The first modal current at 1.123 GHz is along the two long edges of slotted bowtie, which

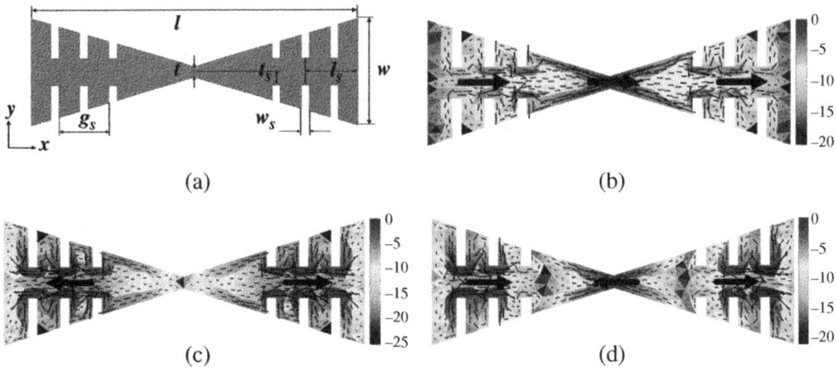

Figure 5.32 Planar slotted bowtie and the modal current distributions. (a) Dimensions. (b) Modal current at 1.123 GHz. (c) Modal current at 2.521 GHz. (d) Modal current at 3.568 GHz.

Table 5.2 Parameters of dual-band bowtie antenna.

Parameter	Value (mm)
W	30
L	90
T	4
l_s	20
w_s	3
t_s	3
g_s	12

is similar to a half-wavelength dipole. The second modal current is similar to a full-wavelength dipole while the third modal current to a one-and-half-wavelength dipole.

The slotted bowtie is now broken into two separate parts and a small gap is introduced in the middle of the bowtie. To excite the first and third modes of the slotted bowtie, a discrete face port with 50 Ω is applied across the gap to form a dual-band bowtie antenna. The fabricated dual-band bowtie antenna is shown in Figure 5.33. The simulated and measured reflection coefficients of the dual-band slotted bowtie antenna are shown in Figure 5.34, where two reflection zeros occur at 1.111 and 3.514 GHz in simulation, very close to the first and third modal resonant

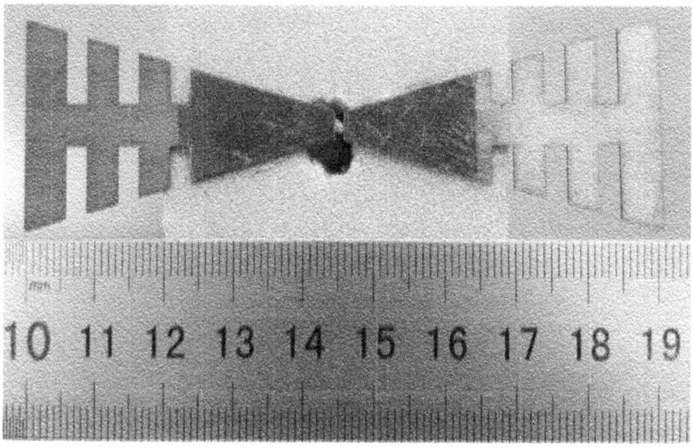

Figure 5.33 Photograph of the fabricated dual-band bowtie antenna.

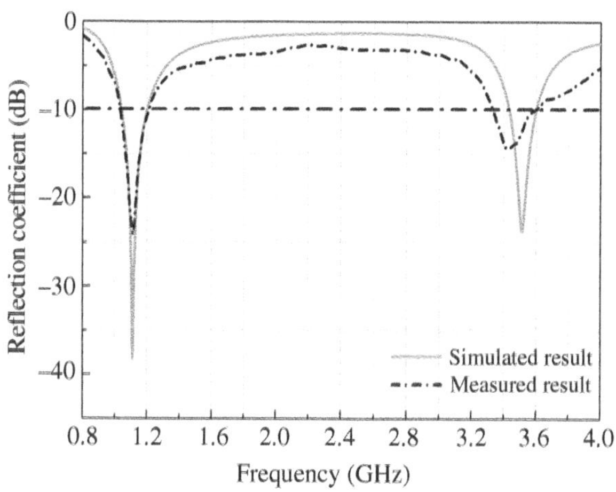

Figure 5.34 Simulated and measured reflection coefficients.

frequencies. The measured reflection coefficient indicates that the first resonant frequency is identical with simulated one while there is a small deviation between the measured and the simulated results for the second resonant frequency, which may be caused by fabrication error of the antenna and the feeding gap introduced in order to build the antenna. The simulated current distributions at the center frequencies of the two bands are shown in Figure 5.35, agreeing well with the modal currents demonstrated in Figure 5.32.

The radiation patterns for the modal currents at the first and third resonant frequencies are compared with the CST simulations and the measurements, and are

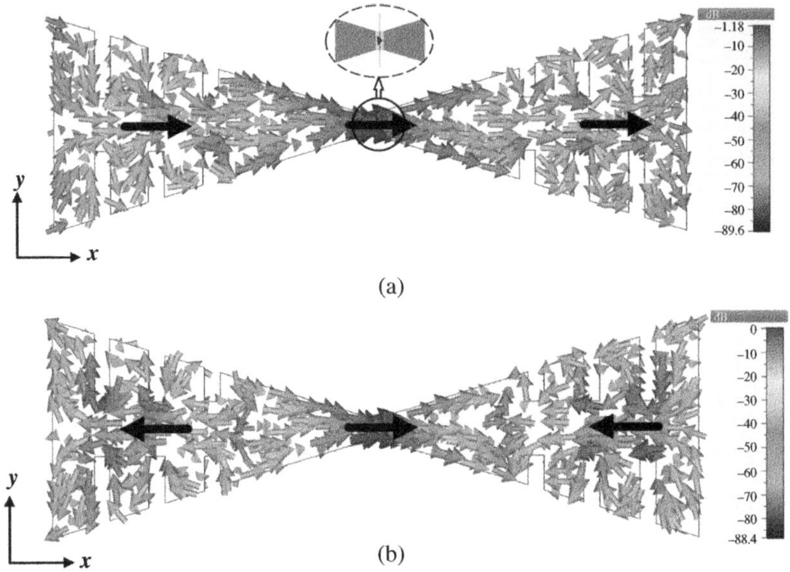

Figure 5.35 Simulated resonant currents. (a) At 1.111 GHz. (b) At 3.514 GHz.

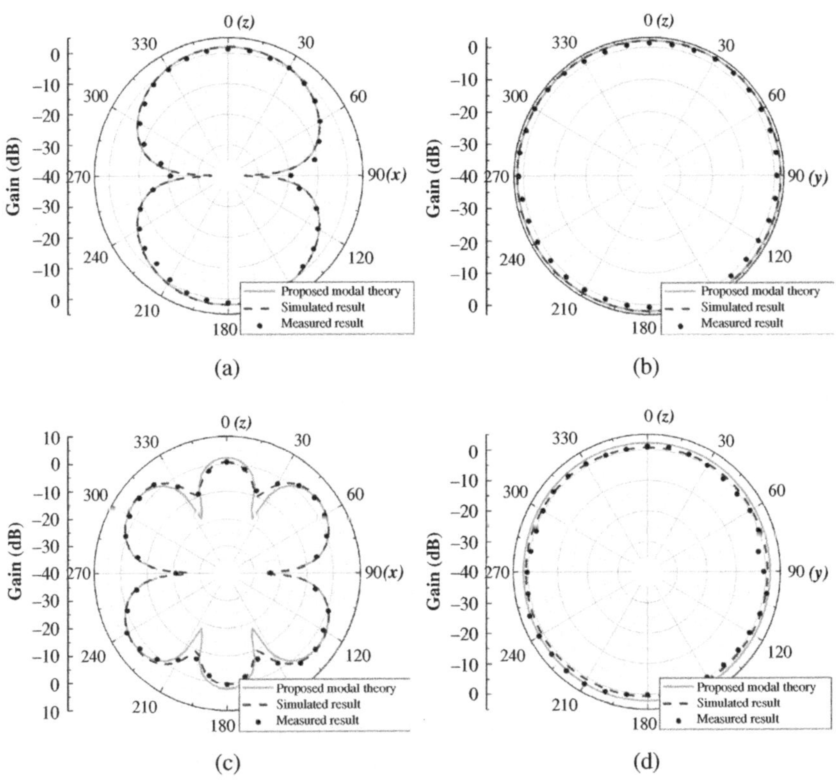

Figure 5.36 Radiation patterns. (a) Patterns at *xoz*-plane at first resonant frequency. (b) Patterns at *yoz*-plane at first resonant frequency. (c) Patterns at *xoz*-plane at third resonant frequency. (d) Patterns at *yoz*-plane at third resonant frequency.

shown in Figure 5.36. The simulated and measured radiation patterns at *xoz*-plane and *yoz*-plane at the first resonant frequency are depicted in Figure 5.36a,b, and agree well with those obtained from the RMT. Similar results can be found in Figure 5.36c,d for the third resonant frequency.

The RMT is different from the previous modal theories that expand the fields by a linear combination of fundamental field patterns (modes) constrained by the boundary conditions. The RMT directly determines the resonant modes for an arbitrary source region by requiring that the stored electric and magnetic field energies of the source are equal, yielding a homogeneous integral equation, from which the resonant frequencies and the corresponding resonant modes can be obtained numerically.

> The more I have learned about physics, the more convinced I am that physics provides, in a sense, the deepest applications of mathematics. The mathematical problems that have been solved, or techniques that have arisen out of physics in the past, have been lifeblood of mathematics. The really deep questions are still in the physical sciences. For the health of mathematics at its research level, I think it is very important to maintain that link as much as possible.
>
> *–Sir Michael Atiyah (British-Lebanese mathematician, 1929–2019)*

References

1 Geyi, W., *Foundations of Applied Electrodynamics*, New York, Wiley, 2010.
2 Geyi, W., *Foundations for Radio Frequency Engineering*, World Scientific, 2015.
3 Chew, W. C., *Waves and Fields in Inhomogeneous Media*, Van Nostrand Reinhold, 1990.
4 Ward, B. J., "Hybrid surface electric field volume magnetic field integral equations for electromagnetic analysis of heterogeneous dielectric bodies with embedded electrically conducting structures", *IEEE Trans. Antennas Propagat.*, Vol. 69, No. 3, pp. 1545–1552, 2021.
5 Bladel, J. V., *Electromagnetic Fields*, IEEE Press, 2007.
6 Felsen, L. B. and N. Marcuvitz, *Radiation and Scattering of Waves*, IEEE Press, 1994.
7 Harrington, R. F., *Time-Harmonic Electromagnetic Fields*, McGraw-Hill Book Company, Inc., 1961.
8 Balanis, C. A., *Antenna Theory: Analysis and Design*, 2nd Ed., John Wiley & Sons, 1997.
9 Geyi, W., "New magnetic field integral equation for antenna system", *Prog. Electromagn. Res.*, Vol. 63, pp. 153–176, 2006.
10 Marcuvitz, N., *Waveguide Handbook*, McGraw-Hill Book Company, Inc., 1951.

11 Deschamps, G.A., "Microstrip microwave antennas", Presented at the Third USAF Symposium on Antennas, 1953.

12 Munson, R. E., "Conformal microstrip antennas and microstrip phased arrays", *IEEE Trans. Antennas Propagat.*, Vol. 22, No. 1, pp. 74–78, 1974.

13 Howell, J. W., "Microstrip antennas", *IEEE Trans. Antennas Propagat.*, Vol. 23, No. 1, pp. 90–93, 1975.

14 Lo, Y. T., D. Solomon, and W. F. Richards, "Theory and experiment on microstrip antennas", *IEEE Trans. Antennas Propagat.*, Vol. 27, No. 2, pp. 137–145, 1979.

15 Derneryd, A. G., "Extended analysis of rectangular microstrip resonator antennas", *IEEE Trans. Antennas Propagat.*, Vol. 27, No. 6, pp. 846–849, 1979.

16 Carver, K. R. and J. W. Mink, "Microstrip antenna technology", *IEEE Trans. Antennas Propagat.*, Vol. 29, No. 1, pp. 2–24, 1981.

17 Elliott, R. S., *Antenna Theory and Design*, New York, Prentice-Hall, 1981.

18 Richards, W. F., Y. T. Lo, and D. D. Harrison, "An improved theory of microstrip antennas with applications", *IEEE Trans. Antennas Propagat.*, Vol. 29, No. 1, pp. 38–46, 1981.

19 James, J. R. and P. S. Hall, *Handbook of Microstrip Antennas*, Vol. 1 and 2, London, UK, Peter Peregrinus, 1989.

20 Bancroft, R., *Microstrip and Printed Antenna Design*, 2nd Ed., SciTech Publishing, Inc., 2009.

21 Fang, D. G., *Antenna Theory and Microstrip Antennas*, CRC Press, 2010.

22 Pozar, D. M. and D. H. Schaubert, *Microstrip Antennas: The Analysis and Design of Microstrip Antennas and Arrays*, Wiley-IEEE Press, 1995.

23 Lee, K. F. and W. Chen, *Advances in Microstrip and Printed Antennas*, John Wiley & Sons, Inc., 1997.

24 Xiao, R., W. Geyi, and W. Wen, "Theory of resonant modes and its application", *IEEE Access*, Vol. 9, pp. 114845–114956, 2021.

25 Rao, S. M., D. R. Wilton, and A. W. Glisson, "Electromagnetic scattering by surfaces of arbitrary shape", *IEEE Trans. Antennas Propagat.*, Vol. 30, No. 3, pp. 409–417, 1982.

6

Radiation in Free Space (III)

Array Analysis and Synthesis

Analysis and synthesis, though commonly treated as two different methods, are, if properly understood, only the two necessary parts of the same method. Each is the relative and correlative of the other. Analysis, without a subsequent synthesis, is incomplete. Synthesis, without a previous analysis, is baseless; for synthesis receives from analysis the elements which it recomposes.
–Sir William Hamilton (9th Baronet Scottish metaphysician, 1788–1856)

The preceding chapter has focused on the analysis of a single antenna to determine the radiation fields from a given current source. The current source can be the actual current distribution on a wire or an equivalent current on an aperture. Antenna synthesis refers to the inverse problem of finding the current source with various optimization methods to achieve a specified electromagnetic (EM) field pattern in the near- or far-field region in an efficient and economical way. Since the realization of a continuous source is not easy in practice, it is often discretized first and then realized by more than one antenna element, called antenna array. Most of the antennas are around one wavelength in size and their radiation patterns often cover a wide angle and thus exhibit poor directivity. In order to enhance the directivity, one has to use electrically large antenna (such as a reflector antenna) or antenna array. Typical array elements include patch, dipole, waveguide slot, and horn [1]. The performances of antenna array are controlled by the relative physical positioning of elements (called the array configuration) and the distribution of excitations. One of the important tools for the array analysis and synthesis is the array factor, which is a complex-valued far-field pattern for an array of isotropic radiators. The conventional methods for pattern synthesis are based on the array factor and are problem specific, relying on certain simplifications in order to reduce the complexity of the optimization process. For example, the Schelkunoff unit circle method [2] is aimed to produce nulls in desired

Foundations of Antenna Radiation Theory: Eigenmode Analysis, First Edition. Wen Geyi.

Published 2023 by John Wiley & Sons, Inc.

directions, the Fourier transform method and Woodward–Lawson method [3] are often used for beam shaping, and the binomial method and the Dolph–Chebyshev method [4] are most effective in generating narrow beams and low sidelobes. However, the conventional array synthesis methods based on the antenna array factor are no longer valid when the array elements are not identical or its surrounding environment is too complicated or the inter-element spacing is very small, which raises a number of challenges for antenna designers.

In order to achieve a desired field pattern, a performance index (target function) must be introduced and optimized during the process of antenna design. For the design of a wireless system intended either for the transmission of information or power, a natural performance index is the power transmission efficiency (PTE) between the transmitting and receiving antennas, which is defined as the ratio of the power delivered to the load of the receiver to the input power of the transmitter. To attain the best possible quality of wireless communication or power transfer, the PTE must be maximized. Motivated by the fact that all antennas are designed to enhance the PTE for a wireless system, the PTE can thus be used as a performance index for the design of antennas.

The power transmission between two antenna arrays was first investigated by the author in [5]. The optimization procedure provides a powerful technique for the synthesis of antenna arrays, which can overcome the challenges with the conventional array design methods. The technique is called the method of maximum power transmission efficiency (MMPTE) and can be used to achieve various field patterns in any complicated environment, either in the near- or far-field region [6]. The conventional array design methods largely rely on field theory, while the MMPTE reduces the field synthesis problem to a circuit analysis problem so that the circuit theory can be applied to solve the original field problem. This feature of MMPTE makes the design process of antenna array more accessible for those who are not very familiar with EM field theory. The circuit parameters can be acquired by simulation or measurement, and for this reason, the MMPTE is applicable to any complicated problem. Whenever the design problem cannot be handled by a state-of-the-art computer, one can resort to measurement to find the circuit parameters. Another important feature of MMPTE is that it contains the information of the environment between Tx and Rx arrays, and therefore can be made adaptive to complicated environment, guaranteeing the best possible performance of the antenna array. The MMPTE has been demonstrated to be superior to most existing array design methods in terms of simplicity, applicability, generality, and design accuracy. It generates an optimized distribution of excitation (ODE) for the antenna array to assure that the gain and efficiency of the array is maximized for a fixed array configuration and is equally applicable for both near- and far-field synthesis problems.

This chapter summarizes several conventional array synthesis methods based on the array factors. The main focus is on the formulations of MMPTE and the extended MMPTE (EMMPTE), and the latter shares the similarity with the MMPTE but does not involve the test receiving antennas, and therefore is actually a field method. Applications of MMPTE and EMMPTE to the designs of focused antenna, smart antenna, beam shaping antenna, multi-beam antenna, end-fire antenna, multi-null steering antennas, as well as the design of WPT system are demonstrated.

6.1 Introduction to Array Analysis

Array analysis refers to the process of understanding how an antenna operates and performs when the distribution of excitations for the array is given. The analysis yields a number of radiation characteristics such as the field distribution around antenna, gain, efficiency, input impedance, bandwidth, and polarization. For complicated antenna structures, various numerical methods have to be adopted to facilitate the analysis.

6.1.1 Array Factor

An **antenna array** is a group of radiators (called elements) excited by currents with different amplitudes and phases. As a result of EM interference, the radiated fields are enhanced in the desired direction and cancelled in the non-desired direction. It will be convenient to introduce two rectangular coordinate systems: the global rectangular coordinate system $\mathbf{r} = (x, y, z)$ and the local rectangular coordinate system $\mathbf{r}_n = (x_n, y_n, z_n)$, which are related by the coordinate transformation $\mathbf{r} = \mathbf{r}_n + \mathbf{d}_n$ shown in Figure 6.1. The vector $\mathbf{d}_n$ starts at the origin of the global coordinate system (x, y, z) and ends at the origin of the local coordinate system (x_n, y_n, z_n). The spherical coordinate systems associated with the rectangular coordinate systems $\mathbf{r}$ and $\mathbf{r}_n$ are denoted by (r, θ, φ) and $(r_n, \theta_n, \varphi_n)$, respectively.

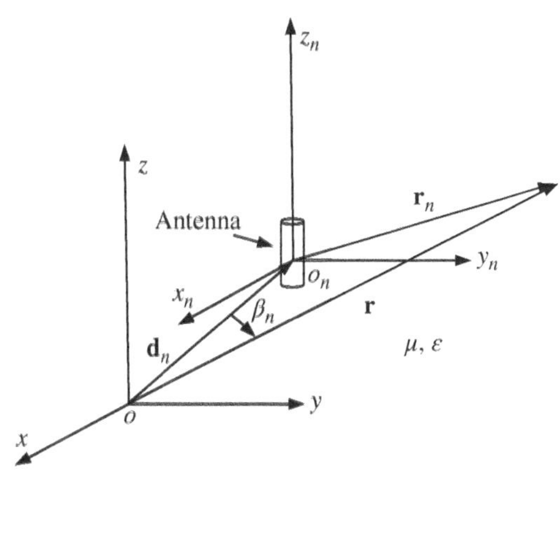

Figure 6.1 Transformation of coordinate systems.

In the local coordinate system (x_n, y_n, z_n), the far-zone electric field of an antenna is given by (4.16):

$$\mathbf{E}_n(\mathbf{r}_n) = -\frac{jk\eta I_n}{4\pi r_n} e^{-jkr_n} \mathbf{L}_n(\mathbf{u}_{r_n}), \tag{6.1}$$

where $\mathbf{L}_n$ is the vector effective length and $I_n = a_n e^{j\alpha_n}$ is the exciting current at the antenna feeding plane. As $r_n \to \infty$, the following relations hold approximately (see Figure 6.1):

$$\mathbf{r}_n \parallel \mathbf{r}, \mathbf{u}_\theta \approx \mathbf{u}_{\theta_n}, \ \theta \approx \theta_n, \ \varphi \approx \varphi_n. \tag{6.2}$$

If $d_n = |\mathbf{d}_n| \ll r_n$ and r_n is sufficiently large, the following approximation can be made:

$$\frac{e^{-jkr_n}}{r_n} = \frac{e^{-jk|\mathbf{r}-\mathbf{d}_n|}}{|\mathbf{r}-\mathbf{d}_n|} \approx \frac{e^{-jkr}}{r} e^{jk\mathbf{u}_r \cdot \mathbf{d}_n} = \frac{e^{-jkr}}{r} e^{jkd_n \cos\beta_n}, \tag{6.3}$$

where β_n is the angle between $\mathbf{d}_n$ and $\mathbf{r}$. By use of (6.2) and (6.3), (6.1) can be written as

$$\mathbf{E}_n(\mathbf{r}) = -\frac{jk\eta I_n e^{-jkr}}{4\pi r} e^{jk\mathbf{u}_r \cdot \mathbf{d}_n} \mathbf{L}_n(\mathbf{u}_r). \tag{6.4}$$

The field expression (6.1) in the local coordinate system is now transformed into that in the global coordinate system (x, y, z).

If there exist N antenna elements such that (6.2) holds for each of them, the total far-zone field can be expressed as the sum of the fields generated by each element

$$\mathbf{E}(\mathbf{r}) = \sum_{n=1}^{N} \mathbf{E}_n(\mathbf{r}) = -\frac{jk\eta}{4\pi r} e^{-jkr} \sum_{n=1}^{N} I_n \mathbf{L}_n(\mathbf{u}_r) e^{jk\mathbf{u}_r \cdot \mathbf{d}_n}, \tag{6.5}$$

where the mutual couplings between the antenna elements have been neglected. If all the antenna elements are identical, such that $\mathbf{L}_n(\mathbf{u}_r) = \mathbf{L}(\mathbf{u}_r)$ $(n = 1, 2, \ldots, N)$, the above equation can be rewritten as a product

$$\mathbf{E}(\mathbf{r}) = \mathbf{E}_e(\mathbf{r}) \times \text{AF}, \tag{6.6}$$

where

$$\text{AF} = \sum_{n=1}^{N} I_n e^{jk\mathbf{u}_r \cdot \mathbf{d}_n} = \sum_{n=1}^{N} a_n e^{j(\alpha_n + k\mathbf{u}_r \cdot \mathbf{d}_n)} \tag{6.7}$$

is known as the **array factor**, and

$$\mathbf{E}_e(\mathbf{r}) = -\frac{jk\eta}{4\pi r} e^{-jkr} \mathbf{L}(\mathbf{u}_r)$$

is the far-zone field of the antenna element located at the origin of the global coordinate system (x, y, z), which is normalized to the exciting current. Equation (6.6) is called the **pattern multiplication** rule for the array of identical elements.

6.1.2 Linear Array

A **linear array** is formed by a group of antenna elements placed along one axis, and is the simplest type of array configuration. The separation among the elements may vary, but it is often assumed to be uniform. Linear arrays can be classified into three categories based on how the axis of the array is related to the direction of radiation. A **broadside array** is a linear array whose main beam points to the direction perpendicular to the array axis. In this case, the antenna elements must be fed in phase. An **end-fire array** is a linear array whose main beam is along the array axis. In this case, the antenna elements must be fed with a constant phase difference related to the separation between the elements. A **phased linear array** refers to a linear array that can direct the main beam to an arbitrary direction.

6.1.2.1 Linear Array with Uniform Amplitude

Consider a linear array of N identical elements equally spaced along the z-axis, as indicated in Figure 6.2. The elements are first assumed to be excited by nonuniform excitation amplitudes with a progressive phase shift α. For convenience, one may set $\alpha_1 = 0$ and $d_1 = 0$ so that $\alpha_n = (n-1)\alpha$, $\mathbf{u}_r \cdot \mathbf{d}_n = (n-1)d\cos\theta$, where d is the separation between two adjacent elements. The array factor (6.7) for a linear array with nonuniform amplitudes can be written as

$$\text{AF} = \sum_{n=1}^{N} a_n e^{j(n-1)\psi}, \tag{6.8}$$

where $\psi = \alpha + kd\cos\theta$. If the excitation currents have equal amplitudes $a_n = a$, (6.8) reduces to

$$\text{AF} = a\frac{1-e^{jN\psi}}{1-e^{j\psi}}. \tag{6.9}$$

The array factor (6.9) is valid for an array with equal excitation amplitude and spacing, and can be written as

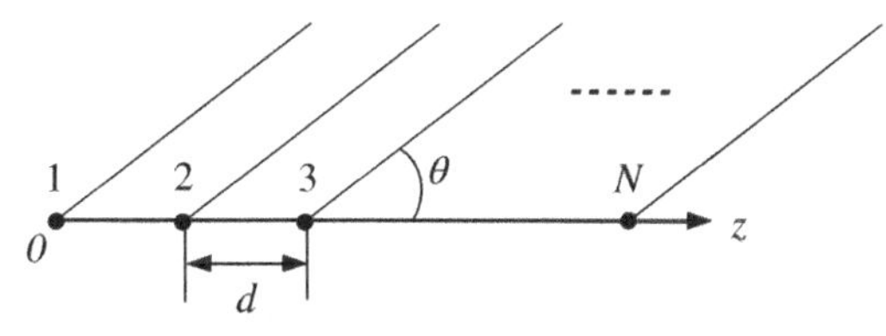

Figure 6.2 Linear array.

$$\mathrm{AF} = ae^{j(N-1)\psi/2}\frac{\sin\dfrac{N}{2}\psi}{\sin\dfrac{\psi}{2}}. \tag{6.10}$$

Dropping the constant factor a and the exponential term, one may introduce the **normalized array factor**

$$\tilde{\mathrm{AF}} = \frac{\sin\dfrac{N}{2}\psi}{N\sin\dfrac{\psi}{2}}. \tag{6.11}$$

Clearly, $\tilde{\mathrm{AF}}$ is a periodic function of ψ with period 2π and its maximum value is unity. The universal radiation patterns based on the normalized array factor $\tilde{\mathrm{AF}}$ for different number of antenna elements are plotted in Figure 6.3. It can be seen that the level of first sidelobe decreases as N increases. Some properties can be derived from the normalized array factor (6.11) and they are summarized below.

1) The normalized array factor $\tilde{\mathrm{AF}}$ attains the maximum value unity, when

$$\frac{1}{2}\psi = \frac{1}{2}(\alpha + kd\cos\theta_n) = \pm n\pi, (n = 0, 1, 2, ...),$$

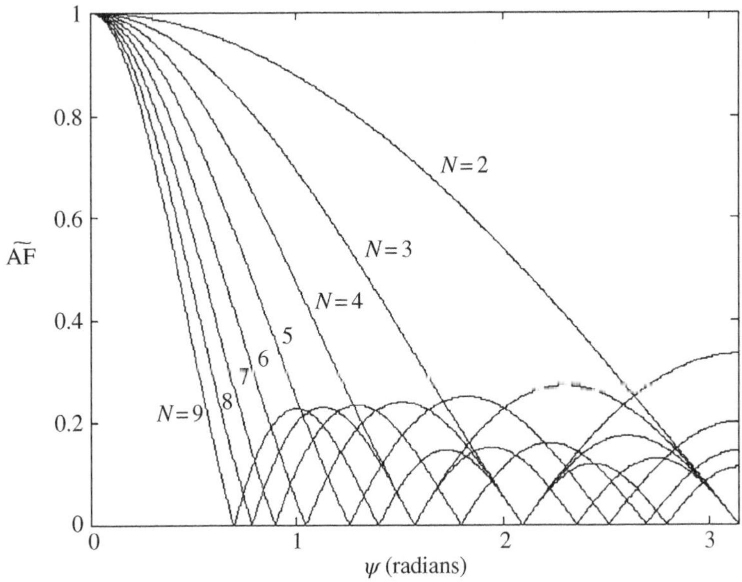

Figure 6.3 Universal pattern for uniform linear array.

or when

$$\theta_n = \arccos\left[\frac{1}{kd}(-\alpha \pm 2n\pi)\right].$$

The main beam occurs at θ_0, and other angles θ_n are where grating lobes take place.

Remark 6.1 (Grating Lobe)
A grating lobe, other than the main lobe, is produced by an array when the inter-element spacing is sufficiently large to permit the in-phase addition of radiated fields in more than one direction. □

2) The 3-dB point of the main beam occurs when

$$\frac{N}{2}\psi = \frac{N}{2}(\alpha + kd\cos\theta_{3\ dB}) = \pm 1.391,$$

or when

$$\theta_{3\ dB} = \arccos\left[\frac{1}{kd}\left(-\alpha \pm \frac{2.782}{N}\right)\right].$$

The half-power beamwidth is determined by

$$\mathrm{BW}_{3dB} = 2|\theta_0 - \theta_{3dB}|.$$

3) The nulls of the normalized array factor $\tilde{\mathrm{AF}}$ occur when

$$\frac{N}{2}\psi = \frac{N}{2}(\alpha + kd\cos\theta_n) = \pm n\pi,$$

or when

$$\theta_n = \arccos\left[\frac{1}{kd}\left(-\alpha \pm \frac{2n\pi}{N}\right)\right], (n = 1, 2, \ldots; n \neq N, 2N, 3N, \ldots).$$

4) The maximum of the first minor lobe occurs (approximately) when

$$\frac{N}{2}\psi = \frac{N}{2}(\alpha + kd\cos\theta_{s1}) = \pm\frac{3\pi}{2},$$

or when

$$\theta_{s1} = \arccos\left[\frac{1}{kd}\left(-\alpha \pm \frac{3\pi}{N}\right)\right].$$

At θ_{s1}, the magnitude of $\tilde{\mathrm{AF}}$ is $(2/3\pi)$, which is $-13.46\ dB$ down from the maximum of the major lobe and is approximately valid for large N.

Example 6.1 (Broadside Array)
To direct the main beam (the first maximum of the array factor) to the direction normal to the axis of the array, i.e. to the direction $\theta = (\pi/2)$, one may let

$$\alpha + kd\cos\theta|_{\theta=\pi/2} = \alpha = 0.$$

The broadside radiation is thus attained if all the elements are excited with equal excitation phases. Figure 6.4a shows the two-dimensional (2D) radiation pattern in the plane $\varphi =$ constant for a five-element broadside array with $d = (\lambda/2)$. □

Example 6.2 (End-Fire Array)
To direct the main beam along the axis of the array, i.e. to the directions $\theta = 0$ or $\theta = \pi$, one may let

$$\alpha + kd\cos\theta|_{\theta=0} = \alpha + kd = 0,$$
$$\alpha + kd\cos\theta|_{\theta=\pi} = \alpha - kd = 0.$$

The end-fire radiation is therefore achieved if the progressive phase shift α satisfies the above equations. Figure 6.4b shows the 2D radiation pattern in the plane $\varphi =$ constant for a five-element end-fire array with $d = \frac{\lambda}{2}$. □

Example 6.3 (Phased Array)
To direct the main beam to an arbitrary direction θ_p $(0 < \theta_p < \pi)$, one may let

$$\alpha + kd\cos\theta_p = 0.$$

The main beam will be directed to the desired direction θ_p if the progressive phase shift is selected as

$$\alpha = -kd\cos\theta_p.$$

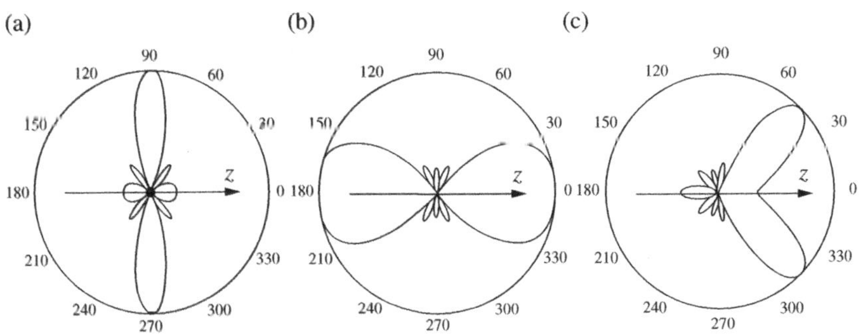

Figure 6.4 Radiation patterns. (a) Broadside. (b) End-fire. (c) Scanning angle $\theta_p = 45°$.

Figure 6.4c shows the 2D radiation pattern in the plane φ = constant for a five-element phased array with $d = (\lambda/2)$ and $\theta_p = 45°$. To avoid the grating lobes, one must have

$$\left|\frac{1}{2}\psi\right| = \left|\frac{1}{2}(kd\cos\theta + \alpha)\right| = \frac{1}{2}kd\left|\cos\theta - \cos\theta_p\right| < \pi$$

according to (6.11). From the above, the separation between two adjacent elements must satisfy

$$d < \frac{\lambda}{1 + \left|\cos\theta_p\right|}. \tag{6.12}$$

The right-hand side of (6.12) is the allowable maximum spacing between elements, exceeding which the grating lobes may occur. □

6.1.2.2 Linear Array with Nonuniform Amplitude

Consider now the linear array with nonuniform amplitude and a progressive phase shift α. For convenience, the origin of the global coordinate system is selected at the center of the equispaced linear array as illustrated in Figure 6.5. It will be assumed that the array is symmetrically excited with a distribution of excitations

$$\left\{a_M e^{-j(2M-1)\alpha/2}, \ldots, a_2 e^{-j3\alpha/2}, a_1 e^{-j\alpha/2}, a_1 e^{j\alpha/2}, a_2 e^{j3\alpha/2}, \ldots, a_M e^{j(2M-1)\alpha/2}\right\}$$

for an array of $2M$ (M is an integer) elements, and

$$\left\{a_{M+1} e^{jM\alpha}, a_M e^{-j(M-1)\alpha}, \ldots, a_2 e^{-j\alpha}, 2a_1, a_2 e^{j\alpha}, \ldots, a_M e^{j(M-1)\alpha}, a_{M+1} e^{jM\alpha}\right\}$$

for an array of $2M + 1$ elements. The array factor for the array of even number of elements is

$$\begin{aligned}\mathrm{AF}_e &= a_1 e^{j\frac{1}{2}(kd\cos\theta+\alpha)} + a_2 e^{j\frac{3}{2}(kd\cos\theta+\alpha)} + \cdots + a_M e^{j\frac{2M-1}{2}(kd\cos\theta+\alpha)} \\ &\quad + a_1 e^{-j\frac{1}{2}(kd\cos\theta+\alpha)} + a_2 e^{-j\frac{3}{2}(kd\cos\theta+\alpha)} + \cdots + a_M e^{-j\frac{2M-1}{2}(kd\cos\theta+\alpha)} \\ &= 2\sum_{n=1}^{M} a_n \cos\left[\frac{2n-1}{2}(kd\cos\theta+\alpha)\right].\end{aligned} \tag{6.13}$$

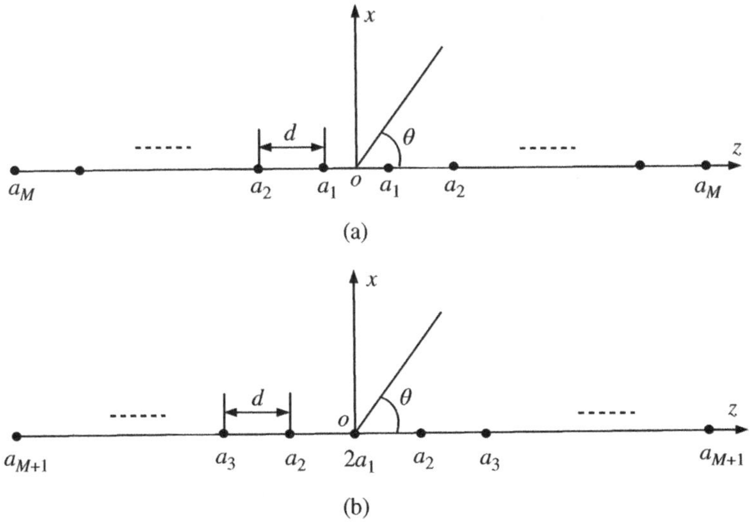

Figure 6.5 Linear arrays. (a) Even number of elements. (b) Odd number of elements.

The array factor for the array of odd number of elements is

$$\begin{aligned} \mathrm{AF}_o &= 2a_1 + a_2 e^{j(kd\cos\theta+\alpha)} + \cdots + a_{M+1}e^{jM(kd\cos\theta+\alpha)} \\ &\quad + a_2 e^{-j(kd\cos\theta+\alpha)} + \cdots + a_{M+1}e^{-jM(kd\cos\theta+\alpha)} \\ &= 2\sum_{n=1}^{M+1} a_n \cos[(n-1)(kd\cos\theta+\alpha)]. \end{aligned} \tag{6.14}$$

Example 6.4 (Binomial Array)
For an integer $m \geq 1$, the binomial theorem is given by

$$(1+x)^{m-1} = \sum_{k=0}^{m-1} C_{m-1}^k x^k, \tag{6.15}$$

where the binomial coefficients are defined by the **combinatory number**

$$C_n^k = \binom{n}{k} = \frac{n!}{(n-k)!k!}.$$

The binomial coefficients can easily be obtained from the **Pascal triangle** shown in Table 6.1. The first and the last number in each row are equal to one and every

Table 6.1 Pascal triangle.

m						Binomial coefficients						
1						1						
2					1		1					
3				1		2		1				
4			1		3		3		1			
5		1		4		6		4		1		
6	1		5		10		10		5		1	
	↑		↑		↑		↑		↑		↑	
	$\binom{5}{0}$		$\binom{5}{1}$		$\binom{5}{2}$		$\binom{5}{3}$		$\binom{5}{4}$		$\binom{5}{5}$	

Table 6.2 Excitation amplitudes.

Number of elements	Amplitude
2	$a_1 = 1$
3	$2a_1 = 2,\ a_2 = 1$
4	$a_1 = 3,\ a_2 = 1$
5	$2a_1 = 6,\ a_2 = 4,\ a_3 = 1$
6	$a_1 = 10,\ a_2 = 5,\ a_3 = 1$

other coefficient is the sum of the two numbers on its both sides in the row above it. The array is referred to as a **binomial array** if its excitation amplitudes are selected as the binomial coefficients listed in Table 6.2.

Figure 6.6 shows the radiation patterns for a five-element broadside binomial array ($\alpha = 0$) with separations of $\lambda/4$, $\lambda/2$, and $3\lambda/5$, respectively. It is observed that there are no minor lobes for the arrays with separations of $\lambda/4$ and $\lambda/2$. In general, a binomial array does not have minor lobes if the separation is equal or less than a half wavelength. □

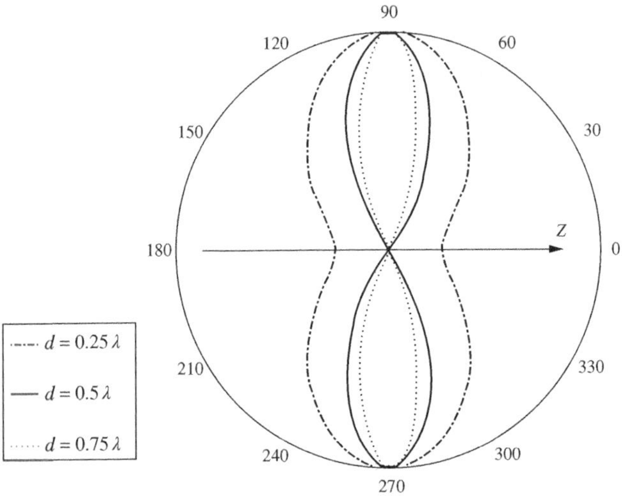

Figure 6.6 Radiation patterns for a five-element broadside binomial array.

6.1.3 Circular Array

For the N-element array placed in a circular ring of radius a shown in Figure 6.7, the position of the nth element is $\mathbf{d}_n = a\mathbf{u}_{\rho n}$, where $\mathbf{u}_{\rho n} = \mathbf{u}_x \cos\varphi_n + \mathbf{u}_y \sin\varphi_n$ is the unit vector along the radial direction of the circle area. One may write

$$\begin{aligned}\mathbf{u}_r \cdot \mathbf{d}_n &= \left(\mathbf{u}_x \sin\theta\cos\varphi + \mathbf{u}_y \sin\theta\sin\varphi + \mathbf{u}_z\cos\theta\right)\cdot a\left(\mathbf{u}_x\cos\varphi_n + \mathbf{u}_y\sin\varphi_n\right)\\ &= a\sin\theta\cos(\varphi-\varphi_n).\end{aligned}$$

The array factor (6.7) becomes

$$\mathrm{AF} = \sum_{n=1}^{N} a_n e^{j[\alpha_n + ka\sin\theta\cos(\varphi-\varphi_n)]}. \tag{6.16}$$

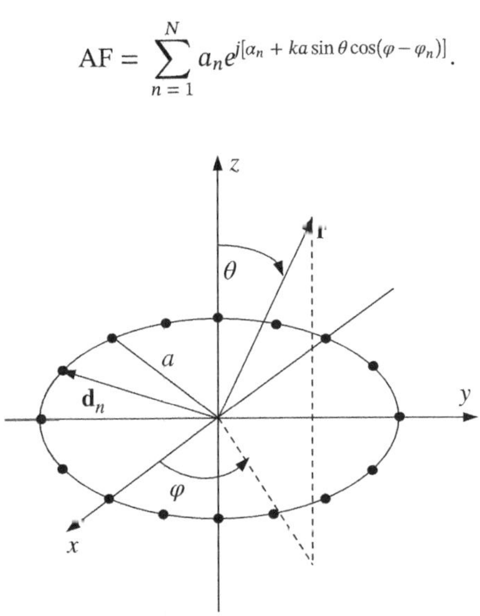

Figure 6.7 Circular array.

If the main beam is in the (θ_0, φ_0) direction, the excitation phases α_n must be selected to be

$$\alpha_n = -ka \sin\theta_0 \cos(\varphi_0 - \varphi_n).$$

The array factor (6.16) thus becomes

$$\mathrm{AF} = \sum_{n=1}^{N} a_n e^{jka[\sin\theta\cos(\varphi-\varphi_n) - \sin\theta_0\cos(\varphi_0-\varphi_n)]}. \tag{6.17}$$

The term in the square bracket can be written as

$$\sin\theta\cos(\varphi-\varphi_n) - \sin\theta_0\cos(\varphi_0-\varphi_n) = \rho_0\cos(\xi-\varphi_n),$$

where

$$\rho_0 = \sqrt{(\sin\theta\cos\varphi - \sin\theta_0\cos\varphi_0)^2 + (\sin\theta\sin\varphi - \sin\theta_0\sin\varphi_0)^2}$$
$$\cos\xi = \frac{1}{\rho_0}(\sin\theta\cos\varphi - \sin\theta_0\cos\varphi_0),$$
$$\sin\xi = \frac{1}{\rho_0}(\sin\theta\sin\varphi - \sin\theta_0\sin\varphi_0).$$

Hence, (6.17) takes a simpler form

$$\mathrm{AF} = \sum_{n=1}^{N} a_n e^{jk\rho\cos(\varphi_n-\xi)}, \tag{6.18}$$

where $\rho = a\rho_0$. The exponential functions in (6.18) can be expanded in terms of Bessel functions

$$e^{jk\rho\cos(\xi-\varphi_n)} = \sum_{m=-\infty}^{+\infty} e^{jm(\varphi_n-\xi+\pi/2)} J_m(k\rho). \tag{6.19}$$

If the array is equispaced in the circular ring $(\varphi_n = (2\pi n/N))$ and excited with uniform excitation $a_n = a_0$, (6.18) can be expressed as

$$\mathrm{AF} = a_0 \sum_{n=1}^{N}\sum_{m=-\infty}^{+\infty} e^{jm\left(\frac{2\pi n}{N}+\frac{\pi}{2}-\xi\right)} J_m(k\rho) = a_0 \sum_{m=-\infty}^{+\infty} J_m(k\rho) e^{jm\left(\frac{\pi}{2}-\xi\right)} \sum_{n=1}^{N} e^{jm\frac{2\pi n}{N}}. \tag{6.20}$$

By use of the summation

$$\sum_{n=1}^{N} e^{jn\frac{2\pi m}{N}} = \frac{e^{j\frac{2\pi m}{N}}(1-e^{j2\pi m})}{1-e^{j\frac{2\pi m}{N}}} = \begin{cases} N, & m = pN, p = 0, 1, 2, \ldots \\ 0, & \text{others} \end{cases},$$

the array factor (6.20) can be written as

$$\mathrm{AF} = a_0 N \sum_{m=-\infty}^{+\infty} J_{mN}(k\rho) e^{jmN\left(\frac{\pi}{2}-\xi\right)}. \tag{6.21}$$

The dominant part comes from $m = 0$ and $J_0(k\rho)$ is called the **principal term**.

6.1.4 Planar Array

A planar array is a 2D array of antennas, which is more flexible to reach better performance. The planar arrays provide more symmetrical patterns with lower side lobes, much higher directivity than linear array and can be used to steer the main beam toward more desired directions. Figure 6.8 shows a rectangular planar array placed in the plane $z = 0$, which consists of $M \times N$ antenna elements equispaced along x and y directions with separations d_x and d_y, respectively. The antenna element positioned at $\mathbf{d}_{mn} = (m-1)d_x\mathbf{u}_x + (n-1)d_y\mathbf{u}_y$ will be labeled as (m, n). One may write

$$\mathbf{u}_r \cdot \mathbf{d}_{mn} = (m-1)d_x \sin\theta\cos\varphi + (n-1)d_y \sin\theta\sin\varphi.$$

The array factor (6.7) for the rectangular array can be written as

$$\mathrm{AF} = \sum_{m=1}^{M}\sum_{n=1}^{N} a_{mn} e^{j\left[\alpha_{mn} + k\left((m-1)d_x \sin\theta\cos\varphi + (n-1)d_y \sin\theta\sin\varphi\right)\right]}. \tag{6.22}$$

If the excitation amplitudes a_{mn} can be decomposed into the product of two constants and the phase angles α_{mn} into the sum of two constants

$$a_{mn} = a_{xm}a_{yn},\ \alpha_{mn} = \alpha_{xm} + \alpha_{yn},$$

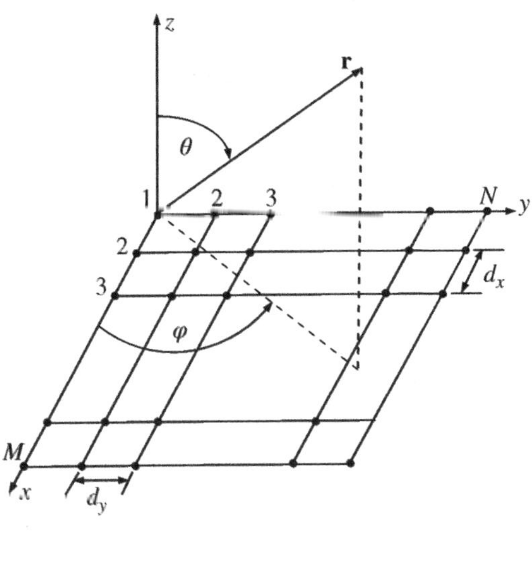

Figure 6.8 Planar array.

the array factor (6.22) can be rewritten as

$$AF = \sum_{m=1}^{M} a_{xm} e^{j[\alpha_{xm} + k(m-1)d_x \sin\theta\cos\varphi]} \sum_{n=1}^{N} a_{yn} e^{j\left[\alpha_{yn} + k(n-1)d_y \sin\theta\sin\varphi\right]}. \tag{6.23}$$

If the excitation amplitudes are constant and the excitations have a constant progressive phase shift

$$a_{xm} = a_x, a_{yn} = a_y, \alpha_{xm} = (m-1)\alpha_x, \alpha_{yn} = (n-1)\alpha_y,$$

the normalized form of (6.23) (see [6.11]) is given by

$$\tilde{AF} = \left(\frac{\sin\frac{M}{2}\psi_x}{M\sin\frac{\psi_x}{2}}\right)\left(\frac{\sin\frac{N}{2}\psi_y}{N\sin\frac{\psi_y}{2}}\right), \tag{6.24}$$

where

$$\psi_x = \alpha_x + kd_x \sin\theta\cos\varphi,$$
$$\psi_y = \alpha_y + kd_y \sin\theta\sin\varphi.$$

Figure 6.9 shows the radiation pattern ($\varphi = 0°$) for a 5×5-element array of isotropic sources with $d_x = d_y = \lambda/2$ and $\alpha_x = \alpha_y = 0$.

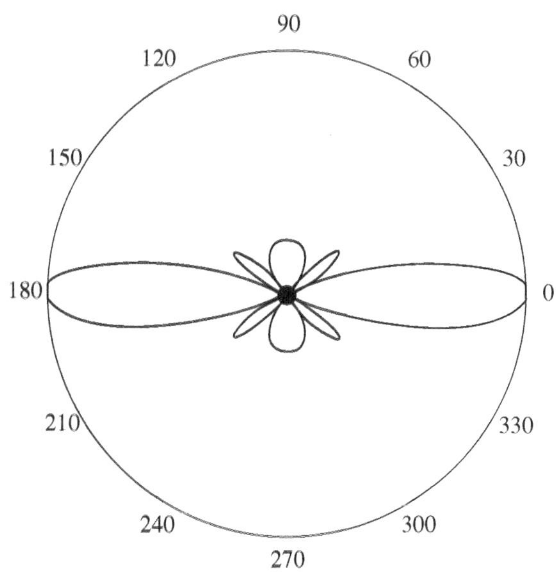

Figure 6.9 Radiation pattern ($\varphi = 0°$) for a 5×5-element rectangular planar array with uniform excitations and $d_x = d_y = (\lambda/2)$.

6.2 Introduction to Array Synthesis with Conventional Methods

For antenna synthesis, it must be understood that an antenna source distribution confined in a finite region may not be unique when a desired radiation pattern is prescribed, or may even not exist. The antenna synthesis provides the best possible solution under certain constraints to achieve the desired radiation pattern. Conventionally, antenna synthesis involves two steps. The first step is to use an analytical expression to represent the desired radiation pattern. The next step is to find the antenna source distribution so that its radiation pattern approximates what is represented by the analytical expression. Two pattern synthesis problems often arise in practice:

1) The width of the main beam and sidelobe levels are specified. The details of the beam shape and sidelobes are not demanded. This is encountered in most applications.
2) In addition to the requirements of beam width and sidelobe levels, the pattern has to be shaped to achieve a desired distribution.

Several conventional methods for antenna pattern synthesis in terms of array factor will be reviewed in this section. More detailed discussions about the conventional pattern synthesis methods can be found in [7–13].

6.2.1 Array Factor and Space Factor for Line Source

When the number of the array elements increases while the length of a linear array remains fixed, the discrete source approaches to a continuous source distribution $I(z)$, and the array factor (6.7) becomes an integral, known as **space factor** denoted by

$$\mathrm{SF}(\xi) = \int_{-l/2}^{l/2} I(z')e^{j\xi z'}dz', \tag{6.25}$$

where l is the length of the array, $\xi = k\cos\theta$, and the origin of the coordinate system is selected at the center of the line source, as illustrated in Figure 6.10. If the source distribution $I(z)$ is a constant I_0, the space factor (6.25) reduces to

$$\mathrm{SF}(\xi) = I_0 l \ \operatorname{sinc}(\pi v), \tag{6.26}$$

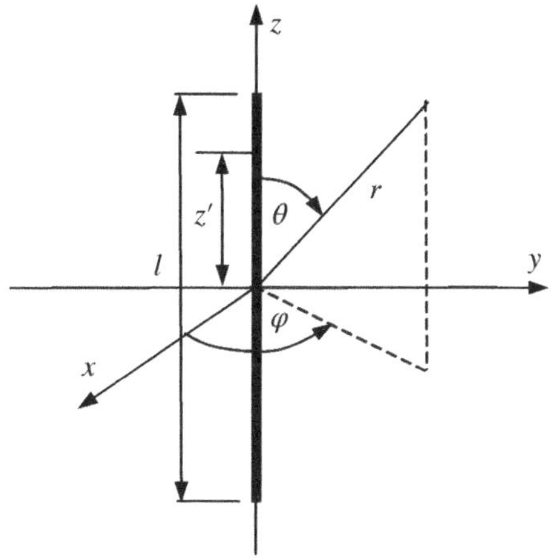

Figure 6.10 A continuous line source.

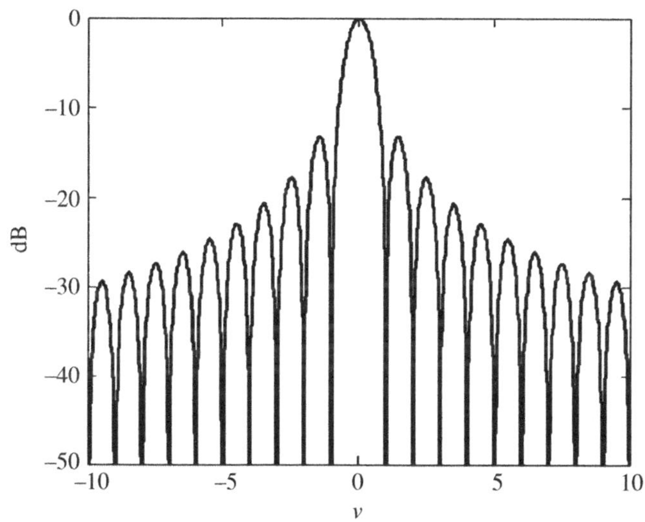

Figure 6.11 Radiation pattern of uniform line source.

with $v = (l/\lambda)\cos\theta$, and

$$\mathrm{sinc}(x) = \frac{\sin x}{x} \tag{6.27}$$

is the **sinc function**. The normalized radiation pattern for the uniform line source is plotted in Figure 6.11. Note that the first minor lobe is −13.3 dB down from the maximum at the major lobe.

6.2.2 Schelkunoff Unit Circle Method

The Schelkunoff unit circle method was first introduced in [2], and is useful in the synthesis problem where radiation nulls need to be generated in desired directions. Consider the linear array shown in Figure 6.2. By introducing a complex variable $z = e^{j\psi}$, the array factor (6.8) can be considered as a polynomial of degree $(N-1)$ and therefore has $(N-1)$ roots denoted by $z_n(n = 1, 2, ..., N-1)$:

$$\mathrm{AF} = \sum_{n=1}^{N} a_n z^{n-1} = a_N \prod_{n=1}^{N-1} (z - z_n). \tag{6.28}$$

As the parameters α, θ, and d change, the complex number z will trace out a path on the unit circle in the complex plane. For fixed α and d, the complex number z will rotate clockwise and the phase angle ψ will vary from the initial phase $\psi_s = \alpha + kd$ to the final phase $\psi_f = \alpha - kd$ as θ changes from 0° to 180°. The variation range of the phase angle ψ is $2kd$, which is called **visible region** as illustrated in Figure 6.12. The visible region is clearly independent of the phase shift α. If all the roots $z_n(n = 1, 2, ..., N-1)$ are located in the visible region, the radiation pattern will have $(N-1)$ zeros. On the other hand, if all the roots are out of the unit circle, the pattern will have no nulls.

Example 6.5 If the excitation currents have equal amplitudes, denoted by a, (6.9) can be written as

$$\mathrm{AF} = a\frac{1-z^N}{1-z}. \tag{6.29}$$

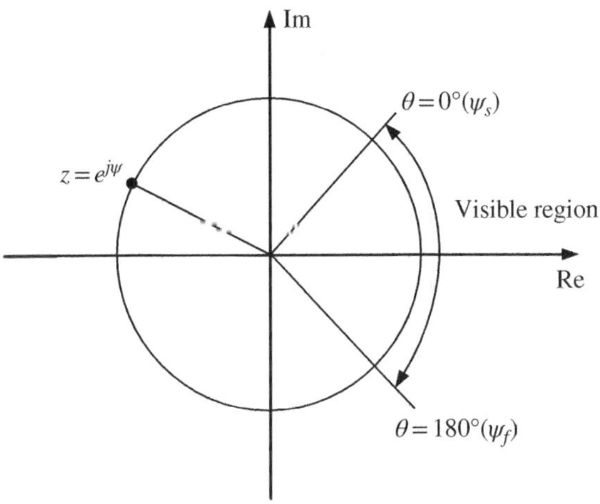

Figure 6.12 Visible region.

Apparently, the equation $\frac{1-z^N}{1-z}=0$ has roots

$$z_m = e^{j2m\pi/N}, m = 1, 2, ..., N-1.$$

Figure 6.13 shows the four roots $z_{1,4} = e^{\pm j(2/5)\pi}$ and $z_{2,3} = e^{\pm j(4/5)\pi}$ for a five-element array. For $d = (\lambda/2)$ and $\alpha = 0$, one finds $\psi_s = \pi$ and $\psi_f = -\pi$, and all the four roots are in the visible region. As z moves clockwise from $e^{j\pi}$ to $e^{-j\pi}$, the array factor traces out half a side lobe, a null at z_2, a full side lobe, a null at z_1, a main beam, a null at z_4, a side lobe, a null at z_3, and finally another half side lobe, as illustrated in Figure 6.13. □

The visible region varies with the separation distance d, and several cases are summarized below.

1) $d < (\lambda/2)$: In this case, $2kd < 2\pi$. As θ changes from 0° to 180°, z moves clockwise and covers the part of unit circle.
2) $d = (\lambda/2)$: In this case, $2kd = 2\pi$. As θ changes from 0° to 180°, z moves clockwise and covers the whole unit circle once.
3) $d > (\lambda/2)$: In this case, $2kd > 2\pi$. As θ changes from 0° to 180°, z moves clockwise and has overlapping paths on the unit circle.

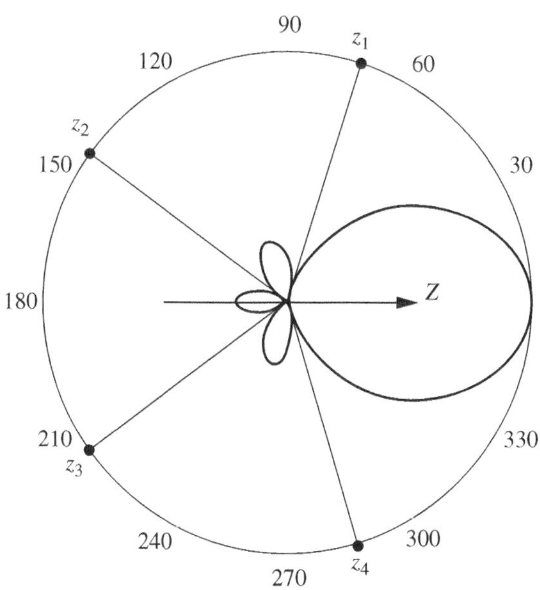

Figure 6.13 Universal pattern for five-element array.

To introduce a null at θ_n, one of the roots of the Schelkunoff polynomial on the unit circle must be

$$z_n = e^{j(kd\cos\theta_n + \alpha)}. \tag{6.30}$$

The Schelkunoff unit circle is best suited for generating nulls [7].

Example 6.6 Design a linear array with $d = \lambda/4$ such that its radiation pattern has zeros at $\theta = 0°$, $90°$ and $180°$.

Solution

For $d = \lambda/4$ and $\alpha = 0$, one may find $\psi_s = \pi/2$ and $\psi_f = -\pi/2$. According to the design requirement, the Schelkunoff polynomial has three roots $z_1 = -j$, $z_2 = 1$, and $z_3 = -j$. The normalized array factor for the design is

$$\tilde{AF} = (z-1)(z-j)(z+j) = z^3 - z^2 + z - 1. \tag{6.31}$$

Therefore, the excitation amplitudes are

$$a_1 = -1, a_2 = 1, a_3 = -1, a_4 = 1.$$

The array factor can be realized by four elements. The radiation pattern for the four-element array is plotted in Figure 6.14, which shows that the radiation pattern has zeros at $\theta = 0°$, $90°$, and $180°$ as required. □

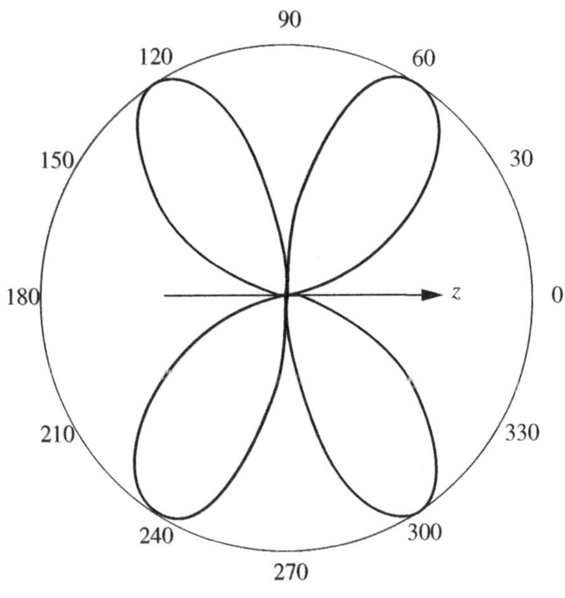

Figure 6.14 Radiation pattern for a four-element array of isotropic sources ($d = (\lambda/4)$, $\alpha = 0$).

6.2.3 Dolph–Chebyshev Method

The method was first introduced by Dolph to the design of broadside array [4]. The **Chebyshev polynomial** of degree n is defined by

$$T_n(z) = \begin{cases} \cos[n\cos^{-1}(z)], & |z| \le 1 \\ \cosh\left[n\cosh^{-1}(z)\right], & |z| > 1 \end{cases} \tag{6.32}$$

or equivalently defined by

$$\begin{aligned} T_1(z) &= z, \\ T_2(z) &= 2z^2 - 1, \\ T_n(z) &= 2zT_{n-1}(z) - T_{n-2}(z), n > 2, \end{aligned} \tag{6.33}$$

and has the following properties (see Figure 6.15):

1) $T_n(z)$ is a polynomial of degree n.
2) $T_n(-z) = (-1)^n T_n(z)$.
3) $T_n(1) = 1$, $T_n(-1) = (-1)^n$.
4) $T_n(z)$ oscillates between ± 1 in $|z| < 1$.
5) $T_n(z)$ has n different simple roots in $|z| < 1$.

The first nine Chebyshev polynomials are listed in Table 6.3.

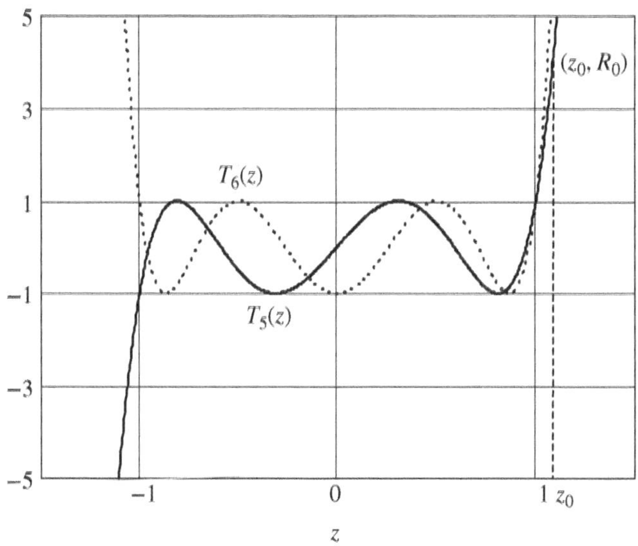

Figure 6.15 Chebyshev polynomials.

Table 6.3 Chebyshev polynomials.

n	$T_n(z)$
0	1
1	z
2	$2z^2 - 1$
3	$4z^3 - 3z$
4	$8z^4 - 8z^2 + 1$
5	$16z^5 - 20z^3 + 5z$
6	$32z^6 - 48z^4 + 18z^2 - 1$
7	$64z^7 - 112z^5 + 56z^3 - 7z$
8	$128z^8 - 256z^6 + 160z^4 - 32z^2 + 1$

The array factor of a linear array is given by (6.13) or (6.14). For a broadside array, one may let $\alpha = 0$ and the array factor can be written as

$$\mathrm{AF}_e = 2\sum_{n=1}^{M} a_n \cos[(2n-1)u], \tag{6.34}$$

$$\mathrm{AF}_o = 2\sum_{n=1}^{M+1} a_n \cos[2(n-1)u], \tag{6.35}$$

respectively, for the arrays of $2M$ elements and $2M+1$ elements with $u = (1/2)kd\cos\theta$. If one introduces

$$z = \cos u, \tag{6.36}$$

(6.34) and (6.35) can be expressed as a linear combination of Chebyshev polynomials

$$\mathrm{AF}_e = 2\sum_{n=1}^{M} a_n T_{2n-1}(z), \tag{6.37}$$

$$\mathrm{AF}_o = 2\sum_{n=1}^{M+1} a_n T_{2(n-1)}(z). \tag{6.38}$$

Therefore, AF_e and AF_o are polynomials of degree $N-1$, where N is the number of antenna elements.

Dolph first applied the Chebyshev polynomial $T_{N-1}(z)$ to obtain an optimum radiation pattern with controllable sidelobe levels for an N-element array. The procedure is summarized below:

1) Let R_0 denote the ratio of the main beam maximum to the minor lobe level. A point z_0 can be then determined such that $T_{N-1}(z_0) = R_0$ (refer to Figure 6.15). Note that $z_0 > 1$ if $R_0 > 1$.
2) Introduce a normalized variable $w = z/z_0$ and replace (6.36) with

$$w = \cos u, \tag{6.39}$$

and substitute this into the array factor (6.34) or (6.35)

$$\mathrm{AF}_e = 2\sum_{n=1}^{M} a_n T_{2n-1}(w), \tag{6.40}$$

$$\mathrm{AF}_o = 2\sum_{n=1}^{M+1} a_n T_{2(n-1)}(w). \tag{6.41}$$

Therefore, the array factor is a polynomial of degree $N-1$.

3) Equate the array factor (6.40) or (6.41) to $T_{N-1}(z)$. This process determines the excitation coefficients a_n.

The Dolph–Chebyshev design yields the smallest possible first-null beamwidth for a given sidelobe level or the smallest sidelobe level for a given first-null beamwidth.

Example 6.7 Design a nine-element broadside array with $d = \lambda/2$ and $R_0 = 20$ (26 dB).

Solution

The point z_0 is determined from the Chebyshev polynomial of degree 8

$$T_8(z_0) = \cosh\left[8\cos\mathrm{h}^{-1}(z_0)\right] = R_0 = 20.$$

This gives

$$z_0 = \cosh\left[\frac{1}{8}\cos\,\mathrm{h}^{-1}(20)\right] = 1.108.$$

Substituting $w = z/z_0$ into (6.41) and dropping the factor 2, one may obtain

$$\begin{aligned}
\mathrm{AF}_o &= \sum_{n=1}^{5} a_n T_{2(n-1)}(w) = a_1 + a_2\left(2w^2 - 1\right) + a_3\left(8w^4 - 8w^2 + 1\right) \\
&\quad + a_4\left(32w^6 - 48w^4 + 18w^2 - 1\right) + a_5\left(128w^8 - 256w^6 + 160w^4 - 32w^2 + 1\right) \\
&= 128a_5\left(\frac{z}{z_0}\right)^8 + (32a_4 - 256a_5)\left(\frac{z}{z_0}\right)^6 + (8a_3 - 48a_4 + 160a_5)\left(\frac{z}{z_0}\right)^4 \\
&\quad + (2a_2 - 8a_3 + 18a_4 - 32a_5)\left(\frac{z}{z_0}\right)^2 + a_1 - a_2 + a_3 - a_4 + a_5.
\end{aligned} \tag{6.42}$$

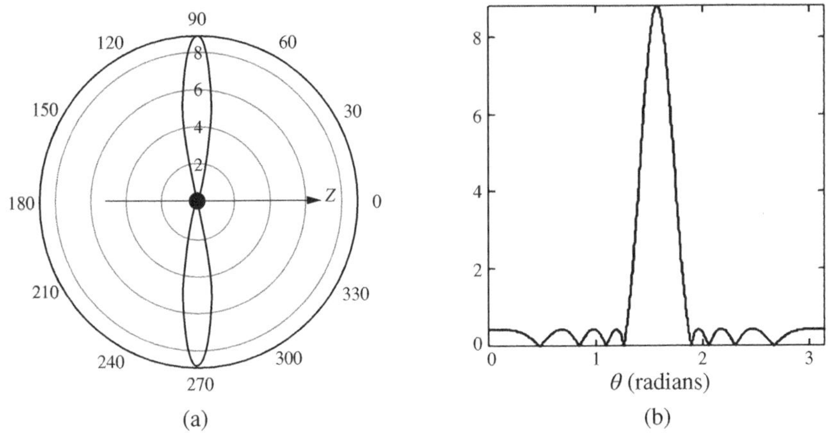

Figure 6.16 Radiation pattern of nine-element broadside array with $d = \lambda/2$. (a) Polar coordinate system. (b) Rectangular coordinate system.

Equating Chebyshev polynomial

$$T_8(z) = 128z^8 - 256z^6 + 160z^4 - 32z^2 + 1$$

to (6.42) and comparing the coefficients of the terms of like degree, one finds

$$\begin{aligned}
a_5 &= z_0^8 = 2.272,\\
a_4 &= 8a_5 - 8z_0^6 = 3.374,\\
a_3 &= 20z_0^4 + 6a_4 - 20a_5 = 4.947,\\
a_2 &= 4a_3 - 9a_4 + 16a_5 - 16z_0^2 = 6.131,\\
a_1 &= 1 + a_2 - a_3 + a_4 - a_5 = 3.286.
\end{aligned}$$

These coefficients can be normalized as follows:

$$\begin{aligned}
a_5 &= 1,\\
a_4 &= 1.485,\\
a_3 &= 2.177,\\
a_2 &= 2.699,\\
a_1 &= 1.446.
\end{aligned}$$

Dropping the factor 2, the array factor (6.35) for the nine-element array can be written as

$$\mathrm{AF}_o = \sum_{n=1}^{5} a_n \cos[2(n-1)u]. \tag{6.43}$$

The radiation pattern (6.43) for the nine-element broadside array is plotted in Figure 6.16a in polar coordinates and Figure 6.16b in rectangular coordinates. It is observed that the roots of the Chebyshev polynomial $T_8(z)$ correspond to the nulls of the radiation pattern. Also note that the ratio of the main beam maximum to the minor lobe level is 20 as specified. □

6.2.4 Fourier Transform Method

The Fourier transform method is based on the fact that the far-field radiation pattern is the Fourier transform of the aperture (source) distribution, and is very useful in pattern synthesis. When the radiation pattern is specified, the source distribution can be directly obtained from the inverse Fourier transform.

6.2.4.1 Continuous Line Source

Since the current distribution $I(z)$ in (6.25) is zero outside the region $|z| \leq l/2$, one can extend integration limits in (6.25) to infinity

$$\mathrm{SF}(\xi) = \int_{-\infty}^{+\infty} I(z')e^{j\xi z'}dz'. \tag{6.44}$$

This implies that the space factor is the Fourier transform of the line source, and its inverse gives the current source distribution

$$I(z) = \frac{1}{2\pi}\int_{-\infty}^{+\infty} \mathrm{SF}(\xi)e^{-j\xi z}d\xi. \tag{6.45}$$

In general, the inverse transform (6.45) is not zero outside the source region $|z| \leq l/2$. Therefore, one has to truncate it at $z = \pm l/2$, yielding an approximate source distribution. The above procedure is called **Fourier transform method**.

Example 6.8 Let the space factor be determined by

$$\mathrm{SF}(\xi) = \begin{cases} 1, & 0 < \theta < \pi \text{ (or } |\xi| < k) \\ 0, & \text{elsewhere} \end{cases}. \tag{6.46}$$

The source distribution is then given by (6.45)

$$I(z) = \frac{1}{2\pi}\int_{-k}^{k} e^{-j\xi z}d\xi = \frac{k}{\pi}\mathrm{sinc}(kz). \tag{6.47}$$

If the source distribution is assumed to be confined in $|z| \leq l/2$, the realized space factor (6.25) becomes

$$\begin{aligned}\mathrm{SF}(\xi) &= \frac{k}{\pi}\int_{-l/2}^{l/2} \mathrm{sinc}(kz')e^{j\xi z'}dz' = \frac{2}{\pi}\int_{0}^{l/2} \frac{\sin kz' \cos \xi z'}{z'}dz' \\ &= \frac{1}{\pi}\int_{0}^{l/2} \frac{\sin(k+\xi)z' + \sin(k-\xi)z'}{z'}dz' \\ &= \frac{1}{\pi}\mathrm{si}(1+\cos\theta)\frac{kl}{2} + \frac{1}{\pi}\mathrm{si}(1-\cos\theta)\frac{kl}{2},\end{aligned} \tag{6.48}$$

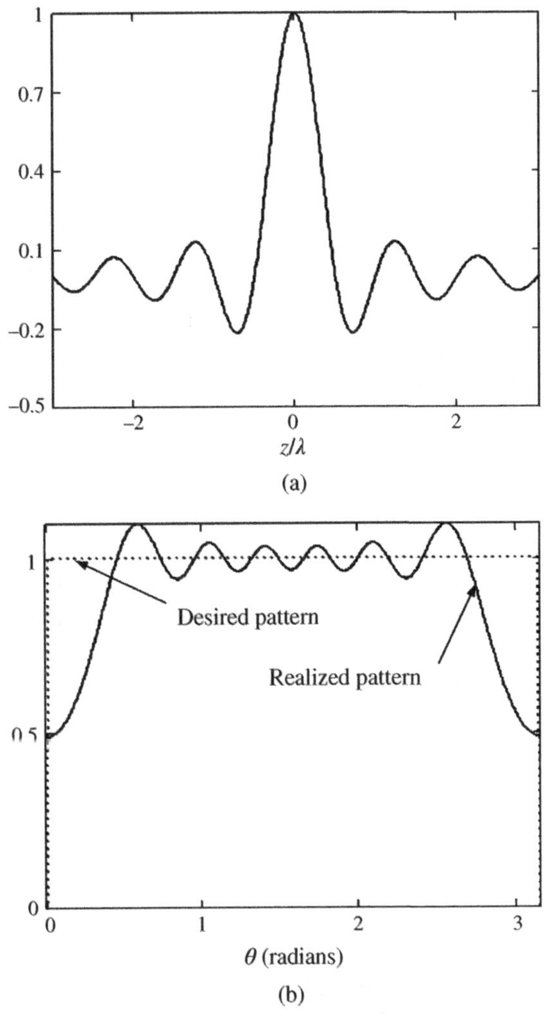

Figure 6.17 Normalized current distribution and realized radiation pattern. (a) Normalized current distribution for $l = 6\lambda$. (b) Realized radiation pattern.

where si is sine integral

$$\mathrm{si}(x) = \int_0^x \mathrm{sinc}(y)dy.$$

The normalized current distribution for the line source of length $l = 6\lambda$ is shown in Figure 6.17a. The realized radiation pattern is shown Figure 6.17b and is compared with the desired radiation pattern. □

6.2.4.2 Linear Array

The Fourier transform method is equally applicable to a linear array and the discussions are similar. Consider an N-element equispaced linear array shown in Figure 6.18. The origin of the global coordinate system is located at the center of the array. Assume that the array is excited with a distribution of excitations

$$\left\{a_{-M}e^{-j(2M-1)\alpha/2}, \ldots, a_{-2}e^{-j3\alpha/2},\ a_{-1}e^{-j\alpha/2}, a_1e^{j\alpha/2}, a_2e^{j3\alpha/2}, \ldots, a_Me^{j(2M-1)\alpha/2}\right\}$$

for an array of $2M$ (M is an integer) elements, and

$$\left\{a_{-M}e^{-jM\alpha}, \ldots, a_{-2}e^{-j2\alpha}, a_{-1}e^{-j\alpha}, a_0, a_1e^{j\alpha}, a_2e^{j2\alpha}, \ldots, a_Me^{jM\alpha}\right\}$$

for an array of $2M + 1$ elements. Correspondingly, the array factors can be written as

$$\begin{aligned} \mathrm{AF}_e &= a_1e^{j\frac{1}{2}(kd\cos\theta+\alpha)} + a_2e^{j\frac{3}{2}(kd\cos\theta+\alpha)} + \cdots + a_Me^{j\frac{2M-1}{2}(kd\cos\theta+\alpha)} \\ &\quad + a_{-1}e^{-j\frac{1}{2}(kd\cos\theta+\alpha)} + a_{-2}e^{-j\frac{3}{2}(kd\cos\theta+\alpha)} + \cdots + a_{-M}e^{-j\frac{2M-1}{2}(kd\cos\theta+\alpha)} \\ &= \sum_{m=-M}^{-1} a_n e^{j\frac{2m+1}{2}\psi} + \sum_{m=1}^{M} a_n e^{j\frac{2m-1}{2}\psi} \end{aligned} \tag{6.49}$$

for $N = 2M$ and

$$\begin{aligned} \mathrm{AF}_o &= a_0 + a_1e^{j(kd\cos\theta+\alpha)} + \cdots + a_Me^{jM(kd\cos\theta+\alpha)} \\ &\quad + a_2e^{-j(kd\cos\theta+\alpha)} + \cdots + a_Me^{-jM(kd\cos\theta+\alpha)} \\ &= \sum_{m=-M}^{M} a_m e^{jm\psi} \end{aligned} \tag{6.50}$$

for $N = 2M + 1$ with $\psi = kd\cos\theta + \alpha$. Expressions (6.49) and (6.50) are the Fourier series expansions of the array factors. Once the array factors are specified, the excitation amplitudes a_n (Fourier coefficients) can be determined by

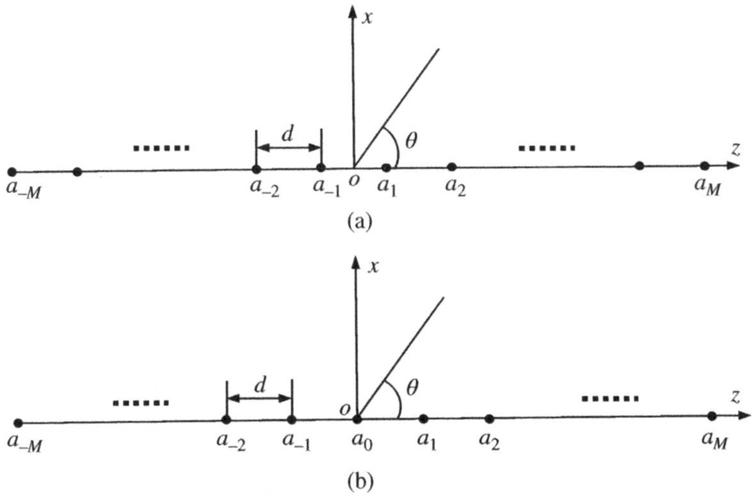

Figure 6.18 Linear arrays. (a) Even number of elements. (b) Odd number of elements.

$$a_m = \begin{cases} \dfrac{1}{2\pi}\displaystyle\int_{-\pi}^{\pi} \mathrm{AF}_e(\psi)e^{-j\frac{2m+1}{2}\psi}d\psi, & -M \leq m \leq -1 \\ \dfrac{1}{2\pi}\displaystyle\int_{-\pi}^{\pi} \mathrm{AF}_e(\psi)e^{-j\frac{2m-1}{2}\psi}d\psi, & 1 \leq m \leq M \end{cases} \tag{6.51}$$

for $N = 2M$ and

$$a_m = \frac{1}{2\pi}\int_{-\pi}^{\pi} \mathrm{AF}_o(\psi)e^{-jm\psi}d\psi, \tag{6.52}$$

for $N = 2M + 1$.

6.3 Power Transmission Between Two Antennas

All wireless systems are designed with the aim to efficiently deliver the power from the transmitter to the receiver. For a power transmission system consisting of antenna 1 and 2, the PTE is defined as the ratio of the power received by the receiving (Rx) antenna 2 to the input power of the transmitting (Tx) antenna 1 and can be expressed by [5, 14, 15]

$$\eta = \frac{\left| \int\limits_{S_1 \text{ or } S_2} (\mathbf{E}_1 \times \mathbf{H}_2 - \mathbf{E}_2 \times \mathbf{H}_1) \cdot \mathbf{u}_n dS \right|^2}{4 \operatorname{Re} \int\limits_{S_1} (\mathbf{E}_1 \times \overline{\mathbf{H}}_1) \cdot \mathbf{u}_n dS \operatorname{Re} \int\limits_{S_2} (\mathbf{E}_2 \times \overline{\mathbf{H}}_2) \cdot \mathbf{u}_n dS}, \tag{6.53}$$

where $\mathbf{E}_i$ and $\mathbf{H}_i$ stand for the fields generated by antenna i when antenna $j(j \neq i)$ is receiving, S_i $(i = 1, 2)$ denotes the closed surface that encloses antenna i only, the overbar represents the complex conjugate operation, and $\mathbf{u}_n$ is unit outward normal to the surface S_i. The PTE gets maximized if the fields satisfy the conjugate matching conditions $\mathbf{E}_1 = \overline{\mathbf{E}}_2$ and $\mathbf{H}_1 = -\overline{\mathbf{H}}_2$ on the closed surface S_1 or S_2. Determining $\mathbf{E}_i$ and $\mathbf{H}_i$ with antenna $j(j \neq i)$ in place is not easy. When the two antennas are in the Fresnel region or far-field region of each other, the problem can be simplified by neglecting the reflections between the antennas. In this case, the fields $\mathbf{E}_i$ and $\mathbf{H}_i$ can be obtained by removing the antenna $j(j \neq i)$. For a wireless power transmission (WPT) system consisting of two planar apertures, the optimization of (6.53) yields an eigenvalue equation. A number of authors have investigated the PTE between two antennas [16–22]. When the Tx antenna and Rx antenna are in the far-field region of each other, the power transmission formula reduces to the well-known far-field range equation obtained by Friis in 1946 [23]. Various applications of the theory of power transmission as well as the related techniques for building the power transmission system were summarized in [24]. It has been demonstrated that, for two planar aperture antennas located in the Fresnel region of each other, the PTE between them reaches maximum if the two planar apertures focus to each other [19–22]. The optimization process yields an eigenvalue problem of an integral operator, which is similar to that found in the study of confocal resonators [25]. The maximum eigenvalue is the maximum PTE, and the corresponding eigenvector is the synthesized optimal aperture distribution.

6.3.1 The General Power Transmission Formula

A simple proof of the power transmission formula (6.53) will be provided below. Consider a two-antenna system contained in a region V_∞ bounded by ∂V_∞. Let V_{0i} denote the source region for antenna i $(i = 1, 2)$. The source region is chosen in such a way that its boundary, denoted by ∂V_{0i}, is coincident with the metal surface of the antennas except for a portion Ω_i, where the boundary crosses the antenna reference plane. Let $V_i^{(j)}$ and $I_i^{(j)}$ $(i, j = 1, 2)$, respectively, represent the modal voltage and modal current at the reference plane of antenna i when antenna j is transmitting. One of the states of operation is illustrated in Figure 6.19a. Figure 6.19b is the corresponding equivalent network representation with

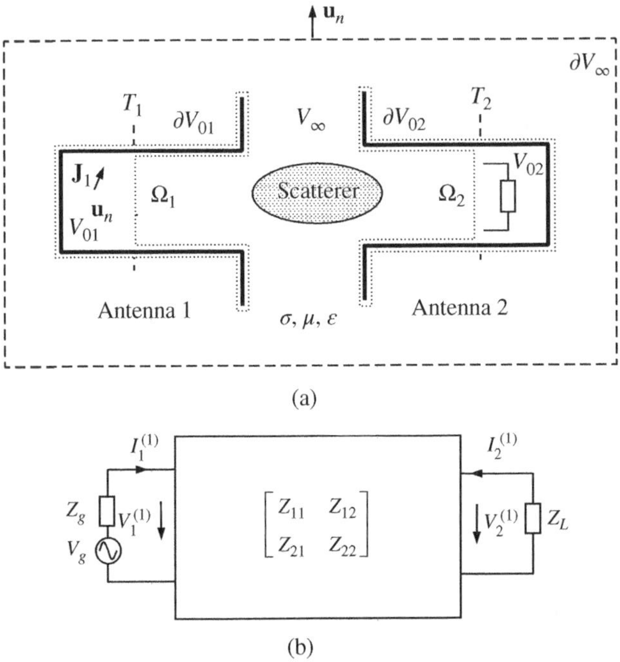

Figure 6.19 (a) A two-antenna system. (b) Equivalent network.

$$\begin{bmatrix} V_1^{(1)} \\ V_2^{(1)} \end{bmatrix} = \begin{bmatrix} Z_{11} & Z_{12} \\ Z_{21} & Z_{22} \end{bmatrix} \begin{bmatrix} I_1^{(1)} \\ I_2^{(1)} \end{bmatrix}.$$

The mutual coupling between the two antennas is characterized by Z_{ij} $(i, j = 1, 2; i \neq j)$, and may be determined by the frequency-domain reciprocity theorem

$$\int_S (\mathbf{E}_1 \times \mathbf{H}_2 - \mathbf{E}_2 \times \mathbf{H}_1) \cdot \mathbf{u}_n dS = 0, \tag{6.54}$$

where S is an arbitrary closed surface that does not contain any impressed sources and $\mathbf{u}_n$ is the outward unit normal. Choosing $S = \partial V_\infty + \partial V_{01} + \partial V_{02}$ in (6.54) yields

$$\int_{\partial V_{01}} (\mathbf{E}_1 \times \mathbf{H}_2 - \mathbf{E}_2 \times \mathbf{H}_1) \cdot \mathbf{u}_n dS + \int_{\partial V_{02}} (\mathbf{E}_1 \times \mathbf{H}_2 - \mathbf{E}_2 \times \mathbf{H}_1) \cdot \mathbf{u}_n dS$$
$$+ \int_{\partial V_\infty} (\mathbf{E}_1 \times \mathbf{H}_2 - \mathbf{E}_2 \times \mathbf{H}_1) \cdot \mathbf{u}_n dS = \sum_{l=1}^{2} \left[V_l^{(2)} I_l^{(1)} - V_l^{(1)} I_l^{(2)} \right] = 0. \tag{6.55}$$

This is the well-known reciprocity theorem in network theory. If one of the antennas is in the state of open circuit when the other is transmitting, the above equation reduces to $V_1^{(2)}I_1^{(1)} = V_2^{(1)}I_2^{(2)}$, or

$$Z_{12} = \left.\frac{V_1^{(2)}}{I_2^{(2)}}\right|_{I_1^{(2)}=0} = \left.\frac{V_2^{(1)}}{I_1^{(1)}}\right|_{I_2^{(1)}=0} = Z_{21}. \tag{6.56}$$

Therefore, the impedance matrix is symmetric. Now choose $S = S_1 + \partial V_{01}$ in (6.54), where S_1 is a closed surface containing antenna 1 only. Then,

$$\int\limits_{\partial V_{01}} (\mathbf{E}_1 \times \mathbf{H}_2 - \mathbf{E}_2 \times \mathbf{H}_1) \cdot \mathbf{u}_n dS + \int\limits_{S_1} (\mathbf{E}_1 \times \mathbf{H}_2 - \mathbf{E}_2 \times \mathbf{H}_1) \cdot \mathbf{u}_n dS = 0.$$

This implies

$$V_1^{(1)}I_1^{(2)} - V_1^{(2)}I_1^{(1)} = \int\limits_{S_1} (\mathbf{E}_1 \times \mathbf{H}_2 - \mathbf{E}_2 \times \mathbf{H}_1) \cdot \mathbf{u}_n dS. \tag{6.57}$$

Similarly,

$$V_2^{(2)}I_2^{(1)} - V_2^{(1)}I_2^{(2)} = \int\limits_{S_2} (\mathbf{E}_2 \times \mathbf{H}_1 - \mathbf{E}_1 \times \mathbf{H}_2) \cdot \mathbf{u}_n dS, \tag{6.58}$$

where S_2 is a closed surface containing antenna 2 only. The right-hand sides of (6.57) and (6.58) can be shown to be equal by choosing $S = S_1 + S_2 + \partial V_\infty$ in (6.54). When one of the antennas is transmitting with the other one being open, one may obtain

$$\begin{aligned} V_1^{(2)}I_1^{(1)} &= -\int\limits_{S_1} (\mathbf{E}_1 \times \mathbf{H}_2 - \mathbf{E}_2 \times \mathbf{H}_1) \cdot \mathbf{u}_n dS \\ &= -\int\limits_{S_2} (\mathbf{E}_2 \times \mathbf{H}_1 - \mathbf{E}_1 \times \mathbf{H}_2) \cdot \mathbf{u}_n dS = V_2^{(1)}I_2^{(2)}. \end{aligned} \tag{6.59}$$

By definition, the mutual impedance of the two-antenna system can be written as

$$Z_{12} = \left.\frac{V_1^{(2)}}{I_2^{(2)}}\right|_{I_1^{(2)}=0} = -\frac{\int\limits_{S_1} (\mathbf{E}_1 \times \mathbf{H}_2 - \mathbf{E}_2 \times \mathbf{H}_1) \cdot \mathbf{u}_n dS}{I_1^{(1)}I_2^{(2)}} = -\frac{\int\limits_{V_{01}} \mathbf{J}_1 \cdot \mathbf{E}_2 dV}{I_1^{(1)}I_2^{(2)}}, \tag{6.60}$$

where use is made of the following reciprocity theorem:

$$\begin{aligned}\int_{V_{02}} \mathbf{J}_2 \cdot \mathbf{E}_1 dV &= \int_{S_2} (\mathbf{E}_2 \times \mathbf{H}_1 - \mathbf{E}_1 \times \mathbf{H}_2) \cdot \mathbf{u}_n dS \\ &= \int_{S_1} (\mathbf{E}_1 \times \mathbf{H}_2 - \mathbf{E}_2 \times \mathbf{H}_1) \cdot \mathbf{u}_n dS = \int_{V_{01}} \mathbf{J}_1 \cdot \mathbf{E}_2 dV,\end{aligned} \tag{6.61}$$

where $\mathbf{J}_j (j = 1, 2)$ is the current distribution of antenna j. Equation (6.60) may be regarded as an exact expression of Huygens' principle in a symmetrical form, and it is generally applicable to an inhomogeneous medium.

Let Z_{si} be the reference impedance for the input terminal of antenna i ($i = 1, 2$). By definition, the **normalized incident wave** $a_j^{(i)}$ and the **normalized reflected wave** $b_j^{(i)}$ are related to the terminal voltage and current by

$$V_j^{(i)} = \frac{\overline{Z}_{sj} a_j^{(i)}}{\sqrt{\operatorname{Re} Z_{sj}}} + \frac{Z_{sj} b_j^{(i)}}{\sqrt{\operatorname{Re} Z_{sj}}}, I_j^{(i)} = \frac{a_j^{(i)}}{\sqrt{\operatorname{Re} Z_{sj}}} - \frac{b_j^{(i)}}{\sqrt{\operatorname{Re} Z_{sj}}}, (i, j = 1, 2). \tag{6.62}$$

Insertion of (6.62) into (6.55) yields

$$\sum_{l=1}^{2} \left[a_l^{(1)} b_l^{(2)} - a_l^{(2)} b_l^{(1)} \right] = 0. \tag{6.63}$$

If one of the antennas is matched when the other one is transmitting, (6.63) reduces to $a_1^{(1)} b_1^{(2)} = a_2^{(2)} b_2^{(1)}$, which gives the symmetric property of scattering matrix

$$S_{12} = \left. \frac{b_1^{(2)}}{a_2^{(2)}} \right|_{a_1^{(2)} = 0,} = \left. \frac{b_2^{(1)}}{a_1^{(1)}} \right|_{a_2^{(1)} = 0} = S_{21}.$$

In terms of the normalized incident and reflected waves, (6.57) and (6.58) can be written as

$$b_1^{(1)} a_1^{(2)} - b_1^{(2)} a_1^{(1)} = \frac{1}{2} \int_{S_1} (\mathbf{E}_1 \times \mathbf{H}_2 - \mathbf{E}_2 \times \mathbf{H}_1) \cdot \mathbf{u}_n dS, \tag{6.64}$$

$$b_2^{(2)} a_2^{(1)} - b_2^{(1)} a_2^{(2)} = \frac{1}{2} \int_{S_2} (\mathbf{E}_2 \times \mathbf{H}_1 - \mathbf{E}_1 \times \mathbf{H}_2) \cdot \mathbf{u}_n dS. \tag{6.65}$$

If one of the antennas is matched when the other one is transmitting, (6.64) and (6.65) imply

$$S_{12} = \left.\frac{b_1^{(2)}}{a_2^{(2)}}\right|_{a_1^{(2)}=0} = -\frac{1}{2a_1^{(1)}a_2^{(2)}}\int_{S_1}(\mathbf{E}_1 \times \mathbf{H}_2 - \mathbf{E}_2 \times \mathbf{H}_1)\cdot\mathbf{u}_n dS, \tag{6.66}$$

$$S_{21} = \left.\frac{b_2^{(1)}}{a_1^{(1)}}\right|_{a_2^{(1)}=0} = -\frac{1}{2a_1^{(1)}a_2^{(2)}}\int_{S_2}(\mathbf{E}_2 \times \mathbf{H}_1 - \mathbf{E}_1 \times \mathbf{H}_2)\cdot\mathbf{u}_n dS. \tag{6.67}$$

The power transmission formula (6.53) is obtained by taking the squared modulus of S_{12} or S_{21}.

6.3.2 Power Transmission Between Two Planar Apertures

As an application of the power transmission formula (6.53), let us consider the power transfer between two planar apertures. The configuration of a two-planar aperture system in free space is shown in Figure 6.20, where both apertures are assumed to be in an infinite conducting screen so that the tangential electric field outside the aperture is zero. When the aperture i ($i = 1, 2$) is used as a transmitting antenna, the aperture field is assumed to be

$$\mathbf{E}_i = \mathbf{u}_x E_i, \mathbf{H}_i = \mathbf{u}_y \frac{1}{\eta} E_i,$$

where $\eta = \sqrt{\mu/\varepsilon}$ is the wave impedance in free space. The same notations will be used for the aperture field distribution and the field produced by the aperture, and this will not cause any confusion. By means of equivalence theorem and image principle, the electric field produced by aperture 1 may be represented by

$$\mathbf{E}_1(\mathbf{r}) = \frac{1}{2\pi}\int_{T_1}\mathbf{u}_y \times \mathbf{u}_R\left(jk + \frac{1}{|\mathbf{r}-\mathbf{r}'|}\right)e^{-jk|\mathbf{r}-\mathbf{r}'|}E_1(\mathbf{r}')dx'dy', \tag{6.68}$$

where $\mathbf{u}_R = (\mathbf{r}-\mathbf{r}')/|\mathbf{r}-\mathbf{r}'|$, $k = \omega\sqrt{\mu\varepsilon}$. In deriving the above expression, the multiple scattering between the apertures has been neglected. If the apertures are

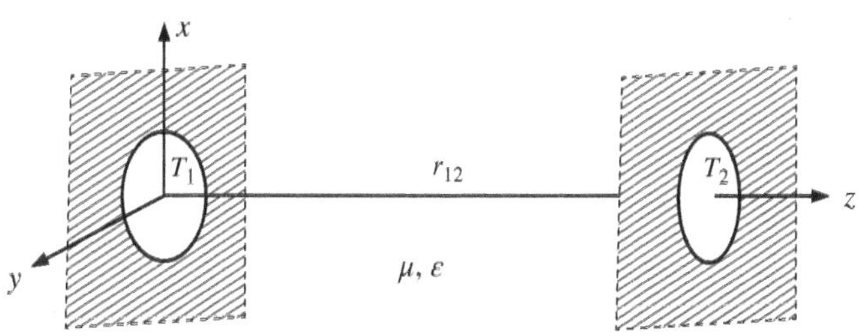

Figure 6.20 Two-planar aperture system.

located in the Fresnel region of each other and the observation point **r** is on the aperture 2, one may make the small angle approximation $\mathbf{u}_y \times \mathbf{u}_R \approx \mathbf{u}_x$ and the following approximations

$$e^{-jk|\mathbf{r}-\mathbf{r}'|} \approx e^{-jk\left\{r_{12} + \frac{1}{2r_{12}}\left[(x-x')^2 + (y-y')^2\right]\right\}}, \quad \frac{1}{|\mathbf{r}-\mathbf{r}'|} \approx \frac{1}{r_{12}}$$

in the integrand of (6.68). Then, (6.68) can be written as

$$\mathbf{E}_1(\mathbf{r}) = \mathbf{u}_x E_1(\mathbf{r}) \approx \mathbf{u}_x \frac{je^{-jkr_{12}}}{\lambda r_{12}} \int_{T_1} E_1 e^{-jk\left[(x-x')^2 + (y-y')^2\right]/2r_{12}} dx' dy'. \tag{6.69}$$

The magnetic field is given by

$$\mathbf{H}_1(\mathbf{r}) = \mathbf{u}_y \frac{1}{\eta} E_1(\mathbf{r}). \tag{6.70}$$

By use of (6.69) and (6.70), the power transmission formula (6.53) can be expressed as

$$T_{12} = \frac{\left|\int_{T_2} E_1 E_2 dxdy\right|^2}{\int_{T_1} |E_1|^2 dxdy \int_{T_2} |E_2|^2 dxdy}. \tag{6.71}$$

Note that the field E_1 in the numerator represents the field at aperture 2, which is produced by the aperture 1. Substituting (6.69) into (6.71) and simplifying, one may find

$$T_{12} = \left(\frac{1}{\lambda r_{12}}\right)^2 \frac{\left|\int_{T_2} \tilde{m}_1 m_2 dxdy\right|^2}{\int_{T_1} |m_1|^2 dxdy \int_{T_2} |m_2|^2 dxdy}, \tag{6.72}$$

where

$$\begin{aligned}
m_1(x,y) &= E_1 e^{-jk(x^2+y^2)/2r_{12}}, \\
m_2(x,y) &= E_2 e^{-jk(x^2+y^2)/2r_{12}}, \\
\tilde{m}_1(x,y) &= \int_{T_1} m_1(x',y') e^{jk(xx'+yy')/r_{12}} dx' dy', \\
\tilde{m}_2(x,y) &= \int_{T_2} m_2(x',y') e^{jk(xx'+yy')/r_{12}} dx' dy'.
\end{aligned}$$

Note that

$$\int_{T_1} m_1 \tilde{m}_2 dxdy = \int_{T_2} \tilde{m}_1 m_2 dxdy.$$

This is equivalent to the reciprocity relation $T_{12} = T_{21}$. Equation (6.72) may be rewritten as

$$T_{12} = T_{12}^{ideal} \cdot U,$$

where

$$U = \frac{\left| \int_{T_2} \tilde{m}_1 m_2 dxdy \right|^2}{\int_{T_2} |\tilde{m}_1|^2 dxdy \int_{T_2} |m_2|^2 dxdy}, \tag{6.73}$$

$$T_{12}^{ideal} = \frac{\text{Re} \int_{T_2} \left(\mathbf{E}_1 \times \overline{\mathbf{H}}_1\right) \cdot \mathbf{u}_z dxdy}{\text{Re} \int_{T_1} \left(\mathbf{E}_1 \times \overline{\mathbf{H}}_1\right) \cdot \mathbf{u}_z dxdy} = \left(\frac{1}{\lambda r_{12}}\right)^2 \frac{\int_{T_2} |\tilde{m}_1|^2 dxdy}{\int_{T_1} |m_1|^2 dxdy}. \tag{6.74}$$

Expression (6.74) is the PTE between two ideal apertures. The PTE T_{12} reaches maximum if both T_{12}^{ideal} and U are maximized. From Schwartz inequality, one may find $\max U = 1$, which can be reached by letting $m_2(x,y) = c_1 \overline{\tilde{m}}_1(x,y)$, $(x, y) \in T_2$, i.e.

$$E_2(x,y) = c_2 \overline{E}_1(x,y), (x,y) \in T_2. \tag{6.75}$$

Both c_1 and c_2 in the above are arbitrary complex numbers. Equation (6.75) implies that the aperture distribution of antenna 2 is equal to the complex conjugate of the field produced by antenna 1 at antenna 2. To find the condition for maximizing T_{12}^{ideal}, one may rewrite (6.74) into the form of the Rayleigh quotient

$$T_{12}^{ideal} = \frac{(\hat{T} m_1, m_1)}{(m_1, m_1)},$$

where $(\cdot, \cdot)$ denotes the inner product defined by $(u, v) = \int_{T_1} u\bar{v} dxdy$ for two arbitrary functions u and v, and $\hat{T}$ is a self-adjoint operator defined by

$$\hat{T} m_1(\xi', \varsigma') = \int_{T_1} K_2(\xi, \varsigma; \xi', \varsigma') m_1(\xi, \varsigma) d\xi d\varsigma$$

with the integral kernel function

$$K_2(\xi, \varsigma; \xi', \varsigma') = \left(\frac{1}{\lambda r_{12}}\right)^2 \int_{T_2} e^{jk[(\xi - \xi')x + (\varsigma - \varsigma')y]/r_{12}} dxdy. \tag{6.76}$$

If the condition (6.75) is met, one may write

$$T_{12} = T_{12}^{ideal} = \frac{(\hat{T}m_1, m_1)}{(m_1, m_1)}. \tag{6.77}$$

The above Rayleigh quotient attains an extremum when m_1 satisfies

$$\hat{T}m_1 = T_{12}m_1. \tag{6.78}$$

Therefore, the power transmission between planar apertures is maximized if the aperture field distributions satisfy (6.75) and (6.78) simultaneously. Equation (6.78) is an eigenvalue problem and its largest eigenvalue is the maximum possible value for the PTE. It may be used first to determine the aperture distribution of antenna 1, and the aperture distribution of antenna 2 can then be determined from (6.75).

Example 6.9 Consider the power transmission between two rectangular apertures $T_1 = [-a_1, a_1] \times [-a_2, a_2]$ and $T_2 = [-b_1, b_1] \times [-b_2, b_2]$. The kernel function (6.76) becomes

$$\begin{aligned} K_2(\xi, \varsigma; \xi', \varsigma') &= \left(\frac{1}{\lambda r_{12}}\right)^2 \int_{-b_1}^{b_1} \int_{-b_2}^{b_2} e^{jk[(\xi - \xi')x + (\varsigma - \varsigma')y]/r_{12}} dxdy \\ &= \left(\frac{k}{\pi r_{12}}\right)^2 \frac{\sin\dfrac{k(\xi - \xi')b_1}{r_{12}} \sin\dfrac{k(\varsigma - \varsigma')b_2}{r_{12}}}{\dfrac{k(\xi - \xi')}{r_{12}} \dfrac{k(\varsigma - \varsigma')}{r_{12}}}. \end{aligned} \tag{6.79}$$

Thus, the eigenvalue equation (6.78) turns out to be

$$\left(\frac{k}{\pi r_{12}}\right)^2 \int_{-a_1}^{a_1} \int_{-a_2}^{a_2} \frac{\sin\dfrac{k(\xi - x)b_1}{r_{12}} \sin\dfrac{k(\varsigma - y)b_2}{r_{12}}}{\dfrac{k(\xi - x)}{r_{12}} \dfrac{k(\varsigma - y)}{r_{12}}} m_1(\xi, \varsigma) d\xi d\varsigma = T_{12}m_1(x, y). \tag{6.80}$$

If $m_1(x, y)$ is assumed to be a separable function of x and y,

$$m_1(x, y) = m_{1x}(x)m_{1y}(y),$$

the eigenvalue equation (6.80) can be divided into two separate eigenvalue equations

$$\begin{aligned}&\int_{-a_1}^{a_1}\frac{\sin\dfrac{kb_1}{r_{12}}(\xi-x)}{\pi(\xi-x)}m_{1x}(\xi)d\xi = T_{12}^x m_{1x}(x),\\&\int_{-a_2}^{a_2}\frac{\sin\dfrac{kb_2}{r_{12}}(\varsigma-y)}{\pi(\varsigma-y)}m_{1y}(\varsigma)d\varsigma = T_{12}^y m_{1y}(y),\end{aligned}\tag{6.81}$$

with $T_{12} = T_{12}^x T_{12}^y$. By use of the coordinate transformations

$$\xi' = \frac{\xi}{a_1}, x' = \frac{x}{a_1}, \varsigma' = \frac{\varsigma}{a_2}, y' = \frac{y}{a_2},$$

(6.81) can be rewritten as

$$\begin{aligned}&\int_{-1}^{1}\frac{\sin c_1(\xi'-x')}{\pi(\xi'-x')}m_{1x}(\xi')d\xi' = T_{12}^x m_{1x}(x'),\\&\int_{-1}^{1}\frac{\sin c_2(\varsigma'-y')}{\pi(\varsigma'-y')}m_{1y}(\varsigma')d\varsigma' = T_{12}^y m_{1y}(y'),\end{aligned}\tag{6.82}$$

where

$$c_1 = \frac{ka_1b_1}{r_{12}},\ c_2 = \frac{ka_2b_2}{r_{12}}.\tag{6.83}$$

The eigenvalue problems in (6.82) often appear in signal theory and have been solved by Slepian and Pollak [26]. The largest eigenvalues are

$$T_{12}^x = \frac{2c_1}{\pi}\left[R_{00}^{(1)}(c_1,1)\right]^2,\ T_{12}^y = \frac{2c_2}{\pi}\left[R_{00}^{(1)}(c_2,1)\right]^2,\tag{6.84}$$

where $R_{00}^{(1)}$ is the **radial prolate spheroidal function**. The eigenfunctions corresponding to (6.84) are, respectively, given by the **angular prolate spheroidal wave functions** $S_{00}(c_1, x/a_1)$ and $S_{00}(c_2, y/a_2)$. Some values of T_{12}^x are listed in Table 6.4. Note that the PTE of 100% can be achieved by increasing the parameter c_1. The maximum PTE and the optimal distribution for aperture T_1 are, respectively, given by

$$\begin{aligned}&T_{12} = \frac{2c_1}{\pi}\left[R_{00}^{(1)}(c_1,1)\right]^2\frac{2c_2}{\pi}\left[R_{00}^{(1)}(c_2,1)\right]^2,\\&E_1(x,y) = S_{00}\left(c_1,\frac{x}{a_1}\right)S_{00}\left(c_2,\frac{y}{a_2}\right)e^{\frac{jk(x^2+y^2)}{2r_{1,2}}}.\end{aligned}\tag{6.85}$$

Table 6.4 Largest eigenvalue.

c_1	T_{12}^{x}
0.5	0.30969
1	0.57258
2	0.88056
4	0.99589
8	1.00000

The optimal distribution for aperture T_2 is determined by (6.75).

Equation (6.85) indicates that the transmitting aperture must be focused to the receiving aperture in order to achieve a maximum power transfer between the two apertures, and the optimization process yields a focused transmitting aperture. In addition, the transverse amplitude distribution has no sidelobes. Some important properties of the focused antenna aperture are illustrated in Figure 6.21. The range between the axial −3 dB points about the maximum intensity point (called principal focal point) is called the focused region and its extension is defined as the **depth of focus.**

In practice, it is difficult to realize the continuous distribution (6.85) with a single aperture. For this reason, an antenna array has to be adopted. □

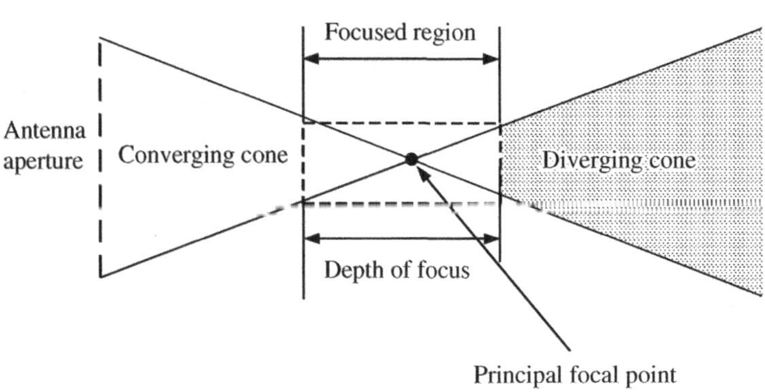

Figure 6.21 Properties of focused antenna aperture.

6.3.3 Power Transmission Between Two Antennas with Large Separation

If the antennas are located in the far-field region of each other, the power transmission formula (6.53) can be simplified. Especially, the calculation of fields $\mathbf{E}_i$ and $\mathbf{H}_i$ can be carried out with the antennas j $(j \neq i)$ removed. Physically, this implies that one can neglect the reflections between the two antennas. Two different coordinate systems $\mathbf{r}_1 = (x_1, y_1, z_1)$ and $\mathbf{r}_2 = (x_2, y_2, z_2)$ for antenna 1 and antenna 2 will be used. The origins of the coordinate systems are chosen to be the geometrical center of the current distributions and the separation between antenna 1 and antenna 2 satisfies $kr_2 \gg 1$, $r_2 \gg d_2$, $r_2 \gg d_1$, where $r_2 = |\mathbf{r}_2|$ is the distance between antenna 2 and an arbitrary point of the circumscribing sphere of antenna 1 (denoted by S_1), as shown in Figure 6.22. Let $\mathbf{r}_1'$ be a point on the circumscribing sphere of antenna 1, and $\mathbf{r}_{12} = r_{12}\mathbf{u}_{r_{12}}$, where r_{12} is the distance between the two origins and $\mathbf{u}_{r_{12}}$ is a unit vector directed from antenna 1 to antenna 2. The far-zone field of antenna 2 at antenna 1 can be expressed as

$$\mathbf{E}_2(\mathbf{r}_2) \approx -\frac{jk\eta I_2^{(2)} e^{-jkr_2}}{4\pi r_2} \mathbf{L}_2(\mathbf{u}_{r_2}), \mathbf{H}_2(\mathbf{r}_2) \approx \frac{1}{\eta} \mathbf{u}_{r_2} \times \mathbf{E}_2(\mathbf{r}_2), \tag{6.86}$$

where $\mathbf{r}_2 = \mathbf{r}_1' - \mathbf{r}_{12}$ is assumed to be a point on the sphere S_1 and

$$\mathbf{L}_2(\mathbf{u}_{r_2}) = \frac{1}{I_2^{(2)}} \int_{V_{02}} [\mathbf{J}_2 - (\mathbf{J}_2 \cdot \mathbf{u}_{r_2})\mathbf{u}_{r_2}] e^{jk\mathbf{r}_2' \cdot \mathbf{u}_{r_2}} dV(\mathbf{r}_2')$$

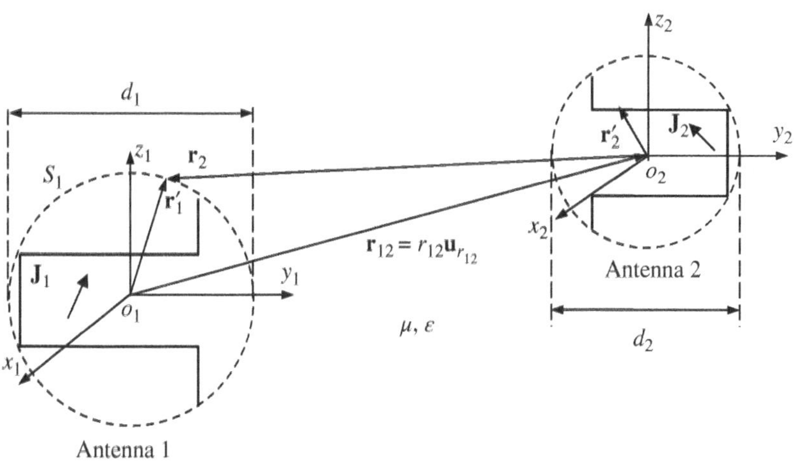

Figure 6.22 Two distant antennas.

is the antenna effective vector length. Since $\mathbf{r}'_1$ is very small compared to r_{12} in magnitude, one can make the approximation $r_2 = |\mathbf{r}'_1 - \mathbf{r}_{12}| \approx r_{12} - \mathbf{u}_{r_{12}} \cdot \mathbf{r}'_1$. The field $\mathbf{E}_2$ in the coordinate system (x_1, y_1, z_1) can then be represented by

$$\begin{aligned} \mathbf{E}_2(\mathbf{r}_2) &\approx -\frac{jk\eta I_2^{(2)} e^{-jkr_{12}} e^{jk\mathbf{u}_{r_{12}} \cdot \mathbf{r}'_1}}{4\pi r_{12}} \mathbf{L}_2(-\mathbf{u}_{r_{12}}), \\ \mathbf{H}_2(\mathbf{r}_2) &\approx -\frac{1}{\eta} \mathbf{u}_{r_{12}} \times \mathbf{E}_2(\mathbf{r}_2). \end{aligned} \tag{6.87}$$

Then,

$$\begin{aligned} \int_{S_1} (\mathbf{E}_1 \times \mathbf{H}_2 - \mathbf{E}_2 \times \mathbf{H}_1) \cdot \mathbf{u}_n dS &= \int_{S_1} \left[-\eta^{-1} \mathbf{E}_1 \times (\mathbf{u}_{r_{12}} \times \mathbf{E}_2) - \mathbf{E}_2 \times \mathbf{H}_1 \right] \cdot \mathbf{u}_n dS \\ &= \int_{S_1} \mathbf{E}_2 \cdot \left[-\eta^{-1} \mathbf{u}_{r_{12}} \times (\mathbf{E}_1 \times \mathbf{u}_n) - \mathbf{H}_1 \times \mathbf{u}_n \right] dS \\ &= \int_{S_1} \mathbf{E}_2 \cdot (\mathbf{J}_{1s} - \eta^{-1} \mathbf{u}_{r_{12}} \times \mathbf{J}_{1ms}) dS, \end{aligned} \tag{6.88}$$

where $\mathbf{J}_{1s} = \mathbf{u}_n \times \mathbf{H}_1$ and $\mathbf{J}_{1ms} = -\mathbf{u}_n \times \mathbf{E}_1$ are the equivalent electric current and magnetic current on the surface S_1, respectively. Substituting (6.87) into (6.88), one may obtain

$$\begin{aligned} \int_{S_1} (\mathbf{E}_1 \times \mathbf{H}_2 - \mathbf{E}_2 \times \mathbf{H}_1) \cdot \mathbf{u}_n dS &\approx -\frac{4\pi r_{12} e^{jkr_{12}}}{jk\eta} \mathbf{E}_1(\mathbf{u}_{r_{12}}) \cdot \mathbf{E}_2(-\mathbf{u}_{r_{12}}) \\ &= \frac{-jk\eta I_1^{(1)} I_2^{(2)} e^{-jkr_{12}}}{4\pi r_{12}} \mathbf{L}_1(\mathbf{u}_{r_{12}}) \cdot \mathbf{L}_2(-\mathbf{u}_{r_{12}}). \end{aligned} \tag{6.89}$$

Here, the following far-field expression of antenna 1 at antenna 2 has been used

$$\begin{aligned} \mathbf{E}_1(\mathbf{r}_{12}) &= \frac{-jk\eta e^{-jkr_{12}}}{4\pi r_{12}} \int_{S'_1} e^{jk\mathbf{u}_{r_{12}} \cdot \mathbf{r}_1} \left[\mathbf{J}_{1s}(\mathbf{r}'_1) - \mathbf{u}_d \times \eta^{-1} \mathbf{J}_{1ms}(\mathbf{r}'_1) \right] dS \\ &= \frac{-jk\eta I_1^{(1)} e^{-jkr_{12}}}{4\pi r_{12}} \mathbf{L}_1(\mathbf{u}_{r_{12}}). \end{aligned} \tag{6.90}$$

It follows from (6.89) that the mutual impedance Z_{12} is given by

$$Z_{12} = \left. \frac{V_1^{(2)}}{I_2^{(2)}} \right|_{I_1^{(2)} = 0} = \frac{jk\eta e^{-jkr_{12}}}{4\pi r_{12}} \mathbf{L}_1(\mathbf{u}_{r_{12}}) \cdot \mathbf{L}_2(-\mathbf{u}_{r_{12}}). \tag{6.91}$$

Note that mutual impedance Z_{12} vanishes if the two effective vector lengths $\mathbf{L}_1$ and $\mathbf{L}_2$ are orthogonal. Furthermore, from (6.89), one may find

$$\left|\int_{S_1}(\mathbf{E}_1\times\mathbf{H}_2-\mathbf{E}_2\times\mathbf{H}_1)\cdot\mathbf{u}_n dS\right|^2 \approx \left(\frac{4\pi r_{12}}{k\eta}\right)^2\left|\mathbf{E}_1(\mathbf{u}_{r_{12}})\right|^2\cdot|\mathbf{E}_2(-\mathbf{u}_{r_{12}})|^2\cos^2\theta_{12}$$
$$=\left(\frac{4\lambda}{r_{12}}\right)^2 U_1(\mathbf{u}_{r_{12}})\cdot U_2(-\mathbf{u}_{r_{12}})\cos^2\theta_{12}, \tag{6.92}$$

where U_1 and U_2 are the radiation intensity of antenna 1 and 2, respectively, and θ_{12} is the angle between $\mathbf{E}_1(\mathbf{r}_{12})$ and $\mathbf{E}_2(-\mathbf{r}_{12})$. From (6.53) and (6.92), one may obtain the well-known **Friis transmission formula**

$$\frac{P_2^{(1)}}{P_1^{(1)}}=\left(\frac{\lambda}{4\pi r_{12}}\right)^2\frac{4\pi U_1(\mathbf{u}_{r_{12}})4\pi U_2(-\mathbf{u}_{r_{12}})\cos^2\theta_{12}}{\frac{1}{2}\operatorname{Re}\int_{S_1'}(\mathbf{E}_1\times\overline{\mathbf{H}}_1)\cdot\mathbf{u}_n dS\frac{1}{2}\operatorname{Re}\int_{S_2'}(\mathbf{E}_2\times\overline{\mathbf{H}}_2)\cdot\mathbf{u}_n dS}$$
$$=\left(\frac{\lambda}{4\pi r_{12}}\right)^2 G_1(\mathbf{u}_{r_{12}})G_2(-\mathbf{u}_{r_{12}})\cos^2\theta_{12}, \tag{6.93}$$

where G_1 and G_2 are the gains of the antenna 1 and antenna 2, respectively. Equation (6.93) may be rewritten as

$$P_2^{(1)}=\frac{\text{EIRP}}{L_s}G_2(-\mathbf{u}_{r_{12}})\cos^2\theta_{12}, \tag{6.94}$$

where $L_s=(4\pi r_{12}/\lambda)^2$ is known as **free-space path loss**, and EIRP stands for the **effective isotropic radiated power** defined by $\text{EIRP}=P_1^{(1)}G_1(\mathbf{u}_{r_{12}})$. The **received isotropic power** is defined as EIRP$/L_s$, which is the power received by an isotropic antenna ($G_2=1$).

6.4 Synthesis of Arrays with MMPTE

Realizing the optimized aperture distribution by a single aperture is impossible in most cases. In addition, the optimization of PTE based on (6.53) gets very complicated when two antennas are in close proximity. To overcome these difficulties, one can instead consider the optimization of the power transmission between two antenna arrays from the circuit point of view. The optimization technique is called the MMPTE and can be applied to the design of both antenna arrays and WPT systems [6].

6.4.1 Power Transmission Between Two Antenna Arrays

Consider a generic WPT system located in an arbitrary scattering environment, as illustrated by Figure 6.23. The system consists of m Tx antennas and n Rx antennas, and each Rx antenna is terminated by a load of reflection coefficient $\Gamma_i(i = m + 1, m + 2, ..., m + n)$. The EM waves emanated from the Tx antenna array may bump into a cluster of obstacles and scatterers in the propagation medium and get reflected, refracted, and diffracted before they reach the Rx antenna array. The separation between the Tx and Rx antenna arrays is assumed to be arbitrary. The system forms an $(m + n)$-port network and may be characterized by scattering parameters. The normalized reflective waves and incident waves at the ports are related by

$$\begin{bmatrix} [b_t] \\ [b_r] \end{bmatrix} = \begin{bmatrix} [S_{tt}] & [S_{tr}] \\ [S_{rt}] & [S_{rr}] \end{bmatrix} \begin{bmatrix} [a_t] \\ [a_r] \end{bmatrix}, \tag{6.95}$$

where

$$[a_t] = [a_1, a_2, ..., a_m]^T, [b_t] = [b_1, b_2, ..., b_m]^T$$

are, respectively, the normalized incident and reflected waves for the Tx array and

$$[a_r] = [a_{m+1}, a_{m+2}, ..., a_{m+n}]^T, [b_r] = [b_{m+1}, b_{m+2}, ..., b_{m+n}]^T$$

are, respectively, the normalized incident and reflected waves for the Rx array. The superscript T denotes the transpose operation. For the Rx array, the normalized incident and reflective waves are related by the condition imposed by the loads

$$[a_r] = [\Gamma_L][b_r], \tag{6.96}$$

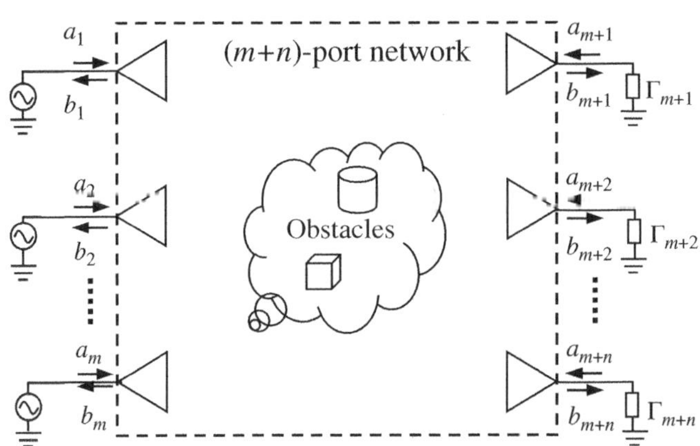

Figure 6.23 A generic WPT system.

where $[\Gamma_L] = \text{diag}[\Gamma_{m+1}, \Gamma_{m+2}, \ldots, \Gamma_{m+n}]$ is the **load reflection coefficient matrix**. Combining (6.95) and (6.96) yields

$$[b_t] = [S_{tt}][a_t] + [S_{tr}][\Gamma_L][b_r],$$
$$[b_r] = [S_{rt}][a_t] + [S_{rr}][\Gamma_L][b_r].$$

By eliminating $[b_r]$ from the first equation, the normalized reflected waves can be expressed in terms of the incident waves $[a_t]$ as follows:

$$[b_t] = [\Gamma_{in}][a_t], \quad [b_r] = [T][a_t], \tag{6.97}$$

where $[\Gamma_{in}]$ and $[T]$ are the input reflection coefficient matrix and the transmission coefficient of the system, respectively, given by

$$\begin{aligned} &[\Gamma_{in}] = [S_{tt}] + [S_{tr}][\Gamma_L][T], \\ &[T] = ([1] - [S_{rr}][\Gamma_L])^{-1}[S_{rt}], \end{aligned} \tag{6.98}$$

and [1] denotes the identity matrix of dimension n. The input power of the WPT system is given by

$$P_{in} = \frac{1}{2}\left(\|[a_t]\|^2 - \|[b_t]\|^2\right) = \frac{1}{2}[a_t]^H[B][a_t],$$

where the superscript H denotes the conjugate transpose of matrix, and

$$[B] = [1] - [\Gamma_{in}]^H[\Gamma_{in}]. \tag{6.99}$$

Here, [1] denotes the identity matrix of dimension m. The received power of the WPT system can be obtained from (6.96) and the second equation of (6.97)

$$\begin{aligned} P_{rec} &= \frac{1}{2}\left(\|[b_r]\|^2 - \|[a_r]\|^2\right) = \frac{1}{2}[b_r]^H\left([1] - [\Gamma_L]^H[\Gamma_L]\right)[b_r] \\ &= \frac{1}{2}[a_t]^H[T]^H\left([1] - [\Gamma_L]^H[\Gamma_L]\right)[T][a_t] = \frac{1}{2}[a_t]^H[A][a_t], \end{aligned}$$

where

$$[A] = [T]^H\left([1] - [\Gamma_L]^H[\Gamma_L]\right)[T].$$

The PTE for the WPT system can then expressed by

$$\eta = \frac{P_{rec}}{P_{in}} = \frac{([A][a_t], [a_t])}{([B][a_t], [a_t])}, \tag{6.100}$$

where $(\cdot, \cdot)$ denotes the usual inner product between two complex column vectors. Equation (6.100) is a generalized Rayleigh quotient. The PTE (6.100) can be optimized subject to constraints. Also note that the optimization of (6.100) is equivalent to optimizing the PTE subject to the constraint that the input power of the

WPT system is kept at a constant. The matrices $[A]$ and $[B]$ are solely determined by the scattering parameters and the load reflection coefficients of the WPT system, and the latter can be obtained either by simulation or by measurement. This flexibility offers great convenience when the system gets too complicated to be modeled by a computer. It is noted that the above derivation is applicable to any WPT system. In other words, the surrounding environment of the system, the separation between the Tx and Rx arrays, the array elements, and the array configuration are all assumed to be arbitrary.

6.4.1.1 Unconstrained Optimization

To optimize the PTE expressed by the Rayleigh quotient (6.100), one can use the variational method. If the quotient (6.100) is required to be stationary at $[a_t]$, one may obtain the following generalized algebraic eigenvalue equation:

$$[A][a_t] = \eta[B][a_t]. \tag{6.101}$$

Since both the matrix $[A]$ and $[B]$ are positive, the eigenvalues of (6.101) are all non-negative. In view of the fact that the number of positive eigenvalues of (6.101) is equal to the rank of the matrix $[A]$ as well as the following inequality [27]

$$\text{rank}[A] \le \text{rank}[S_{rt}] \le \min\{m, n\},$$

the number of positive eigenvalues of (6.101) is less than or equal to $\min\{m, n\}$. The maximum eigenvalue gives the maximum PTE and the corresponding eigenvector gives the ODE for the Tx array. The eigenvector corresponding to the zero eigenvalue is also useful and it represents the ODE for the Tx array, which generates a null in the direction of the Rx array. Based on the ODE, a feeding network for the Tx array can be designed by the theory of transmission line.

Once the ODE for the Tx array is known, the feeding network for the Rx array may be built from the relationship

$$[b_r] = [T][a_t]. \tag{6.102}$$

The above optimization procedure for the WPT system is called **unconstrained MMPTE** (UMMPTE) since no essential constraint is involved.

6.4.1.2 Weighted Optimization

Sometimes constraints have to be imposed to the design of WPT system to realize a specific distribution of the received power among the Rx array elements. To this end, the UMMPTE may be modified by introducing a weighting matrix

$$[W] = \text{diag}[w_{m+1}, w_{m+2}, \ldots, w_{m+n}] \tag{6.103}$$

for the received power $[b_r]$:

$$[b'_r] = [W][b_r] = [w_{m+1}b_{m+1}, w_{m+2}b_{m+2}, \ldots, w_{m+n}b_{m+n}]^T.$$

Accordingly, (6.100) becomes

$$\eta = \frac{\left(\|[b'_r]\|^2 - \|[a_r]\|^2\right)/2}{\left(\|[a_t]\|^2 - \|[b_t]\|^2\right)/2} = \frac{([A'][a_t], [a_t])}{([B][a_t], [a_t])}, \tag{6.104}$$

where $[A'] = [T']^H([1] - [\Gamma_L]^H[\Gamma_L])[T']$ with $[T'] = [W][T]$. Equation (6.101) is then modified to

$$[A'][a_t] = \eta[B][a_t]. \tag{6.105}$$

The MMPTE imposed with the weighting matrix is called **weighted MMPTE** (WMMPTE). The above procedure not only achieves the specific distribution of received power but also guarantees that the received power of the Rx array is maximized for fixed input power.

6.4.1.3 Constrained Optimization

Different constraints may be enforced on the design of WPT systems. For example, one may require that the received power of the Rx array is equally distributed among its elements

$$|b_{m+1}|^2 = |b_{m+2}|^2 = \cdots = |b_{m+n}|^2.$$

In addition to the above constraints, the received power is further required to be maximized. Mathematically one needs to solve a quadratically constrained quadratic programing (QCQP) problem with equality constraints. The solution of this QCQP is not easy [28, 29]. If the Rx array is in the far-field region and positioned on a sphere centered at the Tx array, the Tx array can then be regarded as a point source. In this case, the phase difference between any two Rx array elements is negligible. Since only the phase differences among the antenna elements are relevant, one may set

$$[S_{rt}][a_t] = [b_r] = [c]e^{j\varphi}, \tag{6.106}$$

where $[c]$ is an n-dimensional real constant vector, and φ is also a constant phase. Note that the constraint (6.106) is valid for any distribution of excitations $[a_t]$. The denominator of (6.100) can be normalized by assuming that $[B] = [1]$ and the QCQP problem may be simplified to a linearly constrained quadratic programing (LCQP) problem as follows:

$$\begin{aligned} \max \quad & ([A][a_t], [a_t]) = [a_t]^H[A][a_t] \\ \text{s.t.} \quad & [S_{rt}][a_t] = [c], \end{aligned} \tag{6.107}$$

where the exponential term $e^{j\varphi}$ in (6.106) has been factored in the excitation vector $[a_t]$. The problem (6.107) may be solved by the method of Lagrangian multipliers with a Lagrangian function defined by

$$L([a_t],[\lambda]) = [a_t]^H[A][a_t] - [\lambda]^T([S_{rt}][a_t] - [c]).$$

The extremum point ($[a_t]$, $[\lambda]$) of (6.107) satisfies the following Lagrangian equations:

$$\begin{aligned} \frac{\delta L([a_t],[\lambda])}{\delta[a_t]} &= 2[A][a_t] - [S_{rt}]^H[\lambda] = 0, \\ \frac{\delta L([a_t],[\lambda])}{\delta[\lambda]} &= [S_{rt}][a_t] - [c] = 0, \end{aligned} \tag{6.108}$$

where $\delta/\delta[a_t]$ and $\delta/\delta[\lambda]$ denote the functional derivatives with respect to $[a_t]$ and $[\lambda]$, respectively. It follows from (6.108) that the optimized solution of (6.107) is given by

$$\begin{aligned} [a_t] &= [A]^{-1}[S_{rt}]^H\left([S_{rt}][A]^{-1}[S_{rt}]^H\right)^{-1}[c], \\ [\lambda] &= 2\left([S_{rt}][A]^{-1}[S_{rt}]^H\right)^{-1}[c]. \end{aligned} \tag{6.109}$$

If the Rx array is located in the near-field region of the Tx array, one cannot ignore the phase differences among the Rx elements. If one still follows the above procedure in this case, the distribution of the received power along the Rx array will no longer be flat. Instead, it will oscillate around a constant (the average distribution) with peaks and troughs. To avoid solving the QCQP problem directly, a weighting matrix $[W]$ as defined by (6.103) may be introduced to bring the peaks and troughs back to their average. The weighting matrix can be determined by simulations. The weighting coefficients must be less than one for the Rx elements at the peaks and larger than one for the troughs, as illustrated in Figure 6.24. A few iterations may be needed in order to finalize the weighting matrix $[W]$. One may

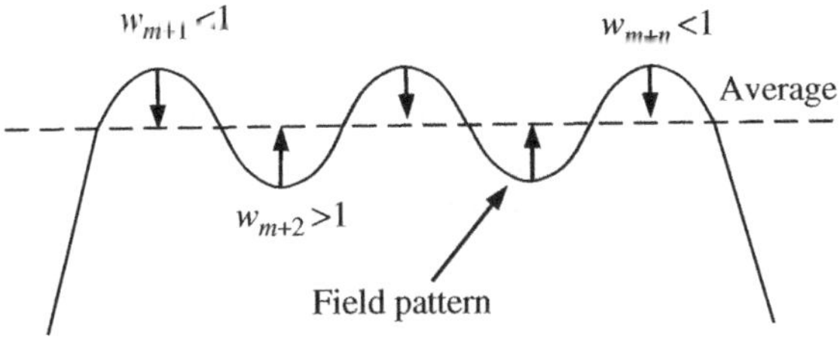

Figure 6.24 Selection of weighting coefficients.

replace the constant vector $[c]$ in (6.107) with a new weighted vector $[W][c]$ to solve a new weighted LCQP problem expressed by

$$\begin{array}{ll} \max & [a_t]^H[A][a_t] \\ \text{s.t.} & [S_{rt}][a_t] = [W][c]. \end{array} \tag{6.110}$$

The solution of (6.110) is then given by

$$\begin{aligned} [a_t] &= [A]^{-1}[S_{rt}]^H\left([S_{rt}][A]^{-1}[S_{rt}]^H\right)^{-1}[W][c], \\ [\lambda] &= 2\left([S_{rt}][A]^{-1}[S_{rt}]^H\right)^{-1}[W][c]. \end{aligned} \tag{6.111}$$

The MMPTE imposed with the linear constraints is called **constrained MMPTE** (CMMPTE). Compared with the UMMPTE and WMMPTE, the optimal solution from the CMMPTE can be determined analytically.

6.4.2 Applications

Since the final goal of antenna design is to maximize the PTE between the Tx and Rx antennas, the MMPTE can also be applied to the design of antenna arrays. In fact, the PTE may be used as a performance index or an objective function to be optimized for all array designs. For this purpose, the array under design may be set as the Tx and a test array may be introduced and set as the Rx so that they form a WPT system, as illustrated in Figure 6.25. Therefore, by properly introducing a test Rx array, the optimal design of an antenna array is transformed into the optimal design of a WPT system and the best possible antenna performance (gain and

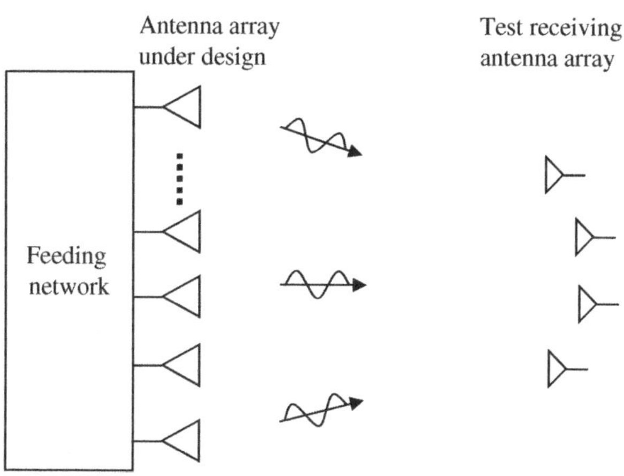

Figure 6.25 Design of antenna array with test receiving antennas.

efficiency) is guaranteed. Different from the design of WPT systems, the Rx array introduced in the design of an antenna array is a virtual tool set for determining the scattering parameters only. There is no need to fabricate it eventually.

The design of antenna arrays for applications in far- or near-field regions can be accomplished in terms of MMPTE by following the same procedures listed below:

1) Set up the WPT system: A WPT system may be set up by assuming that the array under design is used as Tx and a test array is introduced as Rx. The type and the number of the test Rx elements and their configuration depend on the field pattern to be realized on the Rx side. The polarization of the test Rx array elements must match that of the Tx array elements. For example, the design of multi-beam antennas requires a test Rx array with its elements being positioned in the directions where the beam needs to be generated.
2) Determine the scattering parameters of the WPT system: The scattering parameters of the WPT system can be acquired by simulation or by measurement if the surrounding environment is too complicated or unknown.
3) Find the ODE for the array: The ODE can be obtained from (6.101), (6.105), (6.109), or (6.111), depending on which formulation of MMPTE is used.
4) Design the feeding network: Based on the ODE, a feeding network can be designed by the theory of transmission line. The power dividers in the feeding networks control the amplitude distribution of the excitations while the lengths of the feeding lines control the phase distribution of excitations. The feeding network can also be realized by attenuators and phase shifters available in the market. The excitations with negligible amplitudes in the ODE can be ignored to reduce the number of array elements (to generate a sparse array) and simplify the design of the feeding network.

The scattering parameters in MMPTE contain information about the environment between the Tx and Rx arrays and can be obtained by measurement in real time. This is convenient for the design of antennas in a changing environment. For example, the targeted hyperthermia treatment requires accurate knowledge of the electrical properties of the human body, which are usually acquired at the diagnostic stage through medical imaging. In hyperthermia treatment, the patient and properties of the body tissues may change. In these cases, an adaptive approach in real time for hyperthermia treatment is in high demand.

Typically, the MMPTE can be applied to the design of focused antenna arrays, smart (beam steering) antenna arrays, end-fire antenna arrays, multi-beam antenna arrays, and the shaped beam antenna arrays. The MMPTE is applicable to the design of antenna arrays in the near- or far-field region and the design procedures are exactly the same. By introducing a single test Rx antenna in the direction where the radiation intensity needs to be maximized, the UMMPTE can be used to design the focused antennas [30–36], smart antennas for base station and handset applications [30, 37–41], end-fire antennas [42, 43], and polarization and pattern reconfigurable antennas [6, 44]. By introducing a test Rx array, the CMMPTE and WMMPTE can be used to shape beam patterns [6, 45–47], and design antenna arrays with multiple focal points [48] and multiple nulls [49], and multiple beam antennas [30, 50, 51].

The application of MMPTE to the design of WPT system is straightforward. The PTE is a key performance index for the design of WPT system, and the MMPTE maximizes the PTE and therefore provides the best possible system performance. There are three application scenarios for the WPT systems, respectively, corresponding to UMMPTE, WMMPTE, and CMMPTE [6, 52–56].

Example 6.10 (Antenna Array Focused on a Single Point)
In order to illustrate the design process with MMPTE, let us consider a 4×4 microstrip patch antenna array built on FR-4 substrate with thickness of 3 mm, operating at 2.45 GHz. The antenna element is a rectangular microstrip patch with an inset-feed with the length and the width of the patch being 29 mm and 28 mm, respectively, as shown in Figure 6.26a. The length of the feed line is 12 mm and the width is 3 mm which is combined with the 6 mm inset to achieve 50 Ω characteristic impedance. By properly selecting the depth of the inset, a good matching

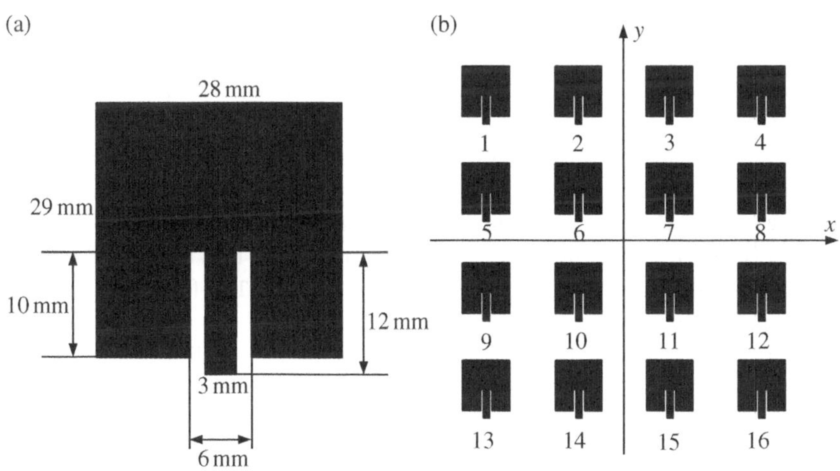

Figure 6.26 (a) Element of microstrip array. (b) Arrangement of array elements.

Table 6.5 ODE.

Port no.	Excitation of port	Port no.	Excitation of port
1	$0.13\angle 0°$	9	$0.22\angle -76°$
2	$0.2\angle -69°$	10	$0.38\angle -138°$
3	$0.2\angle -69°$	11	$0.38\angle -138°$
4	$0.13\angle 0°$	12	$0.22\angle -76°$
5	$0.22\angle -76°$	13	$0.13\angle 0°$
6	$0.38\angle -138°$	14	$0.2\angle -69°$
7	$0.38\angle -138°$	15	$0.2\angle -69°$
8	$0.22\angle -76°$	16	$0.13\angle 0°$

can be achieved without additional matching elements. The distance between the neighboring elements is chosen as 55 mm for the 4×4 array. The 4×4 array with the port number is shown in Figure 6.26b.

If the 4×4 array is required to be focused at $z = 100$ mm, one may introduce a test receiving antenna at $(r, \theta, \varphi) = (100 \text{ mm}, 0, 0)$ in the Fresnel region of the array. The 4×4 array and the test receiving antenna form a WPT system. The WPT system is simulated with one port being active and rest terminated in 50 Ω, which generates the scattering parameters for the system. The ODE is then obtained from eigenvalue equation (6.101) and is listed in Table 6.5. It is easy to find that the optimized phases obey spherical distribution. The feeding network may be modeled by simulation to achieve the ODE at the outputs of the feeding network [30, 31]. The phase distribution can be realized by adjusting the length of the feeding line, and the amplitude distribution can be achieved by power dividers with different choices of width for the feeding lines. During the simulation of the feeding network, each antenna element connected to the feeding network is replaced by a 50 Ω termination. Finally, the feeding network and the antenna array are joined together and simulated as a whole to ensure that the outputs of the feeding network agree well with the optimized values. The final design of the 4×4 microstrip array is displayed in Figure 6.27.

Figure 6.28a shows the measured and simulated normalized distributions of electric field along the z-axis. It can be seen that the measured and simulated results are in a good agreement. The maximum intensity of the electric field occurs at $z = 80$ mm, while the target focal distance is at $z = 100$ mm. The measured and simulated normalized electrical field distribution at the maximum field intensity plane along x-axis is depicted in Figure 6.28b. It can be seen that the measured and simulated results agree very well around the main beams and no sidelobes occur. □

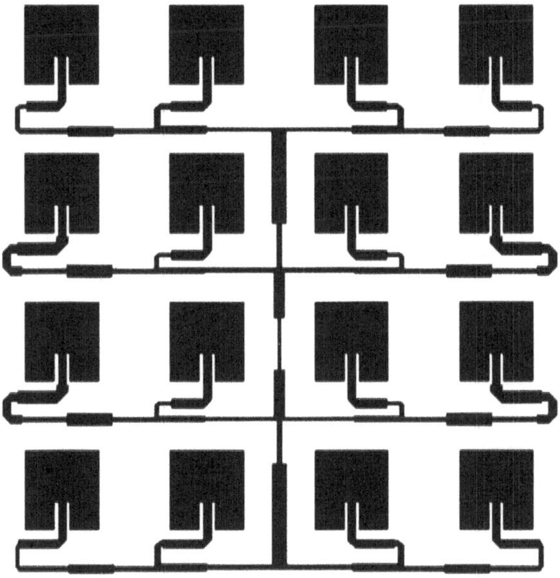

Figure 6.27 A 4 × 4 focused antenna array.

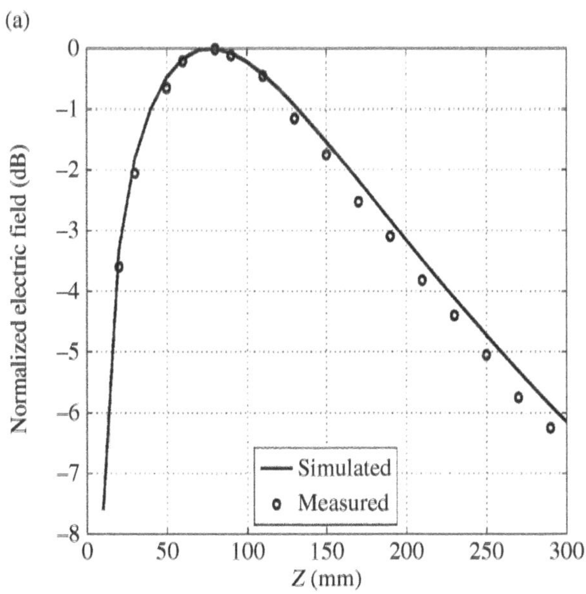

Figure 6.28 (a) Normalized electric field along the z-axis for microstrip array. (b) Normalized electric field distributions at the maximum field intensity plane along x-axis.

(b)

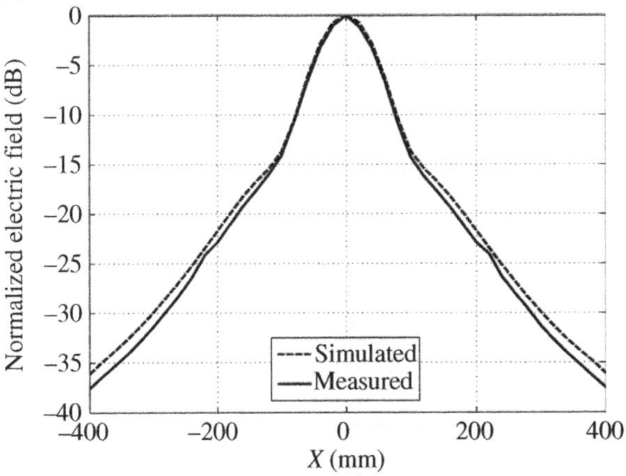

Figure 6.28 (Continued)

Example 6.11 (Antenna Arrays Focused on Multiple Points)
In some circumstances, a prescribed power distribution among multiple focal points may be required to achieve, for example, simultaneous WPT to different devices with different input power levels. Two antenna arrays operating at 2.45 GHz have been designed in [48] by CMMPTE to focus the EM field energy to multiple targets in both closed and open regions, as shown in Figure 6.29. Microstrip patch antenna is selected as the array element for its simple structure, low cost, and easy fabrication. The first antenna array consists of 28 elements and is configured as a square as illustrated in Figure 6.29a. Each side of the square includes seven patch elements. The patch elements are printed on FR4 dielectric substrate. In order to achieve different focusing field patterns, nine test Rx dipoles are introduced and evenly placed inside the square as a 3×3 array. The dipole is chosen as the test receiving antenna for its omnidirectional radiation pattern in the horizontal plane. As a result, a WPT system using the 28-element array as Tx and the nine-element array as Rx is set up, from which the scattering parameters of the whole system can be obtained by ANSYS HFSS (High Frequency Structural Simulator). It is noted that the full wave simulation of the scattering matrix needs to be done just for once for multiple focusing field patterns. The same microstrip patch antenna element is used to build the 6×6 antenna array shown in Figure 6.29b, which are arranged as the Tx array in a square shape. The test Rx array is formed with 3×3 dipoles. The distance between the Tx array and the test Rx array is set to be 150 mm, which is in the Fresnel region of the Tx array. The elements in both Tx and Rx arrays are equally spaced with the inter-element space being 61.2 mm and 153 mm, respectively.

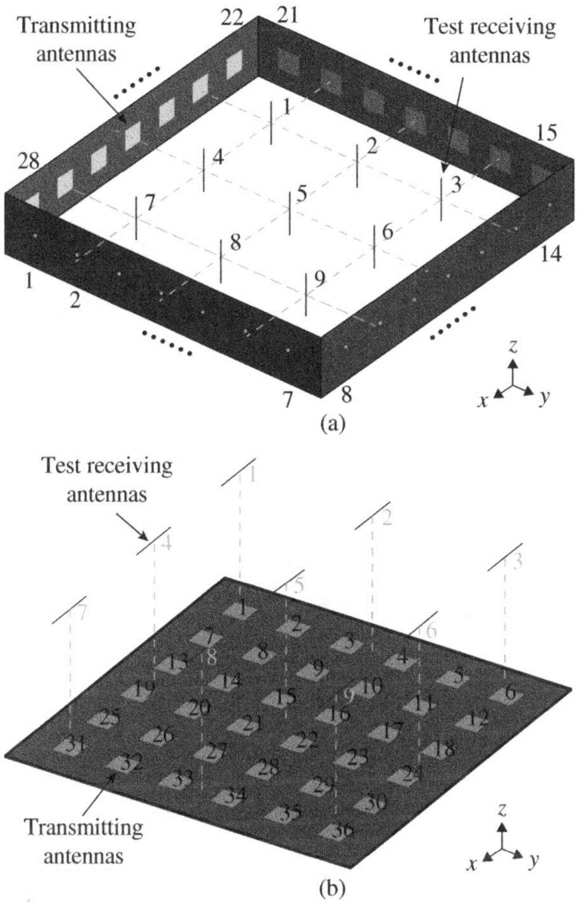

Figure 6.29 Antenna arrays focused on multiple targets. (a) Closed region. (b) Open space. *Source:* Cai et al. [48]/Reproduced with permission of IEEE.

In the design process, the test Rx array elements are divided into two sets. One set denotes the test elements whose received power must be maximized and the other set the test elements whose received power must be minimized. The number of focal points and the power intensity at each focal point are all manageable by regulating the weighting matrix. As a demonstration, the X-shaped, Y-shaped, and L-shaped focusing field patterns are achieved in both closed and open regions and are shown in Figure 6.30. The three focusing field patterns are generated by different weighted coefficients. The weights are initially estimated and then adjusted to make the electric field intensities at each focal point equal. The ODEs for the closed and open arrays are obtained from (6.111). □

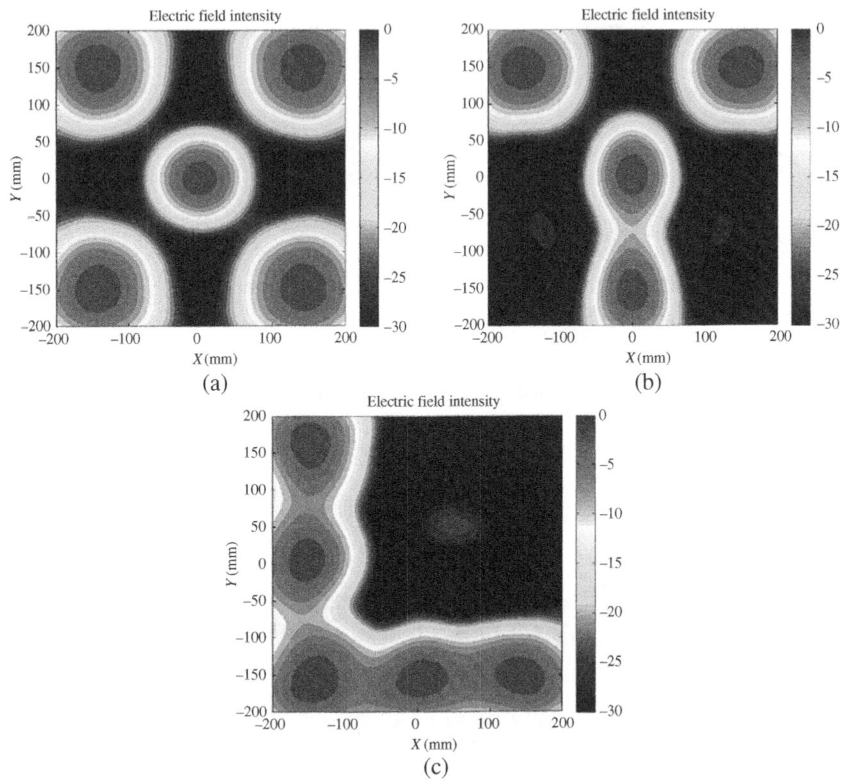

Figure 6.30 Focused field patterns. (a) *X*-shaped pattern. (b) *Y*-shaped pattern. (c) *L*-shaped pattern. *Source:* Cai et al. [48]/Reproduced with permission of IEEE.

Example 6.12 (Beam Shaping in Free Space)

In order to generate a square flat-top radiation pattern in the far-field region, nine test receiving antennas are used and placed on a sphere in the far-field region of a 4×4 patch array operating at 5.8 GHz, and arranged as a square shape, as illustrated in Figure 6.31a. The inter-element spacing is set as half wavelength in free space. The bottom side of the FR4 substrate is the ground plane with 16 feeding ports connected to the patch elements through copper wires. The dimensions of the array are $W = L = 129.4$ mm. The CMMPTE will be used to realize the prescribed pattern. The ODE determine from (6.111) is listed in the left column of Table 6.6. It can be seen that the optimized excitations for ports 2, 3, 5, 8, 9, 12, 14, and 15 are all negligible and can be discarded. An eight-element array is thus obtained from the original 4×4 square array by discarding those elements (shown as dashed squares) with negligible amplitudes of excitation in the ODE, as shown

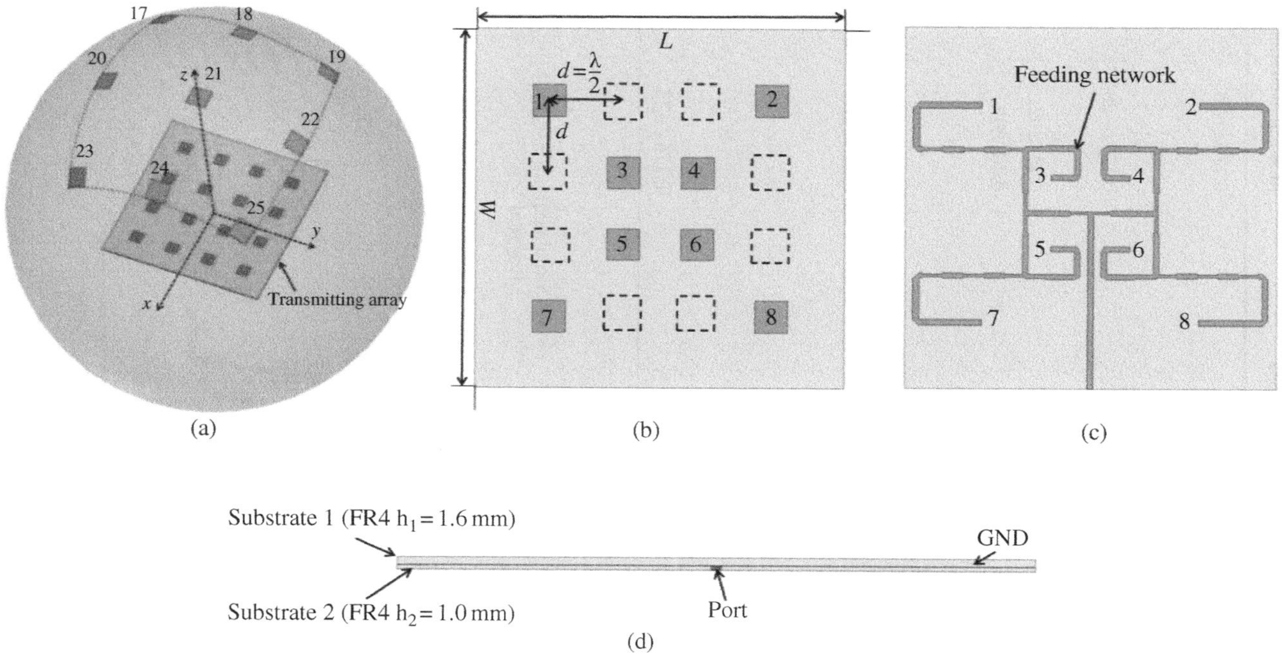

Figure 6.31 Eight-element array. (a) WPT system. (b) Top view. (c) Bottom view. (d) Side view.

Table 6.6 ODEs.

Port number	Optimized excitations	Realized excitations
1	0.176∠−28	0.175∠−28
2	0.005∠31	
3	0.005∠31	
4	0.176∠−28	0.176∠−28
5	0.005∠31	
6	0.468∠93	0.467∠92
7	0.468∠93	0.468∠92
8	0.005∠31	
9	0.005∠31	
10	0.468∠93	0.467∠90
11	0.468∠93	0.466∠91
12	0.005∠31	
13	0.176∠−28	0.176∠−27
14	0.005∠31	
15	0.005∠31	
16	0.176∠−28	0.176∠−27

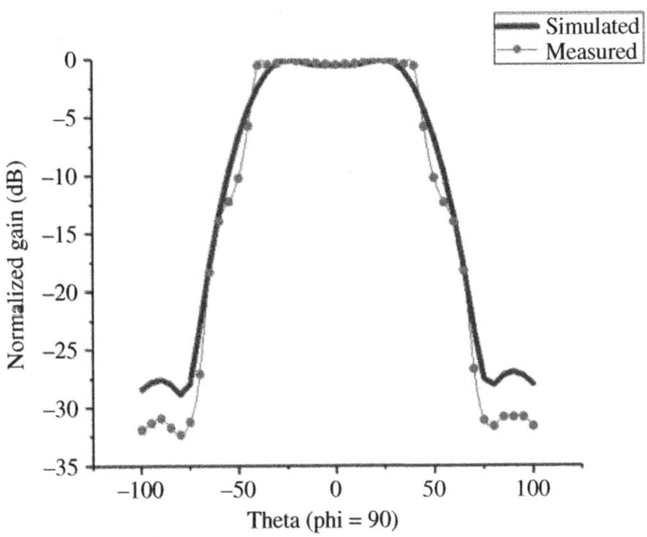

(a)

Figure 6.32 (a) 2D flat top radiation pattern. (b) 3D square flat top radiation pattern.

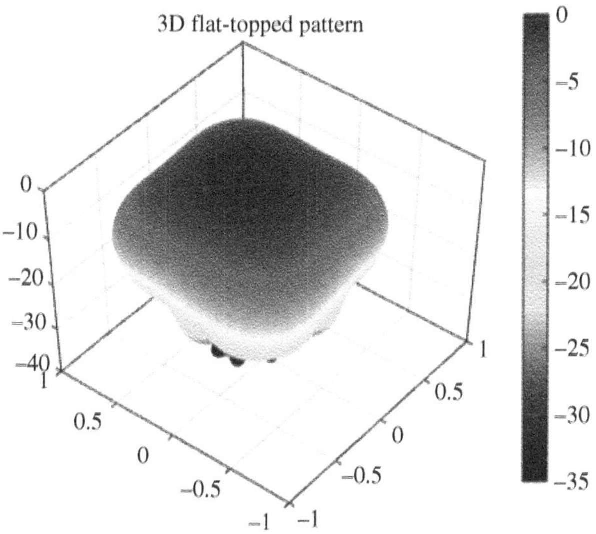

(b)

Figure 6.32 (Continued)

in Figure 6.31b. Figure 6.31c is the feeding network with eight feeding ports, which is designed according to the ODE determined from CMMPTE. Figure 6.32a,b are the 2D and three-dimensional (3D) shaped radiation patterns. It can be seen that field drops quickly off the flat top region and the measured side lobe levels are below −30 dB. □

Example 6.13 (Beam Shaping in Complicated Environment)
Most reader antennas for library applications are designed in free space without considering the influences of the surrounding books. A five-element RFID reader antenna array, operating at 922.5 MHz, for a smart bookshelf is designed with CMMPTE and is shown in Figure 6.33 [47].

The RFID reader antenna is fabricated on an FR4 substrate with thickness of h_1, while the feeding network is fabricated on an FR4 substrate with thickness of h_2. In order to make it more practical, the length L and width W are from a real bookshelf. In order to achieve a flat field pattern across the bookshelf, 11 test Rx antennas are positioned in a row with equal spacing inside a book model of dimensions 850 × 176 × 250 mm whose dielectric constant and loss tangent change with frequency, as illustrated in Figure 6.34. The five-element reader antenna array and the 11 test Rx antennas constitute a 16-port WPT system.

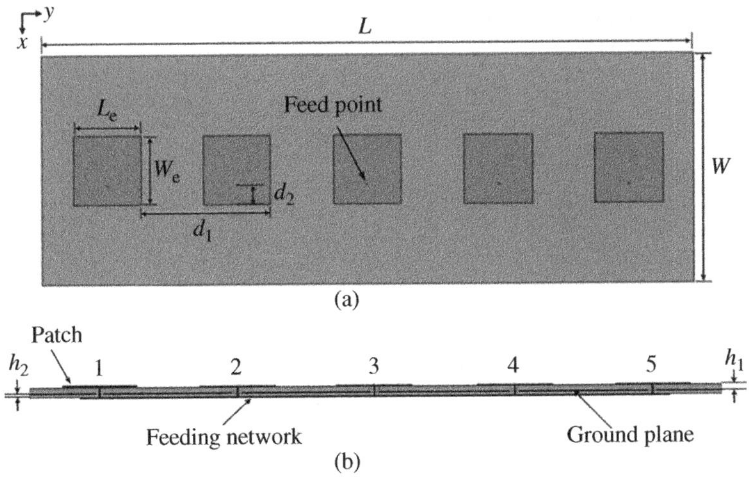

Figure 6.33 Geometry of the RFID reader antenna. (a) Top view. (b) Side view. *Source:* Cai and Geyi [47]/Reproduced with permission of IEEE.

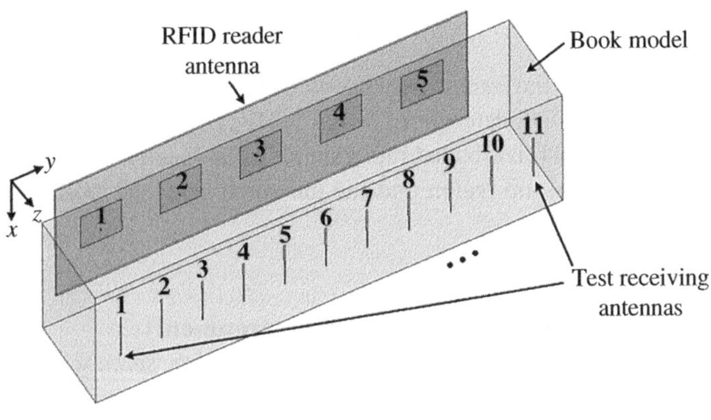

Figure 6.34 Diagram for the near-field power transmission system. *Source:* Cai and Geyi [47]/Reproduced with permission of IEEE.

The ODE for the reader antenna array is obtained from CMMPTE through properly selecting the weighting matrix to generate a flat top radiation pattern inside the book model with the coverage measured at 3 dB being exactly 730 mm. Figure 6.35 shows the wide and flat electric field intensity at $z = 220$ mm in the y-direction achieved after the optimization. The electric field intensity drops abruptly on both left and right side of the bookshelf, which is enabled by placing the outermost test Rx antennas right above the edge of the reader antenna. The

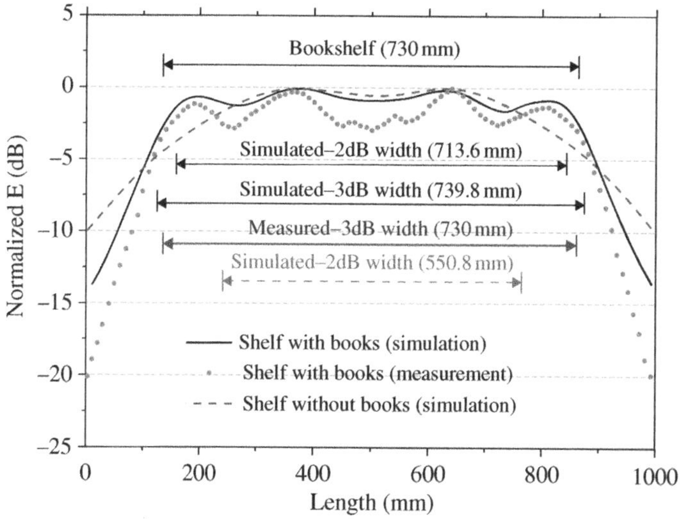

Figure 6.35 Normalized electric field intensity at $z = 220$ mm in y-direction. *Source:* Cai and Geyi [47]/Reproduced with permission of IEEE.

rapid cutoff of the field intensity is helpful in reducing the probability of misreading the books on the adjacent bookshelf.

Many previously reported RFID reader antennas for near-field applications suffer from large fluctuations in field distribution. As a result, the input power may not meet the demands in the lowest electric field areas and may thus cause reading failures. In contrast, the reader antenna designed by the CMMPTE guarantees the reading accuracy for its wide and flat field distribution which is also optimally enhanced in the reading area. □

Example 6.14 (Multi-Beam Antenna Array)
Multi-beam antennas simultaneously generate multiple beams and have found applications in satellite and mobile communications. The WMMPTE and CMMPTE can be applied to the array design for multi-beam applications. By introducing multiple test Rx antennas in the directions where the beams are to be generated, the WMMPTE and CMMPTE produce the ODE to achieve multiple beams in the desired directions.

In order to achieve a comprehensive coverage, the beamwidth needs to be enhanced without sacrificing too much gain. To this end, the array element must be properly selected. An E-shaped microstrip patch antenna is a good candidate for its capability of generating dual beams when it is set to operate in a higher-order mode [57]. The two adjacent beams are expected to be merged together to form a

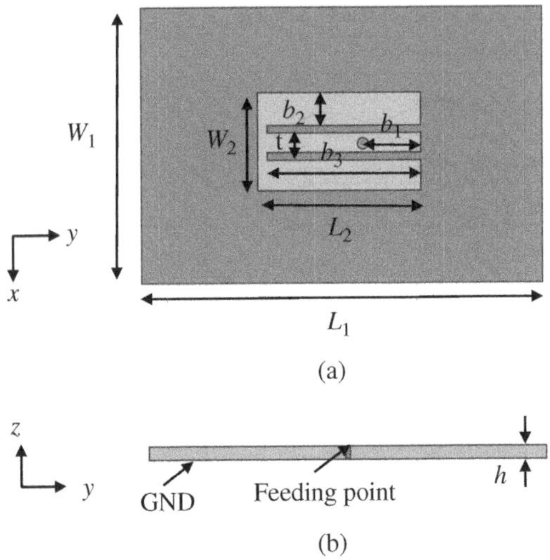

Figure 6.36 Geometry of array element. (a) Top view. (b) Side view.

Table 6.7 Parameters of the element.

Parameter	W_1	W_2	L_1	L_2	b_1	b_2	b_3	t	h
Value (mm)	60	21	80	32.4	11.4	6.75	30.8	4.5	1.57

wide beam after the optimization with MMPTE. The geometry of the E-shaped patch element is shown in Figure 6.36, which is printed on an Arlon Diclad 880 substrate with thickness of 62 mil, relative dielectric constant of 2.2, and loss tangent of 0.0009. The patch element is optimized at the center frequency of 6.0 GHz, and the optimized geometric parameters are obtained by Ansys HFSS and are listed in Table 6.7. By setting the value of L_2 to about 2/3 wavelength, two symmetrical adjacent beams can be generated by adjusting the position of the feeding point, as well as the values of b_2, b_3, and t. The simulated 3D radiation pattern of the E-shaped patch element is shown in Figure 6.37. It can be seen that the patch element generates two adjacent beams, and the maximum gain deviates from the center z-axis by about 30°.

The multi-beam antenna array is formed by four identical E-shaped patch elements with the inter-element spacing along x-axis and y-axis being set as 0.5λ and λ, respectively, as illustrated in Figure 6.38. Four test receiving antennas are introduced in the far-field region and are placed in the directions of the beams to be generated as illustrated in Figure 6.39. The 3D radiation pattern of the antenna array is shown in Figure 6.40, and the realized beam directions, the gains in the beam directions as well as the half-power beamwidth for the four squint beams

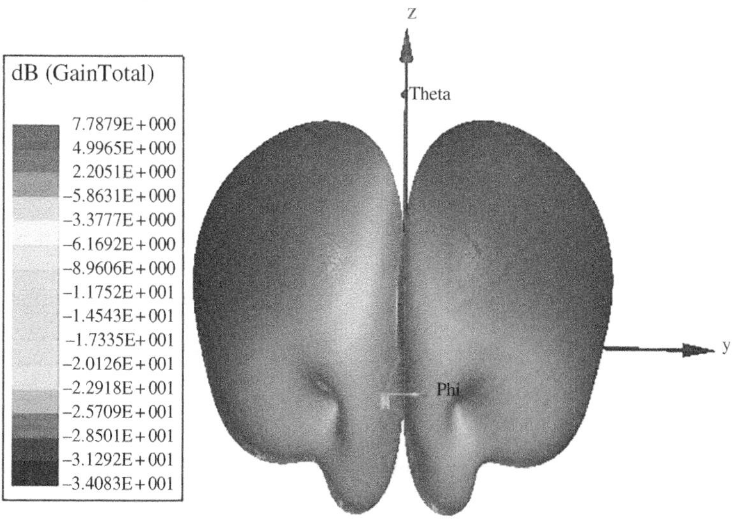

Figure 6.37 Simulated 3D radiation pattern.

Figure 6.38 Configuration of antenna array (W_s = 80 mm, L_s = 100 mm, d_1 = 49.6 mm, d_2 = 25 mm).

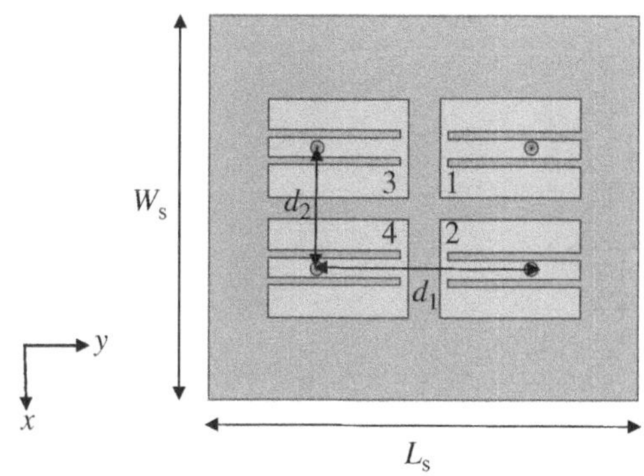

are listed in Table 6.8. It can be seen that the four beams are symmetric about the x-axis and y-axis and are almost identical. Interestingly, the two adjacent beams shown in Figure 6.37 are now merged together into one wide beam after optimization with MMPTE. The fabricated antenna array and the feeding network share a common ground and are shown in Figure 6.41. The feeding network is printed on a Rogers 4003 substrate with thickness of 32 mil, relative dielectric constant of 3.55, and loss tangent of 0.0027. The multi-beam antenna array can be used as an indoor wireless base station installed on the ceiling or wall. In comparison with other similar applications, the feeding network obtained from MMPTE is much simpler, and the highest gain for each beam is guaranteed. □

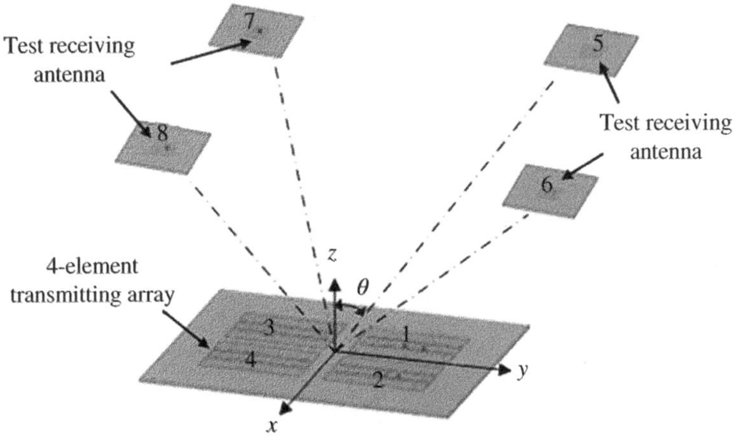

Figure 6.39 WPT system for the design of multi-beam antenna array ($\theta = 45°$).

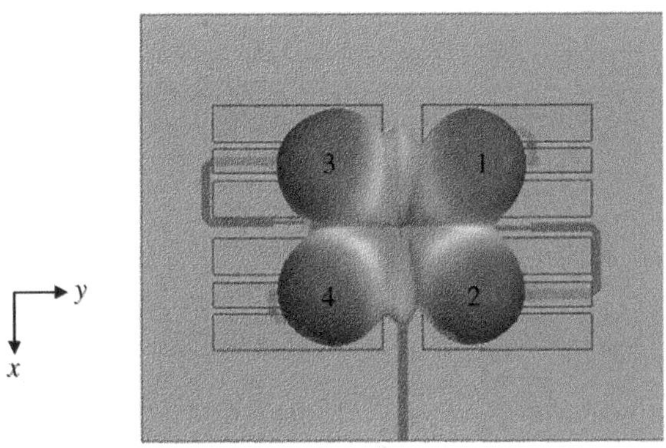

Figure 6.40 3D radiation pattern of antenna array.

Table 6.8 Beam directions, gains, and HPBW.

	Beam 1	Beam 2	Beam 3	Beam 4
θ	45°	45°	45°	45°
φ	45°	135°	315°	225°
Gain (dBi)	8.06	8.17	8.18	8.07
HPBW	44°	43°	43°	44°

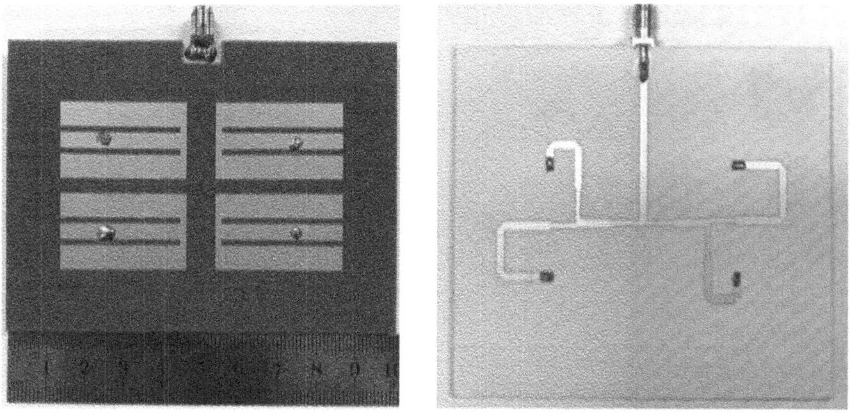

(a) (b)

Figure 6.41 Four-element multi-beam antenna array. (a) Top view. (b) Bottom view.

Example 6.15 (Generalized Yagi–Uda Antenna)
A novel design for end-fire antenna, which generalizes the conventional Yagi–Uda antenna by introducing multiple driven elements, has been presented in [43]. Figure 6.42 shows an eight-element generalized Yagi–Uda antenna with four driven elements, three directors, and one reflector, operating at 2.45 GHz. The four driven elements plus the test receiving antenna placed in the end-fire direction constitute a five-port WPT system, which is used to determine the ODE. The measured and simulated results for radiation patterns on the E-plane and H-plane are plotted in Figure 6.43a,b, respectively. The measured end-fire gain and front-to-back ratio reach 13.4 dBi and 16.4 dB, respectively (with radiation efficiency of 96.5%), which agree well with the simulated results. A comparison with conventional Yagi–Uda dipole array with the same number of elements is also made in Figure 6.43, which indicates that the end-fire gain and front-to-back ratio of the generalized Yagi–Uda design are, respectively, 1.8 and 6.8 dB higher than the conventional Yagi–Uda dipole array. □

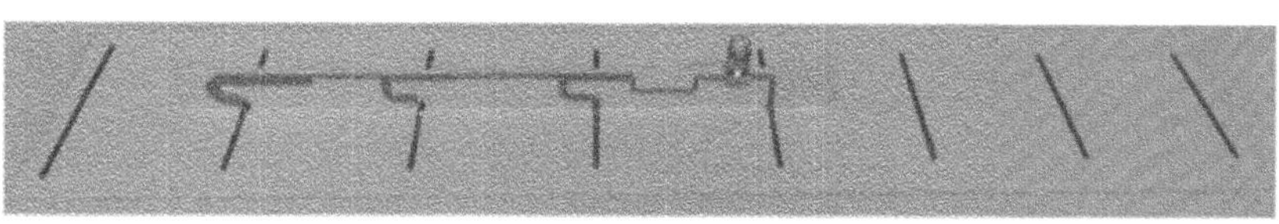

Figure 6.42 The configuration of generalized Yagi–Uda dipole antenna array operating at 2.45 GHz.

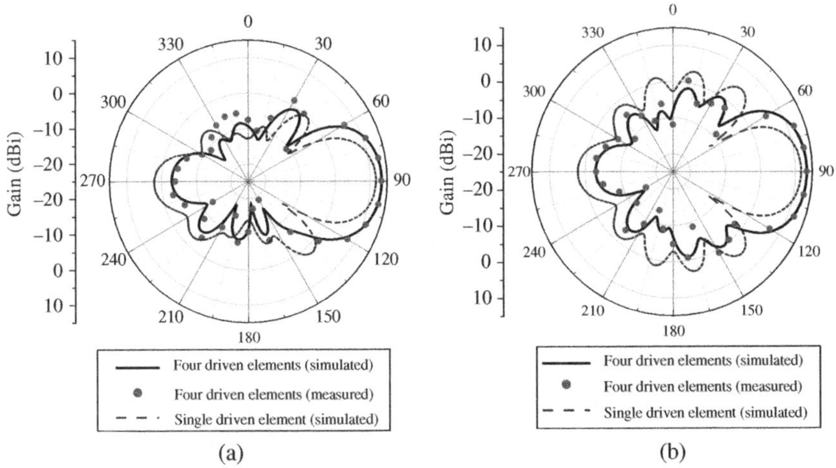

Figure 6.43 Radiation patterns of the generalized Yagi–Uda array. (a) E-plane. (b) H-plane.

Example 6.16 (Smart Antenna for Handheld Device)

An eight-element MIMO smart antenna system consisting of two different array modules for handheld device is shown in Figure 6.44. The first module is a six-element array placed in the middle of the printed circuit board (PCB), operating in N78 (3.3–3.8 GHz) band for 5G, which achieves MIMO functions for receiving

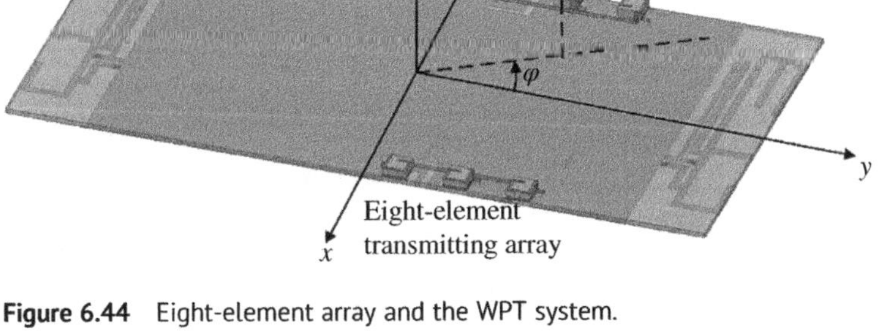

Figure 6.44 Eight-element array and the WPT system.

and beam scanning for transmitting. The second module is a two-element antenna array placed at top and bottom part of the PCB, which operates in LTE/WWAN/N78 (0.7–0.91 GHz, 1.63–2.61 GHz, 3.3–3.8 GHz) bands. To take full advantage of the existing antenna resources in the mobile device, the six antenna elements in the first module are combined with the two antenna elements in the second module to form an eight-element array in the overlapping N78 band. The eight-element array under design is set as the transmitting antenna and a test receiving antenna is introduced and placed in the desired direction in which antenna gain must be maximized, forming a WPT system. The ODE for the combined array can be obtained from solving the eigenvalue equation (6.101). Figure 6.45 shows the simulated and measured 2D radiation patterns of the eight-element array operating at 3.45 GHz. The realized peak gains in positive x- and y-directions are 4.6 dBi and 4.4 dBi, respectively, which are significantly higher than those (less than 0 dBi) radiated from a single antenna element [40].

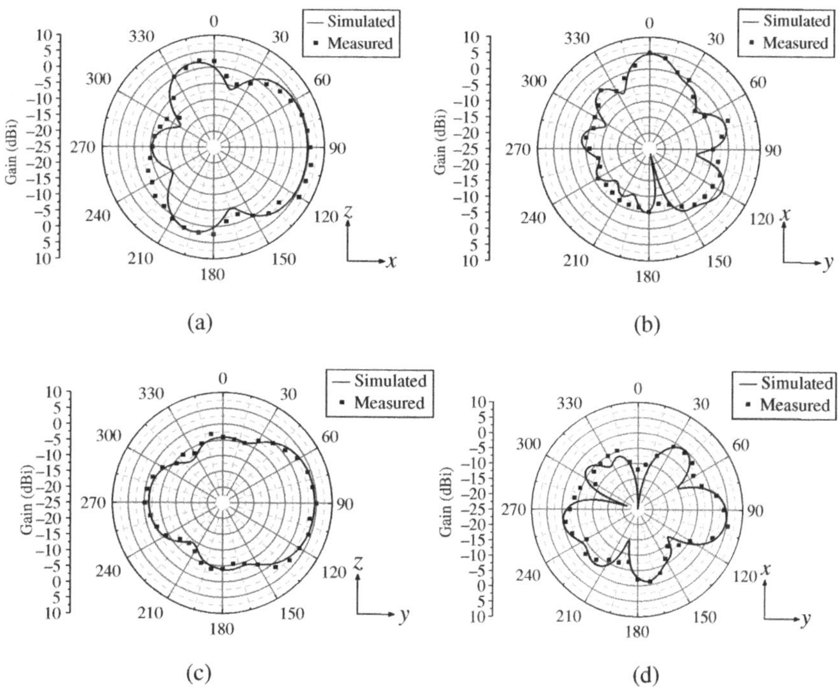

Figure 6.45 Measured and simulated radiation patterns of eight-element antenna array at 3.45 GHz. (a) Directed to positive x axis (xoz-plane). (b) Directed to positive x axis (xoy-plane). (c) Directed to positive y axis (yoz-plane). (d) Directed to positive y axis (xoy-plane).

Example 6.17 (Design of WPT System)
A WPT system generally consists of a transmitting array and a receiving array, and the PTE naturally becomes the performance index for the design of antenna arrays. Both WMMPTE and CMMPTE can be applied to achieve a specified distribution of received power among the Rx array elements. The WMMPTE is most suitable for the situation where the power levels are required to be different among the Rx array elements (e.g. different electronic devices are simultaneously powered). The CMMPTE is most applicable to the scenario where the distribution of received power must be flat along the Rx array elements (e.g. several closely spaced identical electronic devices are wirelessly powered) [56]. The UMMPTE, WMMPTE, and CMMPTE have been applied to the design of a same WPT system operating at 2.45 GHz, in which the Tx array consists of 36 square patch elements and arranged as a square; and the Rx array consists of 5 square patch elements arranged as an L shape. The Tx and Rx arrays are separated by a distance of 15 cm. Figure 6.46 shows the simulation model for the WPT system. The Tx and Rx arrays are built on a 3 mm-thick FR4 substrate. The three formulations of MMPTE yield three feeding schemes for both Tx and Rx arrays, corresponding to three different application scenarios. Simulation and experiment indicate that the PTEs corresponding to UMMPTE, WMMPTE, and CMMPTE are 22.4%, 15.1% and 11.4% respectively. The highest PTE is given by UMMPTE since it does not

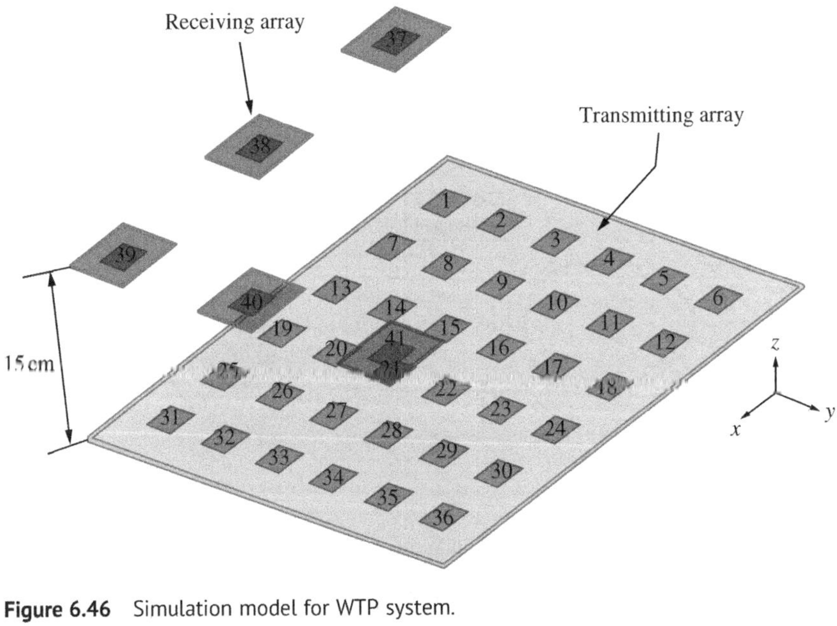

Figure 6.46 Simulation model for WTP system.

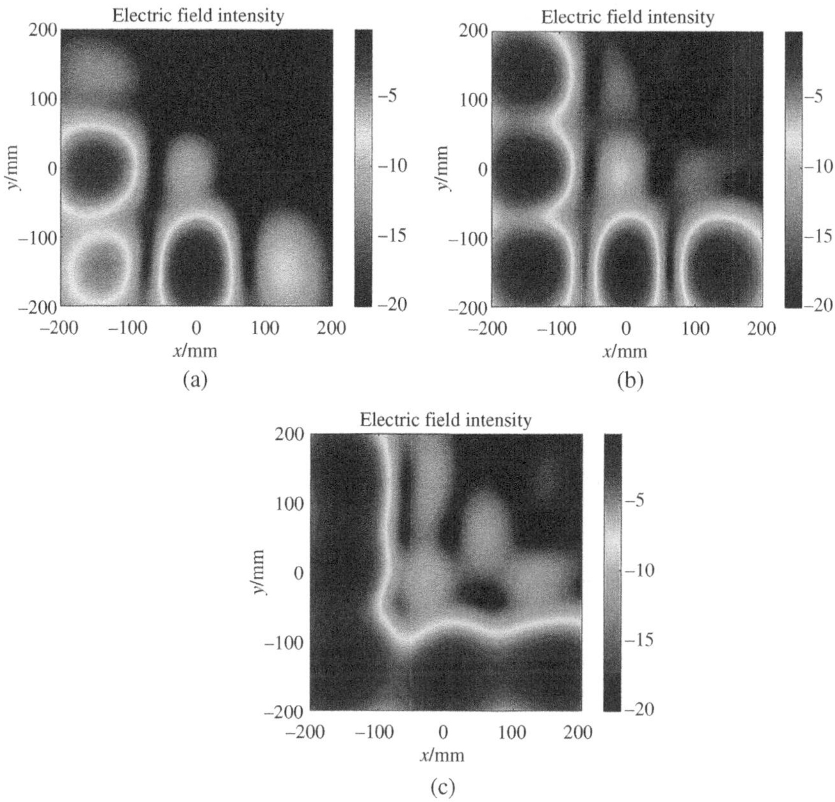

Figure 6.47 Simulated electric field patterns. (a) UMMPTE. (b) WMMPTE. (c) CMMPTE.

involve any constraints. The CMMPTE yields the lowest PTE as it requires that the received energy is equally distributed along the whole path of the L-shape while the WMMPTE only requires that the radiated energy is equally distributed at the locations where the Rx elements are placed. The normalized field patterns generated by the three Tx arrays are, respectively, plotted in Figure 6.47.

6.5 Synthesis of Arrays with EMMPTE

The MMPTE discussed in previous section is based on the circuit theory for a multiport network where both Tx and Tx arrays are involved. One can also get rid of the Rx arrays and introduce other performance indices to achieve various field patterns. The technique will be called the EMMPTE.

6.5.1 Arrays with Specified Energy Distribution

If the transmitting antenna is required to focus the EM field energy to multiple regions designated by $\Omega_p(p=1,2,\ldots,n)$ as illustrated in Figure 6.48, one may introduce the ratio of the weighted sum of the radiated energies in the regions $\Omega_p(p=1,2,\ldots,n)$ over the total input power into the Tx array as the performance index

$$\eta = \frac{\sum_{p=1}^{n}\int_{\Omega_p} W_p(\mathbf{r})|\mathbf{E}(\mathbf{r})|^2 d\Omega(\mathbf{r})}{P_{in}}, \tag{6.112}$$

where $W_p(p=1,2,\ldots,n)$ are weighting functions that regulate the energy distribution among the designated regions $\Omega_p(p=1,2,\ldots,n)$. It is noted that one can make the ratio (6.112) dimensionless by properly selecting the weighting functions although this may not be necessary. Assuming that the transmitting array is well matched, the fields radiated from the transmitting antenna array can then be written as

$$\mathbf{E}(\mathbf{r}) = \sum_{j=1}^{m} a_j \mathbf{E}_j(\mathbf{r}), \mathbf{H}(\mathbf{r}) = \sum_{j=1}^{m} a_j \mathbf{H}_j(\mathbf{r}), \tag{6.113}$$

where $\mathbf{E}_j(\mathbf{r})$ and $\mathbf{H}_j(\mathbf{r})$ are the fields generated by the jth antenna element of the transmitting array when the jth element is excited by $a_j = 1$ and the rest are terminated in a matched load, i.e. $a_i = 0\ (i \neq j)$. Thus, one may write

$$\begin{aligned}\int_{\Omega_p} W_p(\mathbf{r})|\mathbf{E}(\mathbf{r})|^2 d\Omega(\mathbf{r}) &= \sum_{i=1}^{m}\bar{a}_i \sum_{j=1}^{m} a_j \int_{\Omega_p} W_p(\mathbf{r})\mathbf{E}_j(\mathbf{r})\cdot\bar{\mathbf{E}}_i(\mathbf{r})d\Omega(\mathbf{r})\\ &= ([A^p][a_t],[a_t]),\end{aligned}$$

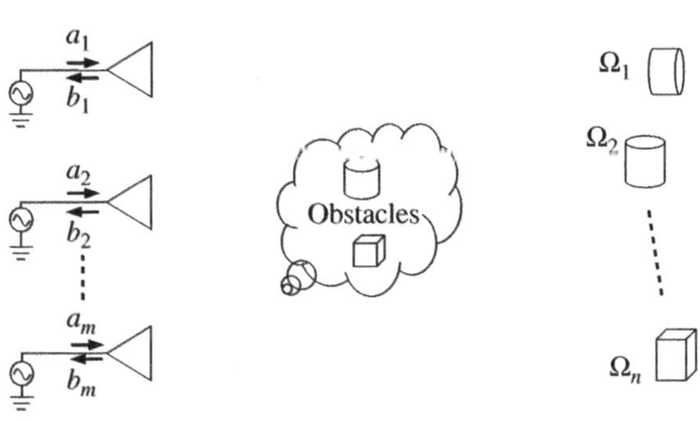

Figure 6.48 A extended WPT system.

where $[A^p]$ is an $m \times m$ matrix with the (i, j) elements given by

$$A_{ij}^p = \int_{\Omega_p} W_p(\mathbf{r})\mathbf{E}_j(\mathbf{r}) \cdot \overline{\mathbf{E}}_i(\mathbf{r})d\Omega(\mathbf{r}). \tag{6.114}$$

If the transmitting antenna elements are all matched, one gets $P_{in} = ([a_t], [a_t])/2$, and (6.112) can thus be written as

$$\eta = \frac{\sum_{p=1}^{n} ([A^p][a_t], [a_t])}{([a_t], [a_t])/2} = \frac{([A][a_t], [a_t])}{([a_t], [a_t])}, \tag{6.115}$$

with $[A] = 2\sum_{p=1}^{n}[A^p]$. Similarly, the optimized solution of (6.115) for the transmitting antenna array can be determined by the eigenvalue equation (6.101) with $[B]$ set to the identity matrix. The transmitting antenna array excited by the ODE obtained from maximizing (6.115) focuses the EM field energy to the multiple target regions designated by $\Omega_p(p = 1, 2, \ldots, n)$. Especially, if the weighting functions are selected to be $W_p = w_p\delta(\mathbf{r} - \mathbf{r}_p)$ $(p = 1, 2, \ldots, n)$, the optimization of (6.115) gives the ODE for the transmitting antenna array focused on multiple points $\mathbf{r}_p$ $(p = 1, 2, \ldots, n)$. The above strategy can be used to design focused antenna array, smart antenna array, and multi-beam antenna array.

One can also introduce the ratio of the total energy absorbed in the region $\sum_{p=1}^{l}\Omega_p$ $(l < n)$ over that absorbed in the region $\sum_{p=l+1}^{n}\Omega_p$ $(l < n)$ as the performance index

$$\eta = \frac{\sum_{p=1}^{l} \int_{\Omega_p} W_p(\mathbf{r})|\mathbf{E}(\mathbf{r})|^2 d\Omega(\mathbf{r})}{\sum_{p=l+1}^{n} \int_{\Omega_p} W_p(\mathbf{r})|\mathbf{E}(\mathbf{r})|^2 d\Omega(\mathbf{r})}. \tag{6.116}$$

Inserting (6.113) into (6.116) leads to

$$\eta = \frac{([A][a_t], [a_t])}{([B][a_t], [a_t])}, \tag{6.117}$$

where

$$[A] = \sum_{p=1}^{l}[A^p], [B] = \sum_{p=l+1}^{n}[A^p].$$

The optimized solution of (6.117) can be determined by the eigenvalue equation (6.101). An optimization criterion similar to (6.116) was introduced in [58] to find the optimal excitations for the multi-antenna applicators in regional hyperthermia.

6.5.2 Arrays with Specified Power Distribution

Let $S_p(p = 1, 2, \ldots, n)$ denote the surface elements located in various directions as illustrated in Figure 6.49, one can introduce the performance index

$$\eta = \frac{\sum_{p=1}^{n} \int_{S_p} W_p(\mathbf{r}) \frac{1}{2} \operatorname{Re} \mathbf{E}(\mathbf{r}) \times \overline{\mathbf{H}}(\mathbf{r}) \cdot \mathbf{u}_n dS(\mathbf{r})}{P_{in}}, \tag{6.118}$$

which is the ratio of the weighted sum of the radiated power in the directions specified by $S_p(p = 1, 2, \ldots, n)$ over the total input power into the transmitting antenna array. In the above, $W_p(\mathbf{r})$ $(p = 1, 2, \ldots, n)$ are the weighting functions that regulate the power distribution in the directions designated by S_p $(p = 1, 2, \ldots, n)$ and $\mathbf{u}_n$ is the unit normal vector to the surface element. Considering (6.113), one may write

$$\begin{aligned}
&\int_{\Omega_l} W_p(\mathbf{r}) \frac{1}{2} \operatorname{Re} \mathbf{E}(\mathbf{r}) \times \overline{\mathbf{H}}(\mathbf{r}) \cdot \mathbf{u}_n dS(\mathbf{r}) \\
&= \frac{1}{2} \operatorname{Re} \sum_{i=1}^{m} \bar{a}_i \sum_{j=1}^{m} a_j \int_{\Omega_p} W_p(\mathbf{r}) \mathbf{E}_j(\mathbf{r}) \times \overline{\mathbf{H}}_i(\mathbf{r}) \cdot \mathbf{u}_n dS(\mathbf{r}) \\
&= \frac{1}{2} \operatorname{Re}([A^p][a_t], [a_t]),
\end{aligned}$$

where $[A^p]$ is an $m \times m$ matrix with the elements given by

$$A_{ij}^p = \int_{\Omega_p} W_p(\mathbf{r}) \mathbf{E}_j(\mathbf{r}) \times \overline{\mathbf{H}}_i(\mathbf{r}) \cdot \mathbf{u}_n dS(\mathbf{r}). \tag{6.119}$$

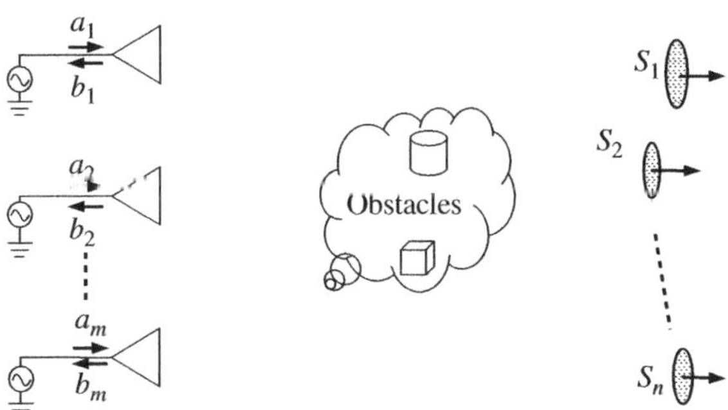

Figure 6.49 An extended WPT system.

If the transmitting antenna elements are all matched, (6.118) can thus be expressed by

$$\eta = \text{Re}\,\frac{([A][a_t], [a_t])}{([a_t], [a_t])} = \frac{([A_c][a_t], [a_t])}{([a_t], [a_t])}, \tag{6.120}$$

with $[A] = \sum_{p=1}^{n} [A^p]$ and $[A_c] = (1/2)([A] + [A]^\dagger)$. The above strategy can be applied to the design of focused antenna array, smart antenna array, and multi-beam antenna array.

Similar to (6.116), one can also introduce the ratio of the total power radiated in the directions $S_1, S_2, \ldots,$ and S_l over that radiated in the directions $S_{l+1}, S_{l+2}, \ldots,$ and S_n as the performance index

$$\eta = \frac{\sum\limits_{p=1}^{l} \int\limits_{S_p} W_p(\mathbf{r})\frac{1}{2}\,\text{Re}\,\mathbf{E}(\mathbf{r}) \times \overline{\mathbf{H}}(\mathbf{r}) \cdot \mathbf{u}_n dS(\mathbf{r})}{\sum\limits_{p=l+1}^{n} \int\limits_{S_p} W_p(\mathbf{r})\frac{1}{2}\,\text{Re}\,\mathbf{E}(\mathbf{r}) \times \overline{\mathbf{H}}(\mathbf{r}) \cdot \mathbf{u}_n dS(\mathbf{r})}. \tag{6.121}$$

Introducing (6.113) into (6.121) yields

$$\eta = \frac{\text{Re}([A][a_t], [a_t])}{\text{Re}([B][a_t], [a_t])} = \frac{([A_c][a_t], [a_t])}{([B_c][a_t], [a_t])}, \tag{6.122}$$

where

$$[A] = \sum_{p=1}^{l} [A^p], \quad [A_c] = \frac{1}{2}\left([A] + [A]^\dagger\right),$$

$$[B] = \sum_{p=l+1}^{n} [A^p], [B_c] = \frac{1}{2}\left([B] + [B]^\dagger\right),$$

and the matrix elements of $[A^p]$ are given by (6.119).

6.5.3 Applications

For most applications, such as the design of focused antenna array, smart antenna array, and multi-beam antenna, it has been verified that that the EMMPTE without using the test receiving antennas yields the same results as MMPTE. Constraints can also be introduced with EMMPTE and the discussions can be carried out in a similar way.

Example 6.18 (Multi-Null Steering Antenna Array)
In wireless communication systems, null steering is vital to decrease overall interference level and to increase throughput by transmitting the signal power toward user ends and minimizing the signal power in unwanted directions, such as the directions of co-channel or co-site base stations. Multi-null patterns can be achieved by MMPTE or EMMPTE. The EMMPTE will be used to illustrate the design process [49].

As shown in Figure 6.50, the null directions are designated by the points on an arc in the far-field region (the dashed line). The weighting functions in (6.112) are selected as $W_p = w_p\delta(\mathbf{r} - \mathbf{r}_p)$ $(p = 1, 2, \ldots, n)$ and the optimization of (6.115) yields eigenvalue equation (6.101). The eigenvector corresponding to the zero eigenvalues are the ODE for generating the radiation pattern with the desired nulls. Note that the electric field generated by each element of the antenna array on the far-field arc can be simulated or measured in advance. To steer the multiple nulls of the radiation pattern, one can choose different position combinations to build the matrix elements (6.114). Because the matrix is usually very small, the computational effort involved is trivial.

To demonstrate the multi-null steering, an eight-element linear patch array is designed and fabricated, as illustrated in Figures 6.50 and 6.51. The patch element is fabricated on a 1.6-mm-thick FR4 substrate, and is fed with coaxial cable. The antenna is designed to operate at 2.45 GHz and linearly polarized along x-axis. Based on the EMMPTE, the ODE for the patch array can be obtained to realize a radiation pattern with desired multiple nulls. Figure 6.52 shows the 3D radiation patterns with multiple nulls that are required to point to $\pm 10°$ (two-direction case), $-5°$ and $-25°$ (two-direction case), $0°$ and $\pm 30°$ (three-direction case), and $\pm 10°$ and $\pm 30°$ (four-direction case), respectively. □

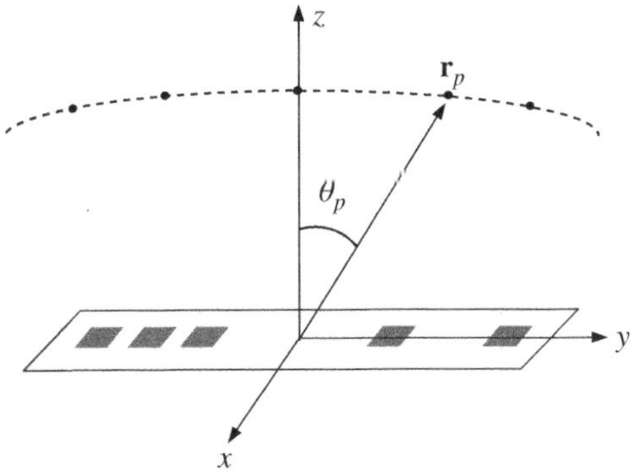

Figure 6.50 Multi-null steering of linear antenna array.

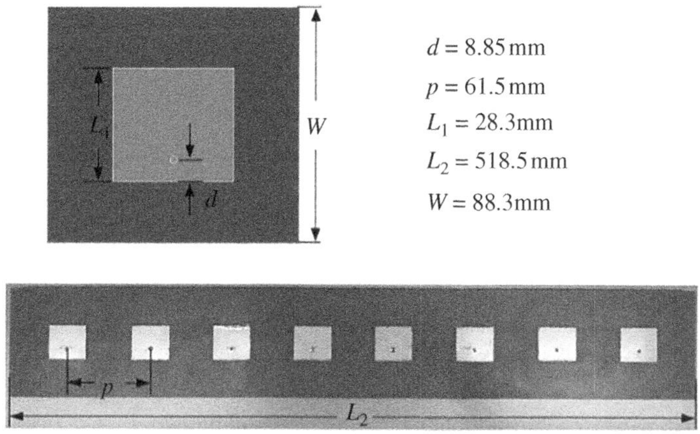

Figure 6.51 Patch element and array configuration.

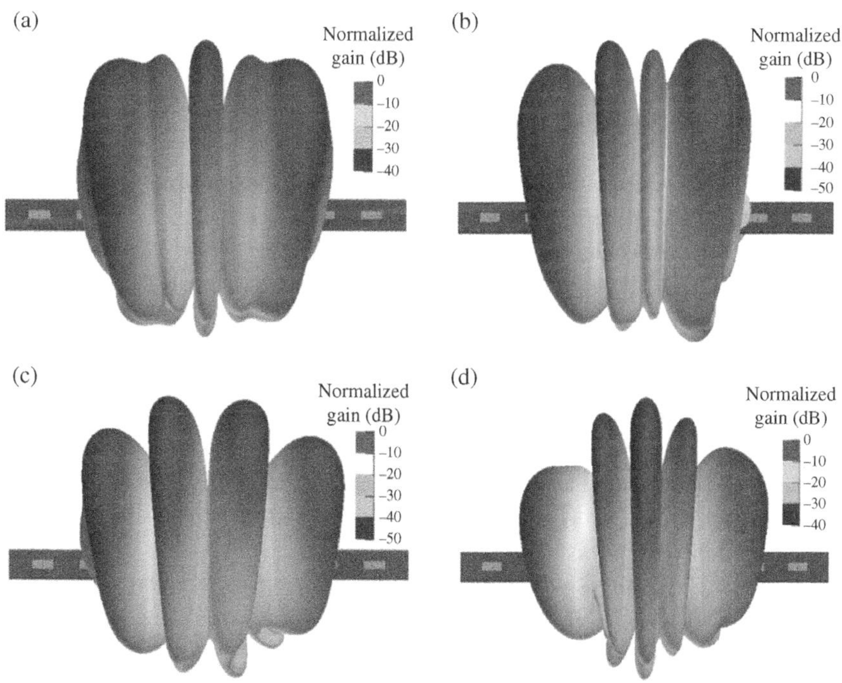

Figure 6.52 Simulated 3D normalized gain pattern with multiple nulls at elevation angles: (a) ±10°, (b) −5° and −25°, (c) 0° and ±30°, and (d) ±10°and ±30°.

Everything should be built top-down, except the first time. Simplicity does not precede complexity, but follows it.

Alan Jay Perlis (American computer scientist, 1922–1990)

References

1 Hansen, R. C., *Phased Array Antennas*, John Wiley & Sons, Inc., 2001.
2 Schelkunoff, S. A., "A mathematical theory of linear arrays", *Bell Sys. Tech. J.*, Vol. 22, pp. 80–107, 1943.
3 Woodward, P. M. and Lawson, J. D., "The theoretical precision with which an arbitrary radiation pattern may be obtained from a source of finite size", *J. IEE*, Vol. 95, pt. III, No. 37, pp. 363–370, 1948.
4 Dolph, C. L., "A current distribution for broadside arrays which optimizes the relationship between beamwidth and sidelobe level", *Proc. IEEE*, Vol. 34, pp. 335–348, 1946.
5 Geyi, W., *Foundations of Applied Electrodynamics*, New York, Wiley, 2010.
6 Geyi, W., "The method of maximum power transmission efficiency for the design of antenna arrays", *IEEE Open J. Antennas Propag.*, Vol. 2, pp. 412–430, 2021.
7 Balanis, C. A., *Antenna Theory: Analysis and Design*, 2nd Ed., John Wiley & Sons, Inc., 1997.
8 Elliott, R. S., *Antenna Theory and Design*, New York, Prentice-Hall, 1981.
9 Milligan, T. A., *Modern Antenna Design*, 2nd Ed., Wiley, 2005.
10 Haupt, R. L., *Antenna Arrays: A Computational Approach*, Wiley-IEEE Press, 2010.
11 Collin, R. E., *Antennas and Radio Wave Propagation*, McGraw-Hill, 1985.
12 Kraus, J. D. and Marhefka, R. J., *Antennas for All Applications*, McGraw-Hill, 2001.
13 Stutzman, W. L. and Thiele, G. A., *Antenna Theory and Design*, New York, John Wiley & Sons, 1981.
14 Hu, M. K., "Near zone power transmission formulae", *IRE Nat'l Conv. Rec.*, Vol. 6, No. 8, pp. 128–135, 1958.
15 Geyi, W., *Foundations for Radio Frequency Engineering*, World Scientific, 2015.
16 Bickmore, R. W., "On focusing electromagnetic radiators", *Can. J. Phys.*, Vol. 35, No. 11, pp. 1292–1298, 1957.
17 Brown, J., "A theoretical analysis of some errors in aerials measurements", *Proc. IEE*, Vol. 156, pp. 343–351, 1958.
18 Robieux, J., "Near-zone power transmission formulas", *Proc. IRE*, Vol. 47, pp. 1161–1163, 1959.
19 Kay, A. F., "Near field gain of aperture antennas", *IRE Trans. Antennas Propagat.*, Vol. 8, pp. 586–593, 1960.

20 Jacobs, E., "Fresnel region power transfer", in *Electromagnetic Theory and Antennas*, New York, Pergamon, pp. 1051–1074, 1963.

21 Soejima, T., "Fresenel gain of aperture aerials", *Proc. IEE*, Vol. 110, pp. 1021–1027, 1963.

22 Borgiotti, G. V., "Maximum power transfer between two planner apertures in the Fresnel zone", *IEEE Trans. Antennas Propag.*, Vol. 14, pp. 158–163, 1966.

23 Friis, H. T., "A note on a simple transmission formula", *IRE Proc.*, Vol. 34, No. 5, pp. 254–256, 1946.

24 Okress, E. C., *Microwave Power Engineering, Vols. 1–2*, Academic Press, 1968.

25 Boyd, G. D. and Kogelnik, H., "Generahzed confocal resonator theory", *Bell Sys. Tech. J.*, Vol. 41, pp. 1347–1369, 1962.

26 Slepian, D. and Pollak, H. O., "Prolate spheroidal wave functions, Fourier analysis and uncertainty-I", *Bell Sys. Tech. J.*, Vol. 40, pp. 43–63, 1961.

27 Horn, R. A. and Johnson, C. R., *Matrix Analysis*, 2nd Ed., Cambridge University Press, 2013.

28 Bar-on, J. R. and Grasse, K. A., "Global optimization of a quadratic functional with quadratic equality constraints, part 2", *J. Optim. Theory Appl.*, Vol. 93, No. 3, pp. 547–556, 1997.

29 Jia, Z. H., Cai, X. J., and Han, D. R., "Comparison of several fast algorithms for projection onto an ellipsoid", *J. Comput. Appl. Math.*, Vol. 319, pp. 320–337, 2017.

30 Geyi, W., "Optimal design of antenna arrays (invited)", in *International Workshop on Antenna Technology*, Sydney, IEEE, 2014.

31 Shan, L. and Geyi, W., "Optimal design of focused antenna arrays", *IEEE Trans. Antennas Propag.*, Vol. 62, No. 11, pp. 5565–5571, 2014.

32 Jiang, Y. H., Geyi, W., and Sun, H., "A new focused antenna array with circular polarization", *Microw. Opt. Technol. Lett.*, Vol. 57, No. 12, pp. 2936–2939, 2015.

33 Jiang, Y. H., Geyi, W., Yang, L. S., and Sun, H., "Circularly-polarized focused microstrip antenna arrays", *IEEE Antennas Wirel. Propag. Lett.*, Vol. 15, pp. 52–57, 2016.

34 He, X. P., Geyi, W., and Wang, S. Y., "Optimal design of focused arrays for microwave-induced hyperthermia", *IET Microw. Antennas Propag.*, Vol. 9, No. 14, pp. 1605–1611, 2015.

35 He, X. P., Geyi, W., and Wang, S. Y., "A hexagonal focused array for microwave hyperthermia: optimal design and experiment", *IEEE Antennas Wirel. Propag. Lett.*, Vol. 15, pp. 1605–1611, 2016.

36 Gao, P., Wang, S. Y., and Geyi, W., "A blind adaptive focusing scheme for the unknown target in lossy medium", *IEEE Antennas Wirel. Propag. Lett.*, Vol. 21, No. 5, pp. 983–987, 2022.

37 Tong, H. and Geyi, W., "Optimal design of smart antenna systems for handheld devices", *IET Microw. Antennas Propag.*, Vol. 10, No. 6, pp. 617–623, 2016.

38 Wan, W., Geyi, W., and Gao, S., "Optimum design of low-cost dual-mode beam-steerable arrays for customer-premises equipment applications", *IEEE Access*, Vol. 6, pp. 16092–16098, 2018.

39 Miao, X., Wan, W., Duan, Z., and Geyi, W., "Design of dual-mode arc-shaped dipole arrays for indoor base-station applications", *IEEE Antennas Wirel. Propag. Lett.*, Vol. 18, No. 4, pp. 752–756, 2019.

40 Wei, P. and Geyi, W., "Design of MIMO/smart antenna arrays using different array modules for handheld device", *Prog. Electromagn. Res. C.*, Vol. 115, pp. 111–126, 2021.

41 Li, T. and Geyi, W., "Design of MIMO beamforming antenna array for mobile handsets", *Prog. Electromagn. Res. C.*, Vol. 94, pp. 13–28, 2019.

42 Cai, X., Geyi, W., and Sun, H. C., "A printed dipole array with high gain and end-fire radiation", *IEEE Antennas Wirel. Propag. Lett.*, Vol. 16, pp. 1512–1515, 2017.

43 Guo, H. and Geyi, W., "Design of Yagi-Uda antenna with multiple driven elements", *Prog. Electromagn. Res. C.*, Vol. 92, pp. 101–112, 2019.

44 Duan, Z., Wang, J., Xu, L.-J., Wang, F., and Geyi, W., "Pattern and polarization reconfigurable loop antenna with circular high-impedance surface for 5G micro base station application", *Int. J. RF Microw. Comput. Aided Eng.*, 2022, DOI:10.1002/mmce.23359.

45 Gu, X. and Geyi, W., "Design of a near-field RFID antenna array in metal cabinet environment", *IEEE Antennas Wirel. Propag. Lett.*, Vol. 18, No. 1, pp. 79–83, 2019.

46 Dong, Y., Cai, X., and Geyi, W., "Circularly polarized antenna array with suppressed sidelobes for electronic toll collection", *IEEE Antennas Wirel. Propag. Lett.*, Vol. 21, No. 5, pp. 988–992, 2022.

47 Cai, X. and Geyi, W., "An optimization method for the synthesis of flat-top radiation patterns in the near- and far-field regions", *IEEE Trans. Antennas Propag.*, Vol. 67, No. 2, pp. 980–987, 2019.

48 Cai, X., Gu, X., and Geyi, W., "Optimal design of antenna arrays focused on multiple targets", *IEEE Trans. Antennas Propag.*, Vol. 68, No. 6, pp. 4593–4603, 2020.

49 Wang, S. Y., Jin, X. Y., Liu, P., and Geyi, W., "Fast multi-null steering of antenna array", *IEEE Antennas Wirel. Propag. Lett.*, Vol. 21, 2022, DOI:10.1109/LAWP.2022.3199078.

50 Sun, H. C. and Geyi, W., "A new rectenna using beamwidth-enhanced antenna array for RF power harvesting applications", *IEEE Antennas Wirel. Propag. Lett.*, Vol. 16, pp. 1451–1454, 2017.

51 Guo, H. and Geyi, W., "Design of bidirectional antenna array with adjustable end-fire gains", *IEEE Antennas Wirel. Propag. Lett.*, Vol. 18, No. 8, pp. 1656–1660, 2019.

52 Sun, H. C. and Geyi, W., "Optimum design of wireless power transmission systems in unknown electromagnetic environments", *IEEE Access*, Vol. 5, pp. 20198–20206, 2017.

53 Xie, F., Yang, G., and Geyi, W., "Optimal design of an antenna array for energy harvesting", *IEEE Antennas Wirel. Propag. Lett.*, Vol. 12, pp. 155–158, 2013.

54 Yang, X. D., Geyi, W., and Sun, H. C., "Optimum design of wireless power transmission system using microstrip patch antenna arrays", *IEEE Antennas Wirel. Propag. Lett.*, Vol. 16, pp. 1824–1827, 2017.

55 Chen, Z. Z., Sun, H. C., and Geyi, W., "Maximum wireless power transfer to the implantable device in the radiative near field", *IEEE Antennas Wirel. Propag. Lett.*, Vol. 16, pp. 1780–1783, 2017.

56 Sun, S. Y. and Geyi, W., "Optimal design of wireless power transmission systems using antenna arrays", *ZTE Commun.*, Vol. 20, No. 2, pp. 19–27, 2022.

57 Khidre, A., Lee, K. F., Elsherbeni, A. Z., and Yang, F., "Wide band dual-beam U-slot microstrip antenna", *IEEE Trans. Antennas Propag.*, Vol. 61, No. 3, pp. 1415–1418, 2013.

58 Böhm, M., Kremer, J., and Louis, A. K., "Efficient algorithm for computing optimal control of antennas in hyperthermia", *Surv. Math. Ind.*, Vol. 3, No. 4, pp. 233–251, 1993.

Appendix A

Vector Analysis

Vector analysis studies the differentiation and integration of vector fields. It plays an important role in electromagnetic field theory. By definition, the gradient of a scalar function $\phi(\mathbf{r})$ at $\mathbf{r}$ is

$$\nabla\phi(\mathbf{r}) = \lim_{V\to 0}\frac{1}{V}\int_S \mathbf{u}_n\phi dS,$$

where V is a volume containing the point $\mathbf{r}$, S is its boundary, and $\mathbf{u}_n$ is the unit outward normal of S. The gradient measures the rate and direction of change in a scalar field. The divergence of a vector function $\mathbf{A}$ is defined by

$$\nabla\cdot\mathbf{A} = \lim_{V\to 0}\frac{1}{V}\int_S \mathbf{u}_n\cdot\mathbf{A}dS.$$

The divergence measures the magnitude of the source of the vector field at a point. The rotation of a vector function $\mathbf{A}$ is defined by

$$\nabla\times\mathbf{A} = \lim_{V\to 0}\frac{1}{V}\int_S \mathbf{u}_n\times\mathbf{A}dS.$$

The rotation measures the tendency of the vector field to rotate about a point. Let $\mathbf{a}$, $\mathbf{b}$, $\mathbf{c}$, and $\mathbf{d}$ be vector functions; and ϕ and ψ be scalar functions. Then,

1) $\mathbf{a}\cdot\mathbf{b}\times\mathbf{c} = \mathbf{b}\cdot\mathbf{c}\times\mathbf{a} = \mathbf{c}\cdot\mathbf{a}\times\mathbf{b}$.
2) $\mathbf{a}\times(\mathbf{b}\times\mathbf{c}) = (\mathbf{a}\cdot\mathbf{c})\mathbf{b} - (\mathbf{a}\cdot\mathbf{b})\mathbf{c}$.
3) $(\mathbf{a}\times\mathbf{b})\cdot(\mathbf{c}\times\mathbf{d}) = (\mathbf{a}\cdot\mathbf{c})(\mathbf{b}\cdot\mathbf{d}) - (\mathbf{a}\cdot\mathbf{d})(\mathbf{b}\cdot\mathbf{c})$.
4) $\nabla(\phi\psi) = \phi\nabla\psi + \psi\nabla\phi$.
5) $\nabla(\mathbf{a}\cdot\mathbf{b}) = (\mathbf{a}\cdot\nabla)\mathbf{b} + (\mathbf{b}\cdot\nabla)\mathbf{a} + \mathbf{a}\times(\nabla\times\mathbf{b}) + \mathbf{b}\times(\nabla\times\mathbf{a})$.
6) $\nabla\cdot(\phi\mathbf{a}) = \mathbf{a}\cdot\nabla\phi + \phi\nabla\cdot\mathbf{a}$.
7) $\nabla\cdot(\mathbf{a}\times\mathbf{b}) = \mathbf{b}\cdot\nabla\times\mathbf{a} - \mathbf{a}\cdot\nabla\times\mathbf{b}$.

Foundations of Antenna Radiation Theory: Eigenmode Analysis, First Edition. Wen Geyi.

Published 2023 by John Wiley & Sons, Inc.

8) $\nabla \times (\phi \mathbf{a}) = \nabla \phi \times \mathbf{a} + \phi \nabla \times \mathbf{a}$.

9) $\nabla \times (\mathbf{a} \times \mathbf{b}) = \mathbf{a} \nabla \cdot \mathbf{b} - \mathbf{b} \nabla \cdot \mathbf{a} + (\mathbf{b} \cdot \nabla)\mathbf{a} - (\mathbf{a} \cdot \nabla)\mathbf{b}$.

10) $\nabla \times \nabla \times \mathbf{a} = \nabla \nabla \cdot \mathbf{a} - \nabla^2 \mathbf{a}$.

Let V be a volume bounded by the closed surface S and $\mathbf{u}_n$ be the unit outward normal of S. Then, the Gauss theorems hold

1) $\int_V \nabla \phi dV = \int_S \phi \mathbf{u}_n dS$.

2) $\int_V \nabla \cdot \mathbf{a} dV = \int_S \mathbf{u}_n \cdot \mathbf{a} dS$.

3) $\int_V \nabla \times \mathbf{a} dV = \int_S \mathbf{u}_n \times \mathbf{a} dS$.

Let S be an unclosed surface bounded by the contour Γ. Then, Stokes theorems hold

1) $\int_S \mathbf{u}_n \times \nabla \phi dS = \int_\Gamma \phi \mathbf{u}_l d\Gamma$,

2) $\int_S \mathbf{u}_n \cdot \nabla \times \mathbf{a} dS = \int_\Gamma \mathbf{a} \cdot \mathbf{u}_l d\Gamma$,

where $\mathbf{u}_l$ is the unit tangent vector along Γ in the positive sense with respect to $\mathbf{u}_n$.

Appendix B

Dyadic Analysis

A **dyad** is the dyadic product of two vectors **a** and **b**, denoted by **ab**. A **dyadic** $\overleftrightarrow{\mathbf{A}}$ is the sum of some dyads. A dyadic $\overleftrightarrow{\mathbf{A}}$ has nine components, and in the rectangular coordinate system, it can be written as

$$\begin{aligned}\overleftrightarrow{\mathbf{A}} &= a_{xx}\mathbf{u}_x\mathbf{u}_x + a_{xy}\mathbf{u}_x\mathbf{u}_y + a_{xz}\mathbf{u}_x\mathbf{u}_z + a_{yx}\mathbf{u}_y\mathbf{u}_x \\ &\quad + a_{yy}\mathbf{u}_y\mathbf{u}_y + a_{yz}\mathbf{u}_y\mathbf{u}_z + a_{zx}\mathbf{u}_z\mathbf{u}_x + a_{zy}\mathbf{u}_z\mathbf{u}_y + a_{zz}\mathbf{u}_z\mathbf{u}_z,\end{aligned}$$

where $\mathbf{u}_i (i = x, y, z)$ denote the unit vectors along the coordinate axis. The above can be rewritten as

$$\begin{aligned}\overleftrightarrow{\mathbf{A}} &= \left(a_{xx}\mathbf{u}_x + a_{yx}\mathbf{u}_y + a_{zx}\mathbf{u}_z\right)\mathbf{u}_x \\ &\quad + \left(a_{xy}\mathbf{u}_x + a_{yy}\mathbf{u}_y + a_{zy}\mathbf{u}_z\right)\mathbf{u}_y \\ &\quad + \left(a_{xz}\mathbf{u}_x + a_{yz}\mathbf{u}_y + a_{zz}\mathbf{u}_z\right)\mathbf{u}_z \\ &= \mathbf{A}'_x\mathbf{u}_x + \mathbf{A}'_y\mathbf{u}_y + \mathbf{A}'_z\mathbf{u}_z,\end{aligned} \quad \text{or} \quad \begin{aligned}\overleftrightarrow{\mathbf{A}} &= \mathbf{u}_x\left(a_{xx}\mathbf{u}_x + a_{xy}\mathbf{u}_y + a_{xz}\mathbf{u}_z\right) \\ &\quad + \mathbf{u}_y\left(a_{yx}\mathbf{u}_x + a_{yy}\mathbf{u}_y + a_{yz}\mathbf{u}_z\right) \\ &\quad + \mathbf{u}_z\left(a_{zx}\mathbf{u}_x + a_{zy}\mathbf{u}_y + a_{zz}\mathbf{u}_z\right) \\ &= \mathbf{u}_x\mathbf{A}_x + \mathbf{u}_y\mathbf{A}_y + \mathbf{u}_z\mathbf{A}_z.\end{aligned}$$

The operations between vector and dyad are defined by

1) $(\mathbf{ab})\cdot\mathbf{c} = \mathbf{a}(\mathbf{b}\cdot\mathbf{c})$.
2) $\mathbf{c}\cdot(\mathbf{ab}) = (\mathbf{c}\cdot\mathbf{a})\mathbf{b}$.
3) $\mathbf{ab}\cdot\mathbf{cd} = \mathbf{a}(\mathbf{b}\cdot\mathbf{c})\mathbf{d}$.
4) $(\mathbf{ab})\times\mathbf{c} = \mathbf{a}(\mathbf{b}\times\mathbf{c})$.
5) $\mathbf{a}\times(\mathbf{bc}) = (\mathbf{a}\times\mathbf{b})\mathbf{c}$.

The **double product** of two dyads is defined by $\mathbf{ab}:\mathbf{cd} = (\mathbf{a}\cdot\mathbf{d})(\mathbf{b}\cdot\mathbf{c})$. The **transpose** of a dyadic $\overleftrightarrow{\mathbf{A}}$ is denoted by $\overleftrightarrow{\mathbf{A}}^T$, and is defined by $\overleftrightarrow{\mathbf{A}}\cdot\mathbf{b} = \mathbf{b}\cdot\overleftrightarrow{\mathbf{A}}^T$. A dyadic is called **symmetric** (or **antisymmetric**) if $\overleftrightarrow{\mathbf{A}}\cdot\mathbf{b} = \mathbf{b}\cdot\overleftrightarrow{\mathbf{A}}$ (or $\overleftrightarrow{\mathbf{A}}\cdot\mathbf{b} = -\mathbf{b}\cdot\overleftrightarrow{\mathbf{A}}$) for any vector **b**. The **identity dyadic** $\overleftrightarrow{\mathbf{I}}$ is defined by $\overleftrightarrow{\mathbf{I}}\cdot\mathbf{b} = \mathbf{b}\cdot\overleftrightarrow{\mathbf{I}} = \mathbf{b}$, which is given by $\overleftrightarrow{\mathbf{I}} = \mathbf{u}_x\mathbf{u}_x + \mathbf{u}_y\mathbf{u}_y + \mathbf{u}_z\mathbf{u}_z$ in the rectangular coordinate system.

Foundations of Antenna Radiation Theory: Eigenmode Analysis, First Edition. Wen Geyi.

Published 2023 by John Wiley & Sons, Inc.

The multiplicative relations are given by

1) $(\mathbf{a}\times\mathbf{b})\cdot\overleftrightarrow{\mathbf{A}} = \mathbf{a}\cdot\left(\mathbf{b}\times\overleftrightarrow{\mathbf{A}}\right) = -\mathbf{b}\cdot\left(\mathbf{a}\times\overleftrightarrow{\mathbf{A}}\right).$

2) $\left(\overleftrightarrow{\mathbf{A}}\times\mathbf{a}\right)\cdot\mathbf{b} = \overleftrightarrow{\mathbf{A}}\cdot(\mathbf{a}\times\mathbf{b}) = -\left(\overleftrightarrow{\mathbf{A}}\times\mathbf{b}\right)\cdot\mathbf{a}.$

3) $\mathbf{a}\times\left(\mathbf{b}\times\overleftrightarrow{\mathbf{A}}\right) = \mathbf{b}\left(\mathbf{a}\cdot\overleftrightarrow{\mathbf{A}}\right) - \overleftrightarrow{\mathbf{A}}(\mathbf{a}\cdot\mathbf{b}).$

4) $(\mathbf{ab}-\mathbf{ba})\cdot\mathbf{c} = (\mathbf{b}\times\mathbf{a})\times\mathbf{c}.$

The basic differential operations are defined by

1) $\nabla = \mathbf{u}_x\frac{\partial}{\partial x} + \mathbf{u}_y\frac{\partial}{\partial y} + \mathbf{u}_z\frac{\partial}{\partial z}.$

2) $\nabla\mathbf{a} = \mathbf{u}_x\frac{\partial\mathbf{a}}{\partial x} + \mathbf{u}_y\frac{\partial\mathbf{a}}{\partial y} + \mathbf{u}_z\frac{\partial\mathbf{a}}{\partial z} = \nabla a_x\mathbf{u}_x + \nabla a_y\mathbf{u}_y + \nabla a_z\mathbf{u}_z.$

3) $\mathbf{a}\nabla = \frac{\partial\mathbf{a}}{\partial x}\mathbf{u}_x + \frac{\partial\mathbf{a}}{\partial y}\mathbf{u}_y + \frac{\partial\mathbf{a}}{\partial z}\mathbf{u}_z.$

4) $\nabla\cdot\overleftrightarrow{\mathbf{A}} = (\nabla\cdot\mathbf{A}'_x)\mathbf{u}_x + \left(\nabla\cdot\mathbf{A}'_y\right)\mathbf{u}_y + (\nabla\cdot\mathbf{A}'_z)\mathbf{u}_z = \frac{\partial\mathbf{A}_x}{\partial x} + \frac{\partial\mathbf{A}_y}{\partial y} + \frac{\partial\mathbf{A}_z}{\partial z}.$

5) $\overleftrightarrow{\mathbf{A}}\cdot\nabla = \nabla\cdot\overleftrightarrow{\mathbf{A}}^T = \mathbf{u}_x\nabla\cdot\mathbf{A}_x + \mathbf{u}_y\nabla\cdot\mathbf{A}_y + \mathbf{u}_z\nabla\cdot\mathbf{A}_z.$

6) $$\nabla\times\overleftrightarrow{\mathbf{A}} = (\nabla\times\mathbf{A}'_x)\mathbf{u}_x + \left(\nabla\times\mathbf{A}'_y\right)\mathbf{u}_y + (\nabla\times\mathbf{A}'_z)\mathbf{u}_z$$
$$= \mathbf{u}_x\left(\frac{\partial\mathbf{A}_z}{\partial y} - \frac{\partial\mathbf{A}_y}{\partial z}\right) + \mathbf{u}_y\left(\frac{\partial\mathbf{A}_x}{\partial z} - \frac{\partial\mathbf{A}_z}{\partial x}\right) + \mathbf{u}_z\left(\frac{\partial\mathbf{A}_y}{\partial x} - \frac{\partial\mathbf{A}_x}{\partial y}\right).$$

7) $\nabla^2\overleftrightarrow{\mathbf{A}} = \frac{\partial^2\overleftrightarrow{\mathbf{A}}}{\partial x^2} + \frac{\partial^2\overleftrightarrow{\mathbf{A}}}{\partial y^2} + \frac{\partial^2\overleftrightarrow{\mathbf{A}}}{\partial z^2} = \nabla\nabla\cdot\overleftrightarrow{\mathbf{A}} - \nabla\times\nabla\times\overleftrightarrow{\mathbf{A}}.$

For a scalar φ, the following differential relations hold

1) $\nabla(\varphi\mathbf{b}) = \nabla\varphi\mathbf{b} - \varphi\nabla\mathbf{b}.$

2) $\nabla\cdot\left(\varphi\overleftrightarrow{\mathbf{A}}\right) = \nabla\varphi\cdot\overleftrightarrow{\mathbf{A}} + \varphi\nabla\cdot\overleftrightarrow{\mathbf{A}}.$

3) $\nabla\times\left(\varphi\overleftrightarrow{\mathbf{A}}\right) = \nabla\varphi\times\overleftrightarrow{\mathbf{A}} + \varphi\nabla\times\overleftrightarrow{\mathbf{A}}.$

For a finite region V bounded by S, the following integral relations are valid

1) $\int_V \nabla\mathbf{a}dV = \int_S \mathbf{u}_n\mathbf{a}dS.$

2) $\int_V \nabla\cdot\overleftrightarrow{\mathbf{A}}dV = \int_S \mathbf{u}_n\cdot\overleftrightarrow{\mathbf{A}}dS.$

3) $\int_V \nabla\times\overleftrightarrow{\mathbf{A}}dV = \int_S \mathbf{u}_n\times\overleftrightarrow{\mathbf{A}}dS.$

4) $\int_V \left[(\nabla\times\nabla\times\mathbf{b})\cdot\overleftrightarrow{\mathbf{A}} - \mathbf{b}\cdot\left(\nabla\times\nabla\times\overleftrightarrow{\mathbf{A}}\right)\right]dV = \int_S \mathbf{u}_n\cdot\left[\mathbf{b}\times\nabla\times\overleftrightarrow{\mathbf{A}} - (\nabla\times\mathbf{b})\times\overleftrightarrow{\mathbf{A}}\right]dS.$

Appendix C

SI Unit System

This book uses SI unit system in which mass M is measured in kilograms, length L in meters, time T in seconds, and charge Q in coulombs. Table C.1 lists the electromagnetic quantities, their symbols, dimensions, and SI unit.

Table C.1 Electromagnetic quantities.

Quantity	Symbol	SI unit	Dimensions
Charge	q	Coulomb	Q
Current	I	Ampere	Q/T
Resistance	R	Ohm	ML^2/TQ^2
Inductance	L	Henry	ML^2/Q^2
Capacitance	C	Farad	Q^2T^2/ML^2
Charge density	ρ	Coulomb/cubic meter	Q/L^3
Current density	$\mathbf{J}$	Ampere/square meter	Q/TL^2
Electric field intensity	$\mathbf{E}$	Volt/meter	ML/QT^2
Electric displacement	$\mathbf{D}$	Coulomb/square meter	Q/L^2
Electric dipole moment	$\mathbf{p}$	Coulomb-meter	QL
Polarization vector	$\mathbf{P}$	Coulomb/square meter	Q/L^2
Magnetic field intensity	$\mathbf{H}$	Ampere/meter	Q/TL
Magnetic induction	$\mathbf{B}$	Weber/square meter	M/QT
Magnetic dipole moment	$\mathbf{m}$	Ampere-square meter	QL^2/T
Magnetization vector	$\mathbf{M}$	Ampere/meter	Q/TL
Vector potential	$\mathbf{A}$	Weber/Henry	ML/QT
Scalar potential	ϕ	Volt	ML^2/QT^2

(*Continued*)

Foundations of Antenna Radiation Theory: Eigenmode Analysis, First Edition. Wen Geyi.

Published 2023 by John Wiley & Sons, Inc.

Table C.1 (Continued)

Quantity	Symbol	SI unit	Dimensions
Conductivity	σ	Mho/meter	Q^2T/ML^3
Permeability	μ	Henry/meter	ML/Q^2
Permittivity	ε	Farad/meter	Q^2T^2/ML^3
Frequency	f	Hertz	$1/T$
Force	$\mathbf{F}$	Newton	ML/T^2
Energy	W	Joule	ML^2/T^2
Power	P	Watt	ML^2/T^3
Poynting vector	$\mathbf{S}$	Watt/square meter	M/T^3

Table C.2 shows the physical constants.

Table C.2 Physical constants.

Quantity	Symbol	Value
Speed of light	c	3.00×10^8 meter/second
Elementary charge	e	1.60×10^{-19} Coulomb
Electron mass	m_e	9.11×10^{-31} kilogram
Proton mass	m_p	1.67×10^{-27} kilogram
Permeability constant	μ_0	1.26×10^{-6} Henry/meter
Permittivity constant	ε_0	8.85×10^{-12} Farad/meter
Gravitational constant	G	6.67×10^{-11} Newton · square meter/square kilogram
Planck's constant	h	6.63×10^{-34} Joule · second
Boltzmann constant	k	1.3807×10^{-23} Joule/Kelvin

Appendix D

Unified Theory for Fields (UTF)

From the outset Maxwell's theory excelled all others in elegance and in the abundance of the relations between the various phenomena which it included.
– Heinrich Hertz (German physicist, 1857–1894)

Although Maxwell's equations are relatively simple, they daringly reorganize our perception of nature, unifying electricity and magnetism and linking geometry, topology and physics. They are essential to understanding the surrounding world. And as the first field equations, they not only showed scientists a new way of approaching physics but also took them on the first step towards a unification of the fundamental forces of nature.
Robert P. Crease (American author, 1957–)

One of the cornerstones of modern physics is special relativity, which has its origins deeply rooted in electrodynamics. Although the theory of electrodynamics preceded that of special relativity, tremendous efforts have been made in deriving Maxwell equations from special relativity and a few assumptions. A brief account of the work on the derivation of Maxwell equations has been given by Heras and Pierce [1, 2]. Page derived Maxwell equations from Coulomb's law and special relativity [3] and later refined his work with Adams in [4]. In their approach, only the electric field is fundamental, and the magnetic field is just a derived quantity. Their work was criticized by Frisch and Wilets for using an overspecialized model based on an emission theory of lines of forces [5]. Frisch and Wilets derived Maxwell equations by applying the force transformations of special relativity to Gauss's law with the assumption that the electromagnetic (EM) signals propagate with light speed and the electric force on a test charge is velocity independent and does

Foundations of Antenna Radiation Theory: Eigenmode Analysis, First Edition. Wen Geyi.

Published 2023 by John Wiley & Sons, Inc.

not depend on the time derivatives of the source position of order higher than the acceleration. Elliott derived Maxwell equations from the Lorentz transformation and the laws for both electric and magnetic statistic fields [6, 7]. Rosser showed that the EM fields of a moving charge with constant velocity can be obtained by using Coulomb's law and the transformations for coordinates, velocities, and forces in the special relativity [8, 9]. Tessman generalized Rosser's work to an accelerating charge [10]. Pedagogical derivations of Maxwell equations were given by Krefetz [11] and Kobe [12, 13], who noted that the deduction of Maxwell equations solely from the Lorentz transformation and Coulomb's law was criticized by Feynman et al. and Jackson as some additional assumptions were always necessary [14, 15]. A different derivation of Maxwell equations was tried by Schwinger et al., in which Maxwell equations were derived from Coulomb's law, the Galilean transformation, and an assumption that the electric field satisfies the wave equation [16].

A different approach for deriving Maxwell equations is to use Newton's law as the starting point. Dyson published a proof, showed to him by Feynman in 1948, that the Maxwell equations could be derived from Newton's law and the commutation relations between coordinates and velocities [17]. Feynman's proof has triggered many discussions and comments [18, 19].

The similarity between Newton's law of gravity and Coulomb's law of electrodynamics motivates that Maxwell-like equations exist for gravity and can be rigorously derived from Newton's law of gravity and special relativity. The major differences are that the sources of EM fields consist of positive and negative charges, which may attract or repel each other, while the sources of the gravitational field are masses which are always attractive. As early as 1893, Heaviside tried to generalize Newton's static gravity to a time-dependent system by postulating a gravitational analog of magnetic field, called **gravitomagnetic field**, to obtain the time-dependent Maxwell-like equations for gravity [20]. Heaviside's work was further explored by Jefimenko, where many interesting results, in analogy to the Maxwell equations, have been derived from the time-dependent equations for gravitational field [21]. Some of these results were believed to be the exclusive consequence of the theory of general relativity. The existence of a gravitational analog of the magnetic field can be easily demonstrated in many cases. By considering a point mass moving perpendicularly toward a long line of rest mass and using the transformation for forces in special relativity, Bedford and Krumm showed that, in addition to the radial gravitational attraction on the point mass due to the line of rest mass, there is another acting force on the point mass, which is the gravitomagnetic field and is parallel to the line mass [22]. The existence of gravitomagnetic field from a uniformly moving mass was also predicted by Kolbenstvedt from special relativity and gravitational time dilation [23]. Campbell and Morgan introduced an electric-type tidal gravitational field strength and a magnetic-type gravitational field strength, both being tensors which satisfy Maxwell-like equations [24]. In fact, the introduction of a gravitomagnetic field is inevitable if Newtonian gravitation is assumed to be compatible with the special relativity. By using analogies with EM radiation, the properties of gravitational radiation emitted by orbiting binary objects were studied by Hilborn, which produces the same results as those obtained from the linear version of general relativity [25].

In this appendix, a unified theory for fields (UTF) will be presented. The UTF unveils that an ontological field (scalar or vectorial), defined as a static field in an inertial system (static system), will emerge as two vector fields **U** and **V** in an inertial system in relative motion with respect to the static system, which satisfy the Maxwell-like equations. This implies that an EM-like radiation, traveling at light speed in free space, will be perceived by an observer who is in motion relative to the ontological field. The UTF is built on the basis of the theory of special relativity. The line of reasoning of UTF is inspired by the Helmholtz theorem, which states that a vector field is determined by its curl and divergence. In order to find how a vector field changes in different inertial systems, one only needs to examine how the curl and divergence of the vector field transform. The Maxwell-like equations are a set of four equations that, respectively, define the curls and divergences for the vector fields **U** and **V** in an inertial system. As applications, the Maxwell equations for EM field and the Maxwell-like equations for the gravitational field are derived from the UTF, and the latter are also derived from the Einstein field equations in the theory of general relativity. Some universal laws of nature are shown to be derivable from the UTF.

D.1 Lorentz Transformation

Consider two reference systems S and S', respectively, called **laboratory system** and **co-moving system** (or **static system**). The coordinate axes of the two systems are parallel and oriented so that the frame S' is moving with a velocity $\mathbf{v}$ as viewed from S, as illustrated in Figure D.1. The two origins o and o' are assumed to coincide at $t = t' = 0$. The two reference systems are then related by the **Lorentz transformation** [26]

$$\mathbf{r} = \overleftrightarrow{\boldsymbol{\alpha}} \cdot \mathbf{r}' + \gamma \boldsymbol{\beta} ct', \quad ct = \gamma(ct' + \boldsymbol{\beta} \cdot \mathbf{r}'), \tag{D.1}$$

where $\boldsymbol{\beta} = \mathbf{v}/c$, $\gamma = 1/\sqrt{1-\beta^2}$, $\beta = |\mathbf{v}|/c$, and $\overleftrightarrow{\boldsymbol{\alpha}}$ is a dyadic defined by

$$\overleftrightarrow{\boldsymbol{\alpha}} = \overleftrightarrow{\mathbf{I}} + (\gamma - 1)\frac{\boldsymbol{\beta}\boldsymbol{\beta}}{\beta^2}, \tag{D.2}$$

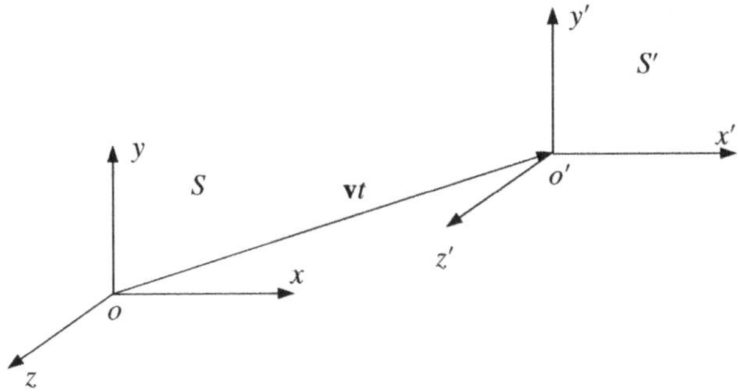

Figure D.1 Two reference systems in relative motion.

and $\overleftrightarrow{\mathbf{I}}$ is the identity dyadic. It can be seen from (D.1) that it is the relative velocity $\boldsymbol{\beta}$ (relative motion) that connects the space and time. The inverse transformation of (D.1) is

$$\mathbf{r}' = \overleftrightarrow{\boldsymbol{\alpha}} \cdot \mathbf{r} - \gamma \boldsymbol{\beta} ct, \quad ct' = \gamma(ct - \boldsymbol{\beta} \cdot \mathbf{r}). \tag{D.3}$$

The Lorentz transformation relates the positions and times of occurrence of a single event, which are measured by two observers in different reference systems. As a first-order approximation, if $\beta^2 << 1$, one may have $\gamma \approx 1$ and $\overleftrightarrow{\boldsymbol{\alpha}} \approx \overleftrightarrow{\mathbf{I}}$. In this case, the Lorentz transformation (D.3) reduces to

$$\mathbf{r}' = \mathbf{r} - \mathbf{v}t, \quad ct' = ct - \boldsymbol{\beta} \cdot \mathbf{r}. \tag{D.4}$$

This becomes the **Galilean transform**. A function $f'(\mathbf{r}', t')$ in the static system S' can be viewed as a function of $(\mathbf{r}, t)$ in the laboratory system S through the Lorentz transformation (D.3). It is easy to show that the time and spatial derivatives are transformed according to

$$\frac{\partial}{\partial ct'} = \gamma \left(\frac{\partial}{\partial ct} + \boldsymbol{\beta} \cdot \nabla \right), \tag{D.5}$$

$$\nabla' = \left(\overleftrightarrow{\boldsymbol{\alpha}} \cdot \nabla + \gamma \boldsymbol{\beta} \frac{\partial}{\partial ct} \right). \tag{D.6}$$

For the Galilean transform, (D.5) and (D.6) reduce to

$$\frac{\partial}{\partial t'} = \frac{\partial}{\partial t} + \mathbf{v} \cdot \nabla, \quad \nabla' = \nabla. \tag{D.7}$$

Some important properties of the dyadic $\overleftrightarrow{\boldsymbol{\alpha}}$ that will be used later are listed below

$$\begin{aligned}
&\overleftrightarrow{\boldsymbol{\alpha}}^{-1} = \overleftrightarrow{\boldsymbol{\alpha}} - \gamma\boldsymbol{\beta}\boldsymbol{\beta}, && \overleftrightarrow{\boldsymbol{\alpha}}^{2} = \overleftrightarrow{\mathbf{I}} + \gamma^2\boldsymbol{\beta}\boldsymbol{\beta},\\
&\left(\overleftrightarrow{\boldsymbol{\alpha}}^{-1}\right)^2 = \overleftrightarrow{\mathbf{I}} - \boldsymbol{\beta}\boldsymbol{\beta}, && \overleftrightarrow{\boldsymbol{\alpha}}\cdot\boldsymbol{\beta} = \boldsymbol{\beta}\cdot\overleftrightarrow{\boldsymbol{\alpha}} = \gamma\boldsymbol{\beta},\\
&\overleftrightarrow{\boldsymbol{\alpha}}^{-1}\cdot\boldsymbol{\beta} = \boldsymbol{\beta}\cdot\overleftrightarrow{\boldsymbol{\alpha}}^{-1} = \gamma^{-1}\boldsymbol{\beta}, && |\overleftrightarrow{\boldsymbol{\alpha}}\cdot\mathbf{B}|^2 = |\mathbf{B}|^2 + \gamma^2|\boldsymbol{\beta}\cdot\mathbf{B}|^2.
\end{aligned} \tag{D.8}$$

D.2 Maxwell-Like Equations Derived from an Ontological Vector Field

In this appendix, the primed and unprimed symbols will be used to designate the field quantities and operations in the static system S' and laboratory system S, respectively. For convenience, the time variable t' will be explicitly included even for the time-independent (static) quantities in S'. Suppose that $\mathbf{U}'(\mathbf{r}', t')$ is an arbitrary static vector field in the system S'. According to Helmholtz theorem (see Section 1.7), the vector field $\mathbf{U}'(\mathbf{r}', t')$ is uniquely determined by its curl and divergence

$$\mathbf{U}'(\mathbf{r}',t') = -\nabla'\int_{R^3}\frac{\nabla''\cdot\mathbf{U}'(\mathbf{r}'',t')}{4\pi|\mathbf{r}'-\mathbf{r}''|}dV(\mathbf{r}'') + \nabla'\times\int_{R^3}\frac{\nabla''\times\mathbf{U}'(\mathbf{r}'',t')}{4\pi|\mathbf{r}'-\mathbf{r}''|}dV(\mathbf{r}'') \tag{D.9}$$

if $\mathbf{U}'$ decreases faster than $1/r'$ as $r' \to \infty$. The static field $\mathbf{U}'$ will be referred to as an **ontological vector field**. In order to find out how a vector field changes in different reference systems, it will be adequate to consider how the curl and divergence of the vector field transform under the Lorentz transformation (D.1). The curl and divergence of the ontological field $\mathbf{U}'$ will be, respectively, denoted by $-\mathbf{j}'_m$ and $\chi^{-1}\rho'$, and they are simply regarded as **ontological sources**. The constant χ is introduced for the applications to be discussed later. Then,

$$\nabla'\times\mathbf{U}'(\mathbf{r}',t') = -\mathbf{j}'_m(\mathbf{r}',t'), \tag{D.10}$$

$$\nabla'\cdot\mathbf{U}'(\mathbf{r}',t') = \frac{1}{\chi}\rho'(\mathbf{r}',t'). \tag{D.11}$$

First, the transformation of the curl equation (D.10) will be considered. Combining (D.6) and (D.10) gives

$$\left(\overleftrightarrow{\boldsymbol{\alpha}}\cdot\nabla + \gamma\boldsymbol{\beta}\frac{\partial}{\partial ct}\right)\times\mathbf{U}' = -\mathbf{j}'_m.$$

Dot-multiplying both sides by $\overleftrightarrow{\boldsymbol{\alpha}}$ yields

$$\overleftrightarrow{\boldsymbol{\alpha}} \cdot \left(\overleftrightarrow{\boldsymbol{\alpha}} \cdot \nabla \times \mathbf{U}'\right) + \gamma \overleftrightarrow{\boldsymbol{\alpha}} \cdot \left(\boldsymbol{\beta} \frac{\partial}{\partial ct} \times \mathbf{U}'\right) = -\overleftrightarrow{\boldsymbol{\alpha}} \cdot \mathbf{j}'_m. \tag{D.12}$$

The first term on the left-hand side can be expanded as follows:

$$\begin{aligned}\overleftrightarrow{\boldsymbol{\alpha}} \cdot \left(\overleftrightarrow{\boldsymbol{\alpha}} \cdot \nabla \times \mathbf{U}'\right) &= \left[\overleftrightarrow{\mathbf{I}} + (\gamma - 1)\frac{\boldsymbol{\beta\beta}}{\beta^2}\right] \cdot \left\{\left[\overleftrightarrow{\mathbf{I}} + (\gamma - 1)\frac{\boldsymbol{\beta\beta}}{\beta^2}\right] \cdot \nabla\right\} \times \mathbf{U}' \\ &= \nabla \times \mathbf{U}' + \frac{\gamma - 1}{\beta^2}(\boldsymbol{\beta} \cdot \nabla)(\boldsymbol{\beta} \times \mathbf{U}') + \frac{(\gamma - 1)}{\beta^2}\boldsymbol{\beta\beta} \cdot \nabla \times \mathbf{U}' \\ &\quad + \frac{(\gamma - 1)^2}{\beta^4}\boldsymbol{\beta\beta} \cdot (\boldsymbol{\beta} \cdot \nabla)(\boldsymbol{\beta} \times \mathbf{U}').\end{aligned} \tag{D.13}$$

Considering $\boldsymbol{\beta} \cdot (\boldsymbol{\beta} \cdot \nabla)(\boldsymbol{\beta} \times \mathbf{U}') = 0$ and $\boldsymbol{\beta} \cdot \nabla \times \mathbf{U}' = -\nabla \cdot (\boldsymbol{\beta} \times \mathbf{U}')$, (D.13) becomes

$$\overleftrightarrow{\boldsymbol{\alpha}} \cdot \left(\overleftrightarrow{\boldsymbol{\alpha}} \cdot \nabla \times \mathbf{U}'\right) = \nabla \times \left[\mathbf{U}' - \frac{\gamma - 1}{\beta^2}\left(\boldsymbol{\beta\beta} \cdot \mathbf{U}' - \beta^2 \mathbf{U}'\right)\right] = \nabla \times \left(\gamma \overleftrightarrow{\boldsymbol{\alpha}}^{-1} \cdot \mathbf{U}'\right). \tag{D.14}$$

The second term on the left-hand side of (D.12) can be written as

$$\gamma \overleftrightarrow{\boldsymbol{\alpha}} \cdot \left(\boldsymbol{\beta} \frac{\partial}{\partial ct} \times \mathbf{U}'\right) = \gamma \frac{\partial}{\partial ct}\left[\overleftrightarrow{\boldsymbol{\alpha}} \cdot (\boldsymbol{\beta} \times \mathbf{U}')\right] = \gamma \frac{\partial}{\partial ct}\boldsymbol{\beta} \times \mathbf{U}'. \tag{D.15}$$

On account of (D.14) and (D.15), (D.12) turns out to be

$$\nabla \times \left(\gamma \overleftrightarrow{\boldsymbol{\alpha}}^{-1} \cdot \mathbf{U}'\right) = -\frac{\partial}{\partial t}\left(\frac{\gamma}{c}\boldsymbol{\beta} \times \mathbf{U}'\right) - \overleftrightarrow{\boldsymbol{\alpha}} \cdot \mathbf{j}'_m. \tag{D.16}$$

Introducing the new fields

$$\mathbf{U}(\mathbf{r}, t) = \gamma \overleftrightarrow{\boldsymbol{\alpha}}^{-1} \cdot \mathbf{U}'(\mathbf{r}', t'), \tag{D.17}$$

$$\mathbf{V}(\mathbf{r}, t) = \chi c \boldsymbol{\beta} \times \mathbf{U}(\mathbf{r}, t), \tag{D.18}$$

and the new source

$$\mathbf{j}_m(\mathbf{r}, t) = \overleftrightarrow{\boldsymbol{\alpha}} \cdot \mathbf{j}'_m(\mathbf{r}', t'), \tag{D.19}$$

in the laboratory system S, (D.16) gives the curl of the vector field $\mathbf{U}$:

$$\nabla \times \mathbf{U}(\mathbf{r}, t) = -\tau \frac{\partial \mathbf{V}(\mathbf{r}, t)}{\partial t} - \mathbf{j}_m(\mathbf{r}, t), \tag{D.20}$$

where τ is defined by $\tau\chi = 1/c^2$. The curl equation (D.10) in the static system S' is now transformed into (D.20) in the laboratory system S. A new field $\mathbf{V}$ appears in

(D.20) due to relative motion, and it will be called the **co-vector field** of **U**. The vector field **U** will be called the **dominant vector field**. The source term $\mathbf{j}_m$ will be called **V-current density**, which is caused by the ontological source $\mathbf{j}'_m$.

Remark D.1 It follows from (D.17) and (D.18) that the ontological field $\mathbf{U}'$ is related to **U** and **V** by

$$\mathbf{U}'(\mathbf{r}', t') = \gamma \overleftrightarrow{\boldsymbol{\alpha}}^{-1} \cdot \mathbf{U}(\mathbf{r}, t) + \gamma c \tau \boldsymbol{\beta} \times \mathbf{V}(\mathbf{r}, t). \tag{D.21}$$

By (D.17) and the last equation of (D.8), one may find

$$|\mathbf{U}'(\mathbf{r}', t')|^2 = \gamma^{-2} |\mathbf{U}(\mathbf{r}, t)|^2 + c^{-2} |\mathbf{v} \cdot \mathbf{U}(\mathbf{r}, t)|^2. \tag{D.22}$$

□

Second, let us consider the transformation of divergence equation (D.11). Equation (D.11) can be rewritten as

$$\nabla' \cdot \gamma \mathbf{U}' = \frac{1}{\chi} \gamma \rho'. \tag{D.23}$$

On account of the relation

$$\begin{aligned} \left(\overleftrightarrow{\boldsymbol{\alpha}} \cdot \nabla'\right) \cdot \left(\gamma \overleftrightarrow{\boldsymbol{\alpha}}^{-1} \cdot \mathbf{U}'\right) &= \nabla' \cdot \gamma \mathbf{U}' + \gamma(\gamma - 1) \frac{1}{\beta^2} (\boldsymbol{\beta} \cdot \nabla')(\boldsymbol{\beta} \cdot \mathbf{U}') \\ &+ \frac{1-\gamma}{\beta^2} (\boldsymbol{\beta} \cdot \nabla')(\boldsymbol{\beta} \cdot \mathbf{U}') + (\gamma - 1) \frac{1-\gamma}{\beta^2} (\boldsymbol{\beta} \cdot \nabla')(\boldsymbol{\beta} \cdot \mathbf{U}') = \nabla' \cdot \gamma \mathbf{U}', \end{aligned}$$

and $\partial \mathbf{U}'/\partial t' = 0$, (D.23) can be expressed as

$$\left(\overleftrightarrow{\boldsymbol{\alpha}} \cdot \nabla' - \gamma \boldsymbol{\beta} \frac{\partial}{\partial c t'}\right) \cdot \left[\gamma \overleftrightarrow{\boldsymbol{\alpha}}^{-1} \cdot \mathbf{U}'(\mathbf{r}', t')\right] = \frac{1}{\chi} \rho(\mathbf{r}, t), \tag{D.24}$$

where

$$\rho(\mathbf{r}, t) = \gamma \rho'(\mathbf{r}', t') \tag{D.25}$$

is a source in the laboratory system S and will be called **U-charge density**. Equation (D.24) can be written as

$$\nabla \cdot \mathbf{U}(\mathbf{r}, t) = \frac{1}{\chi} \rho(\mathbf{r}, t). \tag{D.26}$$

Therefore, the curl and divergence of the vector field **U** are fully determined by the sources and the co-vector **V**. The next step is to determine the co-vector field **V**. According to Helmholtz theorem, one only needs to identify its curl and divergence. From (D.18), the curl of the co-vector field **V** is given by

$$\begin{aligned}\nabla \times \mathbf{V}(\mathbf{r},t) &= \chi c\boldsymbol{\beta}\nabla \cdot \mathbf{U}(\mathbf{r},t) - \chi c(\boldsymbol{\beta}\cdot\nabla)\mathbf{U}(\mathbf{r},t) \\ &= \mathbf{j}(\mathbf{r},t) - \chi c(\boldsymbol{\beta}\cdot\nabla)\mathbf{U}(\mathbf{r},t),\end{aligned} \tag{D.27}$$

where

$$\mathbf{j}(\mathbf{r},t) = c\boldsymbol{\beta}\rho(\mathbf{r},t) \tag{D.28}$$

is a source in the laboratory system S, and will be called **U-current density**. It follows from (D.5) and $\partial\mathbf{U}'/\partial t' = 0$ that

$$\left(\frac{\partial}{\partial t} + c\boldsymbol{\beta}\cdot\nabla\right)\gamma\mathbf{U}'(\mathbf{r}',t') = 0.$$

Dot-multiplying both sides by $\gamma\overleftrightarrow{\boldsymbol{\alpha}}^{-1}$ gives

$$\frac{\partial\mathbf{U}(\mathbf{r},t)}{\partial t} + (c\boldsymbol{\beta}\cdot\nabla)\mathbf{U}(\mathbf{r},t) = 0. \tag{D.29}$$

Combining (D.27) and (D.29) gives the curl of the co-vector field $\mathbf{V}$ in the laboratory system S:

$$\nabla \times \mathbf{V}(\mathbf{r},t) = \mathbf{j}(\mathbf{r},t) + \chi\frac{\partial\mathbf{U}(\mathbf{r},t)}{\partial t}. \tag{D.30}$$

To find the divergence of the co-vector field $\mathbf{V}$, one may take the divergence of (D.20) to get

$$\nabla\cdot\mathbf{j}_m(\mathbf{r},t) + \tau\frac{\partial\nabla\cdot\mathbf{V}(\mathbf{r},t)}{\partial t} = 0. \tag{D.31}$$

Let the divergence of the co-vector field $\mathbf{V}$ be denoted by an unknown quantity $\rho_m(\mathbf{r}, t)$:

$$\nabla\cdot\mathbf{V}(\mathbf{r},t) = \frac{1}{\tau}\rho_m(\mathbf{r},t). \tag{D.32}$$

Insertion of (D.32) into (D.31) yields

$$\nabla\cdot\mathbf{j}_m(\mathbf{r},t) = -\frac{\partial}{\partial t}\rho_m(\mathbf{r},t). \tag{D.33}$$

Therefore, the quantity ρ_m can be determined by the **V**-current density $\mathbf{j}_m$, and will be called **V-charge density**. The curl and divergence of the co-vector field **V** are now fully determined. Equation (D.33) will be referred to as **V-continuity equations**. The **V**-charge density ρ_m arises as result of the ontological source $\mathbf{j}'_m$. If the ontological field $\mathbf{U}'$ is curl free, the **V**-charge density ρ_m will not occur and the divergence of the co-vector field **V** will be zero.

In summary, an ontological vector field $\mathbf{U}'$ in a static system generates a dominant vector field $\mathbf{U}$, and a co-vector field $\mathbf{V}$ in the laboratory system S, which satisfy the Maxwell-like equations

$$\begin{aligned}
\nabla \times \mathbf{U}(\mathbf{r},t) &= -\mathbf{j}_m(\mathbf{r},t) - \tau\frac{\partial \mathbf{V}(\mathbf{r},t)}{\partial t},\\
\nabla \times \mathbf{V}(\mathbf{r},t) &= \mathbf{j}(\mathbf{r},t) + \chi\frac{\partial \mathbf{U}(\mathbf{r},t)}{\partial t},\\
\nabla \cdot \mathbf{U}(\mathbf{r},t) &= \frac{1}{\chi}\rho(\mathbf{r},t),\\
\nabla \cdot \mathbf{V}(\mathbf{r},t) &= \frac{1}{\tau}\rho_m(\mathbf{r},t),
\end{aligned} \tag{D.34}$$

where $\tau\chi = (1/c^2)$. From (D.34), it is readily found that

$$\nabla \cdot \mathbf{j}(\mathbf{r},t) = -\frac{\partial}{\partial t}\rho(\mathbf{r},t). \tag{D.35}$$

This will be called **U-continuity equations**. The **U-charge** contained in space is defined by

$$q = \int_{R^3} \rho dV. \tag{D.36}$$

Similarly, the **ontological charge** may be defined by

$$q' = \int_{R^3} \rho' dV'. \tag{D.37}$$

It is easy to show that charge is conserved in different coordinate systems

$$q = \int_{R^3} \rho dV = \int_{R^3} \gamma\rho'(\mathbf{r}',t')\frac{1}{\gamma}dV' = q', \tag{D.38}$$

where the relation $dV' = \gamma dV$ has been used. The solution of the Maxwell-like equations (D.34) will be called a **(U, V)-field**. The Maxwell-like equations (D.34) set up the rules of how the curls and divergences of the vector fields $\mathbf{U}$ and $\mathbf{V}$ are determined by each other as well as the sources.

Remark D.2 (Generalized Lenz's Law)
As stated by the Maxell-like equation (D.34), an increasing vector field $\mathbf{U}$ will generate the vector field $\mathbf{V}$ according to the right-hand rule (see Figure D.2a) and the latter will induce a vector field $\mathbf{U}$ according to the left-hand rule so that the

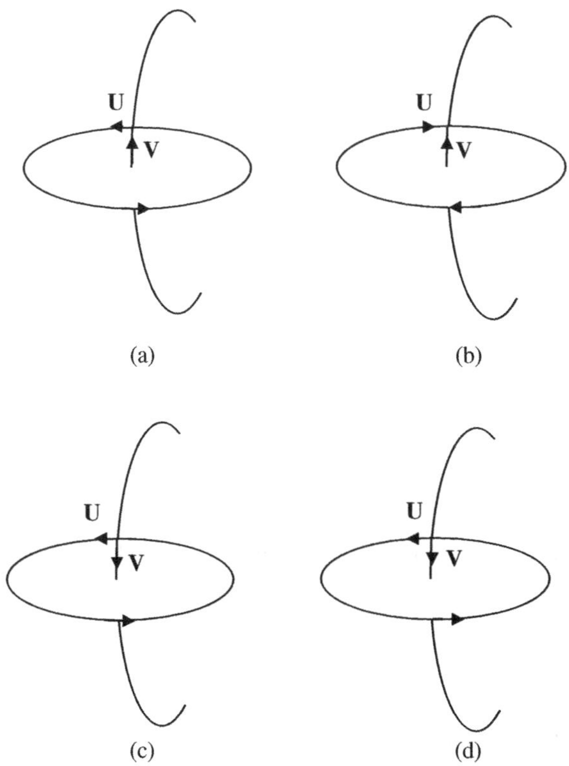

Figure D.2 (a) An increasing **U** generates **V**. (b) The increasing **V** induced by the increasing **U** generates a field that is opposite to the original **U**. (c) A decreasing **U** generates **V**. (d) The increasing **V** induced by the decreasing **U** generates a field that is in the same direction of the original **U**.

direction of the induced **U** is opposite to that of the original **U** that creates **V** (see Figure D.2b); a decreasing vector field **U** will generate the vector field **V** according to the left-hand rule (see Figure D.2c) and the latter will induce a vector field **U** according to the left-hand rule so that the direction of the induced **U** is in the same direction as the original **U** that creates **V** (see Figure D.2d). Similarly, an increasing vector field **V** will generate the vector field **U** according to the left-hand rule and the latter will induce a vector field **V** according to the right-hand rule so that the direction of the induced **V** is opposite to that of the original **V** that creates **U**; a decreasing vector field **V** will generate the vector field **U** according to the right-hand rule and the latter will induce a vector field **V** according to the right-hand rule so that the direction of the induced **V** is in the same direction of the original **V** that creates **U**. □

D.3 Field Equations Derived from an Ontological Scalar Field

A static scalar ϕ' in the system S' will be referred to as an **ontological scalar field**. Suppose that the scalar field ϕ' decreases faster than $1/r'$ as $r' \to \infty$. By means of the identity

$$\nabla^2 \frac{1}{4\pi|\mathbf{r}-\mathbf{r}'|} = -\delta(\mathbf{r}-\mathbf{r}'),$$

one may arrive at

$$\begin{aligned} \phi'(\mathbf{r}',t') &= \int_{R^3} \phi'(\mathbf{r}'',t')\delta(\mathbf{r}'-\mathbf{r}'')dV(\mathbf{r}'') \\ &= -\int_{R^3} \phi'(\mathbf{r}'',t')\nabla''^2 \frac{1}{4\pi|\mathbf{r}''-\mathbf{r}'|}dV(\mathbf{r}'') \\ &= \int_{R^3} \frac{\nabla''^2\phi'(\mathbf{r}'',t')}{4\pi|\mathbf{r}'-\mathbf{r}''|}dV(\mathbf{r}''). \end{aligned} \tag{D.39}$$

This indicates that an ontological scalar field ϕ' is determined by its Laplacian $\nabla'^2\phi$. A source function ρ' (ontological source) and a constant χ may be introduced such that

$$\nabla'^2\phi'(\mathbf{r}',t') = -\frac{1}{\chi}\rho'(\mathbf{r}',t'), \tag{D.40}$$

with

$$\frac{\partial\phi'(\mathbf{r}',t')}{\partial t'} = 0. \tag{D.41}$$

D.3.1 Wave Equation Derived from an Ontological Scalar Field

Substituting (D.6) into (D.40) yields

$$\left(\overleftrightarrow{\boldsymbol{\alpha}}\cdot\nabla + \gamma\boldsymbol{\beta}\frac{\partial}{\partial ct}\right)\cdot\left(\overleftrightarrow{\boldsymbol{\alpha}}\cdot\nabla\phi'(\mathbf{r}',t') + \gamma\boldsymbol{\beta}\frac{\partial}{\partial ct}\phi'(\mathbf{r}',t')\right) = -\frac{1}{\chi}\rho'(\mathbf{r}',t').$$

Expanding the left-hand side, the above equation becomes

$$\begin{aligned} &\left(\overleftrightarrow{\boldsymbol{\alpha}}\cdot\nabla\right)\cdot\overleftrightarrow{\boldsymbol{\alpha}}\cdot\nabla\phi' + \left(\overleftrightarrow{\boldsymbol{\alpha}}\cdot\nabla\right)\cdot\gamma\boldsymbol{\beta}\frac{\partial}{\partial ct}\phi' + \gamma\frac{\partial}{\partial ct}\boldsymbol{\beta}\cdot\overleftrightarrow{\boldsymbol{\alpha}}\cdot\nabla\phi' \\ &+ \gamma^2\frac{\partial}{\partial ct}\boldsymbol{\beta}\cdot\boldsymbol{\beta}\frac{\partial}{\partial ct}\phi' = -\frac{1}{\chi}\rho'. \end{aligned} \tag{D.42}$$

The first, the second, and the third terms on the left-hand side of (D.42) can be, respectively, written as

$$\begin{aligned}\left(\overleftrightarrow{\boldsymbol{\alpha}}\cdot\nabla\right)\cdot\overleftrightarrow{\boldsymbol{\alpha}}\cdot\nabla\phi' &= \left[\overleftrightarrow{\mathbf{I}}+(\gamma-1)\frac{\boldsymbol{\beta}\boldsymbol{\beta}}{\beta^2}\right]\cdot\nabla\cdot\left[\overleftrightarrow{\mathbf{I}}+(\gamma-1)\frac{\boldsymbol{\beta}\boldsymbol{\beta}}{\beta^2}\right]\cdot\nabla\phi' \\ &= \nabla^2\phi' + 2(\gamma-1)\frac{1}{\beta^2}(\boldsymbol{\beta}\cdot\nabla)^2\phi' + (\gamma-1)^2\frac{1}{\beta^2}(\boldsymbol{\beta}\cdot\nabla)^2\phi'.\end{aligned} \tag{D.43}$$

$$\begin{aligned}\left(\overleftrightarrow{\boldsymbol{\alpha}}\cdot\nabla\right)\cdot\gamma\boldsymbol{\beta}\frac{\partial}{\partial ct}\phi' &= \left[\nabla+(\gamma-1)\frac{\boldsymbol{\beta}}{\beta^2}\boldsymbol{\beta}\cdot\nabla\right]\cdot\gamma\boldsymbol{\beta}\frac{\partial}{\partial ct}\phi' \\ &= (\boldsymbol{\beta}\cdot\nabla)\gamma\frac{\partial}{\partial ct}\phi' + \gamma(\gamma-1)(\boldsymbol{\beta}\cdot\nabla)\frac{\partial}{\partial ct}\phi'.\end{aligned} \tag{D.44}$$

$$\begin{aligned}\gamma\frac{\partial}{\partial ct}\boldsymbol{\beta}\cdot\overleftrightarrow{\boldsymbol{\alpha}}\cdot\nabla\phi' &= \gamma\frac{\partial}{\partial ct}\boldsymbol{\beta}\cdot\left[\nabla\phi'+(\gamma-1)\frac{\boldsymbol{\beta}}{\beta^2}(\boldsymbol{\beta}\cdot\nabla)\phi'\right] \\ &= (\boldsymbol{\beta}\cdot\nabla)\gamma\frac{\partial\phi'}{\partial ct} + \gamma(\gamma-1)(\boldsymbol{\beta}\cdot\nabla)\frac{\partial\phi'}{\partial ct}.\end{aligned} \tag{D.45}$$

Upon substituting (D.43)–(D.45) into (D.42), one may find

$$\begin{aligned}&\nabla^2\phi' + 2(\gamma-1)\frac{1}{\beta^2}(\boldsymbol{\beta}\cdot\nabla)^2\phi' + (\gamma-1)^2\frac{1}{\beta^2}(\boldsymbol{\beta}\cdot\nabla)^2\phi' + 2(\boldsymbol{\beta}\cdot\nabla)\gamma\frac{\partial\phi'}{\partial ct} \\ &\quad + 2\gamma(\gamma-1)(\boldsymbol{\beta}\cdot\nabla)\frac{\partial\phi'}{\partial ct} + \gamma^2\frac{\partial}{\partial ct}\boldsymbol{\beta}\cdot\boldsymbol{\beta}\frac{\partial\phi'}{\partial ct} = -\frac{1}{\chi}\rho'.\end{aligned} \tag{D.46}$$

To simplify (D.46), one needs to use (D.41), which, after applying (D.5), can be written as

$$\left(\frac{\partial}{\partial ct}+\boldsymbol{\beta}\cdot\nabla\right)\phi'(\mathbf{r}',t') = 0. \tag{D.47}$$

By use of (D.47), (D.46) reduces to

$$\begin{aligned}&\nabla^2\phi' + 2(\gamma-1)\frac{1}{\beta^2c^2}\frac{\partial^2\phi'}{\partial t^2} + (\gamma-1)^2\frac{1}{\beta^2c^2}\frac{\partial^2\phi'}{\partial t^2} \\ &\quad - \frac{2\gamma}{c^2}\frac{\partial^2\phi'}{\partial t^2} - \frac{2\gamma(\gamma-1)}{c^2}\frac{\partial^2\phi'}{\partial t^2} + \frac{\gamma^2-1}{c^2}\frac{\partial^2\phi'}{\partial t^2} = -\frac{1}{\chi}\rho'(\mathbf{r}',t').\end{aligned} \tag{D.48}$$

After simplification, one may find that the scalar field $\phi(\mathbf{r}, t) \equiv \gamma\phi'(\mathbf{r}', t')$ in the laboratory system S satisfies the wave equation

$$\nabla^2\phi(\mathbf{r},t) - \frac{1}{c^2}\frac{\partial^2\phi(\mathbf{r},t)}{\partial t^2} = -\frac{1}{\chi}\rho(\mathbf{r},t), \tag{D.49}$$

where $\rho(\mathbf{r}, t) = \gamma\rho'(\mathbf{r}', t')$ represents the source in the system S.

D.3.2 Maxwell-Like Equations Derived from an Ontological Scalar Field

Maxwell-like equations can also be derived from an ontological scalar field ϕ'. A vector $\mathbf{U}'$ (ontological vector field) may be introduced such that

$$\mathbf{U}'(\mathbf{r}', t') = -\nabla'\phi'(\mathbf{r}', t'). \tag{D.50}$$

The divergence of the ontological vector field is then given by

$$\nabla \cdot \mathbf{U}'(\mathbf{r}', t') = \frac{1}{\chi}\rho'(\mathbf{r}', t'). \tag{D.51}$$

Similar to the derivation of (D.26), one may introduce the dominant vector field $\mathbf{U}(\mathbf{r}, t) = \gamma\overleftrightarrow{\boldsymbol{\alpha}}^{-1} \cdot \mathbf{U}'(\mathbf{r}', t')$, and the $\mathbf{U}$-charge density $\rho(\mathbf{r}, t) = \gamma\rho(\mathbf{r}', t')$ in the laboratory system S to get the divergence of the vector field $\mathbf{U}$:

$$\nabla \cdot \mathbf{U}(\mathbf{r}, t) = \frac{1}{\chi}\rho(\mathbf{r}, t). \tag{D.52}$$

Considering (D.6), (D.50) can be written as

$$\overleftrightarrow{\boldsymbol{\alpha}} \cdot \nabla\phi'(\mathbf{r}', t') + \gamma\boldsymbol{\beta}\frac{\partial}{\partial ct}\phi'(\mathbf{r}', t') = -\mathbf{U}'(\mathbf{r}', t'). \tag{D.53}$$

Multiplying both sides of (D.53) by $\overleftrightarrow{\boldsymbol{\alpha}}^{-1}$ gives

$$-\nabla\gamma\phi'(\mathbf{r}', t') - \gamma^2\overleftrightarrow{\boldsymbol{\alpha}}^{-1} \cdot \boldsymbol{\beta}\frac{\partial}{\partial ct}\phi'(\mathbf{r}', t') = \gamma\overleftrightarrow{\boldsymbol{\alpha}}^{-1} \cdot \mathbf{U}'(\mathbf{r}', t'). \tag{D.54}$$

By inserting the field quantities introduced in the laboratory system S:

$$\begin{aligned} &\phi(\mathbf{r}, t) = \gamma\phi'(\mathbf{r}', t'), \\ &\mathbf{A}(\mathbf{r}, t) = \frac{\gamma^2}{c}\overleftrightarrow{\boldsymbol{\alpha}}^{-1} \cdot \boldsymbol{\beta}\phi'(\mathbf{r}', t') = \frac{\gamma}{c}\boldsymbol{\beta}\phi'(\mathbf{r}', t'), \\ &\mathbf{U}(\mathbf{r}, t) = \gamma\overleftrightarrow{\boldsymbol{\alpha}}^{-1} \cdot \mathbf{U}'(\mathbf{r}', t'), \end{aligned} \tag{D.55}$$

(D.54) can be expressed by

$$\mathbf{U}(\mathbf{r}, t) = -\nabla\phi(\mathbf{r}, t) - \frac{\partial \mathbf{A}(\mathbf{r}, t)}{\partial t}. \tag{D.56}$$

Note that

$$\begin{aligned} \nabla \times \mathbf{A}(\mathbf{r}, t) &= \frac{\gamma}{c}\nabla \times [\boldsymbol{\beta}\varphi(\mathbf{r}', t')] \\ &= \frac{\gamma}{c}\left(\overleftrightarrow{\boldsymbol{\alpha}} \cdot \nabla' - \gamma\boldsymbol{\beta}\frac{\partial}{\partial ct'}\right)\phi'(\mathbf{r}', t') \times \boldsymbol{\beta} \\ &= \frac{\gamma}{c}\nabla'\phi'(\mathbf{r}', t') \times \boldsymbol{\beta} = \frac{\gamma}{c}\boldsymbol{\beta} \times \mathbf{U}'(\mathbf{r}', t'). \end{aligned}$$

The curl of the vector field **U** may then be determined from (D.56)

$$\nabla \times \mathbf{U}(\mathbf{r},t) = -\frac{\partial \nabla \times \mathbf{A}(\mathbf{r},t)}{\partial t} = -\tau\frac{\partial}{\partial t}\mathbf{V}(\mathbf{r},t), \tag{D.57}$$

where

$$\mathbf{V}(\mathbf{r},t) = \frac{1}{\tau}\nabla \times \mathbf{A}(\mathbf{r},t) = \frac{\gamma}{c\tau}\boldsymbol{\beta} \times \mathbf{U}'(\mathbf{r}',t') = \chi c\boldsymbol{\beta} \times \mathbf{U}(\mathbf{r},t) \tag{D.58}$$

is the co-vector field of **U**. It is easy to show that the curl of the co-vector field is given by

$$\nabla \times \mathbf{V}(\mathbf{r},t) = \mathbf{j}(\mathbf{r},t) + \chi\frac{\partial \mathbf{U}(\mathbf{r},t)}{\partial t}, \tag{D.59}$$

where $\mathbf{j}(\mathbf{r},t) = c\boldsymbol{\beta}\rho(\mathbf{r},t)$ represents the **U**-current density in the laboratory system S. Evidently, the divergence of the co-vector field zero: $\nabla \cdot \mathbf{V}(\mathbf{r},\ t) = 0$. Equations (D.52), (D.57), and (D.59) constitute the Maxwell-like equations

$$\begin{aligned}
\nabla \times \mathbf{U}(\mathbf{r},t) &= -\tau\frac{\partial}{\partial t}\mathbf{V}(\mathbf{r},t),\\
\nabla \times \mathbf{V}(\mathbf{r},t) &= \mathbf{j}(\mathbf{r},t) + \chi\frac{\partial \mathbf{U}(\mathbf{r},t)}{\partial t},\\
\nabla \cdot \mathbf{U}(\mathbf{r},t) &= \frac{1}{\chi}\rho(\mathbf{r},t).
\end{aligned} \tag{D.60}$$

Note that the **V**-current density and charge do not appear in (D.60) because the ontological field **U′** defined by (D.50) is curl free.

Example D.1 To show how an ontological vector field **U′** emerges as the dominant vector field **U** and the co-vector field **V** in the laboratory system S, let us consider an ontological charge q' located at the origin of the static system S'. From the Helmholtz theorem (D.9), the ontological vector field **U′** generated by the charge q' is given by

$$\mathbf{U}'(\mathbf{r}',t') = \frac{q'}{4\pi\chi r'^2}\mathbf{u}_{r'}, \tag{D.61}$$

where $\mathbf{r}' = \overleftrightarrow{\boldsymbol{\alpha}} \cdot \mathbf{r} - \gamma\boldsymbol{\beta}ct = \mathbf{r} + (\gamma-1)(\boldsymbol{\beta}/\beta^2)\boldsymbol{\beta}\cdot\mathbf{r} - \gamma\boldsymbol{\beta}ct$, with

$$\begin{aligned}
r'^2 &= \left[\mathbf{r} + (\gamma-1)\frac{\boldsymbol{\beta}}{\beta^2}\boldsymbol{\beta}\cdot\mathbf{r} - \gamma\boldsymbol{\beta}ct\right] \cdot \left[\mathbf{r} + (\gamma-1)\frac{\boldsymbol{\beta}}{\beta^2}\boldsymbol{\beta}\cdot\mathbf{r} - \gamma\boldsymbol{\beta}ct\right]\\
&= r^2 + (\gamma^2-1)\frac{1}{\beta^2}(\boldsymbol{\beta}\cdot\mathbf{r})^2 - 2\gamma^2 ct\boldsymbol{\beta}\cdot\mathbf{r} + (\gamma\beta ct)^2.
\end{aligned}$$

The unit vector along the radial direction is given by

$$\mathbf{u}_{r'} = \frac{1}{r'}\left[\mathbf{r}+ (\gamma-1)\frac{\boldsymbol{\beta}}{\beta^2}\boldsymbol{\beta}\cdot\mathbf{r}-\gamma\boldsymbol{\beta}ct\right].$$

Therefore, the ontological vector field $\mathbf{U}'$ can be expressed as

$$\mathbf{U}'(\mathbf{r}',t') = \frac{q'}{4\pi\chi r'^3}\left[\mathbf{r}+ (\gamma-1)\frac{\boldsymbol{\beta}}{\beta^2}\boldsymbol{\beta}\cdot\mathbf{r}-\gamma\boldsymbol{\beta}ct\right].$$

By definition, the corresponding dominant vector field $\mathbf{U}$ in the laboratory system S is

$$\begin{aligned}\mathbf{U}(\mathbf{r},t) &= \gamma\overleftrightarrow{\boldsymbol{\alpha}}^{-1}\cdot\mathbf{U}'(\mathbf{r}',t') \\ &= \frac{q'}{4\pi\chi r'^3}\left[\gamma\overleftrightarrow{\mathbf{I}} + (1-\gamma)\frac{\boldsymbol{\beta}\boldsymbol{\beta}}{\beta^2}\right]\cdot\left[\mathbf{r} + (\gamma-1)\frac{\boldsymbol{\beta}}{\beta^2}\boldsymbol{\beta}\cdot\mathbf{r}-\gamma\boldsymbol{\beta}ct\right] \\ &= \frac{q\gamma}{4\pi\chi r'^3}(\mathbf{r}-\boldsymbol{\beta}ct),\end{aligned} \tag{D.62}$$

where q has been set to q' from (D.38). The co-vector field $\mathbf{V}$ is then given by

$$\begin{aligned}\mathbf{V}(\mathbf{r},t) &= \chi\mathbf{v}\times\mathbf{U}(\mathbf{r},t) \\ &= \frac{q\gamma}{4\pi r'^3}\mathbf{v}\times(\mathbf{r}-\boldsymbol{\beta}ct) = \frac{q\gamma c}{4\pi r'^3}\boldsymbol{\beta}\times\mathbf{r}.\end{aligned} \tag{D.63}$$

□

D.3.3 Conservation Laws

So far, no physical meanings have been given to the vector fields **U** and **V**. Since they satisfy the Maxwell-like equations, the conservation laws that hold for the EM field are also applicable for the (**U**, **V**)-field.

D.3.3.1 Conservation of Energy

Let $\mathbf{S} = \mathbf{U}\times\mathbf{V}$. From the Maxwell-like equations (D.34), one may find

$$\begin{aligned}\nabla\cdot\mathbf{S} &= -\mathbf{U}\cdot\nabla\times\mathbf{V}+\mathbf{V}\cdot\nabla\times\mathbf{U} \\ &= -\mathbf{j}\cdot\mathbf{U}-\mathbf{j}_m\cdot\mathbf{V}-\chi\mathbf{U}\cdot\frac{\partial\mathbf{U}}{\partial t}-\tau\mathbf{V}\cdot\frac{\partial\mathbf{V}}{\partial t} \\ &= -\mathbf{j}\cdot\mathbf{U}-\mathbf{j}_m\cdot\mathbf{V}-\frac{1}{2}\frac{\partial}{\partial t}(\chi|\mathbf{U}|^2+\tau|\mathbf{V}|^2).\end{aligned} \tag{D.64}$$

The integral form of (D.64) over a finite region V bounded by S is

$$-\int_V(\mathbf{j}\cdot\mathbf{U}+\mathbf{j}_m\cdot\mathbf{V})dV = \int_S\mathbf{S}\cdot\mathbf{u}_n dS + \frac{\partial}{\partial t}\int_V\frac{1}{2}(\chi|\mathbf{U}|^2+\tau|\mathbf{V}|^2)dV, \tag{D.65}$$

where $\mathbf{u}_n$ is the unit outward normal of S. To give (D.65) a physical meaning, let us assume that the vector **S**, which is also called **Poynting vector**, represents the power flow density of the (**U**, **V**)-field measured in watts per square meter (W/m^2). The first term on the right-hand side then denotes the power flowing out of S.The left-hand side can thus be interpreted as the power supplied by the current sources **j** and $\mathbf{j}_m$, while the second term on the right-hand side can be interpreted as the work done per second by the current sources to establish the (**U**, **V**)-field. As a result, one may introduce the total (**U**, **V**)-field energy density w:

$$w = w_U + w_V, \tag{D.66}$$

where w_U and w_V will be called the **U-field energy density** and **V-field energy density**, respectively, defined by

$$w_U = \frac{1}{2}\chi|\mathbf{U}|^2, \quad w_V = \frac{1}{2}\tau|\mathbf{V}|^2. \tag{D.67}$$

D.3.3.2 Conservation of Momentum

Taking the partial derivative of the Poynting vector **S** with respect to time and making use of the Maxwell-like equations (D.34) yield

$$\frac{\partial \mathbf{S}}{\partial t} = \frac{\partial \mathbf{U}}{\partial t} \times \mathbf{V} + \mathbf{U} \times \frac{\partial \mathbf{V}}{\partial t} = \frac{1}{\chi}(\nabla \times \mathbf{V} - \mathbf{j}) \times \mathbf{V} - \frac{1}{\tau}\mathbf{U} \times (\nabla \times \mathbf{U} + \mathbf{j}_m).$$

This can be written as

$$\frac{1}{c^2}\frac{\partial \mathbf{S}}{\partial t} + \tau\mathbf{j} \times \mathbf{V} + \chi\mathbf{U} \times \mathbf{j}_m = \tau(\nabla \times \mathbf{V}) \times \mathbf{V} + \chi(\nabla \times \mathbf{U}) \times \mathbf{U}.$$

By using the vector identity $(\nabla \times \mathbf{A}) \times \mathbf{A} = -(1/2)\nabla|\mathbf{A}|^2 + (\mathbf{A} \cdot \nabla)\mathbf{A}$, the above equation may be rewritten as

$$\begin{aligned} &\frac{1}{c^2}\frac{\partial \mathbf{S}}{\partial t} + \tau\mathbf{j} \times \mathbf{V} + \chi\mathbf{U} \times \mathbf{j}_m \\ &= \tau\left[-\frac{1}{2}\nabla|\mathbf{V}|^2 + (\mathbf{V} \cdot \nabla)\mathbf{V}\right] + \chi\left[-\frac{1}{2}\nabla|\mathbf{U}|^2 + (\mathbf{U} \cdot \nabla)\mathbf{U}\right]. \end{aligned} \tag{D.68}$$

It follows from (D.26) and (D.31) that

$$\begin{aligned} &\frac{1}{c^2}\frac{\partial \mathbf{S}}{\partial t} + \rho\mathbf{U} + \tau\mathbf{j} \times \mathbf{V} + \rho_m\mathbf{V} + \chi\mathbf{U} \times \mathbf{j}_m \\ &= \tau\left[-\frac{1}{2}\nabla|\mathbf{V}|^2 + (\mathbf{V} \cdot \nabla)\mathbf{V} + \mathbf{V}(\nabla \cdot \mathbf{V})\right] + \chi\left[-\frac{1}{2}\nabla|\mathbf{U}|^2 + (\mathbf{U} \cdot \nabla)\mathbf{U} + \mathbf{U}(\nabla \cdot \mathbf{U})\right]. \end{aligned} \tag{D.69}$$

By use of the identities

$$\nabla \cdot \left(\varphi \overleftrightarrow{\mathbf{I}}\right) = \nabla \varphi, \ \ \nabla \cdot (\mathbf{AB}) = (\nabla \cdot \mathbf{A})\mathbf{B} + \mathbf{A} \cdot \nabla \mathbf{B},$$

(D.69) can be expressed in dyadic notations as

$$\nabla \cdot \overleftrightarrow{\mathbf{T}} - \frac{\partial \mathbf{g}}{\partial t} = \mathbf{f}, \tag{D.70}$$

where

$$\begin{aligned}
\mathbf{g} &= \frac{1}{c^2}\mathbf{S}, \\
\mathbf{f} &= \rho \mathbf{U} + \tau \mathbf{j} \times \mathbf{V} + \rho_m \mathbf{V} + \chi \mathbf{U} \times \mathbf{j}_m, \\
\overleftrightarrow{\mathbf{T}} &= \chi \mathbf{UU} + \tau \mathbf{VV} - \frac{1}{2}\tau |\mathbf{V}|^2 \overleftrightarrow{\mathbf{I}} - \frac{1}{2}\chi |\mathbf{U}|^2 \overleftrightarrow{\mathbf{I}}.
\end{aligned}$$

The integral form of (D.70) over a finite region V bounded by S is

$$\int_S \mathbf{u}_n \cdot \overleftrightarrow{\mathbf{T}} dS - \int_V \mathbf{f} dV = \frac{\partial}{\partial t} \int_V \mathbf{g} dV. \tag{D.71}$$

Since the Poynting vector **S** has been interpreted as power flow density, the vector **g** can be interpreted as the **momentum density** of the field. As a result, **f** can be interpreted as the **force density** acting on the sources by the field; $\mathbf{u}_n \cdot \overleftrightarrow{\mathbf{T}}$ can be interpreted as the force per unit area acting on the surface S. $\overleftrightarrow{\mathbf{T}}$ will be called **stress tensor**.

D.4 Interaction Between Two Systems

Consider two field systems $(\mathbf{U}_1, \mathbf{V}_1)$ and $(\mathbf{U}_2, \mathbf{V}_2)$, and their composite field will be denoted by $(\mathbf{U}, \mathbf{V})$ with $\mathbf{U} = \mathbf{U}_1 + \mathbf{U}_2$, $\mathbf{V} = \mathbf{V}_1 + \mathbf{V}_2$. The stress tensor for the composite field can be decomposed into

$$\overleftrightarrow{\mathbf{T}} = \chi \mathbf{UU} + \tau \mathbf{VV} - \frac{1}{2}\tau |\mathbf{V}|^2 \overleftrightarrow{\mathbf{I}} - \frac{1}{2}\chi |\mathbf{U}|^2 \overleftrightarrow{\mathbf{I}} = \overleftrightarrow{\mathbf{T}}_1 + \overleftrightarrow{\mathbf{T}}_2 + \overleftrightarrow{\mathbf{T}}_{12},$$

where $\overleftrightarrow{\mathbf{T}}_1$ and $\overleftrightarrow{\mathbf{T}}_2$ are the stress tensors for the systems $(\mathbf{U}_1, \mathbf{V}_1)$ and $(\mathbf{U}_2, \mathbf{V}_2)$, respectively, and

$$\overleftrightarrow{\mathbf{T}}_{12} = \chi(\mathbf{U}_1\mathbf{U}_2 + \mathbf{U}_2\mathbf{U}_1) + \tau(\mathbf{V}_1\mathbf{V}_2 + \mathbf{V}_2\mathbf{V}_1) - (\chi \mathbf{U}_1 \cdot \mathbf{U}_2 + \tau \mathbf{V}_1 \cdot \mathbf{V}_2)\overleftrightarrow{\mathbf{I}} \tag{D.72}$$

is the stress tensor for the interaction between the two systems, called **interaction stress tensor**. The interaction force between the two systems can be found by the integration of (D.72) over a closed surface that encloses one of the systems

$$\begin{aligned}\mathbf{F}_{12} &= \int_S \mathbf{u}_n \cdot \overleftrightarrow{\mathbf{T}}_{12} dS \\ &= \chi \int_S [\mathbf{U}_2(\mathbf{U}_1 \cdot \mathbf{u}_n) - \mathbf{U}_2 \times (\mathbf{u}_n \times \mathbf{U}_1)] dS \\ &+ \tau \int_S [\mathbf{V}_2(\mathbf{V}_1 \cdot \mathbf{u}_n) - \mathbf{V}_2 \times (\mathbf{u}_n \times \mathbf{V}_1)] dS.\end{aligned} \quad \text{(D.73)}$$

Example D.2 (Derivation of Inverse Square Law)

Consider the interaction between two static vector fields $\mathbf{U}_1$ and $\mathbf{U}_2$ (the prime for the ontological field will be dropped in this example), which are generated by the point ontological charges q_1 and q_2. The two charges are, respectively, placed at the origin of the coordinate system $\mathbf{r}_1 = (x_1, y_1, z_1)$ and $\mathbf{r}_2 = (x_2, y_2, z_2)$, as illustrated in Figure D.3. It will be assumed that both $\mathbf{U}_1$ and $\mathbf{U}_2$ are curl free $\nabla \times \mathbf{U}_1 = \nabla \times \mathbf{U}_2 = 0$. According to the Helmholtz theorem (D.9), the expressions for the fields $\mathbf{U}_1$ and $\mathbf{U}_2$ are given by

$$\mathbf{U}_1 = \frac{q_1 \mathbf{r}_1}{4\pi\chi r_1^3}, \quad \mathbf{U}_2 = \frac{q_2 \mathbf{r}_2}{4\pi\chi r_2^3}. \quad \text{(D.74)}$$

In order to find the interaction between the two charges, the closed surface S in (D.73) may be selected as an infinitely small sphere that encloses the ontological charge q_1. From (D.73), the force acting on q_1 by the charge q_2 can be written as

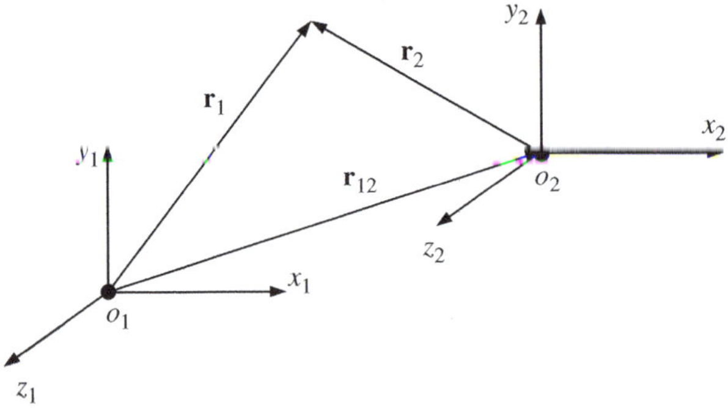

Figure D.3 Two coordinate systems.

$$\begin{aligned}\mathbf{F}_{12} &= \int_S \chi[\mathbf{U}_2(\mathbf{U}_1 \cdot \mathbf{u}_n) - \mathbf{U}_2 \times (\mathbf{u}_n \times \mathbf{U}_1)]dS \\ &= \int_S \chi \mathbf{U}_2 \frac{q_1}{4\pi\chi r_1^2} dS = q_1 \mathbf{U}_2|_{\text{at } o_1} = -\frac{q_1 q_2 \mathbf{r}_{12}}{4\pi\chi r_2^3}.\end{aligned} \tag{D.75}$$

This turns out to be the well-known **Coulomb's law** if q_1 and q_2 are interpreted as the electric charges and χ is set to ε_0, or **Newton's law of gravitation** if q_1 and q_2 are interpreted as masses and χ is set to $-1/4\pi G$ (G is gravitational constant), both becoming a derived result instead of an experimental fact. □

According to interaction force given by (D.75), a test charge q' in the ontological vector field $\mathbf{U}'$ will be affected by a force

$$\mathbf{F}'(\mathbf{r}', t') = q'\mathbf{U}'(\mathbf{r}', t'). \tag{D.76}$$

Since charge is conserved in different coordinate systems, one may set $q = q'$. Insertion of (D.21) into (D.76) gives

$$\begin{aligned}\frac{1}{\gamma}\mathbf{F}'(\mathbf{r}', t') &= q\left[\overleftrightarrow{\boldsymbol{\alpha}}^{-1} \cdot \mathbf{U}(\mathbf{r}, t) + \mathbf{v} \times \tau\mathbf{V}(\mathbf{r}, t)\right] \\ &= q[\mathbf{U}(\mathbf{r}, t) + \mathbf{v} \times \tau\mathbf{V}(\mathbf{r}, t)] + q\left(\frac{1}{\gamma} - 1\right)\frac{\boldsymbol{\beta}[\boldsymbol{\beta} \cdot \mathbf{U}(\mathbf{r}, t)]}{\beta^2}.\end{aligned} \tag{D.77}$$

By use of $\boldsymbol{\beta}[\boldsymbol{\beta} \cdot \mathbf{E}(\mathbf{r}, t)] = \boldsymbol{\beta}\left[\boldsymbol{\beta} \cdot \gamma\overleftrightarrow{\boldsymbol{\alpha}}^{-1} \cdot \mathbf{E}'(\mathbf{r}', t')\right] = \boldsymbol{\beta}[\boldsymbol{\beta} \cdot \mathbf{E}'(\mathbf{r}', t')]$, (D.77) can be written as

$$\mathbf{F}(\mathbf{r}, t) = q[\mathbf{U}(\mathbf{r}, t) + \mathbf{v} \times \tau\mathbf{V}(\mathbf{r}, t)], \tag{D.78}$$

where

$$\begin{aligned}\mathbf{F}(\mathbf{r}, t) &= \frac{1}{\gamma}\mathbf{F}'(\mathbf{r}', t') + \left(1 - \frac{1}{\gamma}\right)\frac{\boldsymbol{\beta}[\boldsymbol{\beta} \cdot \mathbf{F}'(\mathbf{r}', t')]}{\beta^2} \\ &= \frac{1}{\gamma}\left\{\mathbf{F}'(\mathbf{r}', t') + (\gamma - 1)\frac{\boldsymbol{\beta}[\boldsymbol{\beta} \cdot \mathbf{F}'(\mathbf{r}', t')]}{\beta^2}\right\} = \frac{1}{\gamma}\overleftrightarrow{\boldsymbol{\alpha}} \cdot \mathbf{F}'(\mathbf{r}', t')\end{aligned} \tag{D.79}$$

may be interpreted as the force acting on the charge q, observed in the laboratory system S. If $\mathbf{U}'$ stands for a static electric field, (D.78) stands for the well-known **Lorentz force equation**.

D.5 Applications

The theory presented above is applicable to any physical field and will be called the **unified theory for fields** (UTF). The UTF will now be applied to the EM field and gravitational field. Let us first consider the electric field generated by an electric

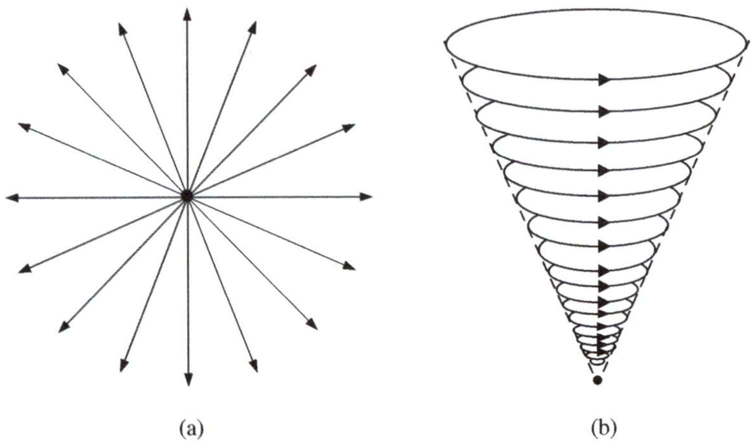

Figure D.4 (a) Field pattern of the ontological field. (b) Helical structure.

charge distribution. An observer, who is stationary with respect to the electric charge distribution, will observe an electric field but no magnetic field; while an observer, who is moving at a constant speed with respect to the charge distribution, will observe both an electric field and a magnetic field. This observation implies that the magnetic field is an effect of relativity caused by the movement of the charge distribution. It has been experimentally verified (as stated by Coulomb's law) that there exists an inertial reference system in which an electric field can be considered static. For this reason, the Maxwell equations are derivable from a static electric field or a static scalar field via the theory of special relativity. In other words, the Maxwell equations are implied by the theory of special relativity. Similarly, Newton's law of universal gravitation implies the existence of a static gravitational field, and as a result, a gravitational wave exists and obeys the Maxwell-like equations. It is noted that both the static electric field and the static gravitational field are curl free, and it is relative motion that causes curls or rotations. This explains why the nature has so many circular movements and helical structures. Figure D.4 shows the typical field pattern for the static (ontological) fields and the helical structure.

D.5.1 Maxwell Equations Derived from a Static Electric Field

By experiment, there exists a static system S' in which the following facts hold

$$\begin{aligned} &\nabla' \times \mathbf{E}'(\mathbf{r}', t') = 0, \\ &\nabla' \cdot \mathbf{E}'(\mathbf{r}', t') = \frac{1}{\varepsilon_0}\rho'(\mathbf{r}', t'), \end{aligned} \tag{D.80}$$

with $\partial \mathbf{E}'/\partial t' = \partial \rho'/\partial t' = 0$. The vector field $\mathbf{E}'$ is the static electric field measured in volts per meter (V/m), the scalar field ρ' is the electric charge density measured in coulombs per cubic meter (C/m^3) and is real, and $\varepsilon_0 = 8.85 \times 10^{-12}$ (F/m) is the permittivity in free space. Corresponding to the static electric field $\mathbf{E}'$, the Maxwell-like equations (D.34) can be written as

$$\begin{aligned}
\nabla \times \mathbf{H}(\mathbf{r}, t) &= \varepsilon_0 \frac{\partial \mathbf{E}(\mathbf{r}, t)}{\partial t} + \mathbf{J}(\mathbf{r}, t), \\
\nabla \times \mathbf{E}(\mathbf{r}, t) &= -\mu_0 \frac{\partial \mathbf{H}(\mathbf{r}, t)}{\partial t}, \\
\nabla \cdot \mathbf{E}(\mathbf{r}, t) &= \frac{1}{\varepsilon_0} \rho(\mathbf{r}, t),
\end{aligned} \tag{D.81}$$

where $\mu_0 = 1.26 \times 10^{-6}$ is defined by $\mu_0 \varepsilon_0 = 1/c^2$ and

$$\begin{aligned}
\mathbf{E}(\mathbf{r}, t) &= \gamma \overleftrightarrow{\boldsymbol{\alpha}}^{-1} \cdot \mathbf{E}'(\mathbf{r}', t'), \\
\mathbf{H}(\mathbf{r}, t) &= \varepsilon_0 \mathbf{v} \times \mathbf{E}(\mathbf{r}, t), \\
\rho(\mathbf{r}, t) &= \gamma \rho'(\mathbf{r}', t'), \\
\mathbf{J}(\mathbf{r}, t) &= \rho(\mathbf{r}, t) \mathbf{v}.
\end{aligned}$$

The co-vector $\mathbf{H}$ is defined as the **magnetic field**. Equation (D.81) is the well-known Maxwell equations.

Remark D.3 A heuristic derivation of Maxwell equations starting from Coulomb's law together with the imposition of Galilean invariance was given in [16]. In order to obtain the second equation of (D.34), one has to use an additional assumption that the electric field $\mathbf{E}$ must satisfy the wave equation. □

D.5.2 Maxwell-Like Equations Derived from a Static Gravitational Field

By experiment, there exists an inertial system S' in which the following facts hold

$$\begin{aligned}
\nabla' \times \mathbf{E}_g'(\mathbf{r}', t') &= 0, \\
\nabla' \cdot \mathbf{E}_g'(\mathbf{r}', t') &= \frac{1}{\varepsilon_g} \rho_g'(\mathbf{r}', t'),
\end{aligned} \tag{D.82}$$

with $\partial \mathbf{E}_g'/\partial t' = \partial \rho_g'/\partial t' = 0$. The vector field $\mathbf{E}_g'$ is the static gravitational field measured in newton per kilogram (N/kg), the scalar field ρ_g' is the mass density measured in kilogram per cubic meter (kg/m^3) and is always positive, $\varepsilon_g = -1/4\pi G = 1.19 \times 10^9$ (kg^2/N $\cdot$ m^3), and $G = 6.67 \times 10^{-11}$ (N $\cdot$ m^2/kg^2) is the gravitational constant. For the static gravitational field $\mathbf{E}_g'(\mathbf{r}', t')$ defined by (D.82), the Maxwell-like equations (D.34) become

$$\nabla \times \mathbf{H}_g(\mathbf{r},t) = \varepsilon_g \frac{\partial \mathbf{E}_g(\mathbf{r},t)}{\partial t} + \mathbf{J}_g(\mathbf{r},t),$$
$$\nabla \times \mathbf{E}_g(\mathbf{r},t) + \mu_g \frac{\partial \mathbf{H}_g(\mathbf{r},t)}{\partial t} = 0, \tag{D.83}$$
$$\nabla \cdot \mathbf{E}_g(\mathbf{r},t) = \frac{1}{\varepsilon_g} \rho_g(\mathbf{r},t),$$

where $\mu_g = -9.3 \times 10^{-27}$ is a very small number defined by $\mu_g \varepsilon_g = 1/c^2$, and

$$\mathbf{E}_g(\mathbf{r},t) = \gamma \overleftrightarrow{\boldsymbol{\alpha}}^{-1} \cdot \mathbf{E}'_g(\mathbf{r}',t'),$$
$$\mathbf{H}_g(\mathbf{r},t) = \varepsilon_g \mathbf{v} \times \mathbf{E}_g(\mathbf{r},t),$$
$$\rho_g(\mathbf{r},t) = \gamma \rho'_g(\mathbf{r}',t'),$$
$$\mathbf{J}_g(\mathbf{r},t) = \rho_g(\mathbf{r},t)\mathbf{v}.$$

The co-vector field $\mathbf{H}_g(\mathbf{r}, t)$ is referred to as **gravitomagnetic field**, which is similar to what was first introduced by Heaviside in 1893 as the analog of magnetic force, more than a decade before the invention of special relativity by Einstein [20]. The linear Maxwell-like equations (D.83) give a field theory for the gravity, which was either introduced earlier as an assumption [20, 21] or derived from the Einstein field equation using a weak field approximation as will be discussed below. They clearly indicate that the gravitational interaction propagates at light speed. In a weak gravitational field, the theory of general relativity also predicts that the gravitational wave propagates at light speed. Note that the minus sign before μ_g and ε_g can be made to disappear by changing the sign of $\mathbf{E}_g$.

D.6 Maxwell-Like Equations Derived from Einstein Field Equations

The Maxwell-like equations (D.83) will now be re-derived from the Einstein field equation. According to the general principle of relativity, all the reference systems are equivalent with respect to the formulation of the fundamental laws of physics. In an accelerated system relative to an inertial system, fictitious forces occur and are interpreted as gravitational forces. Such an interpretation is validated by the fact that the fictitious forces and gravitational forces all act on the mass particles and are independent of the mass of particles. The fictitious forces are very much like the magnetic field caused by relative motion as discussed before. The term "gravito-electromagnetism (GEM)" has been used to describe the Maxwell-like equations obtained from Einstein's famous field equations in the weak gravitational field approximation. The GEM has been discussed by a number of authors and a review can be found in [27]. For a brief introduction to the theory of general relativity and the notations used below, please refer to [28]. The Einstein's field equation is nonlinear and given by

$$R_{\mu\nu} - \frac{1}{2} g_{\mu\nu} R = \frac{8\pi G}{c^4} T_{\mu\nu}, \tag{D.84}$$

where $R_{\mu\nu}$ is Ricci tensor, R is scalar curvature, $g_{\mu\nu}$ is the metric tensor, and $T_{\mu\nu}$ is the stress-energy (or energy-momentum) tensor. The tensor $T_{\mu\nu}$ acts as the sources of the gravitational field and contains all types of mass densities and energy densities except for the energy density of the gravitational field itself. Equation (D.84) can be used to re-derive the Maxwell-like equations (D.83) based on the weak gravitational field approximation. In fact, the metric tensor $g_{\mu\nu}$ may be written as

$$g_{\mu\nu} = \eta_{\mu\nu} + h_{\mu\nu}, \tag{D.85}$$

where $\eta_{\mu\nu} = \text{diag}(1, 1, 1, -1)$ is the Minkowski flat metric. By a weak gravitational field approximation or linearized gravity, it means that only the terms linear in $h_{\mu\nu}$ are retained when (D.85) is substituted into Einstein's field equation. The Christoffel symbol $\Gamma^{\alpha}{}_{\mu\nu}$ may be approximated by

$$\begin{aligned} \Gamma^{\alpha}{}_{\mu\nu} &= \frac{1}{2} g^{\alpha\beta} \left(\partial_\mu g_{\nu\beta} + \partial_\nu g_{\mu\beta} - \partial_\beta g_{\mu\nu} \right) \\ &\approx \frac{1}{2} \eta^{\alpha\beta} \left(\partial_\mu h_{\nu\beta} + \partial_\nu h_{\mu\beta} - \partial_\beta h_{\mu\nu} \right), \end{aligned} \tag{D.86}$$

where $\partial_\mu = \partial / \partial x^\mu$ denotes the derivative operator associated with the flat metric $\eta_{\mu\nu}$, and Einstein summation convention is used. Retaining the linear terms in $h_{\mu\nu}$, the Ricci tensor $R_{\mu\nu}$ is given by

$$\begin{aligned} R_{\mu\nu} &= \partial_\alpha \Gamma^{\alpha}{}_{\mu\nu} - \partial_\mu \Gamma^{\alpha}{}_{\alpha\nu} + \Gamma^{\alpha}{}_{\mu\nu} \Gamma^{\beta}{}_{\alpha\beta} - \Gamma^{\alpha}{}_{\mu\beta} \Gamma^{\beta}{}_{\nu\alpha} \\ &\approx \partial_\alpha \Gamma^{\alpha}{}_{\mu\nu} - \partial_\mu \Gamma^{\alpha}{}_{\alpha\nu}. \end{aligned} \tag{D.87}$$

It follows from (D.86) that the partial derivatives of the Christoffel symbol are given by

$$\begin{aligned} \partial_\alpha \Gamma^{\alpha}{}_{\mu\nu} &= \partial_\alpha \left[\frac{1}{2} \eta^{\alpha\beta} \left(\partial_\mu h_{\nu\beta} + \partial_\nu h_{\mu\beta} - \partial_\beta h_{\mu\nu} \right) \right] \\ &= \frac{1}{2} \eta^{\alpha\beta} \left(\partial_\alpha \partial_\mu h_{\nu\beta} + \partial_\alpha \partial_\nu h_{\mu\beta} - \partial_\alpha \partial_\beta h_{\mu\nu} \right), \end{aligned} \tag{D.88}$$

$$\begin{aligned} \partial_\mu \Gamma^{\alpha}{}_{\alpha\nu} &= \partial_\mu \left[\frac{1}{2} \eta^{\alpha\beta} \left(\partial_\alpha h_{\nu\beta} + \partial_\nu h_{\alpha\beta} - \partial_\beta h_{\alpha\nu} \right) \right] \\ &= \frac{1}{2} \eta^{\alpha\beta} \left(\partial_\mu \partial_\alpha h_{\nu\beta} + \partial_\mu \partial_\nu h_{\alpha\beta} - \partial_\mu \partial_\beta h_{\alpha\nu} \right). \end{aligned} \tag{D.89}$$

Substituting (D.88) and (D.89) into (D.87) gives the Ricci tensor

$$
\begin{aligned}
R_{\mu\nu} &= \frac{1}{2}\eta^{\alpha\beta}\left(\partial_\alpha\partial_\nu h_{\mu\beta} - \partial_\alpha\partial_\beta h_{\mu\nu}\right) - \frac{1}{2}\eta^{\alpha\beta}\left(\partial_\mu\partial_\nu h_{\alpha\beta} - \partial_\mu\partial_\beta h_{\alpha\nu}\right) \\
&= \frac{1}{2}\left(\partial^\beta\partial_\nu h_{\mu\beta} - \partial^\beta\partial_\beta h_{\mu\nu}\right) - \frac{1}{2}\left(\partial_\mu\partial_\nu h - \partial_\mu\partial^\alpha h_{\alpha\nu}\right),
\end{aligned}
$$

where $h = \eta^{\mu\nu}h_{\mu\nu} = h^\beta{}_\beta$. The scalar curvature is then given by

$$
\begin{aligned}
R &= \eta^{\mu\nu}R_{\mu\nu} \\
&= \frac{1}{2}\left(\partial^\beta\partial^\mu h_{\mu\beta} - \partial^\beta\partial_\beta h\right) - \frac{1}{2}\left(\partial^\nu\partial_\nu h - \partial^\nu\partial^\alpha h_{\alpha\nu}\right) \\
&= \partial^\alpha\partial^\beta h_{\alpha\beta} - \partial^\beta\partial_\beta h.
\end{aligned}
$$

The Einstein tensor can thus be approximated by

$$
\begin{aligned}
G_{\mu\nu} &= R_{\mu\nu} - \frac{1}{2}g_{\mu\nu}R \\
&= \frac{1}{2}\left(\partial^\beta\partial_\nu h_{\mu\beta} - \partial^\beta\partial_\beta h_{\mu\nu}\right) - \frac{1}{2}\left(\partial_\mu\partial_\nu h - \partial_\mu\partial^\alpha h_{\alpha\nu}\right) - \frac{1}{2}\eta_{\mu\nu}\left(\partial^\alpha\partial^\beta h_{\alpha\beta} - \partial^\beta\partial_\beta h\right).
\end{aligned}
\tag{D.90}
$$

The above expression can be simplified by introducing

$$
\overline{h}_{\mu\nu} = h_{\mu\nu} - \frac{1}{2}\eta_{\mu\nu}h,
$$

and the Einstein tensor (D.90) becomes

$$
\begin{aligned}
G_{\mu\nu} &= \frac{1}{2}\left(\partial^\beta\partial_\nu\overline{h}_{\mu\beta} - \partial^\beta\partial_\beta\overline{h}_{\mu\nu}\right) - \frac{1}{2}\left(\partial_\mu\partial_\nu h - \partial_\mu\partial^\alpha\overline{h}_{\alpha\nu}\right) \\
&\quad - \frac{1}{2}\eta_{\mu\nu}\left(\partial^\alpha\partial^\beta\overline{h}_{\alpha\beta} - \partial^\beta\partial_\beta h\right) \\
&\quad + \frac{1}{4}\left(\partial_\mu\partial_\nu h - \partial^\beta\partial_\beta h\eta_{\mu\nu}\right) + \frac{1}{4}\partial_\mu\partial_\nu h - \frac{1}{4}\eta_{\mu\nu}\partial^\alpha\partial_\alpha h \\
&= \frac{1}{2}\left(\partial^\beta\partial_\nu\overline{h}_{\mu\beta} - \partial^\beta\partial_\beta\overline{h}_{\mu\nu}\right) + \frac{1}{2}\partial_\mu\partial^\alpha\overline{h}_{\alpha\nu} - \frac{1}{2}\eta_{\mu\nu}\partial^\alpha\partial^\beta\overline{h}_{\alpha\beta} \\
&= -\frac{1}{2}\partial^\beta\partial_\beta\overline{h}_{\mu\nu} + \frac{1}{2}\partial^\beta\partial_\nu\overline{h}_{\mu\beta} + \frac{1}{2}\partial_\mu\partial^\beta\overline{h}_{\beta\nu} - \frac{1}{2}\eta_{\mu\nu}\partial^\alpha\partial^\beta\overline{h}_{\alpha\beta}.
\end{aligned}
\tag{D.91}
$$

After enforcing the transverse gauge condition

$$
\partial^\alpha\overline{h}_{\alpha\beta} = 0,
$$

one obtains

$$
G_{\mu\nu} = -\frac{1}{2}\partial^\beta\partial_\beta\overline{h}_{\mu\nu}. \tag{D.92}
$$

As a result, Einstein's field equation (D.84) reduces to the wave equation

$$\Box \bar{h}_{\mu\nu} = -\frac{16\pi G}{c^4} T_{\mu\nu}, \tag{D.93}$$

where $\Box = \eta^{\mu\nu}\partial_\mu\partial_\nu$ is the flat-space-time wave operator. The wave equation (D.93) admits the retarded solution

$$\bar{h}_{\mu\nu} = \frac{4G}{c^4} \int_V \frac{T_{\mu\nu}(\mathbf{r}', t - R/c)}{R} dV(\mathbf{r}'), \tag{D.94}$$

where $R = |\mathbf{r} - \mathbf{r}'|$. Note that

$$T_{\mu\nu} = \begin{bmatrix} \tilde{\rho}_{g0}u_1u_1 & \tilde{\rho}_{g0}u_1u_2 & \tilde{\rho}_{g0}u_1u_3 & \tilde{\rho}_{g0}u_1u_4 \\ \tilde{\rho}_{g0}u_2u_1 & \tilde{\rho}_{g0}u_2u_2 & \tilde{\rho}_{g0}u_2u_3 & \tilde{\rho}_{g0}u_2u_4 \\ \tilde{\rho}_{g0}u_3u_1 & \tilde{\rho}_{g0}u_3u_2 & \tilde{\rho}_{g0}u_3u_3 & \tilde{\rho}_{g0}u_3u_4 \\ \tilde{\rho}_{g0}u_4u_1 & \tilde{\rho}_{g0}u_4u_2 & \tilde{\rho}_{g0}u_4u_3 & \tilde{\rho}_{g0}u_4u_4 \end{bmatrix} = \begin{bmatrix} \rho_g v_1 v_1 & \rho_g v_1 v_2 & \rho_g v_1 v_3 & -\rho_g c v_1 \\ \rho_g v_2 v_1 & \rho_g v_2 v_2 & \rho_g v_2 v_3 & -\rho_g c v_2 \\ \rho_g v_3 v_1 & \rho_g v_3 v_2 & \rho_g v_3 v_3 & -\rho_g c v_3 \\ -\rho_g c v_1 & -\rho_g c v_2 & -\rho_g c v_3 & \rho_g c^2 \end{bmatrix},$$

where $\tilde{\rho}_{g0}$ is the rest mass density in the co-moving system, $\rho_g = \tilde{\rho}_{g0}/\left[1-(v/c)^2\right]$ is the mass density in the laboratory system, u_μ is four velocity, and $\mathbf{v} = (v_1, v_2, v_3)$ is the speed of the matter distribution $\tilde{\rho}_{g0}$. One can now introduce the GEM potentials ϕ_g and $\mathbf{A}_g = (A_{g1}, A_{g2}, A_{g3})$ defined by

$$\begin{aligned} \bar{h}_{44} &= -\frac{4}{c^2}\phi_g = \frac{4G}{c^4} \int_V \frac{c^2\rho_g\left(\mathbf{r}', t - \frac{R}{c}\right)}{R} dV(\mathbf{r}'), \\ \bar{h}_{4i} &= \frac{4A_{gi}}{c} = \frac{4G}{c^4} \int_V \frac{-cJ_{gi}\left(\mathbf{r}', t - \frac{R}{c}\right)}{R} dV(\mathbf{r}'), \quad (i = 1, 2, 3), \end{aligned} \tag{D.95}$$

where $\mathbf{J}_g = \rho_g \mathbf{v} = (J_{g1}, J_{g2}, J_{g3})$. Thus,

$$\begin{aligned} \phi_g &= \int_V \frac{\rho_g\left(\mathbf{r}', t - \frac{R}{c}\right)}{4\pi\varepsilon_g R} dV(\mathbf{r}'), \\ \mathbf{A}_g &= \int_V \frac{\mu_g \mathbf{J}_g\left(\mathbf{r}', t - \frac{R}{c}\right)}{4\pi R} dV(\mathbf{r}'), \quad (i = 1, 2, 3), \end{aligned} \tag{D.96}$$

where $\varepsilon_g = -1/4\pi G$, $\mu_g = -4\pi G/c^2$. The scalar and vector potentials satisfy the Lorenz gauge condition

$$\nabla \cdot \mathbf{A}_g + \frac{1}{c^2}\frac{\partial \phi_g}{\partial t} = 0, \tag{D.97}$$

which is equivalent to the continuity equation

$$\nabla \cdot \mathbf{J}_g + \frac{\partial \rho_g}{\partial t} = 0. \tag{D.98}$$

One can now define the GEM fields through

$$\mathbf{E}_g = -\nabla\phi_g - \frac{\partial \mathbf{A}_g}{\partial t}, \mu_g \mathbf{H}_g = \nabla \times \mathbf{A}_g.$$

It is easy to show that these fields satisfy the Maxwell-like equations

$$\begin{aligned} \nabla \times \mathbf{H}_g &= \varepsilon_g \frac{\partial \mathbf{E}_g}{\partial t} + \mathbf{J}_g, \\ \nabla \times \mathbf{E}_g &= -\mu_g \frac{\partial \mathbf{H}_g}{\partial t}. \end{aligned} \tag{D.99}$$

Consequently, there is an analogy between EM wave and gravitational wave. The gravitomagnetic field $\mathbf{H}_g$ is usually resulted from the rotation of mass, and has the dimension of angular velocity. Compared with the gravitational attraction, the gravitomagnetic field is often too weak to be observed [29].

D.7 Universal Laws of Nature Derived from UTF

Maxwell equations are the crown jewel of physics, and they have a long history and have accumulated a rich, deep knowledge base that can be borrowed and applied to the study of the Maxwell-like equations (D.34). The Maxwell-like equations are not only suited for the EM fields, but also for any physical system of fields which has a time-independent state. For this reason, many areas have yet to be explored. The UTF indicates that an observer moving relative to an ontological vector field $\mathbf{U}'$ (or a scalar field ϕ') will detect a dominant vector field $\mathbf{U}$ and a co-vector field $\mathbf{V}$, the latter being originated from the ontological field $\mathbf{U}'$ and the relative motion. The two vector fields $\mathbf{U}$ and $\mathbf{V}$ satisfy the Maxwell-like equations and constitute an EM-like field that travels at light speed. It is the relative motion that causes the ontological vector field $\mathbf{U}'$ (or a scalar field ϕ') to emerge as two vector fields $\mathbf{U}$ and $\mathbf{V}$ with respect to a moving observer. This discloses that the light and thus all the phenomena that can be observed are originated from relative motion. The UTF also shows that if the curl of the ontological field $\mathbf{U}'$ is zero or has a scalar potential function, the **V**-charge density ρ_m will not occur relative to a moving observer. Especially, if the ontological field is the static electric field, this implies that the magnetic charge will not occur, explaining why the experimental search for magnetic monopole particles has been in vain. Since the (**U**, **V**)-field in the UTF stand for any physical system of fields that is generated by an arbitrary ontological (or static) field, the following universal laws of nature can be derived from the UTF. They govern all the phenomena that can be observed and are philosophically important.

1) **The Law of "Light"**: Relative motions generate "light." The "light" here means any field that travels at the light speed relative to the observer, including the EM field and gravitational wave that have been discovered. If all relative motions are stopped, there was nothing to be seen.
2) **The Law of Attraction and Repulsion**: All sources (such as matters or charges) in nature intend to attract or repel each other due to the interactions among the fields generated by the sources.
3) **The Law of Opposite and Complementary**: A phenomenon is defined as a (**U**, **V**)-field, and is composed of two aspects **U** and **V**, which are opposite and complementary to each other. If one of them is getting stronger (or weaker), the other will start to oppose (or support) it.
4) **The Law of Cause and Effect**: One phenomenon can cause another if there exists interaction among them. Essentially, the ontological fields cause all phenomena due to relative motions.
5) **The Law of Phenomenology and Ontology**: A phenomenon is a manifestation of an ontological field due to relative motion. Phenomenon and ontological field are an integral whole, like waves and water, and they are not two distinct things.
6) **The Law of Conservation of Energy and Momentum**: Phenomenon has energy and momentum, both being conserved.

> There is nothing in the world except empty curved space. Matter, charge, electromagnetism, and other fields are only manifestations of the curvature of space.
>
> – *John Archibald Wheeler (American theoretical physicist, 1911–2008)*

> Reality is merely an illusion.
>
> *Albert Einstein*

References

1 Heras, J. A., "An axiomatic approach to Maxwell's equations", *Eur. J. Phys.*, Vol. 37, 055204, 2016.

2 Pierce, A. D., "Derivation of Maxwell's equations via the covariance requirements of the special theory of relativity, starting with Newton's law", *Journal of Applied Mathematics and Physics*, Vol. 7, No. 9, 2019, doi: 10.4236/jamp.2019.79141.

3 Page, L., "A derivation of the fundamental relations of electrodynamics from those of electrostatics", *Am. J. Sci.*, Vol. 44, pp. 57–68, 1912.

4 Page, L. and N. I. Adams Jr., *Electrodynamics*, New York, Van Nostrand, pp. 129–154, 1940.

5 Frisch, D. H. and L. Wilets, "Development of the Maxwell-Lorentz equations from special relativity and Gauss's law", *Am. J. Phys.*, Vol. 24, pp. 574–579, 1956.

6 Elliott, R. S., "Relativity and electricity", *IEEE Spectr.*, Vol. 3, pp. 140–152, 1966.

7 Elliott, R. S., *Electromagnetics-History, Theory and Applications*, IEEE Press, 1993.

8 Rosser, W. G. V., "Electromagnetism as a second-order effect", *Contemp. Phys.*, Vol. 1, No. 6, pp. 453–466, 1960.

9 Rosser, W. G. V., *The Special Theory of Relativity*, Butterworth and Go Ltd., 1964.

10 Tessman, J. R., "Maxwell-Out of Newton, Coulomb, and Einstein", *Am. J. Phys.*, Vol. 34, pp. 1048–1055, 1966.

11 Krefetz, E., "A "derivation" of Maxwell's equations", *Am. J. Phys.*, Vol. 38, No. 4, pp. 513–516, 1970.

12 Kobe, D. H., "Derivation of Maxwell's equations from the local gauge invariance of quantum mechanics", *Am. J. Phys.*, Vol. 46, pp. 342–348, 1978.

13 Kobe, D. H., "Generalization of Coulomb's law to Maxwell's equations using special relativity", *Am. J. Phys.*, Vol. 54, pp. 631–636, 1986.

14 Feynman, R. P., R. B. Leighton, and M. Sands (eds.), *The Feynman Lectures on Physics, Mainly Electromagnetism and Matter*, Reading, Addison-Wesley, 26–2, 1964.

15 Jackson, J. D., *Classical Electrodynamics*, 2nd Ed., Wiley, pp. 578–581, 1975.

16 Schwinger, J., L. L. DeRaad Jr., K. A. Milton, and W. Tsai, *Classical Electrodynamics*, Perseus Books, 1998.

17 Dyson, F. J., "Feynman's proof of the Maxwell equations", *Am. J. Phys.*, Vol. 58, pp. 209–211, 1990.

18 Hughes, R. J., "On Feynman's proof of the Maxwell equations", *Am. J. Phys.*, Vol. 60, pp. 301–306, 1992.

19 Lee, C. R., "The Feynman-Dyson proof of the gauge field equations", *Phys. Lett. A*, Vol. 148, No. 3-4, pp. 146–148, 1990.

20 Heaviside, O., "A gravitational and electromagnetic analogy", *The Electrician*, Vol. 31, pp. 281–282, 1893.

21 Jefimenko, O. D., *Causality electromagnetic induction and gravitation*, Star City, Electret Scientific Company, 2000.

22 Bedford, D. and P. Krumm, "On relativistic gravitation", *Am. J. Phys.*, Vol. 53, pp. 889–890, 1985.

23 Kolbenstvedt, H., "Gravomagnetism in special relativity", *Am. J. Phys.*, Vol. 56, pp. 523–524, 1988.

24 Campbell, W. B. and T. A. Morgan, "Maxwell form of the linear theory of gravitation", *Am. J. Phys.*, Vol. 44, pp. 356–365, 1976.

25 Hilborn, R. C., "Gravitational waves from orbiting binaries without general relativity", *Am. J. Phys.*, Vol. 86, pp. 186–197, 2018.

26 Kong, J. A., *Electromagnetic Wave Theory*, New York, Wiley-Interscience, 1990.

27 Mashhoon, B., "Gravitoelectromagnetism: a brief review", in *The Measurement of Gravitomagnetism: A Challenging Enterprise*, edited by L. Iorio, New York, Nova Science, 2007.

28 Geyi, W., *Foundations of Applied Electrodynamics*, New York, Wiley, 2010.

29 Forward, R. L., "General relativity for the experimentalist", *Proc. IRE*, pp. 892–904, 1961.

Index

Foundations of Antenna Radiation Theory: Eigenmode Analysis, First Edition. Wen Geyi.

Published 2023 by John Wiley & Sons, Inc.

IEEE Press Series on Electromagnetic Wave Theory

Field Theory of Guided Waves, Second Edition
Robert E. Collin

Field Computation by Moment Methods
Roger F. Harrington

Radiation and Scattering of Waves
Leopold B. Felsen, Nathan Marcuvitz

Methods in Electromagnetic Wave Propagation, Second Edition
D. S. J. Jones

Mathematical Foundations for Electromagnetic Theory
Donald G. Dudley

The Transmission-Line Modeling Method: TLM
Christos Christopoulos

Methods for Electromagnetic Field Analysis
Ismo V. Lindell

Singular Electromagnetic Fields and Sources
Jean G. Van Bladel

The Plane Wave Spectrum Representation of Electromagnetic Fields, Revised Edition
P. C. Clemmow

General Vector and Dyadic Analysis: Applied Mathematics in Field Theory
Chen-To Tai

Computational Methods for Electromagnetics
Andrew F. Peterson

Finite Element Method Electromagnetics: Antennas, Microwave Circuits, and Scattering Applications
John L. Volakis, Arindam Chatterjee, Leo C. Kempel

Waves and Fields in Inhomogenous Media
Weng Cho Chew

Electromagnetics: History, Theory, and Applications
Robert S. Elliott

Plane-Wave Theory of Time-Domain Fields: Near-Field Scanning Applications
Thorkild B. Hansen

Foundations for Microwave Engineering, Second Edition
Robert Collin

Time-Harmonic Electromagnetic Fields
Robert F. Harrington

Antenna Theory & Design, Revised Edition
Robert S. Elliott

Differential Forms in Electromagnetics
Ismo V. Lindell

Conformal Array Antenna Theory and Design
Lars Josefsson, Patrik Persson

Multigrid Finite Element Methods for Electromagnetic Field Modeling
Yu Zhu, Andreas C. Cangellaris

Electromagnetic Theory
Julius Adams Stratton

Electromagnetic Fields, Second Edition
Jean G. Van Bladel

Electromagnetic Fields in Cavities: Deterministic and Statistical Theories
David A. Hill

Discontinuities in the Electromagnetic Field
M. Mithat Idemen

Understanding Geometric Algebra for Electromagnetic Theory
John W. Arthur

The Power and Beauty of Electromagnetic Theory
Frederic R. Morgenthaler

Electromagnetic Modeling and Simulation
Levent Sevgi

Multiforms, Dyadics, and Electromagnetic Media
Ismo V. Lindell

Low-Profile Natural and Metamaterial Antennas: Analysis Methods and Applications
Hisamatsu Nakano

From ER to E.T.: How Electromagnetic Technologies Are Changing Our Lives
Rajeev Bansal

Electromagnetic Wave Propagation, Radiation, and Scattering: From Fundamentals to Applications, Second Edition
Akira Ishimaru

Time-Domain Electromagnetic Reciprocity in Antenna Modeling
Martin Štumpf

Boundary Conditions in Electromagnetics
Ismo V. Lindell, Ari Sihvola

Substrate-Integrated Millimeter-Wave Antennas for Next-Generation Communication and Radar Systems
Zhi Ning Chen, Xianming Qing

Electromagnetic Radiation, Scattering, and Diffraction
Prabhakar H. Pathak, Robert J. Burkholder

Electromagnetic Vortices: Wave Phenomena and Engineering Applications
Zhi Hao Jiang, Douglas H. Werner

Advances in Time-Domain Computational Electromagnetic Methods
Qiang Ren, Su Yan, Atef Z. Elsherbeni

Foundations of Antenna Radiation Theory: Eigenmode Analysis
Wen Geyi

Printed and bound by CPI Group (UK) Ltd, Croydon, CR0 4YY

17/09/2020

14964308-0008